Contents

The History of Geoconservation

Geological Society books refereeing procedures

The Society makes every effort to ensure that the scientific and production quality of its books matches that of its journals. Since 1997, all book proposals have been refereed by specialist reviewers as well as by the Society's Books Editorial Committee. If the referees identify weaknesses in the proposal, these must be addressed before the proposal is accepted.

Once the book is accepted, the Society Book Editors ensure that the volume editors follow strict guidelines on refereeing and quality control. We insist that individual papers can only be accepted after satisfactory review by two independent referees. The questions on the review forms are similar to those for *Journal of the Geological Society*. The referees' forms and comments must be available to the Society's Book Editors on request.

Although many of the books result from meetings, the editors are expected to commission papers that were not presented at the meeting to ensure that the book provides a balanced coverage of the subject. Being accepted for presentation at the meeting does not guarantee inclusion in the book.

More information about submitting a proposal and producing a book for the Society can be found on its web site: www.geolsoc.org.uk.

It is recommended that reference to all or part of this book should be made in one of the following ways:

BUREK, C. V. & PROSSER, C. D. (eds) 2008. *The History of Geoconservation*. The Geological Society, London, Special Publications, **300**.

HOUSHOLD, I. & SHARPLES, C. 2008. Geodiversity in the wilderness: a brief history of geoconservation in Tasmania. *In*: BUREK, C. V. & PROSSER, C. D. (eds) *The History of Geoconservation*. The Geological Society, London, Special Publications, **300**, 257–272.

GEOLOGICAL SOCIETY SPECIAL PUBLICATION NO. 300

The History of Geoconservation

EDITED BY

C. V. BUREK
University of Chester, UK

and

C. D. PROSSER
Natural England, UK

2008
Published by
The Geological Society
London

THE GEOLOGICAL SOCIETY

The Geological Society of London (GSL) was founded in 1807. It is the oldest national geological society in the world and the largest in Europe. It was incorporated under Royal Charter in 1825 and is Registered Charity 210161.

The Society is the UK national learned and professional society for geology with a worldwide Fellowship (FGS) of over 9000. The Society has the power to confer Chartered status on suitably qualified Fellows, and about 2000 of the Fellowship carry the title (CGeol). Chartered Geologists may also obtain the equivalent European title, European Geologist (EurGeol). One fifth of the Society's fellowship resides outside the UK. To find out more about the Society, log on to www.geolsoc.org.uk.

The Geological Society Publishing House (Bath, UK) produces the Society's international journals and books, and acts as European distributor for selected publications of the American Association of Petroleum Geologists (AAPG), the Indonesian Petroleum Association (IPA), the Geological Society of America (GSA), the Society for Sedimentary Geology (SEPM) and the Geologists' Association (GA). Joint marketing agreements ensure that GSL Fellows may purchase these societies' publications at a discount. The Society's online bookshop (accessible from www.geolsoc.org.uk) offers secure book purchasing with your credit or debit card.

To find out about joining the Society and benefiting from substantial discounts on publications of GSL and other societies worldwide, consult www.geolsoc.org.uk, or contact the Fellowship Department at: The Geological Society, Burlington House, Piccadilly, London W1J 0BG: Tel. +44 (0)20 7434 9944; Fax +44 (0)20 7439 8975; E-mail: enquiries@geolsoc.org.uk.

For information about the Society's meetings, consult *Events* on www.geolsoc.org.uk. To find out more about the Society's Corporate Affiliates Scheme, write to enquiries@geolsoc.org.uk.

Published by The Geological Society from:
The Geological Society Publishing House, Unit 7, Brassmill Enterprise Centre, Brassmill Lane, Bath BA1 3JN, UK

Orders: Tel. +44 (0)1225 445046, Fax +44 (0)1225 442836
Online bookshop: www.geolsoc.org.uk/bookshop

The publishers make no representation, express or implied, with regard to the accuracy of the information contained in this book and cannot accept any legal responsibility for any errors or omissions that may be made.

British Library Cataloguing in Publication Data

A catalogue record for this book is available from the British Library.

ISBN 978-1-86239-254-0

Typeset by Techset Composition Ltd., Salisbury, UK

Printed by MPG Books Ltd., Bodmin, UK

Distributors

North America
For trade and institutional orders:
The Geological Society, c/o AIDC, 82 Winter Sport Lane, Williston, VT 05495, USA
Orders: Tel. +1 800-972-9892
            Fax +1 802-864-7626
            E-mail: gsl.orders@aidcvt.com

For individual and corporate orders:
AAPG Bookstore, PO Box 979, Tulsa, OK 74101-0979, USA
Orders: Tel. +1 918-584-2555
            Fax +1 918-560-2652
            E-mail: bookstore@aapg.org
            Website: http://bookstore.aapg.org

India
Affiliated East-West Press Private Ltd, Marketing Division, G-1/16 Ansari Road, Darya Ganj, New Delhi 110 002, India
Orders: Tel. +91 11 2327-9113/2326-4180
            Fax +91 11 2326-0538
            E-mail: affiliat@vsnl.com

Preface

The idea of a conference on The History of Geoconservation was first aired at a History of Geology Group of the Geological Society, London (HOGG) committee meeting on 14 January 2004 by Cynthia Burek (O'Connor, A. 2004. HOGG committee Minutes (36,2). HOGG is an affiliated group of the Geological Society of London. It was inaugurated on 4th October 1994 to encourage interest in the lives and work of those scientists and philosophers who influenced both the study and practice of geology. The first chairman was John Thackray, secretary John Martin and treasurer John Fuller. It is run by a small committee that meets regularly in the Geological Society of London to plan conferences and other activities. Having attended a Geoconservation Commission meeting in late 2003 followed shortly by the January HOGG meeting, the timeliness of linking the two subjects through a conference became apparent. The HOGG committee agreed that a proposal to hold a conference should be taken to the next meeting of the Geoconservation Commission (GCC). This GCC meeting included a standing item on 'Conferences involving the Commission', and the idea of a History of Geoconservation conference was warmly welcomed. In order to take the conference forward, Colin Prosser (now of Natural England) offered to represent the GCC and Natural England whilst Cynthia Burek agreed to represent HOGG and UKRIGS.

Negotiations with HOGG, GCC and other interested parties led to a decision to hold the conference in Dudley, a town with a long association with geology and geoconservation, and to include the Black Country Geological Society (BCGS) alongside the HOGG and the GCC as organizers. An added benefit of Dudley as a venue was the opportunity to link the event with a series of other events celebrating the fiftieth anniversary of the declaration of Wren's Nest National Nature Reserve, one of the oldest and best known geological reserves in the UK.

An organizing committee consisting of Alan Cutler (BCGS), Graham Worton (Dudley Museum and Art Gallery), as well as Cynthia Burek and Colin Prosser was set up with additional occasional input from Jonathan Larwood on behalf of the Geologists' Association and Hannah Townley (Natural England). Enthusiasm for the conference was high amongst all partners with the conference being seen as chance both to learn from the past and to celebrate achievements to date. Key questions that the conference hoped to address included: what is geoconservation, when did it start, and how did we get to where we are today? The conference aimed to answer these questions within three themes:

- The origins of geoconservation;
- Geoconservation in the British Isles; and
- Geoconservation on an international scale.

The conference was held in Dudley Museum on 24–25 November 2006, with a conference dinner in the limestone caverns beneath Dudley (that were once visited by Roderick Murchison) which included an address from a modern 'Dr William Buckland' (Patrick Boylan in disguise). The following day Graham Worton led an excursion to the Wren's Nest National Nature Reserve.

The first theme of the conference, 'The origins of geoconservation', included four read papers by Phil Doughty, Barry Thomas, Tom Hose and Murray Gray. The second theme, 'Geoconservation in the British Isles' comprised papers from Cynthia Burek, Chris Green, Neil Ellis, Colin Prosser and Graham Worton, whilst Cheryl Jones, Lars Erikstad and Patrick Boylan presented against the third theme, 'Geoconservation on an international scale'. The success of this conference demonstrated a strong interest in the history of geoconservation and enhanced by additional papers developing the local and international angle, has led to this publication. Although focused on the UK and Europe it also includes papers exploring the history of geoconservation in the USA, Australia and all parts of the world engaged with World Heritage sites and Geoparks. The book demonstrates the importance of looking backwards in order to push forward with the conservation and promotion of features, processes, sites and specimens needed to contribute to the sustainable development of our natural environment. After all we are only the custodians of the Earth for future generations.

The Geological Society wishes to acknowledge the financial support of Natural England.

Reference

O'CONNOR, A. 2004. Minutes of the HOGG Committee, **36**, 2.

CYNTHIA BUREK & COLIN PROSSER

UKRIGS

GEOCONSERVATION
ASSOCIATION

The history of geoconservation: an introduction

C. V. BUREK[1] & C. D. PROSSER[2]

[1]*Centre for Science Communications, Department of Biological Sciences, University of Chester, Parkgate Road, Chester CH1 4BJ, UK (e-mail: c.burek@chester.ac.uk)*

[2]*Natural England, Northminster House, Peterborough PE1 1UA, UK*

In many parts of the world, the regeneration, economic growth and social changes that took place in the two decades that followed the Second World War, led to increased leisure time and tourism and a greater awareness of the world around us. In addition, the realization of our ability to destroy both ourselves and the environment in which we live, clearly evident during the Cold War years, led to a greater appreciation of the fragile nature of the natural environment. By the late 1960s, increasing loss of countryside to development, and the ability to see our planet from space, led to an enhanced regard of the fragility of the environment in which we live. By the 1970s an environmental revolution, with conservation at its core, was in full swing, highlighted by the pioneering 1972 United Nations Conference on the Human Environment held in Stockholm. By the 1990s the Earth Summit, held in Rio in 1992, had placed the environment, through its role in achieving sustainable development, on the global political and social agenda. Today, it is climate change that reminds us that we have the power to do irreparable damage to the natural environment that supports us.

This book provides the first collection of papers to address the history of geoconservation. It seeks to explore the origins of the subject and the concepts that helped to define it; it describes the history of geoconservation in the UK, looks more widely to the Republic of Ireland, mainland Europe and Australia and explores the evolution and impact of global conservation initiatives including World Heritage sites and Geoparks. In doing this, it highlights the invaluable contributions to geoconservation made by academics, geological societies, governments, conservationists, volunteers and local communities. The papers demonstrate that the origin and development of this subject is interesting and informative in itself but more importantly, through revealing the history of geoconservation successes and failures, they provide us with an increased understanding of how we got to where we are now; invaluable knowledge in helping geoconservation meet the challenges that lie in the future.

Geoconservation is now a growing and widespread activity that is well established in the UK, Europe and many other parts of the world. Prior to the conference held in Dudley, England, in November 2006, there had been little thought or material published on the history of geoconservation. There are a number of reasons for this. The first is that geoconservation is a relatively new discipline that has had a low profile until the last couple of decades during which it has grown rapidly. Another is that this expansion has been sustained by a forward looking approach rather than on looking back at the history of the subject. This pattern of slow steady growth, with more recent rapid expansion, is well illustrated in the UK. Here, a few early but isolated examples of geoconservation can be identified prior to the twentieth century; conservation legislation and a nationally coordinated and structured approach to geoconservation was in place by 1950; and the rise of the voluntary sector in the form of Regionally Important Geological/geomorphological Sites (RIGS) groups (Regionally Important Geodiversity Sites in Wales) boosted activity levels and participation in geoconservation by the 1990s. By the twenty-first century, the appearance of European Geoparks has led to another step-up in geoconservation activity level. In many other parts of the world, activity levels have risen even more rapidly, jumping from relatively low levels to relatively high levels as Geoparks have opened up new opportunities and enthusiasm for geoconservation.

Geoconservation is undoubtedly an expanding and dynamic activity. It is 'happening today' and through Geoparks is growing a strong international community involving more countries than ever before. It is an exciting time for those interested in geoconservation. It is possible to demonstrate how geoconservation can inform an enlightened public and how geological and geomorphological features, processes, sites and specimens can contribute to the environmental, social and economic

From: BUREK, C. V. & PROSSER, C. D. (eds) *The History of Geoconservation.*
Geological Society, London, Special Publications, **300**, 1–5.
DOI: 10.1144/SP300.1   0305-8719/08/$15.00 © The Geological Society of London 2008.

pillars of sustainable development (Webber *et al.* 2006). There is work to be done, plans to be made, partnerships to be built, funding to be secured, decision makers to be influenced and people to be enthused. With all this going on, it is not surprising that geoconservationists have, until now, been looking forward to the next challenge, rather than backwards into the history of their discipline. However, the past can inform the future and reflection is always valuable.

What is geoconservation?

Defining geoconservation, or geodiversity conservation is a subject in itself, thoughts on which can be read in Sharples (2002), Prosser (2002*a*, *b*) Gray (2004), Prosser *et al.* (2006). However, it is prudent here to highlight the difference between conservation and preservation as applied to the natural environment. Conservation can be taken as meaning the 'active management of something to ensure its quality is retained'. This places the emphasis on management of something to retain a particular quality, rather than on preservation of the feature, site, process etc. with no change at all. Geoconservation, therefore, usually involves working with natural change to retain a feature of interest, for example, maintaining a clear exposure of a stratigraphical sequence in an eroding cliff, despite the erosion. It is not about stopping the erosion and freezing the exposure in time. Preservation on the other hand, can be taken as 'keeping something in the same state, stopping it from changing', i.e. mothballing it and allowing no physical change. However, in some circumstances conservation of a finite and sensitive feature such as a mineral vein may require an approach much more akin to preservation than conservation.

In simple terms, and for the purpose of this paper, geoconservation can be defined as action taken with the intent of conserving and enhancing geological and geomorphological features, processes, sites and specimens. As successful conservation often depends on understanding and valuing the feature, process, site, or specimens to be conserved, the actions taken often also include promotional and awareness raising activities. The need for this awareness raising is captured well in the Local Geodiversity Action Plans (LGAPs) process for example (Burek & Potter 2004, 2006).

Geoconservation today

There is now a general acceptance amongst Earth scientists and conservation practitioners that our geological and geomorphological heritage is an important, and in places threatened, part of our natural heritage and that it is worthy of conservation for future generations. Many practising Earth scientists, including teachers, have first-hand experience of the need for geoconservation. Some may have witnessed the loss of a favourite exposure or have been personally involved in geoconservation activity in some way, such as through advising on the importance, value or management needs of a site with which they have a research or teaching interest or which may be local to where they live or work.

As described above, geoconservation is a growing activity, with more participants and a greater profile now than ever before. Geoconservation is very well established in the UK and increasingly across Europe and Australia, and with the World Heritage List and especially the rapid growth of Geoparks, it is now coming to prominence in many other parts of the world. In addition to the international frameworks there are many national level geoconservation initiatives. These include establishment and use of conservation legislation and government policy to conserve geological and geomorphological features, processes, sites and specimens and to create geological reserves or parks. There are also many geoconservation activities and projects led by geological societies, associations, academic organizations, museums and geological surveys. At a local level, planning authorities and very importantly, voluntary geoconservation groups such as the RIGS movement, are playing a critical role in bringing geoconservation to local people. In the UK, this has been achieved through LGAP partnerships and increasingly through RIGS regional partnerships, Geopark events and Scottish geology week.

The geoconservation activity described above is now established in many places across the world. It started in different places at different times and in different ways and now involves many people from a variety of backgrounds including Earth scientists, conservationists, land managers and landowners. It has a significant role to play in helping to deliver sustainable development through conserving and promoting scientifically, educationally, recreationally and culturally important features, sites and specimens, many of which are important to an individual region or country's economic wealth and cultural identity (Webber *et al.* 2006).

The origins of geoconservation?

Although it may be desirable to identify when and where geoconservation first began, the inevitable range of opinions over which historic activities were, or were not examples of geoconservation, mean that the origin of geoconservation is likely to be a subject of debate, rather than a consensus, for some years to come. One aim of this book is to provide some context and observations on this subject to help take thinking forward. It is argued here that it is relatively easy to identify a series of activities that definitely are geoconservation and a further series of activities that definitely are not (Table 1). The challenge lies in the fact that there are also a number of activities that may or may not be geoconservation. Taking the definition of geoconservation given above, namely 'action taken with the intent of conserving and enhancing geological and geomorphological features, processes, sites and specimens for the future, and generally involving awareness raising activity in support of this aim' it is possible to explore the origins of geoconservation further.

Experience in the UK suggests that geoconservation and the stages leading up to it can be divided into a number of steps (Table 1). First and foremost an understanding of the conservation issue must be engendered. You cannot undertake conservation without first having an appreciation of the value of the item to be conserved. Thus, although not directly geoconservation, some of the steps listed, such as building an awareness of geological/geomorphological features, processes, sites and specimens, describing and auditing them and developing an appreciation of them are clearly not examples of geoconservation as they are carried out without intent to conserve. Other steps, including conservation audits such as the Geological Conservation Review (GCR) or the Welsh Assembly Government RIGS audit, use of

Table 1. *What is geoconservation? Geoconservation and the steps leading up to it*

Activity relating to geological/ geomorphological features, processes, sites and specimens	Examples of activity	Comments
Initial awareness	Appreciation that geological/ geomorphological features, processes, sites and specimens exist	Not geoconservation—just awareness of natural environment or heritage/culture
Examination, description, scientific audit	Specimen collecting for curiosity, visiting and describing features, sites etc., geological mapping/ survey	Not geoconservation—collecting and scientific description. Classification and taxonomy start of scientific thinking
Value/appreciation	Retaining specimens, telling others about features, sites etc., drawing and painting of features, sites etc.	Not geoconservation—but a subconscious state likely to result in support of conservation if a threat is perceived
Awareness of threat/perceived threat	Concern and desire to act	Not geoconservation—but likely to be followed by geoconservation
Unintentional or coincidental activity that leads to a geoconservation benefit	Conservation of valued woodland, including a geological feature that coincidentally benefits from conservation of the woodland	Geoconservation ? 'Grey area' No intent here, likely area for debate
Conservation audit	An assessment of what is important to keep and where it is e.g. the GCR	Geoconservation—action to identifying conservation priorities
Protection through legal/policy means	Conservation legislation or National Park/planning policy	Geoconservation—action to protect through law or practice
Management	Purchase of land or specimen, creation of reserve, securing of a site, enhancement of an exposure	Geoconservation—direct action to protect or manage
Awareness raising of importance of feature	Interpretation, books, media, lobbying of politicians, education, involvement of local community	Geoconservation—indirect action to build support for conservation
Development of a holistic approach to conservation showing the interdependence of all aspects of nature	Integrated landscape scale approaches, integrated biodiversity/geodiversity/ landscape/archaeology conservation	Geoconservation—as part of a strategic, holistic and integrated approach to managing the natural environment

conservation legislation and policies, creation of geological reserves, on-site management work and raising of awareness to generate support, are all examples of geoconservation, as they are actions carried out with the intent to conserve. These are direct and explicit. However, there is a 'grey area' where actions may or may not be geoconservation. This includes 'unconscious' or 'coincidental' actions that may lead to geoconservation taking place. In Table 1, this 'grey area' lies somewhere between 'valuing/appreciating' and 'taking conscious action'. The 'unconscious' action could be buying a site or specimen because it appeals, but without realizing that it is the geological character of the site or specimen that appeals and that through buying it and looking after it, it will be retained for the future whether by an individual or institution. The 'coincidental' action may be through conservation action taken to benefit a different valued feature, such as a wilderness or woodland. Here, action taken to conserve the coastal cliff for bird sanctuary, wilderness or woodland, could result in totally unintentional conservation benefits for geological features such as an exposure of a mineral vein. The debate about the origins of geoconservation revolves around perceptions of these 'unconscious' or 'coincidental' actions.

The discussion of when geoconservation began is taken up by **Doughty, Thomas & Warren, Worton** and **Erikstad**. **Doughty** uses the Giant's Causeway in Northern Ireland and Yosemite National Park in the USA, to explore the interface between geoconservation and curiosity in the natural world, art, literature, tourism and wilderness. The paper also makes the case for considering geological specimens and collections in addition to sites, when seeking to identify the origins of geoconservation. Another key factor that has influenced when, where and how geoconservation came about is land ownership. Conservation is always easier where the land in question is under the control of those wishing to see conservation taking place. **Thomas & Warren** compare and contrast the development of geoconservation and conservation more widely in the USA and the UK. Differing approaches and geoconservation histories are attributed to a large extent, to different situations in terms of ownership of land. The paper demonstrates that National Parks and conservation areas are much easier to establish where large areas of land are under 'state' control, as in the USA, than where land is largely under private ownership, as in the UK. This historical comparison accounts for the present situation where geoconservation in the USA is based on National Parks, whereas in the UK it is based on the Sites of Special Scientific Interest (SSSI) and the RIGS framework where land remains within private ownership. The local

community approach is taken up by **Worton** who uses the seventeenth century observations of Dud Dudley from Dudley as an example, illustrating that the importance of recognizing local contributions in the origins of geoconservation is vital. **Sharples & Houshold** demonstrate the growth of geoconservation in Tasmania, Australia and question the problems of conserving wilderness.

The impact of tourism increasing awareness of 'place' is another strand explored in the origins of geoconservation and developed by **Hose, Doughty**, and **Parkes**.

In terms of advocating the earliest examples of geoconservation, **Thomas & Warren** propose Hutton's Rock, Holyrood Park, Edinburgh, from 1845 and **Erikstad**, investigating geoconservation in Europe, identifies show caves as the pioneers of geoconservation. He cites Baumannshöle, a cave in Germany, which was subject to a nature conservation decree as far back as 1668.

The development of geoconservation

Having explored the origins of geoconservation, the rest of this book is devoted to the history of geoconservation concepts, the development of geoconservation in the UK followed by experience from elsewhere in the world, especially Europe and Tasmania, and finally looks at the history of global initiatives such as the World Heritage Site series. There are many other ways of analysing the history of geoconservation, two of which are briefly discussed below.

The key players

Geoconservation would never have happened without those who advocated and got involved in making it happen. The strength of the discipline today and the opportunities that are available are a consequence of the vision, belief, action and hard work of those that went before. These papers highlight the critical roles that have been played by so many individuals, communities, societies, associations, groups, organizations, governments, landowners and business interests. They illustrate the range of obstacles that have had to be overcome and the innovative solutions and ideas that have moved geoconservation foward. Importantly, they illustrate that geoconservation had developed in many parts of the world, at local, regional, national and international level, through the efforts of geologists, conservationists, politicians, landowners and members of the public.

At the local and regional level, the work of local communities (**Worton**) natural history societies (**Burek**), RIGS groups and Geology Trusts

(**Burek and Radley**) museums (**Munt; Radley**), cave owners, managers and enthusiasts (**Erikstad; Murphy**) and private companies and local authorities (**Doyle**) is vital. Geoconservation is only really effective with local buy-in, and local buy-in can only be secured by local champions and local action. Throughout the history of geoconservation, local action has provided examples of good practice that have been picked up and disseminated at national and international level (**Prosser & Larwood; Sharples & Houshold; Worton**).

Action at a national level has been essential in providing a robust framework in which local delivery can take place. National surveys such as the map making of the British Geological Survey (BGS) (**McMillan**) and the GCR (**Ellis**) have provided the consistent geological information upon which geoconservation is built. National legislation and policy, implemented by national conservation agencies has provided a robust framework for protecting and managing sites and in declaring and managing National Nature Reserves (NNRs) and National Parks (**Prosser, Parkes, Thomas & Warren**). National societies such as the Geologists Association (**Green**) and 'umbrella groups' such as UKRIGS (**Burek**) have played important roles in co-ordinating and supporting the local activities of their member groups working locally.

International initiatives such as Geoparks (**Jones**), World Heritage Sites (**Boylan**) and those led by ProGEO (**Erikstad**), have all helped to ensure that the best of our geological heritage is considered at an international level and that geoconservation has an international profile and is supported by a community of interest.

The success of geoconservation has been attributable to many individuals and bodies, working at all levels. There is no reason to believe that the future of geoconservation will not depend on the same wide range of key players, working wherever there is a need or opportunity for geoconservation, to meet challenges similar to those that have been faced in the past.

The practice of geoconservation

As illustrated in Table 1, there are a number of different activities and stages which make up geoconservation and the stages leading up to it (**Doughty; Thomas & Warren**). The concepts which underpin aspects of geoconservation such as geodiversity (**Gray**) and geotourism (**Hose**) are addressed. The basic foundation for geoconservation, geological audit and mapping is described by **McMillan**, and **Boylan**, **Burek**, **Ellis** and

Prosser consider the history of conservation audits used to underpin a range of designations. The history of protection and management of geological and geomorphological features, processes, sites and specimens is described in each paper whilst the history of activity to promote geology and geoconservation is central to papers by **Hose**, **Green**, **Burek**, **Burek**, **Munt**, **Doyle**, **Sharples & Houshold**, **Jones** and **Boylan**.

Conclusion

This book shows that geoconservation has reached a point where there is enough history to look back and learn from. It also provides answers as to where geoconservation came from and why things are as they are today. Geoconservation is a relatively young discipline and one that can learn from approaches to conservation adopted in archaeological, biological and heritage fields. This book demonstrates that the history of geoconservation has been a series of challenges, set-backs and successes; the future is likely to be the same. Most importantly, this book enables us to share the experience of the past and use this to provide a basis for taking geoconservation forward to meet the challenges of the future. The price for not learning this is high and we owe it to future generations to aim for successful geoconservation.

References

BUREK, C. V. & POTTER, J. 2004. Local Geodiversity Action Plans—Sharing Good Practice Workshop, Peterborough, 3 December 2003. *English Nature Research Report*, **601**, Peterborough, 1–37.

BUREK, C. V. & POTTER, J. 2006. Local Geodiversity Action Plans—Setting the context for geological conservation, *English Nature Research Report*, **560**, Peterborough.

GRAY, M. 2004. *Geodiversity: Valuing and Conserving Abiotic Nature.* John Wiley, Chichester.

PROSSER, C. 2002a. Terms of endearment. *Earth Heritage*, **17**, 12–13.

PROSSER, C. 2002b. Speaking the same language. *Earth Heritage*, **18**, 24–25.

PROSSER, C., MURPHY, M. & LARWOOD, J. 2006. Geological conservation: a guide to good practice. *English Nature*, Peterborough, 1–145.

SHARPLES, C. 2002. *Concepts and Principles of Geoconservation.* PDF Document, Tasmanian Parks and Wildlife Service. Website: www.dpiw.tas.gov.au/inter.nfs/webpages/SJON-57W4FD.

WEBBER, M., CHRISTIE, M. & GLASSER, N. 2006. The social and economic value of the UK's geodiversity. *English Nature Research Reports*, **709**, Peterborough, 1–122.

How things began: the origins of geological conservation

PHILIP DOUGHTY

14 Orchard Rise, Belfast BT8 7DA UK (e-mail: philip.doughty@ntlworld.com)

Abstract: The origins of geoconservation cannot be investigated without first defining its scope. This presents a problem because there is no established working definition of the field. By using the Giant's Causeway as a case study, useful parameters were identified revealing great complexity. They embrace initial curiosity, scientific communication, mythology, access issues, the involvement of national scientific institutions, controversial but ultimately successful iconography, the invention of new artistic conventions, dissemination by engraving, scientific reaction, rekindling of a fundamental geological controversy, tourism, popular literature, modes of transport, commercialization, additions to fundamental science, designation history and historic associations of the site. Other sites are similarly complex and assist in refining the scope. Sites are seen as the principal resource but on analysis achieve their status from what they reveal or the importance of the materials they yield, in turn spotlighting the major museum collections. These are now well documented though not all are secure. It was not the scientific imperative that established the first public designation but an impassioned delight in unspoiled nature which three men, two Americans, Henry Thoreau and George Marsh but especially the Scots environmentalist, John Muir, projected carefully into the attuned ear of the US President. This brief overview closes with the revelation of neglected areas of heritage, paths that geoconservation could have taken and still may and suggests how earlier definitions could be elevated into a more specific and holistic geoconservation strategy.

I accepted this topic reluctantly and the more I have read, the more reluctant I have become. The reason is simple; geoconservation is a nebulous concept. Everyone who has approached the topic has found a major problem in defining its scope. Stevens (1994) was among the first to examine the dilemma. Should it be 'Geological conservation'; 'Geological and landscape'; 'Geological and geomorphological'; 'Earth sciences' or, to push the limit, 'Earth's resources'? In the end he proposed the following definition.

Earth heritage conservation is concerned with sustaining the part of the physical resources of the Earth that represents our cultural heritage, including our geological understanding, and the inspirational response to the resource.

Arid but succinct and while achieving a large measure of inclusiveness, it is thin on practical implications and applications.

Stanley (2002) suggests that even this definition is limiting and a more holistic approach is required, essentially including all things relating to and reliant on the material resources of the planet which he extends to include biodiversity and beyond. Such an approach is contentious, straying onto ground already well cultivated by the biological conservation community and such organizations as Common Ground (see Clifford & King 2006) but it has an undeniable logic. The terminology attaching to all these ideas is thoroughly reviewed by Prosser (2002) and to pursue it further will only add to the fog of words. Suffice it to say that the emerging consensus favours a liberal approach.

Discovery and iconography of the Giant's Causeway: '... a remnant of chaos'

My brief, having loosely established the field, is to explore the first stirrings that led to this modern conservation movement and a single case study with roots in the seventeenth century will be used to establish some parameters. The site is the Giant's Causeway in Northern Ireland described by William Makepeace Thackeray in October 1842 as 'a remnant of chaos' (Thackeray 1843; Watson 1992).

The first mention of the Giant's Causeway in literature was in 1693 in a second-hand account by an anonymous Cambridge graduate describing a visit made in the previous year in the company of the Bishop of Derry. Word reached Sir Richard Bulkeley in Trinity College, Dublin and, through him, Sir Martin Lister, the President of the Royal Society. Lister approved the publication of Bulkeley's letter which appeared in the *Transactions of the Royal Society* in the following year. Before 1693 there is no mention of the Giant's Causeway in Irish topographies or on maps and charts of the time. It must always have been known to the locals but there is no record of what they made of it or called it.

From: BUREK, C. V. & PROSSER, C. D. (eds) *The History of Geoconservation.*
Geological Society, London, Special Publications, **300**, 7–16.
DOI: 10.1144/SP300.2   0305-8719/08/$15.00 © The Geological Society of London 2008.

The name is based on a somewhat flexible Irish 'legend' involving two giants, one Irish, Finn MacCool and his Scottish rival, Benandonner. To facilitate a deciding confrontation Finn constructed a causeway extending to the Scottish island of Staffa and beyond. Here the various accounts become inconsistent but the invariable outcome is that Benandonner escaped to Scotland, ripping up the causeway behind him as he went, leaving remnants in Ireland and at Fingal's Cave. The tale has no deep roots in Irish folk culture and is without mention in the *Mythological, Ulster, Fenian* or *Historical* cycles. It appears to have been an early example of tourism initiative.

From the appearance of the description in the *Transactions of the Royal Society*, curiosity about this seemingly fantastic phenomenon was intense. However, travel through rural Ireland to view it was such a major undertaking that the natural philosophers of the day much preferred to sponsor artists to make accurate representations of the columns for them, obviating the need to undertake the arduous trek to this wild coastline themselves. The earliest known sponsor was Samuel Foley, Bishop of Down and Connor, who employed a local artist, Christopher Cole Foley published an account (1694; Ashworth 2004) illustrated with an engraving of Cole's drawing. Both the drawing and the engraving from it were considered inadequate.

The Royal Dublin Society then intervened through Thomas Molyneux who acted as spokesman for a group of Dublin 'philosophical gentlemen'. They employed one Edwin Sandys 'a good Master in Designing and Drawing Prospects' to take an accurate and genuine representation of the Causeway and its setting. This he produced in 1696 apparently to the delight of Molyneux who, of course, had never seen the Causeway. It was immediately sent to London to be engraved and it was this engraving that was used by Thomas Molyneux's brother, William, as the basis of his description of the Causeway (Molyneux 1697; Ashworth 2004). The original drawing was lost but the engraving proved controversial, indeed it was condemned as a fiasco by the Reverend William Hamilton (1786), an early and influential Vulcanist, who lived and worked on the north coast and who knew the Causeway well.

Neither the talents nor the fidelity of the artist seem to have been at all suited to the purpose of the philosophical landscape .... In this true prospect, the painter has very much indulged his own imagination at the expense of his employers, insomuch that several tall pillars, in the steep banks of this fanciful scene, appear loaded with luxuriant branches, skirting the wild and rocky bay of Port Noffer with the gay exhibition of forest trees. In the background he discovered a parcel of rude and useless materials which his magic pencil soon transformed into comfortable dwelling-houses;

and for chimneys he has happily introduced some detached pillars of basalts, which, from their peculiar situation, and the name given to them by the peasants of the country [*still known today as the chimney tops*], naturally excited the attention of this extraordinary artist.

Later speculation suggested a creative engraver but in 1994 the original drawing surfaced in Sotheby's London auction rooms and it became obvious that Sandys was the culprit. Could this simply have been a case of artistic myopia? The original is now in the collections of the Ulster Museum.

It was a further 44 years before the next and most significant development. The Dublin Society offered an annual premium in open competition and in 1740 it was awarded to an anonymous artist for two remarkably accurate and very fine paintings in gouache representing east and west prospects of the Causeway (Figs 1 and 2) (Anglesea & Preston 1980). They proved to be the work of a little known Dublin artist, Susanna Drury. Mrs Delaney, the noted diarist, records that Susanna spent 3 months at the site while she was preparing them. Their accuracy, even by modern standards, is impressive and they created an immediate excitement when they were exhibited. They have proved to be landmarks both in topographic painting and European scientific illustration. There are two known pairs of Drury originals, both slightly different, the better finished now in the Ulster Museum.

It was decided that they should be engraved but the originals were so loaded with scientific detail that the choice of engraver was critical. A young and rapidly rising London Huguenot engraver, Francois Vivares, was selected. He had learned his trade with Joseph Wagner and by 1739, aged 30, he was producing excellent prints. He was strongly drawn to landscape engraving and became a founder and one of the finest exponents of the English school of landscape engravers. Two superbly detailed and scientifically annotated prints, based on Drury's paintings, the '*East and West Prospects of the Giant's Causway in the County of Antrim in the Kingdom of Ireland*' appeared in 1742/3 (Figs 3 and 4). They were inscribed by Drury to the Right Honourable Alexander McDonnel Earl of Antrim (west prospect) and the Right Honourable John Boyle, Earl of Orrery (east prospect). The prints, of exceptional quality, sold rapidly and new editions were produced steadily. The plates were reworked by Vivares in a crisp new edition and republished by John Boydell of Cheapside, London, as late as 1777. The last known reprinting was in 1837.

As the first accurate representations of basalt columns, enhanced by their spectacular coastal setting, they were widely circulated in polite

Fig. 1. *West Prospect of the Giant's Causeway* by Susanna Drury. Gouache on vellum, 34 × 66 cm, 1740, in the collection of the Ulster Museum, Belfast. This, with the East Prospect (Fig. 2), was the first accurate representation of the columns on the promontories at the Causeway, achieved after almost half a century of failed attempts. It has proved to be a landmark in Irish topographic painting and European scientific illustration.

society throughout Europe and one consequence was the reigniting of the debate on the origin of basalts.

The power of the image is nowhere better shown than here and the difficulty in realizing it was

not due to simple naïvety. As Rudwick (1992) points out:

... artistic conventions do not fall ready-made from heaven, nor are they concocted or decreed at a given moment. They are

Fig. 2. *East Prospect of the Giant's Causeway* by Susanna Drury. Gouache on vellum, 34 × 67 cm, 1740, in the collection of the Ulster Museum, Belfast. For details, see Fig. 1. Both prospects won the £25 premium of the Dublin Society in 1740. Little is known of the artist, who may have been a Mrs Warter, possibly with Huguenot connections. A further, smaller, pair by Drury is in the collection of the Knight of Glin. There may be a third, much larger, pair once owned by Dr John Barrett (1753–1821), Vice-Provost of Trinity College, Dublin.

Fig. 3. Detail from *West Prospect of the Giant's Causeway*. Line engraving by Francois Vivares after gouache by Susanna Drury, 1740. First published by Susanna Drury on 1 February 1743 or 1744. This version is from the refreshed plate, published by John Boydell, Cheapside, London, 1 May, 1777. In the author's collection. Dedicated by Drury to Alexander McDonnel, Earl of Antrim. The quality and accuracy of both the landscape and the column detail is superb. This is the acme of eighteenth century scientific engraving and was never surpassed. The wide distribution of the prints throughout Europe rekindled the Neptunist/Vulcanist debate and allowed Nicolas Desmarest, the French Vulcanist, to confirm the igneous origin of the Causeway. Vivares was a founder of the English school of landscape engravers and produced many fine topographic prints, some after his own drawings such as the exceptional Malham Cove (1753).

the product of *historical* development; they are constructed in the course of artistic practice in specific historical circumstances.

These paintings and the engravings that followed, directly reflect this. They met a specific and urgent demand and are among the earliest, finest and most accurate scientific landscapes in both media. They represent success after several failed earlier attempts.

Sometime before 1786, the Reverend William Hamilton was firmly convinced of the volcanic origin of the Causeway basalts (Hamilton 1786), the first to express this view in writing. It was contrary to the Neptunist opinions of the Dublin school of the day led by Richard Kirwan, a particularly vitriolic protagonist. As a direct response to the Vivares engravings, crushing support for Hamilton's opinion came from Nicolas Desmarest. He was then the leading authority on volcanoes and author of the masterly mapping and analysis of

the extinct volcanoes of the Auvergne where flows containing columnar basalts were directly linked to undeniable volcanic cones. He immediately recognized the Causeway columnar basalts as near-identical with his, strongly fortifying the Vulcanist cause in general.

Tourism

Further features of the paintings and engravings are the figures in the landscape (Figs 3 and 4), gentlemen in frock coats and breeches with tricorn hats and small social groups, including crinolined ladies wearing elaborate bonnets, all obviously in discourse on the features they were so evidently indicating. Clearly tourism was established among well-to-do classes by the 1730s and it has grown and broadened ever since. Early distinguished visitors included John Whitehurst (1773); John Wesley

Fig. 4. Detail from *East Prospect of the Giant's Causeway*. Line engraving by Francois Vivares after gouache by Susanna Drury, 1740. First published by Susanna Drury on 1 February 1743 or 1744. This image from the refreshed plate, published by John Boydell, Cheapside, London, 1st May, 1777. In the author's collection. Dedicated by Drury to John Boyle, Earl of Orrery. Both engravings were extensively annotated for a scientific audience, numbered elements on both prints relating to a descriptive key. Neither of the engravings is faithful to either of the known pairs of paintings and the fidelity of geological detail far exceeds the possibilities of gouache, Drury's chosen medium. Possible explanations include a Drury sketchbook or a visit by Vivares to the site.

(1778); Abraham Mills (1787/8); Humphry Davy (1806) and Jean Francois Berger (1811) (Wyse Jackson 1997).

Enthusiasm for the site was not universally shared. James Boswell (1791) was quite unable to overcome Dr Johnson's aversion to all things Irish when, in 1780, he suggested a tour of Ireland.

Johnson. It is the last place where I would wish to travel.

Boswell. Should you not like to see Dublin, Sir?

Johnson. No, Sir; Dublin is only a worse capital.

Boswell. Is not the Giant's-causeway worth seeing?

Johnson. Worth seeing? yes; but not worth going to see.

Nor was William Thackeray's experience entirely satisfactory. On a stormy October day in 1842 he was tumbled into an open boat by his guide who, 20 minutes later, pointed out the Causeway

through the waves. His reaction: 'Mon Dieu! And have I travelled a hundred and fifty miles to see *that*?' He recovered his equanimity after a good meal and a better bottle of wine in Miss Henry's Inn, one of two then catering for visitors and still surviving as the Causeway Hotel over 170 years later. From Thackeray's *Irish Sketch Book* (1843) it is obvious that there was a mature tourist experience at the Causeway in the early nineteenth century with fleets of jaunting cars, a crowd of guides, half a dozen boats at Portnaboe propelled by an impassive band of oarsmen and a wizened lady on station at the Wishing Well who, for a fee, would add a drop of whiskey to the water, to improve or possibly sterilize it. The guides were capable of a geological résumé of sorts but all visitors were regaled with the 'ancient' tale of the giants.

In 1883 a major and innovative transport development transformed access to the Causeway.

It linked Portrush, a local seaside resort, first to Bushmills, 10 km to the east and by an extension of a further 4 km in 1887 to the terminus at Causeway Head. It was a quaint hydro-electric tramway that continued to operate until 1949 when neglect forced its closure. It provided mass access to the widest public for the very first time. It has now been revived but employs more conventional technology.

An enterprising group of local businessmen recognized the market opportunity offered by this link and formed a syndicate that, after a protracted legal struggle, managed to override local opposition and enclose the site from 1898. A bitterly resented admission charge was imposed from this time and remained in force until 1961 when the National Trust acquired the site.

Two further developments are relevant to this account. The first has international application. Columnar basalts are known from over a hundred sites around the globe but with no viable explanation of the form until 1940. Tomkeieff's definitive account (1940) is based exclusively on the basalts of the Causeway Tholeiite Member as observed at the Causeway and the cliffs to the east and still stands as the primary reference.

The second, linking directly to modern conservation practice, is its history of designation. Surprisingly, the first formal acknowledgement of the site's importance was the World Heritage Convention's when they added it to the list of sites and monuments in 1986. There can be few sites where World Heritage status preceded national recognition. National Nature Reserve designation followed in 1987 and the much broader Area of Outstanding Natural Beauty in 1989.

Despite this high level of recognition there are still unresolved issues at the site, particularly the paucity of geological information on the approach to and at the causeways themselves and the persistence of the crude Health and Safety equipment that defaces the promontories. While there is unfettered freedom to explore the causeways, access to the full sections in the cliffs to the east is still prevented by a failure to maintain the spectacular lower cliff path.

This case study of the Giant's Causeway shows just how complex is the interweaving of scientific and general culture. It also shows how late and almost incidental was the formal recognition and designation that provided a measure of conservation. The understanding of this site demands an appreciation of this rich, colourful and eventful history.

Although not directly related to the Causeway itself, the history is also indelibly tied by guides and guidebooks alike to the final foundering of the Spanish Armada. On 30 October 1588 the galleass *Girona* was driven onto Lacada Point at the foot of 100 m high basalt cliffs 1 km NE of the Causeway. She was heavily overloaded with over 1300 Spanish aristocrats, remnants of soldiery and crews rescued from earlier wrecks. All but five perished. The wreck site was relocated in 1967 and yielded a considerable amount of armament and personal treasure now in the collections of the Ulster Museum.

The significance of sites

The Giant's Causeway is not unique in its concentration of interest. The UK's second Earth Science World Heritage Site, the Jurassic Coast, in Dorset, is similarly rich but with contrasting themes: the history of palaeontology; stratigraphy; building stones; commercial collecting; literary fiction; a seminal woman collector in Mary Anning, all set in a different, largely nineteenth century, time frame. Durlston Downs, rolling uplands on the Purbeck Limestones west of Swanage, teeming with wild flowers, butterflies and birds, add a new and significant biodiversity component underlining and justifying Stanley's (2002) more holistic approach.

Scientific and intellectual quests have also yielded sites of consequence. The measuring and defining of Phanerozoic time is a particular example. In the nineteenth century the United Kingdom of Britain and Ireland exercised a brand of scientific imperialism. Depending on how one partitions the Cenozoic, UK geologists named six of the eleven or ten of the fourteen Phanerozoic divisions and most of them were in England and Wales (Wyse Jackson 2006). When William Phillips and Daniel Conybeare started work in the early 1820s and Adam Sedgwick and Roderick Murchison a decade later, the span of geological time was acknowledged to be vast but how vast and how the pieces fitted together was yet to be determined. William Smith's map of 1815 had laid the foundations but recognizing and characterizing the sequence in time, the business of defining the limits of geological periods, required the minute and meticulous examination of rocks. This new and demanding practice began to throw up sites that were key to the quest. For Murchison such localities as Cavansham Ferry and the Ludlow sections; for Sedgwick the Berwyn Range to Caernarvon and the Bala Limestone; for Charles Lapworth the centimetre by centimetre probing of 75 m of the Dob's Linn section that proved to be the key to resolving the Cambrian–Silurian controversy (Secord 1986; Hallam 1989) defining the Ordovician in the process. Stratigraphic geology is replete with such examples.

For the small coterie of nineteenth century geologists these were vital localities essential to this new and seductive science but I have yet to find a single word from the principal players to suggest that any were deserving of some kind of special recognition and protection.

Localities are often of less interest than the material they have yielded, particularly when they become problematic. Many are worked-out, flooded, infilled, built-over, or inaccessible in some other way, all of which gives a new importance to the material already secured.

In 1912, a local doctor and amateur geologist, William Mackie, was investigating the rocks of Ord Hill south of Elgin in Aberdeenshire. In a dry stone wall he noticed some unusual cherts and collected a few to take for later examination. Fortunately he was able to prepare his own thin sections and immediately recognized perfectly preserved plant stems with their detailed cellular structure intact. And so the famous Rhynie Chert, with its unique early Devonian terrestrial ecosystem, materialized in the geological consciousness. The site has little rock exposure; most material has been collected as loose blocks or from trenching and drilling. The material remains the main focus of interest and so the collections assume prominence, in this case the Kidston and Lang collections and those in the universities of Glasgow, Edinburgh and Aberdeen and equivalent collection in the British Geological Survey. There are relatively few lagerstätten of this type but thousands of lesser sites world-wide which are now uncollectible.

No traveller from the British Isles with scientific interests undertaking the Grand Tour could exclude the Monte Bolca fish mines in northern Italy from their itinerary. The site was discovered in the sixteenth century and opened specifically to extract its Eocene fossil fishes. Over the years it has yielded thousands of specimens of more than 250 species, 150 genera, 90 families and 19 orders. It has a prominent place in the historic literature and figures prominently in Louis Agassiz's classic five volume work *Recherches sur les Poissons Fossilles* (1833–1844). Specimens from the area have filled collectors' cabinets for over a dozen generations and some of the most exquisite specimens are valued as much for their aesthetics as their scientific importance and command high prices.

The Ludford Lane site of the Ludlow Bone Bed was first described by Murchison in 1839. The remarkable three-dimensional fishes of the Brazilian Santana Formation were discovered in 1817 and have been worked ever since (Spix & Martius 1823–31). Other sites, like Monte Bolca, predate scientific study. Fossils were first found at Holzmaden in 1595, long before it became known for

its large marine vertebrates. The Romans were quarrying plates of Solnhofen Limestone for roofing and floor tiles almost two millennia before it became famous for the *Archaeopteryx* sensation.

All of these sites and their collections beg the question of where the origins of geological conservation truly lie. All were being exploited long before there were any formal thoughts of site protection but the very existence of collections meant that specimens were valued and consequently offered some form of protection while in the hands of their collectors. Indeed some areas, particularly the nascent commercial sites, were already unwittingly offered protection by excluding or discouraging the inquisitive, pointing the way to a very different future.

These cultural treasures are now found mostly, though not exclusively, in the greater and lesser museums of the world. If modern geoconservation is to have any meaning at all it has to be accepted that the vital collections are at least as important as their sites of origin. But it should not be assumed that because a collection is in a museum, it has found a secure repository. The Geological Curators' Group collection surveys (Doughty 1981; Fothergill 2005) show just how precarious is their status in many museums. It is not as though the existence and locations of these collections are unknown. There is a relatively little advertised but virtually complete and searchable database of UK geological collections compiled by the Federation for Natural Sciences Collection Research (http://fenscore.man.ac.uk/fenscore/index.htm). Such a vital search tool should be integrated into the geoconservation mainstream as a matter of urgency. A further useful and more detailed guide to the principal UK collections is the Directory of British Geological Museums (Nudds 1994) which, despite the title, also includes the most significant Irish centres.

Objects, ideas and destinations

The earliest stirrings of geological curiosity reach back to the very origins of civilization. Baltic amber jewellery more than 10 000 years old has been found, of such quality that a discerning collector must have been involved. Flat quartz prisms were treasured in prehistory almost certainly because, used as crude lenses, they were able to kindle a flame. This mystic ability to focus the sun's rays endowed them with great power which may explain their introduction into religious ritual. The priest's breastplate of the Old Testament may have its origins here. Minerals were used as amulets in Babylon, Persia and Assyria, carved into scarabs by the Egyptians and facetted in Persia, Greece and Rome. 3000 years ago 125 gem minerals had been named. The skills of the Greeks and Romans in sourcing and selecting stone for building and decoration is still unsurpassed (Price 2007). The importance of such materials is evident in the writings of Theophrastus, Strabo and Pliny the Elder. All these examples show a discrimination informed by long tradition and experience although the chance find, such as a beautifully preserved cycad trunk placed in an Etruscan tomb 2500 years ago, was not passed over.

The search for the true nature of fossils is a vital part of our inheritance and is now well documented. As early as the sixth century the Chinese scholar Li Tao Yuan described a site called Stone Fish Mountain in Hsiang-hsiang Hsien. He described its fossil fish in minute detail and obviously accepted them as true fish although he offered no thoughts on their setting.

Avicenna, the Moslem Aristotelian philosopher, also had a grasp of the origin of sedimentary rocks and their fossils around 1000 AD. As economic minerals assumed growing importance, a great ferment of ideas swirled around the Earth and its processes. By the fifteenth century speculation about the true nature of fossils, in the original sense of 'something dug up', was rife and a new history based on documents began with Alberti, Allesandri, Gesner, Cardano, Palissy and Imperato, all concerned with the nature of fossils (Rudwick 1972). Leonardo da Vinci had realized the true nature of organic fossils long before Gesner but his thoughts remained in the security of his notebooks. The writings of these early authors record halting steps towards a rudimentary grasp of the workings of the planet and as such form the early, swelling archives of Earth science to be joined from the eighteenth century onwards by an ever widening flood of manuscripts, sections, sketches, maps, paintings, engravings, photographs and digital media.

The lives of the greats of geology, such figures as Buffon, Hutton, de Luc, Saussure, Werner, Cuvier, Smith, Lyell, and Darwin are increasingly probed by modern authors stoking a renewed interest in their residences, personalia and memorials, further elements of the geoscience heritage, often neglected, but ripe for recognition.

The fragmented narrative of this paper is a true reflection of our disjointed attitude to many parts of our science that deserve to be protected for posterity. Still, in many areas, we witness vital material fading into the mists of oblivion without any framework for intervention. The holistic view championed by Stanley (2002) demands crisper definition to become an infrastructure for action.

Breakthrough

This paper gives an outline of what had been achieved by the end of the eighteenth and through the nineteenth century. Geoscience was pregnant with a scientific and cultural heritage poised to deliver in some form but how and with what vehicle was far from clear. When action came it was from a surprising quarter.

Anyone who has read Henry Thoreau's account of his time at Walden Pond between 1845 and 1847 will realize that he was not the starry-eyed idealist that many represent him to be. He took enormous and beautifully expressed delight in his surroundings and gave voice to his great concern for the future of pristine nature and its preservation. *Walden* (1854), his book about this period of seclusion, had little immediate impact but became progressively more influential and is now, rightly, considered a masterpiece. The stage was being set for selective preservation of the environment.

A greater sense of urgency was injected by another American, George Perkins Marsh, who demonstrated man's impact on the environment in his book *Man and Nature* (1864) and in doing so took the balance of argument to its tipping point.

It is generally agreed that it was John Muir, the Scottish-born environmentalist, who was the most influential figure in establishing the first formal area of protection, Yosemite. His restless spirit led him to explore some of the great unspoiled landscapes of the American west and to write of them with enormous passion. It also helped that he had the ear of the President, Theodore Roosevelt, himself a great outdoors man, but that only came about as a result of his sustained and rousing enthusiasm. The power of his surroundings often overwhelmed him as in this passage following his discovery of Yosemite.

19 July 1869. Watching the daybreak and sunrise. The pale rose and purple sky changing softly to daffodil yellow and white, sunbeams pouring through the passes between the peaks and over the Yosemite domes making their edges burn; the silver firs in the middle ground catching the glow on the spiry tops, and our camp grove fills and thrills with glorious light. Everything awakened alert and joyful; the birds begin to stir and innumerable insect people. Deer quietly withdraw into leafy hiding places in the chaparral; the dew vanishes, flowers spread their petals, even the pulse beats high, every life cell rejoices, the very rocks seem to thrill with life. The whole landscape glows like a human face in a glory of enthusiasm, and the blue sky, pale around the horizon bends peacefully down over all—like one vast flower. (Muir 1911)

And by contrast:

Only 30 years ago, the Great Central Valley of California, 500 miles long and 50 miles wide, was one bed of golden and purple flowers. Now it is ploughed and pastured out of existence, gone for ever—scarce a memory of it left in fence corners and along the bluffs of the streams. The gardens of the Sierra, also, and the noble forests in both the reserved and unreserved portions are sadly hacked and trampled, notwithstanding the ruggedness of the topography—all except those of the parks guarded by a few soldiers. In the noblest forests of the world, the ground, once divinely beautiful, is desolate and repulsive, like a face ravaged by disease. This is true also of many other Pacific Coast and Rocky Mountain valleys and forests. The same fate, sooner or later, is awaiting them all, unless awakening public opinion comes forward to stop it. Even the great deserts of Arizona, Nevada, Utah and New Mexico, which offer so little to attract settlers, and which a few years ago pioneers were afraid of, as places of desolation and death, are now taken as pastures at the rate of one or two square miles per cow, and of course their plant treasures are passing away—the delicate ambronias, phloxes, gilias etc. Only a few of the bitter, thorny and unbitable shrubs are left, and the sturdy cactuses that defend themselves with bayonets and spears.

The first passage may be too florid for modern tastes but such impassioned descriptions fired the imagination of the President to the point where he joined Muir in Yosemite and ultimately established the first of the public lands. The second passage, more reflective, has a disconcertingly modern ring.

It was not the weight of scientific opinion that turned the tide in the end but the drive and sense of wonderment in the face of nature of a few North American visionaries. Without them it seems unlikely that the more esoteric corners of conservation could ever have gained a foothold.

Once the dam was broken it released a flood of world conservation demands and it is historically interesting that it is site protection that has achieved such a disconcerting dominance in the field of geological conservation. The range of geological and geomorphological conservation as defined by Stevens (1994) has been developed in more specific terms by the GeoConservation Commission of the Geological Society of London and represents a more reflective and rounded view that should better express our origins and heritage (GeoConservation Commission 1998). It begins with the key sites of the UK heritage as defined by the main agencies and RIGS groups and incorporates the World Heritage sites in our sphere as designated by UNESCO. It embraces the Earth science materials from these sites and all other collections of significance reflecting the seminal investigations of UK geology and geologists. Selected landscapes that express the diversity of UK geology and geomorphology are also included as are the residences, memorials, sites, personalia, manuscripts, maps and libraries associated with the major geological personalities. Archives of original art and illustration, portraits and early photographic collections are also essential components as are the materials in the built environment—building and decorative stones and fabricated products. This may not be an exhaustive list but by focusing on the specifics it indicates some elements of our heritage beyond sites alone—the 'origins' of my title but somewhat neglected. We live in an expansive and increasingly trodden world and our movement urgently needs to secure and consolidate its cultural roots before yet more decay into the mists of faded memory.

References

AGASSIZ, L. 1833–1844. *Recherches sur les Poissons Fossiles.* 5 vols, Neuchatel, Switzerland.

ANGLESEA, M. & PRESTON, J. 1980. A philosophical landscape: Susanna Drury and the Giant's Causeway. *Art History*, **3**, 252–273.

ASHWORTH, W. B., JR. 2004. *Vulcan's Forge and Fingal's Cave. Volcanoes, Basalt, and the Discovery of Geological Time.* Linda Hall Library of Science, Engineering and Technology, Kansas City, Missouri.

BOSWELL, J. 1791. *The Life of Samuel Johnson LL.D.* Carter Header & Co., Boston.

CLIFFORD, S. & KING, A. 2006. *England in Particular.* Hodder & Stoughton, London.

DOUGHTY, P. S. 1981. *The State and Status of Geology in UK Museums.* Miscellaneous Paper, **13**. Geological Society, London.

FOLEY, S. 1694. An Account of the Giants Causeway in the North of Ireland. *Philosophical Transactions of the Royal Society of London*, **18** (212), 170–175.

FOTHERGILL, H. 2005. The State and Status of Geology in UK Museums: 2001. *The Geological Curator*, **8**, 55–136.

GeoConservation Commission. 1998. *A Forum for Earth Heritage Conservation In the UK.* Geo Conservation Commission of the Geological Society of London.

HALLAM, A. 1989. *Great Geological Controversies.* Oxford University Press, Oxford.

HAMILTON, W. 1786. *Letters Concerning the North Coast of County Antrim.* Dublin.

MARSH, G. P. 1864. *Man and Nature.* Scribner and Company, New York.

MOLYNEUX, W. 1697. A true prospect of the Giants Cawsway near Pengore-Head in the County of Antrim. *Philosophical Transactions of The Royal Society of London*, **19** (235), 798.

MUIR, J. 1911. *My First Summer in the Sierra.* Houghton Mifflin, Boston.

NUDDS, J. R. 1994. *Directory of British Geological Museums.* Miscellaneous Paper **18**. Geological Society London.

PRICE, M. 2007. *Decorative Stone.* Thames & Hudson, London.

PROSSER, C. 2002. Speaking the same language. *Earth Heritage*, **18**, 24–25.

RUDWICK, M. J. S. 1972. *The Meaning of Fossils.* Macdonald, London.

RUDWICK, M. J. S. 1992. *Scenes From Deep Time.* The University of Chicago Press, Chicago.

SECORD, J. A. 1986. *Controversy in Victorian Geology.* Princeton University Press, NJ, USA.

SPIX, J. B. & MARTIUS, C. F. P. 1823–1831. *Reise im Brasilien.* 3 vols. Vol. 1: Lindauer; Vol. 2: Lentner; Vol. 3: u.beim Verfasser, Munich.

STANLEY, M. 2002. Geodiversity—linking people, landscapes and their culture. *Natural and Cultural Landscapes, The Geological Foundation.* Royal Irish Academy, Dublin.

STEVENS, C. 1994. Defining geological conservation. *In*: O'HALLORAN, D., GREEN, C., HARLEY, M., STANLEY, M. & KNILL, J. (eds) *Geological and Landscape Conservation.* Geological Society, London, 499–502.

THACKERAY, W. M. 1843. *The Irish Sketch Book of 1842.* Chapman Hall, London, 499–502.

THOREAU, H. D. 1854. *Walden.* Ticknor & Fields, Boston.

TOMKEIEFF, S. I. 1940. The Basalt Lavas of the Giant's Causeway district of Northern Ireland. *Bulletin Volcanologique*, Series 2, **6**, 89–143.

WATSON, P. S. 1992. *The Giant's Causeway.* HMSO, Belfast.

WYSE-JACKSON, P. N. 1997. Fluctuations in Fortune: Three Hundred Years of Irish Geology. *Nature in Ireland.* Lilliput, Dublin.

WYSE-JACKSON, P. N. 2006. *The Chronologers' Quest. Episodes in the Search for the Age of the Earth.* Cambridge University Press, Cambridge.

Geological conservation in the nineteenth and early twentieth centuries

BARRY A. THOMAS[1] & LYNDA M. WARREN[2]

[1]Institute of Rural Sciences, Aberystwyth University, Llanbadarn Fawr, Aberystwyth, Ceredigion SY23 3AL, UK (e-mail: ba.thomas@btopenworld.com)

[2]Department of Law and Criminology, Aberystwyth University, Penglais, Aberystwyth, Ceredigion SY23 3DY, UK

Abstract: Before the middle of the twentieth century there were very few geological reserves in Britain and there was no government legislation to protect them. In other countries and especially in the USA, there were many more such sites protected by a number of legislative processes. In nineteenth century Britain most of the land was owned by comparatively few wealthy people and common land was being steadily reduced through increasing numbers of Enclosure Acts. This meant that there were very few opportunities for conservation action especially as there was no legal basis for doing so other than through land ownership. In the USA the situation was completely different. The westward expansion was in full swing resulting in an increasing amount of federal land holdings owned by Congress. This, together with a desire of the federal government to save special sites for future generations, resulted in the extensive National Parks created by statute and the cultural and national monuments protected by the 1906 Preservation of American Antiquities Act. It took another forty years for Britain to have similar legislation.

The reasons for conserving the natural environment are many and various but underlying all of them is the basic belief that the feature is 'worthy' of conservation because it has some special value. It follows that there can be no conservation if there is no interest and thus no sense of value. The interest in geological conservation is a natural development of the interest in geology that burgeoned in the late eighteenth century and nineteenth century. The formalization of geological conservation through the promulgation of laws is the subject of another paper in this volume (Prosser 2008). This paper describes the context for some of the earliest thinking on geological conservation, focusing in particular on Great Britain and the USA because this is where most of the interest is manifest. It does not attempt to provide comprehensive history of geological scientific work because much of this is not relevant to the history of geological conservation but concentrates on those activities that gave rise to moves to conserve features of 'value' that were in danger of being lost.

Great Britain

Early discoveries and the beginnings of public interest in geology

Popular interest in geology was one of the consequences of the process of industrialization, the success of which depended, in part, on detailed scientific knowledge. By the early nineteenth century, quarrying, mining and the construction of canals and railroads were starting to be carried out on a more scientific basis with the increasing knowledge of geology. Books covering a range of geological topics were published in the late eighteenth and early nineteenth centuries, introducing the public to the Earth sciences. Key amongst these was Hutton's (1785) *Theory of the Earth*, in which he sought to explain geological features in scientific rather than biblical terms. The first geological map of England and Wales was published by the surveyor and civil engineer William Smith in 1815 (for an account of Smith's work see Winchester 2001). Later, there were books on animal fossils by Agassiz and Mantell & Owen (see Dean 1999 for an overview) and plant fossils by Artis 1825, Lindley & Hutton 1831–1837, and Bowerbank 1840.

These works prompted and inspired the enthusiasm of fossil and mineral collectors. The collection of fossils of large marine reptiles made by Mary Anning in the 1820s stimulated scientific work on the evolution of these animals and Richard Owen's choice of the name Dinosauria for the largest of the extinct animals guaranteed public interest in these 'terrible lizards'. The extent of this interest is well illustrated by the dinosaur models at Crystal Palace (Doyle 2008). This 'Dinosaur Court', as it was originally called, was the world's first 'geological theme park'. It was an educational attraction with full scale dinosaur

From: BUREK, C. V. & PROSSER, C. D. (eds) *The History of Geoconservation*.
Geological Society, London, Special Publications, **300**, 17–30.
DOI: 10.1144/SP300.3   0305-8719/08/$15.00 © The Geological Society of London 2008.

Fig. 1. Victorian geological models. (**a**) *Megalosaurus*; (**b**) The newly restored 'Coal Formation' complete with coal seam and two faults.

reconstructions placed on rocks of the same age as that in which the fossils of the animals had been found (Fig. 1). Selective plantings were added to reflect the type of vegetation growing at the time these animals were alive. There were also reconstructions of geological sections and a lead mine complete with artificial stalactites and stalagmites. The reconstructed animals and the geological features have recently been restored in a £4 million project, headed by the London Borough of Bromley with financial support from the Heritage Lottery Fund.

Professional and amateur interest was harnessed by the creation of the Geological Society, London, which was established in 1807, and the Geological Association, which was set up in 1858. There were many exciting geological discoveries during the nineteenth century, as a result of which many rocks, minerals and fossils found their way into private collections or into the new museums that were being built around the country. The eighteenth century collections of Hans Sloane and Joseph Banks formed the basis of the original natural history collections in the British Museum. These collections were moved to the British Museum (Natural History) at South Kensington in 1881 (now known as the Natural History Museum). The Museum of Practical Geology (later to become the Geological Survey Museum) was established in 1841. A number of municipal and university museums were also founded in the nineteenth century, often bringing together smaller collections of geological specimens. These include the Natural History Museum in Dublin which opened in 1857, Oxford University Museum in 1860, Manchester Museum in 1890, the Hunterian Museum Glasgow, founded in 1807 but acquiring its main geological collections in the late 1880s, and the Sedgwick Museum, Cambridge in 1904. Although it was considered worthwhile to build up these

vast collections of geological specimens there was never any thought of preserving localities for their scientific value.

'Indestructable' self-preserved sites

Some sites in the UK had a large degree of self preservation built into them. The chert at Rhynie in Scotland is world-famous for its anatomically preserved Devonian land plants, algae and arthropods, and has been studied by many renowned palaeontologists since it was discovered in 1913 by Mackie, a fossil collector to the Geological Survey. It is now an SSSI (Cleal & Thomas 1995; Barclay *et al.* 2005) but this site survived only because there was no natural outcrop and the only way to obtain material was to dig a trench (Fig. 2). In contrast, the fish locality at Cromarty beach, made famous by Hugh Miller (Miller 1841) was effectively destroyed within 20 years by over-collecting. Other geological sites survived because they were simply too large to be totally destroyed, such as the basaltic Fingal's Cave on Staffa (given to the National Trust for Scotland in 1986 and declared a National Nature Reserve in 2001), intrusive dykes like the Whin Sill, County Durham and Northumberland (Loughlin 2003), coastal features like Chesil Beach, Dorset (May 2003) and Gibraltar Point, Lincolnshire (King & May 2003). Caves are also safe up to a point, although many had most of their stalactites and stalagmites taken as souvenirs, removed for commercial gain, or even just smashed by vandals. Some caves were opened as show caves and kept in almost the same condition as when they were found, e.g. Dan yr Ogof, Wookey Hole, Cox's Cave and Gough's Caves in Cheddar Gorge; see Ford (1990) for further information on cave conservation. One major problem here is encapsulated by the legal maxim *cuius est solum eius est usque ad*

Fig. 2. A trench cut into the Rhynie Chert for the Tenth International Botanical Congress Edinburgh, Scotland, in August 1964.

coelum et ad inferos (he who owns land owns everything reaching up to the very heavens and down to the depths of the earth) (Gray 1987). This means that caves belong *de facto* to the owner of the overlying land. As a result large cave systems might then have several owners even if there is only one entrance although, in practice, property rights in the caves have sometimes been sold off separately from the land above them.

Special reasons for conservation

A few sites were considered to be so special as to be thought of as 'natural monuments' and attempts were made to prevent then becoming spoiled though commercial exploitation. In the vicinity of Edinburgh there are several special sites that have survived for a number of reasons. There are the Dinantian volcanic rocks within Edinburgh itself—Castle Rock surmounted by Edinburgh Castle, Calton Hill with the Observatory and Nelson's Monument, and Salisbury Crags and Arthur's Seat in Holyrood Park—that constitute what has been described as 'one of the prime geological sites in Scotland if not in the whole of Great Britain' (Upton 2003). There is also Agassiz Rock, on the south side of Blackford Hill in Edinburgh, where in 1840 Louis Agassiz saw

evidence of ice action in smoothing, striating and undercutting, confirming and demonstrating to the many sceptics his argument that much of northern Britain had been subject to a geologically recent major glaciation (Gordon 1993).

Hutton's Rock in Holyrood Park is a vein of iron ore (hematite) and is of considerable interest because the geologist William Hutton is reputed to have requested that this unusual and interesting geological feature be saved from quarrying of the Salisbury Crags by the Earl of Haddington who was selling most of it to the town council for his own profit. Lord Haddington held the title of Hereditary Keeper of the King's Park which gave him considerable rights over the land in Holyrood Park, including the right to quarry stone for profit. However, it appears he was somewhat excessive in his quarrying efforts. Stone from the Crags had been taken for hundreds of years without complaint but a case was brought against Lord Haddington by the citizens of Edinburgh to test his right to destroy property that had been entrusted to the safe keeping of his ancestors. In 1831, the House of Lords decided against Lord Haddington. In 1843 Parliament passed an Act authorizing the transfer of the land from the Hereditary Keeper of the King's Park to the Commissioner of Woods and Forests. The Earl's interest was duly bought

(a)   **(b)**

Fig. 3. (**a**) Arthur's Seat and the quarried face of Salisbury Crag; (**b**) Hutton's Rock in the Quarry.

out in 1845 and the transfer completed. The site, represents one of the earliest examples of geological conservation (Fig. 3).

In contrast to this idea of preventing commercial exploitation, the Giant's Causeway in Northern Ireland can be singled out as the greatest geological visitor attraction in the area with more than 350 000 people visiting it each year (Fig. 4). The Bishop of Derry visited the Causeway in 1692 and brought it to the notice of the Dublin intelligentsia and then Sir Richard Bukely gave a paper to the Royal Society about it in 1694. It first came to the public's notice in 1740, through sketches made by a Dublin spinster named Susanna Drury that were turned into engravings, and which were widely distributed throughout Europe, North America and the Far East. During the latter part of the nineteenth century and up to the National Trust's acquisition of the site in 1961, the area was highly commercialized with stalls and huts for servicing the increasing numbers of visitors. There was even a house for the custodian appointed to look after the site. In 1883 the world's first hydroelectric tramway (affectionately called the toast rack) was opened to bring the visitors and in 1887 this was extended to the Causeway Head where two hotels and some guesthouses were built (Hose 2008).

Early conservation efforts for scientific preservation

Among the new fossil discoveries made in the nineteenth century were the large, spreading basal portions of what we now know to be arborescent lycophytes such as *Lepidodendron*, *Lepidophloios* and *Sigillaria*. The bases themselves we call *Stigmaria*. Professor W. C. Williamson moved one of the best examples of a *Stigmaria* to the Manchester Museum (Williamson 1896). Earlier, in 1874, H. C. Sorby, Professor of Geology at Sheffield University College, saw a group of these stigmarias uncovered in excavations for the nearby new Wadsley Lunatic Asylum. He believed that these should be preserved where they were originally growing and ensured that three were protected in specially constructed buildings (Sorby 1875) (Fig. 5b). The site has recently been excavated revealing remains of many more stigmarian bases and fallen stems (Boon 2004). Other groups of stigmaria had been found in the Glasgow area (Buckland 1840; Young 1868) but these were not preserved. Then in 1887 another group was uncovered during excavations in the new Victoria Park in Glasgow that had just been opened to honour the Queen's Jubilee (Young & Glen 1888). The Scottish palaeobotanist Robert Kidston, who lived in Stirling, had become involved in the excavation and probably played a part in persuading Glasgow Council to construct the glass-roofed building that now protects this world-famous 'Fossil Grove' (Fig. 5a). For further information on this site see: McGregor & Walton (1848, 1972), McLean (1973), Lawson & Lawson (1976), Gastaldo (1986) and Cleal & Thomas (1995).

Despite this interest in geology there was no legislation for conservation until Part III of the National Parks and Access to the Countryside Act 1949 provided for important localities to be designated for preservation through the creation of

(a)   (b)

Fig. 4. Giants Causeway, Ulster. (**a**) Columnar basalt in the cliffs and (**b**) on the foreshore with a 30 cm bear for scale.

nature reserves for the purposes of, *inter alia*, pro-viding opportunities for the study of geological fea-tures of special interest in the area and/or for preserving them (section 15). In some other countries legislation for geological conservation was brought in much earlier. The greatest contrast is provided by the early history of geological con-servation in the USA, which is described in detail below.

The USA

The beginnings of a country

In the USA, political, social and economic circum-stances in the early nineteenth century were totally different from those prevailing in the UK at the time. In the early 1800s the basic occupation of the North American colonies was farming and as

(a)   (b)

Fig. 5. (**a**) The stigmarian bases in Victoria Park, Glasgow soon after their discovery (left) and covered by the building (right). (**b**) One of the original buildings at Wadsley covering a stigmarian base.

productivity decreased through loss of fertility of the land westward migration began. The expansion westwards was stimulated by the purchase of land from France in 1803. This so-called 'Louisiana Purchase' almost doubled the size of the Colonies taking in roughly a third of the present continental USA including all the present-day states of Arkansas, Oklahoma, Missouri, Kansas, Iowa and Nebraska and parts of Minnesota, South Dakota, North Dakota, Montana, Wyoming, Colorado, New Mexico, Texas and Louisiana (Miller 1931). To find out what they had actually purchased, President Jefferson commissioned Meriwether Lewis (his secretary and Captain 1st US Regt. Infantry) and William Clark (Lewis's friend) to head a four-year transcontinental expedition of 33 military and non-military men, called the Corps of Discovery, up the Missouri River, across the 'great divide', and along the Columbia River to the Pacific Ocean. Their geographical discoveries expanded American knowledge of the continent and promoted settlement and trade (Dayton & Burns 1997). Migration increased enormously after the 1812 War of Independence and several of the states began geological surveys in aid of their failing agriculture. In 1824 Congress passed the General Survey Act that authorised the Army Engineers to make engineering surveys for roads and canals, and national military, commercial and postal purposes. However, geological surveys were not included in the federal remit for another decade.

Federal policy changed and in 1834 the Topographical Bureau of the US Army began to prepare a geological map of the United States, one year before the British Geological Survey was established. But it was too ambitious an aim and was abandoned two years later. More important was the establishment of the US Army Corps of Topographical Engineers in 1838 whose aim was to explore and map the continent. With the westward migration well under way the Topographical Engineers had their work cut out keeping up with the settlers' need to know routes and the possibilities for agriculture. Migration westwards began early in 1834 following army expeditions and fur traders into what is now Oregon, resulting in what was called the Oregon Trail. The increasingly large numbers of people heading west eventually led to the war with Mexico that resulted in the brief Republic of Texas, which was incorporated into the USA in 1845 with the boundary between the two countries being set at the Rio Grande. The war also gave the USA the land consisting of California, Nevada, Utah and part of Arizona. Oregon was then purchased in 1846 at the same time that the Oregon Treaty fixed the boundary with Canada. US Cavalry expeditions were sent into these territories to 'pacify' the native Indians and to provide assistance to Topographical Engineers to map routes for wagon trails and later railroads. The Mormon Trek began in 1845 and the California 'gold rush' of 1848–1855 gave added impetus for finding an east–west route, especially when California became a state in 1850.

Army expeditions and geological discoveries

In 1849 the first US cavalry expedition entered Navajo lands; part of what would eventually become the state of Arizona. Led by Colonel John M. Washington and accompanied by Lieutenant James H. Simpson of the US Army Corps of Topographical Engineers, this expedition was the first to encounter petrified logs (Simpson 1850) (Fig. 7). Other expeditions into the area led by Captain Lorenzo Sitgreaves, in 1851, and Lieutenant Whipple in 1853, found many more petrified logs (Fig. 6) in what was eventually to become the Petrified Forest National Park (Ash 1969, 1972; Thomas 2005). It was Whipple who named Lithodendron Creek (now Lithodendron Wash) because of the enormous numbers of petrified logs he found there. A geologist, Jules Marcou, accompanied the Whipple expedition and in his report (1855) dated the rocks more or less correctly as Triassic and he correctly identified most of the wood as coniferous. A further expedition, led by Lieutenant Joseph Ives, was accompanied by John Newberry who described the discovery of further coniferous wood (Newberry 1861). Then Second Lieutenant John F. C. Hegewald collected and shipped a log to Washington on the orders of General Sherman. It is allegedly this log that is now in the Smithsonian Institution although there is some doubt as to its provenance (Ash 1972).

The motivation for these later expeditions was to survey four possible routes for a transcontinental railway. Then in 1859, silver was discovered in Nevada, prompting a new rush for claims, and the first successful oil well was drilled in Pennsylvania. However, the American Civil War (1861–1865) and the following Indian Wars effectively stopped these army expeditions as the soldiers were either fighting or manning the western army forts in New Mexico and Arizona.

Roughly half a million people moved west during the Civil War, with a third going to Oregon and a third to California. After hostilities finished there was increasing enthusiasm to settle in these new western territories and the army expeditions recommenced taking with them topographers and scientists. The major problems were friction between settlers and the Indian tribes already there and comparative lawlessness. The

(a)   (b)

Fig. 6. (**a**) Simpson; (**b**) Early topographical engineers posing on a rocky outcrop.

US cavalry increased its activities to combat both of these 'problems'. In 1874, Lt Colonel George Armstrong Custer had confirmed gold to be present in the area now known as the Black Hills of South Dakota and Wyoming. Gold miners quickly followed Custer and then Colonel Richard I. Dodge took a scientific team into the area, even though this was in direct violation of Indian treaty rights. In the mountainous area of what is now Wyoming they saw a towering mountain of fluted stone rising 1280 feet from the valley (Fig. 8). Dodge called it the Devil's Tower, taking its name from

Fig. 7. Lithodendron Wash.

Fig. 8. Devil's Tower in Arizona. Note the size of the trees for scale. Courtesy of the US National Parks Service.

one of its Indian names, literally translated as 'Bad God's Tower'. It is now known to be an igneous intrusion into sedimentary rocks of phonolite porphyry in 4 to 7-sided columns. The sedimentary rocks subsequently eroded away leaving the exposed tower. The downside of such expeditionary forces into Indian lands was Indian outrage at having their sacred lands defiled. The US government attempted to buy the Black Hills for 6 million dollars, but their offer was refused. In 1876 the town of Deadwood was established in the area and this provoked the Sioux into action. Following a six hour battle with General George Cook at Rosebud Creek, the Sioux, led by Sitting Bull, and the Cheyenne camped near the Little Bighorn in Montana. Here, Custer and the 7th Cavalry rather foolishly attacked them resulting in the virtual annihilation of his troops. The loss of Custer's 7th Cavalry led to reprisals against the Indians so Sitting Bull took his tribe into Canada. In the end, the dispute over the sacred lands was resolved by the expediency of redrawing the Indian Reservation boundary to leave the Black Hills outside of it! Devil's Tower was now owned by Congress.

Meanwhile, further south, settlers began to establish themselves in the Northern Arizona Territory from the late 1870s. They were encouraged to settle by the Desert Land Act of 1877, which

granted up to 640 acres of public land to any citizen at a cost of $1.25 per acre.

Exploitation or conservation

The petrified forest found by the army expeditions was now within reach of settlers who soon began to take souvenirs. Then George F. Kunz published popular accounts of the fossil forest making much of the large quantities of beautiful silicified wood (Kunz 1885, 1886, 1890; Fig. 9). News spread fast and professional collectors and jewellers came in increasing numbers to take away bits of the wood. Some even started to blow up the larger logs in search of the occasional amethyst crystals that might be found in cavities in the wood. Then the Acheson Topeka and Santa Fe Railroad met up with the Gulf Colorado and Santa Fe Railroad and continued westwards to Flagstaff and eventually to Los Angeles (Berkman 1988). This took the railroad through the middle of the petrified forest so the way was now open to collect and transport the larger logs. The Drake Company in Sioux Falls, South Dakota, transported large loads of logs to their works where they were cut and polished. By this means polished sections from '6 inches to 5 feet in diameter' and from '50 to 2500 pounds in weight' were offered for sale. The Drake Company exhibited specimens at the popular World's Fair in Chicago in 1893, showed them in New York at Davis Callamar & Co. Ltd on Broadway, and exhibited at the Paris Fair in 1898 (Fig. 10). One large section exhibited at Paris was offered for sale to the British Museum (Natural History) for 2000 francs, making the claim that such large sections were never likely to be offered for sale in the future. This was in consequence of the growing disquiet of the locals at the ever-increasing destruction of the petrified forest. There seemed to be no way of stopping the commercial pillage because the forest was on public land and railway land and there were no laws to prevent it. The Mining Law of 1872 had provided for the localization and patent of some mineral deposits such as gold, silver, lead and zinc and it is fortunate that it did not extend to petrified wood or else there would have been a claims rush and all would have been lost to individuals.

In the early 1890s a crushing mill was built in the railroad town of Adamana to turn the petrified logs into abrasives. This was a step too far, so the Legislature of the Arizona Territory petitioned Congress to create a Petrified Forest National Park. Congress had already established some National Parks, on what had been public land, for 'the benefit and enjoyment of the people'. There was Yellowstone National Park in 1872 (US Statutes at Large, Vol. 17, Chap. 324,

(a)

(b)

Fig. 9. Early panoramic views of the Petrified Forest. (**a**) The bridge; (**b**) Scattered logs.

BOIS PÉTRIFIÉS, COMTÉ D'APACHE, ARIZONA, ÉTATS-UNIS.
DRAKE COMPANY, ST. PAUL, MINN., E. U. A.

Fig. 10. The Drake Company's advertisement at the Paris Fair.

pp. 32–33 [S. 392]), Sequoia National Park in 1890 (US Statutes at Large, Vol. 26, Chap. 926, p. 478 [HR. 1570]) and Yosemite National Park also in 1890 (US Statutes at Large, Vol. 26, Chap. 1263, pp. 650–652 [HR. 1263]). There are several petrified forests in the Yellowstone National Park (Amethyst Mountain, Specimen Ridge, Tower Falls, Cache Creek, and several other smaller sites). Knowlton (1928) stated that 'there is hardly a square mile in the north eastern portion of the park that is without its fossil forest, scattered trunks, or erratic fragments' and described them as 'the most remarkable fossil forests known'. Even more encouraging for the Arizona Legislature was the establishment of the Wind Cave National Park in South Dakota for its unusual features (US Statutes at Large, Vol. 32, part 1, Chap. 634, pp. 765–766 [Public Act No. 16]). Congress acted quickly on the petition from the Arizona Legislature and sent Professor Lester F. Ward of the US Geological Survey to survey the area and make recommendations on the need for Congressional interference. After visiting Arizona, Ward recommended, in November 1899, that the area be withdrawn from public use and a National Park established to protect the forests (Ward 1900, 1901). Congress again acted quickly to withdraw the area with the petrified logs from public use. This afforded some protection from exploitation of the logs even though Congress did not establish a National Park for the forest.

Legislation

The real breakthrough for geological conservation came when the Act for the Preservation of American Antiquities (34 Stat. 225) was passed by Congress in 1906 as a means to protect some of America's cultural and scientific resources. The Act was initially intended to protect archaeological records preserved on federal lands, but the terms 'object of antiquity' and 'object of scientific interest' also applied to fossils, although its use in this context was generally limited to controlling the excavation of vertebrates. A system was developed within the meaning of the Act to allow 'qualified' institutions, 'reputable' museums, universities, or other recognized scientific or educational institutes to undertake research on federal lands. The Act made it clear that materials collected on federal land remained public property and any specimens collected through permits issued under the Act must be stored in a museum and be accessible to the public (Clemens 1988). Additional laws and regulations concerning fossils collected on federal lands, especially since the late

1940s, made land management much more complex (see Raup et al. 1987).

The Act for the Preservation of American Antiquities also gave the President of the United States direct authority to set aside areas of federal land of significant scientific or scenic values as National Monuments. Devil's Tower, in Wyoming, was the first National Monument to be declared on 24 September 1906 by President Theodore Roosevelt. This was the world's first legally protected geological site. Then on 8 December in the same year Roosevelt declared the Petrified Forest National Monument on the grounds of the site's 'scientific interest and value' making it the world's first legally protected palaeontological site. Early details of the Petrified Forest National Monument are given in Merrill (1911), Anon (1949) and Broderick (1951). The area of the National Monument has been changed several times since its inception. In December 1962, with an area of 93 500 acres, it was declared a National Park and in 2004 President George W. Bush signed the Petrified Forest National Park Expansion Act adding another 125 000 acres to more than double its size. More information on the park and its fossils can be found at http://www.nps.gov/pefo/. Two other Monuments that conserve palaeontological sites are the Dinosaur National Monument in Utah and the Florissant Fossil Beds in Colorado.

The National Park Service was established in 1916 under the Organic Act with the mission 'to conserve the scenery and the natural and historic objects and the wildlife therein and to provide for the enjoyment of future generations.' Originally the National Park Service was established to administer areas designated as National Parks, Monuments, and Reservations and therefore took over responsibility for both the Devil's Tower and the Petrified Forest. Today, the Service also administers historical/cultural parks, seashores, scenic river ways, recreation areas and a variety of other federal land designations.

State Legislature can be used to save sites. For example, the Ginkgo Petrified Forest State Park, located near the geographical centre of Washington in Kittitas County, was established by Washington State Legislature in 1935 and the initial Interpretive Centre completed in 1936 (Fig. 11). Over 50 species of petrified trees have been identified here, including oak, beech and elm as well as the Maidenhair Tree (Ginkgo).

Legislation can be revoked if there is thought to be sufficient reason. One of the world's greatest concentrations of Cretaceous cycads had been exposed on the surface of a 320 acre site in the Black Hills during the early years of the twentieth century (Weiland 1916). Weiland obtained the

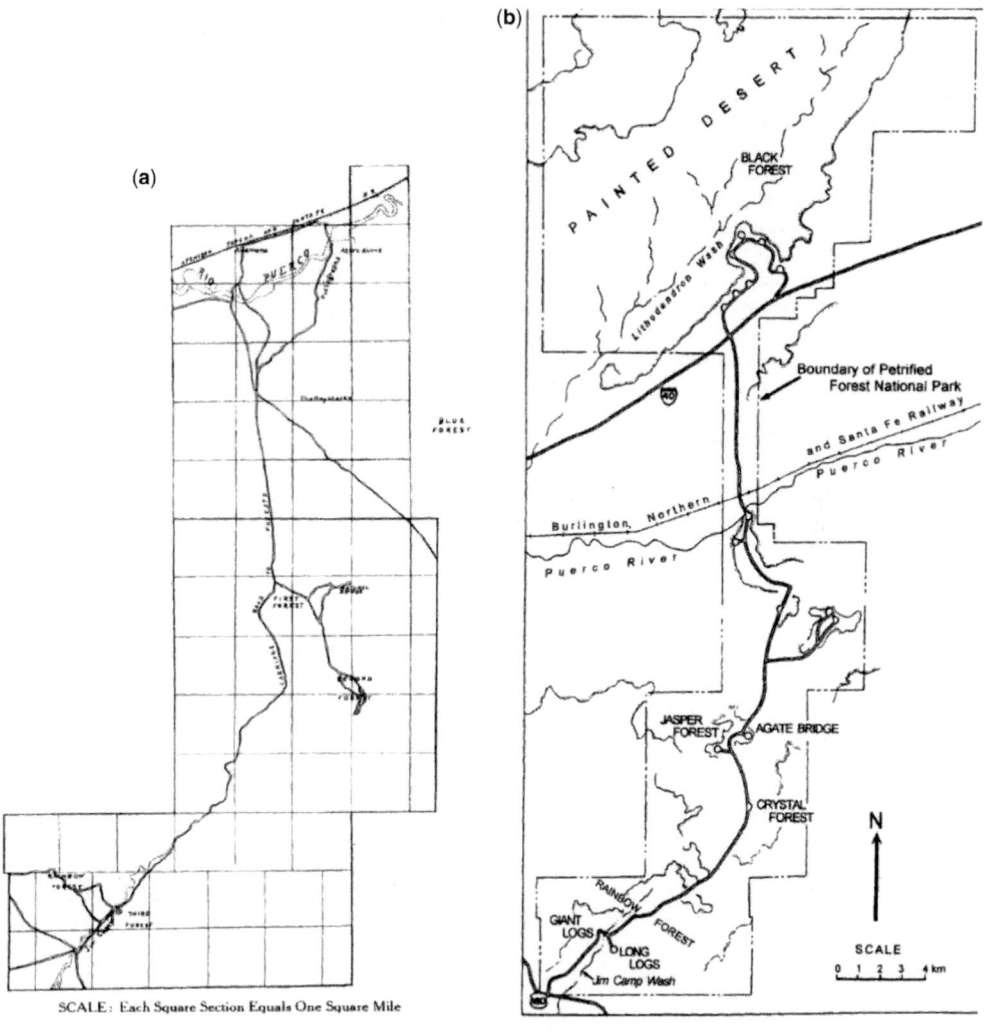

Fig. 11. The changing outline of the Petrified Forest from its creation as a National Monument to the present-day National Park (see Fig. 6 of Thomas 2005).

fossil-rich land under the Homestead Act (37th Congress Session II, Chapter LXXV. An Act to secure Homesteads to actual Settlers on the Public Domain) 'in order that the cycads might not fall into unworthy hands'. He then offered to return the land to the federal government so that a National Monument could be established. President Warren G. Harding proclaimed the Fossil Cycad National Monument on October 1922 under the Antiquities Act (Presidential Proclamation 1641), 'Whereas there are located in section thirty-five, township seven south, range three east of Black Hills Meridian, South Dakota, rich deposits of fossil cycads and other characteristic examples of paleobotany, which are of great scientific interest and value'.

However, neglect, and unregulated collecting had led to a near total loss of the resource that the monument was created for and in 1946 not a single cycad could be located at the site. On the advice of the National Parks Service, it was abolished as a National Monument (Senate Bill 1161). The Bill was signed into law on 1 August 1956 and became effective on 1 September 1957. The legislation included 'That if any excavations on such lands for the recovery of fissionable material or any other minerals should be undertaken, such fossils remains discovered shall become property of the Federal government.' The land was then turned over to the Bureau of Land Management on 6 December 1957 through Public Order 1562 by the Assistant Secretary of Interior. In 1980 construction of a public highway unearthed fossil cycads. Members of the public nominated the Fossil Cycad area for Area of Critical Environmental Concern designation under Regulation 301CMR12.00. The regulations direct the Executive Office of the Environment Affairs Agency to take actions, administer programmes, and revise regulations to preserve, restore or enhance Areas of Critical Concern. Therefore, the Bureau of Land Management published an environmental assessment and prepared a Draft Amendment to the state Resource Management Plan that in summary recommended keeping the land in public ownership, allowing rights of way but restricting activities to protect the area, and recovering any exposed fossils to make them available for research. It appears that the Fossil Cycad National Monument should never have been abolished at all (see Santucci & Hughes' website for further details).

Discussion

There was clearly a major difference in the early approach to geological conservation in the UK and the USA. This was directly related to land ownership.

In the USA, the coastal colonies had originally claimed land to the west of their boundaries, but in 1781 they ceded their claimed lands west of the Allegheny Mountains to Congress. This allowed the Articles of Confederation to be drawn up and then resulted in the Land Ordinance of 1785 that was set up to provide a plan for surveying and disposal of land for revenue and to encourage settlement. It also reserved 'one third part of all gold, silver, lead and copper mines to be disposed of as Congress decided'. By 1879 there were eight classes of public lands with separate regulations for their disposal. Indian Treaty Lands were set aside, although these Treaties were often broken for financial reasons. Much of the land was, therefore, federal until it was disposed of through sale, land allocation for homesteading (e.g. the Great Oklahoma Land Rush in September 1893 when 42 000 parcels of land were opened for settlement) or turned over to the new states when they ceased being territories. This made it easy for Congress to approve the establishment of National Parks which were true wilderness areas (Category I of the IUCN–Anon 1994) for the President to declare National Monuments, and for the states to declare State Parks. The mix of federal land, private land and mixed estate land (where only the surface rights were sold and the federal government retained subsurface ownership i.e. the mineral

rights) varies considerably across the country. About one third of the United States (nearly 740 million acres) is owned by the federal government. Within the 48 contiguous states most federal land is west of the Mississippi in the Great plains, the American far west and the Rocky Mountains. The percentage of federal land in some states is high, comprising for example 86% of Nevada, 64% of Utah and Idaho and over 40% of Arizona, California, Oregon and Wyoming (Clemens 1988). Many of these federal lands include important fossil sites that need to be accessible for research but protected from commercial collecting. The problem here has been that the mix of federal, state and local government agencies responsibilities for land management has sometimes resulted in a need for permits to be issued by several agencies for legitimate research and a confusion about which agency is responsible for policing the system.

In Britain the situation developed along completely different lines. Although under feudal law all of the land under control of the Normans was owned by the monarch and so might be thought to be under state control, this ownership was not linked to any duty to use the land for the benefit of the population at large. Instead parcels of land were granted to loyal followers in return for services performed. In time a system of private ownership developed that differed from the present system mainly in the small number of landowners. Land was held in large estates that passed from generation to generation under a complicated system of inheritance law.

Land ownership brought with it almost complete power and control of the land; a suitably minded landowner could easily protect any feature of the land and he (and only very occasionally she) could equally well destroy it. The power of the landowner increased with the enclosure of open common land under the numerous Enclosure Acts of the eighteenth and nineteenth centuries. By the end of the Georgian era over 7 million acres of land had been taken into private ownership under some 5000 Enclosure Acts. Enclosures continued through the reign of Queen Victoria with the last major Enclosure Act being for Skipwith Common in Yorkshire in 1903. So, while federal land was increasing in the USA, much of the remaining common land in Britain was being parcelled off to individuals who walled and/or hedged it in to exclude the original commoners. In a class-ridden society where politicians were drawn from, and elected by, a land-owning minority, there was no will in Parliament for the introduction of national state-run regimes, for National Parks, reserves or national monuments. It took two world wars and a returning post-war Labour Government before statutory protection could be achieved under Public Act of Parliament (see Prosser 2008 for details of post war conservation in Britain).

The seeds of change, however, were being sown in the closing years of the nineteenth century as the middle classes began to take an interest in science, the arts and philanthropy. Surprisingly, UK legislative history of site designation has its roots in the aesthetics rather than the science of nature. The National Trust for Places of Historic Interest and Natural Beauty was founded in 1895 and its Memorandum of Association under the Companies Act states its purpose to be 'to promote the permanent preservation, for the benefit of the Nation, of lands ... of beauty and historic interest; and ... to preserve (so far as practicable) their natural aspect, features, and animal and plant life.' Here then, is a recognition that sympathetic land ownership is necessary. Although the land would still be in private hands with no state interference, public recognition of the importance of preserving land in this way is indicated by the passing of a Private Act of Parliament in 1907 which gave the National Trust powers to declare inalienable those of its properties deemed to be held 'for the benefit of the nation.' For its time this was a major chink in the armour of the landed gentry.

References

ANON. 1949. *Petrified Forest National Monument Arizona*. United States Department of the Interior, National Park Service.

ANON. 1994. *Guidelines for Protected Area Management Categories and the World Convention*. Monitoring Centre, Gland, Switzerland and Cambridge, UK.

ARTIS, E. T. 1825. *Antidiluvian Phytology*. The author, London.

ASH, S. R. 1969. Ferns from the Chinle Formation (Upper Triassic) in the Fort Wingate Area, New Mexico. *US Geological Survey, Professional Paper*, **613-D**.

ASH, S. R. 1972. The search for plant fossils in the Chinle Formation. *In*: BREED, C. S. & BREED, J. W. (eds) Investigations in the Triassic Chinle Formation. *Museum of North Arizona Bulletin*, **47**, 45–58.

BARCLAY, W. J., STONE, P. & TREWIN, N. H. 2005. Rhynie, Aberdeenshire. *In*: BARCLAY, W. J., BROWNE, M. A. E., McMILLAN, A. A., PICKETT, E. A., STONE, P. & WILBY, P. R. (eds) *The Old Red Sandstone of Great Britain*. Geological Conservation Review Series, No. 31, Joint Nature Conservation Committee, Peterborough, 117–124.

BERKMAN, P. (ed.). 1988. *The History of the Atchison, Topeka and Santa Fe*. Brompton Banks Corp. Greenwich, CT.

BOON, G. 2004. Buried treasure—Sheffield's lost fossil forest laid to rest (again). *Earth Heritage*, **22**, 8–9.

BOWERBANK, J. S. 1840. *A History of the Fossil Fruits and Seeds of the London Clay*. John van Voorst, London.

BRODERICK, H. J. 1951. Agatized Rainbows. *A story of the Petrified Forest*. Popular series No. 3. Petrified Forest Museum Association, Holbrook, Arizona and the Arizona State Highway Department.

BUCKLAND, W. 1840. Anniversary Address to the Geological Society of London. *Proceedings of the Geological Society, London*, **III**, 231.

CLEAL, C. J. & THOMAS, B. A. 1995. *Palaeozoic Palaeobotany of Great Britain*. Geological Conservation Review Series No. 9. Chapman and Hall, London.

CLEMENS, W. A. 1988. Challenges of management of palaeontological site resources in the United States. *In*: CROWTHER, P. & WIMBLEDON, W. A. (eds) *Special Papers in Palaeontology*, **40**, 173–180.

DAYTON, D. & BURNS, K. 1997. *Lewis & Clark. The Journey of the Corps of Discovery: An Illustrated History*. Knopf.

DEAN, D. S. 1999. *Gideon Mantell and the discovery of the Dinosaurs*. Cambridge University Press.

DOYLE, P. 2008. A vision of 'deep time': the 'Geological Illustrations' of Crystal Place Park, London. *In*: BUREK, C. V. & PROSSER, C. D. (eds) *The History of Geoconservation*. The Geological Society, London, Special Publications, **300**, 197–205.

FORD, T. 1990. Caves and Conservation. *Earth Science Review*, **28**, 6–8.

GASTALDO, R. A. 1986. An explanation for lycopod configuration. 'Fossil Grove' Victoria Park, Glasgow. *Scottish Journal of Geology*, **22**, 77–83.

GORDON, J. E. 1993. Agassiz Rock. *In*: GORDON, J. E. & SUTHERLAND, D. G. (eds) *Quaternary of Scotland*. Geological Conservation Review Series, No. 6, Joint Nature Conservation Committee, Peterborough, 565–568.

GRAY, K. 1987. *Elements of Land Law*. Butterworths, London.

GUNNING, R. 1995. *The Fossil Grove*. Glasgow Museum, Glasgow.

HOSE, T. A. 2008. Towards a history of geotourism: definitions, antecedents and the future. *In*: BUREK, C. V. & PROSSER, C. D. (eds) *The History of Geoconservation*. The Geological Society, London, Special Publications, **300**, 37–60.

KING, C. A. M. & MAY, V. J. 2003. Gibraltar Point, Lincolnshire. *In*: MAY, V. J. & HANSOM, J. D. (eds) *Coastal Geomorphology of Great Britain*. Geological Conservation Review Series, No. 28, Joint Nature Conservation Committee, Peterborough, 439–443.

KNOWLTON, F. H. 1928. *Fossil Forests of the Yellowstone National Park*. United States Government Printing Office, Washington.

KUNZ, G. F. 1885. On the agatized woods and the malachite, azurite, etc. from Arizona. *Transactions of the New York Academy of Sciences*, **5**, 9–11.

KUNZ, G. F. 1886. Agatized and jasperised wood of Arizona. *Popular Science Monthly*, **28**, 362–367.

KUNZ, G. F. 1890. *Gems and Precious Stones of North America*. New York Science Publishing Company. (Reprinted by Dover Publications, New York, 1968.)

LAWSON, J. A. & LAWSON, J. D. 1976. *Geology explained around Glasgow and South-west Scotland, including Arran*. David & Charles, Newton Abbott.

LINDLEY, J. & HUTTON, W. 1831–1837. *The Fossil Flora of Great Britain; or, Figures and Descriptions of the Vegetable Remains Found in a Fossil State in this Country.* Volume 1 (1831); Volume 2 (1835): Volume 3 (1837). John Ridgeway, London.

LOUGHLIN, S. C. 2003. Tholeitic sills and dykes of Scotland and northern England. *In*: STEPHENSON, D., LOUGHLIN, S. S., MILLWARD, D., WATERS, C. N. & WILLIAMSON, I. T. (eds) *Carboniferous and Permian Rocks of Great Britain North of the Variscan Front.* Geological Conservation Review Series, No. 27. Joint Nature Conservation Committee, Peterborough, 217–277.

MARCOU, J. 1855. Résumé explicatif d'une carte géologique des États-Unis et des provinces anglaises de l'Amerique du Nord, avec un profil géologique allant de la vallée du Mississippi aux côtes du Pacifique, et une plance de fossils. *Bulletin de la société Géologique de France*, **serie 2, 12**, 813–936.

MAY, V. J. 2003. Chesil Beach, Dorset. *In*: MAY, V. J. & HANSOM, J. D. (eds) *Coastal Geomorphology of Great Britain.* Geological Conservation Review Series, No. 28. Joint Nature Conservation Committee, Peterborough, 254–266.

MCLEAN, A. C. 1973. Excursion 1: Fossil Grove. *In*: BLUCK, B. J. (ed.) *Excursion Guide to the Geology of the Glasgow District.* Geological Society of Glasgow, Glasgow.

MCGREGOR, M. & WALTON, J. 1948. *The Story of the Fossil Grove at Glasgow Public Parks and Botanic Gardens, Glasgow.* Glasgow D.C. Parks Department, Glasgow.

MCGREGOR, M. & WALTON, J. 1972. *The Story of the Fossil Grove at Glasgow Public Parks and Botanic Gardens, Glasgow*, rev. edn. Glasgow D.C. Parks Department, Glasgow.

MERRILL, G. P. 1911. *The Fossil Forest of Arizona.* Arizona Geological Survey, Tucson.

MILLER, H. 1841. *The Old Red Sandstone or New Walks in an Old Field.* John Johnstone, Edinburgh.

MILLER, H. 1931. *Treaties and Other International Acts of the United States of America.* Volume **2**. Documents 1–40: 1818. Government Printing Office, Washington.

NEWBERRY, J. S. 1861. Geological report. *In*: IVES, J. C. *ET AL.* (eds) *Report upon the Colorado River of the West explored in 1857 and 1858.* U.S. 36th Congress, 1st session. Senate Executive Document and House Executive Document 90, 1–156.

PROSSER, C. D. 2008. The history of geoconservation in England: legislative and policy milestones. *In*: BUREK, C. V. & PROSSER, C. D. (eds) *The History of Geoconservation.* The Geological Society, London, Special Publications, **300**, 113–121.

RAUP, D. M., BLACK, C. C., BLACKSTONE, S. *ET AL.* 1987. *Paleontological collecting.* National Academy Press, Washington, DC.

SANTUCCI, V. L. & HUGHES, M. Fossil Cycad National Monument: a case of paleontological resource mismanagement. http://www2.nature.nps.gov/grd/geology/paleo/pub/grd3_3/focy1.htm (accessed on 12th December 2006).

SIMPSON, J. H. 1850. Journal of a military expedition to the Navaho Country, made in 1849. *US Congress, 1st session Senate Executive Document*, **64**, 56–138. (Reprinted and edited by Frank McNite in 1954 and published by University of Oklahoma Press.)

SORBY, H. C. 1875. On the remains of fossil forest in the Coal-measures at Wadsley, near Sheffield. *Quarterly Journal of the Geological Society, London*, **31**, 458–500.

THOMAS, B. A. 2005. The palaeobotanical beginnings of geological conservation: with case studies from the USA, Canada and Great Britain. *In*: BOWDEN, A. J., BUREK, C. V. & WILDING, R. (eds) *History of Palaeobotany: Selected Essays.* Geological Society, London, Special Publications, **241**, 95–110.

UPTON, B. G. J. 2003. Arthur's Seat Volcano, City of Edinburgh. *In*: STEPHENSON, D., KOUGHLIN, S. C., MILLWARD, D., WATERS, C. N. & WILLIAMSON, I. T., *Carboniferous and Permian Igneous Rocks of Great Britain North of the Variscan Front.* Geological Conservation Review Series, No. 27. Joint Nature Conservation Committee, Peterborough, 54–74.

WARD, L. F. 1900. *Report on the Petrified Forests of Arizona.* Washington, US Department of the Interior, Washington, DC.

WARD, L. F. 1901. *The Petrified Forests of Arizona.* Annual Report, Smithsonian Institute for 1899, 189–307.

WEILAND, G. R. 1916. *American Fossil Cycads: Volume 2.* Carnegie Institution of Washington, Washinton DC. Publication no. 34.

WILLIAMSON, W. C. 1896. *Reminiscences of a Yorkshire Naturalist.* George Redway, London (reprinted in 1985 with additions by Watson, J. & Thomas, B. A.).

WINCHESTER, S. 2001. *The Map that Changed the World: the Tale of William Smith and the Birth of Science.* HarperCollins, London.

YOUNG, J. 1868. Note on the section of strata in the Gillmore Quarry and Boulder Clay on the site of the new University buildings. *Transactions of the Geological Society, Glasgow*, **III**, 298.

YOUNG, J. & GLEN, D. C. 1888. Notes on a section of Carboniferous Strata containing erect stems of fossil trees and beds of intrusive dolerite in the old Whinstone Quarry Victoria Park. *Transactions of the Geological Society, Glasgow*, **VIII**, 227–235.

Geodiversity: the origin and evolution of a paradigm

MURRAY GRAY

*Department of Geography, Queen Mary, University of London, Mile End Road,
London E1 4NS, UK (e-mail: j.m.gray@qmul.ac.uk)*

Abstract: 'Geodiversity' can be defined as the range of geological, geomorphological and soil features. Although the word itself was first used only in the early 1990s, the principles behind its application to nature conservation have a longer history. For example, the search for representative sites has been a guiding principle for conservation site selection in the UK since the Second World War, and can also be detected as the basis for new site selection criteria in the USA, Ireland and many other countries. It is also starting to be used as a means of analysing the existing World Heritage Sites list and may become one factor in assessing new site applications. The word was first widely adopted in Tasmania and has a status equal to biodiversity within the Australian Natural Heritage Charter. Despite some opposition, the term is increasingly being used around the world, but has been adopted most enthusiastically in the UK, where many Geodiversity Audits, Local Geodiversity Action Plans and Company Geodiversity Action Plans have been published or are planned. A national Geodiversity Action Plan will be published in 2008. The term has also been adopted in national planning guidance in the UK and, as a result, is finding its way into regional and local planning policies. The paper concludes with some speculations about its future use in geoconservation.

Geodiversity is the abiotic equivalent of biodiversity and has been defined as 'the natural range (diversity) of geological (rocks, minerals, fossils) geomorphological (land form, processes) and soil features. It includes their assemblages, relationships, properties, interpretations and systems' (Gray 2004, p. 8). At the present time there are up to 5000 named minerals, some of which are extremely rare and could easily be lost. In turn, these minerals combine to create thousands of named rock types. Hundreds of thousands of fossil species have been discovered and probably millions more remain to be uncovered. There are 19 000 named soil series in the USA alone. On top of this there is a huge diversity of physical processes (e.g. fluvial, coastal, glacial, periglacial, slope, aeolian, hydrological, volcanic, tectonic, etc.) and a huge variation in land form and landscape character. The conclusion must be that there is as much geodiversity in the world as biodiversity.

Humans have long understood and exploited this diversity of the physical world. Some sites provided better defense than others. Different rock materials could be put to different uses. Some crops grew better in some soils. Some locations provided water supplies. And some landforms were so distinctive that they were imbued with spiritual significance. This understanding of the range of values of the diversity of the physical world has continued to the present day in which a dazzling array of uses are fashioned from the geodiversity of planet Earth (Gray 2004).

Geodiversity as a geoconservation principle

Although the word 'geodiversity' was first used only in the 1990s, the principles behind its application to nature conservation have a longer history. For example, in the UK the Report of the *Wild Life Conservation Special Committee* (Huxley 1947) that led two years later to the establishment of the Nature Conservancy and Sites of Special Scientific Interest (SSSIs), contains the following quote:

Great Britain presents in a small area an extremely wide range of geological phenomena ... the supply of a steady flow of trained geologists for industrial work at home and overseas, requires that there shall be available in this country a sufficient number of representative areas for geological study. (Huxley 1947, Para 64)

For 'range of geological phenomena' in this quote we could easily substitute 'geodiversity', and 'representative areas' must logically mean areas representative of the country's geodiversity.

Similarly, the Geological Conservation Review (GCR) that undertook a major site selection programme in Britain between 1977 and 1990 was intended to 'reflect the range and diversity of Great Britain's Earth heritage' (Ellis *et al.* 1996, p. 45). Site selection was achieved on three main criteria, one of which was 'sites that are nationally important because they are representative of an Earth science feature, event or process which is fundamental to Britain's Earth history' (Ellis *et al.* 1996, p. 45). Note the use of the words 'range', 'diversity' and 'representative' in this quote.

From: BUREK, C. V. & PROSSER, C. D. (eds) *The History of Geoconservation*.
Geological Society, London, Special Publications, **300**, 31–36.
DOI: 10.1144/SP300.4   0305-8719/08/$15.00 © The Geological Society of London 2008.

Similar uses of geodiversity principles in nature conservation site selection can be detected in other countries. For example, the USA has two main conservation programmes. The National Parks network is world famous and new units can be added if they meet certain criteria, one of which is that new units must not represent a feature already adequately represented in the system. Similarly, to be included on the National Natural Landmarks list, units must be 'one of the best examples of a type of biotic community or geological feature'. In other words, in the USA there is an attempt to conserve different types of geological features, i.e. geodiversity.

Ireland has come late to geoconservation but is now selecting sites. The Irish Geological Heritage programme has identified 16 geological themes, e.g. Precambrian, coastal geomorphology. 'Each theme is intended to provide a national network of Natural Heritage Area sites and will include all components of the theme's scientific interest' (Parkes & Morris 2001, p. 82), i.e. the system is intended to establish a representative selection of Ireland's geodiversity.

Until recently, World Heritage Sites (WHS) have been proposed by countries and accepted by UNESCO if they met the criterion of universal heritage value, i.e. UNESCO adopted a reactive role. In the last few years, IUCN/UNESCO have become more proactive and this includes the geological component of the WHS list. For example, Dingwall et al. (2005) have examined the list to determine whether the geological column is represented in the list. They have discovered a significant gap at the Silurian with no sites of this age represented. They have also proposed establishing a list a 13 geothemes to help in assessing future WHS applications and identifying possible gaps in representation (Table 1). There is a sense here of trying to

ensure that the world's geodiversity is represented in the WHS list. However, at present it is an inadequate representation, and in particular, with only two stratigraphic sites listed, it is evident that most Global Stratotypes have no international protection.

Early use of 'geodiversity'

Kevin Kiernan, working for the Tasmanian Forestry Commission in the 1980s, was using the terms 'landform diversity' and 'geomorphic diversity' and was drawing parallels with biological concepts in using such terms as 'landform species' and 'landform communities' (K. Kiernan, pers. comm.). In one seminar paper in 1991, he made the point that 'The diversity among landforms is just as valid a target as the diversity of life when developing nature conservation programs . . .' (Kiernan 1991). This statement was made the year before the adoption of the Convention on Biodiversity (CBD) at the Rio Earth Summit in 1992 and was certainly ahead of its time. Once the CBD had been adopted, it became obvious to several geologists and geomorphologists independently that there was a physical equivalent and so the word 'geodiversity' was born.

Gray (2004) suggested that the first use of the word was by Chris Sharples (1993, p. 7), also then working for the Tasmanian Forestry Commission, who wrote that 'Geoconservation aims at conserving the diversity of Earth features and systems ('Geodiversity') and allowing their ongoing processes to continue to function and evolve in a natural fashion'. However, he has pointed out that this was printed in October 1993, subsequent to its first use by F. W. Wiedenbein in a German

Table 1. *Proposed geothemes for geological World Heritage sites and number of current sites within each theme after Dingwall* et al. (2005). *Some sites fall into more than one theme*

Geotheme	No. of World Heritage sites
Tectonic & structural features	3
Volcanoes/volcanic features	13
Mountain systems	11
Stratigraphic sites	2
Fossil sites	11
Fluvial/lacustrine systems/landscapes	10
Caves & karst	7
Coastal development	8
Reefs, atolls & oceanic islands	·1
Glaciers & ice caps	6
Ice ages	7
Arid & semi-arid landforms & landscapes	4
Meteorite impact	1

publication in April 1993 (see Wiedenbein 1993, 1994). However, Sharples (1993) appears to remain the first use in English.

Subsequently, the concept was accepted rapidly as the basis for many geoconservation projects in Tasmania (e.g. Kiernan 1996, 1997) and, crucially, was adopted in 1996 as a key principle in the Australian Natural Heritage Charter (Australian Heritage Commission 1996, updated 2002). This gave equal weight to biodiversity and geodiversity in assessing nature conservation sites. For example, Article 5 states that 'Conservation is based on respect for biodiversity and geodiversity. It should involve least possible physical intervention to ecological processes, evolutionary processes and Earth processes'.

The use of 'geodiversity' was debated at an international geoconservation conference held at Malvern, England in July, 1993 'but failed to receive significant support' (Joyce 1997, p. 38) though in the conference volume (O'Halloran *et al.* 1994) the word 'geodiversity' was used by Wiedenbein (1994), whereas Erikstad (1994), Harley (1994*a*) and Todorov (1994) used the term 'geological diversity'. Joyce has been very critical of the word 'geodiversity' since it 'may be attempting to draw too strong a parallel between sites, landscape features and processes in biology and geology' (Joyce 1997, p. 39). In answering these criticisms, Gray (2004, p. 347–348) accepted that there is a political dimension to the use of 'geodiversity' but also argued that 'diversity' is a basic guiding principle underlying all nature conservation, not just bioconservation. As such, the emergence of the concept of 'biodiversity' (referring to the variety of living nature), made the use of 'geodiversity' (referring to the variety of non-living nature) obvious and almost unavoidable.

There was also opposition to use of the word from English Nature in the 1990s. Although Harley (1994*b*, p. 2) wrote in an editorial that 'in this issue of *Earth Heritage*, we explore the interface between 'geodiversity' and other aspects of our natural and cultural heritage', the use of the term was subsequently blocked within English Nature as its Council and Management Board felt that its use would confuse their audiences, and perhaps be seen as jumping on the biodiversity bandwagon. Staff were therefore asked to drop the term (Colin Prosser, pers. comm.).

Further criticisms were made by Vincent (2004) who argued that the term 'remains vague' and 'is not particularly helpful', and preferred the concept of 'geotope' which takes into account landscape evolution and spatial assemblages rather than individual landforms. However, we still need to use geodiversity as a principle to conserve representative examples of landform assemblages.

Increasing adoption of the geodiversity paradigm

The use of the word 'geodiversity' started to creep into various publications during the 1990s. Gray (1997, p. 323), for example commented that 'perhaps one day we will see ... a Geodiversity Action Plan for the UK to rank alongside its biological counterpart'. However, adoption of the term was slow until 2000, but has snowballed since then particularly in the UK. An important international milestone was the publication by the Nordic countries (Sweden, Norway, Finland, Denmark and Iceland) of *Geodiversitet i Nordisk Naturvard* (Johansson 2000). This made the case for conservation of the superb geodiversity of these countries and an English summary has helped to make the case more internationally accessible (Nordic Council of Ministers, 2003). In recent years the term 'geodiversity' has been adopted as the title of a book (Gray 2004) that attempted to establish a theoretical framework for the concept. The term has also been used in several other countries including Spain (e.g. Nieto 2001), Portugal (e.g. Brilha 2005; Azevedo 2006), Poland (e.g. Kozlowski 2004), Japan (Watanabe 2005) and the USA (Santucci 2005).

Despite these last two publications, the adoption of the 'geodiversity' concept has so far been greatest in Australia and Europe, and in Europe, nowhere has 'geodiversity' had more impact than in the UK. Although a *Geodiversity Update* newsletter edited by Mick Stanley and published in 2001–02 was terminated when the Royal Society for Nature Conservation decided to cease all Earth science work, there has been considerable activity elsewhere. One field in which this has occurred is the UK Minerals Industry working in partnership with the nature conservation bodies. In particular, in 2003 English Nature in association with the Quarry Products Association and the Silica and Moulding Sands Association published guidance on *Geodiversity and the minerals industry: conserving our geological heritage* (English Nature *et al.* 2003). The important strength of this is the support of the mineral companies who can see the benefits of their work for geodiversity. Other support from the minerals industry in the UK includes various publications and other initiatives (e.g. www. mineralsandnature.org.uk; www.hanson.biz/files/ pdf/CR_BAPGAP.pdf).

By this stage, opposition to the use of the term geodiversity by English Nature had waned and Prosser (2002*a*, *b*), in discussing geoconservation terminology, accepted its validity. Whereas 'geodiversity' is a value-neutral term describing the diversity of abiotic phenomena, 'geoheritage' is a value-laden term used to identify those specific elements of geodiversity that are selected for geoconservation.

Subsequently, a range of English Nature publications have adopted the term (e.g. Stace & Larwood 2006; Webber *et al.* 2006) and the Joint Nature Conservation Committee has also begun to use it on a UK basis (e.g. Larwood & Durham 2005). Particularly significant has been the Local Geodiversity Action Plan (LGAP) initiative (Burek & Potter 2004*a*, *b*) which involves a wide range of local groups, organizations and individuals in agreeing priorities and actions to conserve local geodiversity using the limited resources available. Supported by English Nature and now Natural England, about 30 of these plans are currently published or in preparation over the UK as a whole, which is a remarkable achievement given that the first LGAP in Cheshire was published in 2003. In addition, the first company Geodiversity Action Plan (cGAP) has now been published for the aggregates industry (Thompson *et al.* 2006) and a national Geodiversity Action Plan to which the local plans can relate will be published in 2008. An important early objective of an LGAP should be a geodiversity audit, though in some cases this has been carried out and published as a separate exercise. In the case of Durham, the geodiversity audit was carried out by the British Geological Survey (Lawrence *et al.* 2004), marking its growing interest in geoconservation matters. Scottish Natural Heritage has also started to use the word 'geodiversity' in the last few years, most notably in the beautifully produced book on the evolution of Scotland's geology and landforms *Land of Mountain and Flood* (McKirdy *et al.* 2007) and in the leaflet *Scotland's geodiversity* (Scottish Natural Heritage 2007).

Geodiversity thinking has also now permeated the land-use/spatial planning system in England. Planning Policy Statement 9 on *Biodiversity and Geological Conservation* (DCLG 2005) resisted the use of the word 'geodiversity' in the title but uses both the term 'geological diversity' and 'geodiversity' in the text of the statement, e.g. 'Local authorities should take an integrated approach to planning for biodiversity and geodiversity when preparing local development documents' (Paragraph 4). Similarly, Mineral Policy Statement 1 on *Planning and Minerals* (DCLG 2006) states that Mineral Planning Authorities should 'consider carefully mineral proposals within or likely to affect regional and local sites of biodiversity, geodiversity, landscape, historical and cultural heritage'. These statements are significant since they indicate a UK government acceptance of the term. As a result, regional and local planning bodies are also beginning to adopt the term. For example, in September 2006, a consultation draft of further alterations to the London Plan was published

that included an innovative new policy on 'geological conservation':

The Mayor will work with partners to ensure the protection and promotion of geodiversity. Boroughs should:

- accord the highest protection to nationally designated sites (SSSIs) in accordance with Government guidance;
- give strong protection in their DPDs to Regionally Important Geological Sites (RIGS) which, in addition to nationally designated sites, includes sites of strategic importance for geodiversity across London.

An example of a local authority introducing the term is the Kent County Council Mineral Development Framework whose draft Primary Minerals Development Control Policies DPD (Kent County Council 2006) contains Policy MDC11c:

Proposals for mineral related development will be required to identify biodiversity and geodiversity interests and where appropriate make provision for their safeguarding, retention and enhancement. Proposals for mineral related development will only be permitted where the proposals have been designed to minimise their impact on the biodiversity and geodiversity of the County.

The status and future of 'geodiversity'

The title of this paper makes the claim that the concept of 'geodiversity' has evolved towards the status of a paradigm. This is a fairly major claim but can be justified since it now has the theoretical/conceptual status and the history of usage that means that it meets various dictionary definitions of 'paradigm'. These definitions include 'a theoretical framework of ideas', 'a generally accepted model of how ideas relate to one another, forming a conceptual framework within which scientific research is carried out' and 'a set of assumptions, concepts, values and practices that constitutes a way of viewing reality for a community that shares them, especially in an intellectual discipline'. In this author's view, under any of these definitions, 'geodiversity' unquestionably has attained the status of a significant geological paradigm.

Some predictions can be made about the future of 'geodiversity'. First, a national Geodiversity Action Plan for the UK will be published in 2008. This will need to be updated every few years but will create a national framework to which LGAPs can relate and a national perspective and plan for promoting geoconservation of geodiversity in the UK. Hopefully it will create an example that other countries will follow. Secondly, we need to see as much of the UK as possible covered by LGAPs that are implemented and updated every few years. Thirdly, having created excellent protected site networks in the UK, it is likely that increased attention will be given to protecting geodiversity in the wider landscape following key principle (ii) of PPS9 (DCLG 2005) which refers to the need to

Table 2. *Proposed management aims for different elements of geodiversity*

Element of geodiversity	Rare or common	Management aims
Rocks & minerals	Rare	Maintain integrity of outcrop and subcrop. Remove samples for curation.
	Common	Maintain exposure and encourage responsible collecting. Encourage sustainable use. Value historical and modern uses of geomaterials.
Fossils	Rare	Wherever possible, preserve *in situ*. Otherwise remove for curation.
	Common	Encourage responsible collecting and curation.
Land form		Maintain integrity of land form/landforms. Encourage authentic contouring in restoration work and new landscaping schemes.
Landscape		Maintain contribution of natural land form, rock outcrops and active processes to landscape. Encourage authentic design in restoration work and new landscaping schemes.
Processes		Maintain dynamics and integrity of operation. Encourage restoration of process and form using authentic design principles.
Soil		Maintain soil quality, quantity and function.

consider geological interests in the wider environment. A commissioned report on this topic has been written for Natural England (Gray 2006). Fourthly, it is likely that more sophisticated geoconservation policies and practices will be developed taking into account the very different strategies needed to protect the different elements of geodiversity. Table 2 is a first approximation to what may emerge as geoconservation management aims.

More speculatively, there needs to be increased international adoption of geodiversity and geoconservation principles, particularly in the developing world where geoheritage losses are probably very significant. We certainly need to know more about the losses that are occurring so that perhaps a Geodiversity Red Book can be established to sit alongside its biodiversity equivalent which lists those species that are rare or threatened with extinction and therefore in need of conservation. Finally, we need to try to achieve equal status for biodiversity and geodiversity in as many strategies, policies and plans as possible. This will involve continuing efforts to educate and inform those decision-makers responsible for developing these policies and ensuring that they are implemented. The task is a huge one but geodiversity as a paradigm has made significant strides in the last few years for the simple reason that it is too important to ignore any longer.

I am very grateful to C. Prosser and C. Sharples for their referees' comments on this paper.

References

AUSTRALIAN HERITAGE COMMISSION. 1996. *Australian Natural Heritage Charter*. Australian Heritage Commission, Canberra.

AUSTRALIAN HERITAGE COMMISSION. 2002. *Australian Natural Heritage Charter*, 2nd edn. Australian Heritage Commission. Canberra.

AZEVEDO, M. T. M. 2006. Geodiversidade e geoturismo na bacia do Tejo portugues—uma abordagem preliminar. *Publicacoes da Associacao Portuguesa de Geomorfologos*, **3**, 161–165.

BRILHA, J. 2005. *Património geologico e geoconservacao*. Palimage Editores, Braga.

BUREK, C. & POTTER, J. 2004a. *Local Geodiversity Action Plans: Setting the Context for Geological Conservation*. English Nature Research Report, **506**. English Nature, Peterborough.

BUREK, C. & POTTER, J. 2004b. *Local Geodiversity Action Plans: Sharing Good Practice Workshop*. English Nature Research Report, **601**. English Nature, Peterborough.

DCLG. 2005. *Planning Policy Statement 9: Biodiversity & Geological Conservation*. Department of Communities & Local Government, London (www.communities.gov.uk).

DCLG. 2006. *Minerals Policy Statement 1: Planning and Minerals*. Department of Communities & Local Government. London (www.communities.gov.uk).

DINGWALL, P., WEIGHELL, T. & BADMAN, T. 2005. *Geological World Heritage: a Global Framework. A contribution to the Global Theme Study of World Heritage Natural Sites, Protected Area Programme*. IUCN, Gland.

ELLIS, N. V., BOWEN, D. Q., CAMPBELL, S. *ET AL.* 1996. *An introduction to the Geological Conservation Review*. Joint Nature Conservation Committee, Peterborough.

ENGLISH NATURE, QUARRY PRODUCTS ASSOCIATION & SILICA AND MOULDING SANDS ASSOCIATION. 2003. *Geodiversity and the Minerals Industry: Conserving Our Geological Heritage*. Entec UK Ltd.

ERIKSTAD, L. 1994. The building of an international airport in an area of outstanding geological diversity and quality. *In*: O'HALLORAN, D., GREEN, C., HARLEY, M., STANLEY, M. & KNILL, J. (eds) *Geological and Landscape Conservation*. Geological Society, London, 47–51.

GRAY, M. 1997. Planning and landform: geomorphological authenticity or incongruity in the countryside. *Area*, **29**, 312–324.

GRAY, M. 2004. *Geodiversity: Valuing and Conserving Abiotic Nature.* Wiley, Chichester.

GRAY, M. 2006. *Conserving Geodiversity in the Wider Landscape: Making it Happen.* Unpublished English Nature Research Report.

HARLEY, M. 1994a. The RIGS (Regionally Important Geological/geomorphological Sites) challenge— involving local volunteers in conserving England's geological heritage. *In*: O'HALLORAN, D., GREEN, C., HARLEY, M., STANLEY, M. & KNILL, J. (eds) *Geological and Landscape Conservation.* Geological Society, London, 313–317.

HARLEY, M. 1994b. Editorial. *Earth Heritage*, **2**, 2.

HUXLEY, J. 1947. *Conservation of Nature in England and Wales.* Report of the Wild Life Conservation Special Committee. HMSO, London.

JOHANSSON, C. E. (ed.) 2000. *Geodiversitet i Nordisk Naturvard.* Nordic Council of Ministers, Copenhagen.

JOYCE, E. B. 1997. Assessing geological heritage. *In*: EBERHARD, R. (ed.) *Pattern & Process: Towards a Regional Approach to National Estate Assessment of Geodiversity.* Australian Heritage Commission, Canberra, 35–40.

KENT COUNTY COUNCIL. 2006. *Primary Minerals Development Control Policies DPD.* Kent County Council, Maidstone.

KOZLOWSKI, S. 2004. Geodiversity: the concept and scope of geodiversity. *Przeglad Geologiczny*, **52**, 833–837.

KIERNAN, K. 1991. *Landform Conservation and Protection.* Unpublished paper for CONCOM regional seminar, Tasmania.

KIERNAN, K. 1996. *The Conservation of Glacial Landforms.* Forest Practices Unit, Hobart, Tasmania.

KIERNAN, K. 1997. *The Conservation of Landforms of Coastal Origin.* Forest Practices Unit, Hobart, Tasmania.

LARWOOD, J. G. & DURHAM, E. 2005. *Involving People in Geodiversity.* Joint Nature Conservation Committee, Peterborough.

LAWRENCE, D. J. D., VYE, C. L. & YOUNG, B. 2004. *Durham geodiversity audit.* British Geological Survey & Durham County Council.

MCKIRDY, A., GORDON, J. & CROFTS, R. 2007. *Land of Mountain and Flood: the Geology and Landforms of Scotland.* Birlinn & Scottish Natural Heritage, Edinburgh.

NIETO, L.M. 2001. Geodiversidad: propuesta de una definición integradora. *Boletin Geológico y Minero*, **112**, 3–12.

NORDIC COUNCIL of MINISTERS. 2003. *Diversity in Nature.* Nordic Council of Ministers, Copenhagen.

O'HALLORAN, D., GREEN, C., HARLEY, M., STANLEY, M. & KNILL, J. (eds) 1994. *Geological and Landscape Conservation.* Geological Society, London.

PARKES, M. A. & MORRIS, J. H. 2001. Earth science conservation in Ireland: the Irish Geological Heritage programme. *Irish Journal of Earth Sciences*, **19**, 79–90.

PROSSER, C. 2002a. Terms of endearment. *Earth Heritage*, **17**, 12–13.

PROSSER, C. 2002b. Speaking the same language. *Earth Heritage*, **18**, 24–25.

SANTUCCI, V. L. 2005. Geodiversity & geoconservation. *George Wright Society Forum*, **22**, 4–34.

SCOTTISH NATURAL HERITAGE. 2007. *Scotland's Geodiversity.* Scottish Natural Heritage, Inverness.

SHARPLES, C. 1993. *A Methodology for the Identification of Significant Landforms and Geological Sites for Geoconservation Purposes.* Forestry Commission, Tasmania.

STACE, H. & LARWOOD, J. G. 2006. *Natural Foundations: Geodiversity for People, Places and Nature.* English Nature, Peterborough.

THOMPSON, A., POOLE, J., CARROLL, L., FOWERAKER, M., HARRIS, K. & COX, P. 2006. *Geodiversity Action Plans for Aggregate Companies: a Guide to Good Practice.* Report to the Minerals Industry Research Organisation. Capita Symonds Ltd., East Grinstead.

TODOROV, T. A. 1994. Earth science conservation in Bulgaria. *In*: O'HALLORAN, D., GREEN, C., HARLEY, M., STANLEY, M. & KNILL, J. (eds) *Geological and Landscape Conservation.* Geological Society, London, 247–248.

VINCENT, P. 2004. *What's in a Name?* Earth Heritage, **22**, 7.

WATANABE, T. 2005. Geodiversity: necessity of its conservation and research in Japan. *Global Environmental Research*, **10**, 125–126.

WEBBER, M., CHRISTIE, M. & GLASSER, N. 2006. *The Social and Economic Value of the UK's Geodiversity.* English Nature Research Report, **709**. English Nature, Peterborough.

WIEDENBEIN, F. W. 1993. Ein Geotopschutzkonzept für Deutschland. *In*: QUASTEN, H. (ed.) *Geotopschutz, Probleme der Methodik und der Praktischen Umsetzung.* 1. Jahrestagung der AG Geotopschutz, Otzenhausen/Saarland, 17. University de Saarlandes, Saarbrucken.

WIEDENBEIN, F. W. 1994. Origin and use of the term 'geotope' in German-speaking countries. *In*: O'HALLORAN, D., GREEN, C., HARLEY, M., STANLEY, M. & KNILL, J. (eds) *Geological and Landscape Conservation.* Geological Society, London, 117–120.

Towards a history of geotourism: definitions, antecedents and the future

THOMAS A. HOSE

School of Sport, Leisure and Travel, Faculty of Enterprise & Innovation, Buckinghamshire New University, Wellesbourne Campus, Kingshill Road, High Wycombe, Buckinghamshire HB13 5BB, UK (e-mail: those01@bucks.ac.uk)

Abstract: Geotourism is a relatively new form of tourism with considerable growth potential. Initially researched and defined within the UK, it is a growing field of international academic study. The term passed into general usage in the early 1990s, although its antecedents date back to the seventeenth century. Its resource base includes geosites, museum, library and archive collections and artistic outputs. It has significant social history and industrial archaeology underpinnings. Relatively recently defined, and benefiting from a new appreciation of its historical roots and various outcomes, the concept is already undergoing redefinition and refinement. However, because of an inadequately developed historical perspective and theoretical framework, the rationale for its provision and the societal significance of its resource base is not always fully appreciated by existing and potential stakeholders. This account presents geotourism's historical and theoretical development, especially in Britain from which examples are drawn, and explores its likely future.

This chapter provides a background to the development of geotourism by examining its definition, historical antecedents and modern provision. Geotourism is an aspect of the UK's burgeoning heritage industry, contributing to tourism's success even though: 'The purists may not like it. The cynics may sneer at it. But ... there is a powerful argument for protecting the heritage because it earns its own living' (Ross 1991, p. 175). It may provide a rationale and funding for geoconservation. Recounting geotourism's history might appear straightforward involving a mere examination, following its first definition, of 1990s events and literature. Since many of the activities it encompasses have antecedents in considerably earlier natural science and aesthetic movements, the development of tourism that can be directly attributed to the promotion of landscape and geology must be examined; the former dating from the late seventeenth and the latter from the early nineteenth centuries. Geotourism, allied with interpretative provision and geoconservation, is a late twentieth century development that has burgeoned to become a recognized element within some tourism development strategies, especially with the emergence of geoparks. Encompassing three-hundred years' of landscape-focused tourism is perhaps best approached by examining key areas (see Fig. 1). Two that were seminal in early tourism landscape promotion are the Peak District and the Lake District; and central southern England was significant in geotourism's development.

Geotourism at the participant level is 'recreational geology' that, unlike many forms of countryside recreation, is not limited by the seasons (Hose 1996, p. 211). It could extend the season in many traditional, especially coastal, tourism areas and underpin regeneration strategies in old mining and industrial areas. Geotourism, as tourism associated with geological and geomorphological sites (or geosites) and collections, is a form of 'special interest' tourism. The seminal Australian study noted: 'The quality of the rural landscape with regard to features such as peace and quiet and scenery are almost as essential to on-site fossickers as the activity itself ... a good way of introducing children to the outdoors ...' (Jenkins 1992, p. 134); the appeal of such landscapes can only be maintained through enforced geoconservation measures.

Geotourism defined

The Jenkins Australian study (Jenkins 1992) employed 'fossicking', an Australian 1850s 'gold rush' term, to describe geology-focused tourism. Although some authors make fleeting reference to tourism and geology (e.g. Maini & Carlisle 1974; Jenkins 1992; De Bastion 1994; Martini 1994; Spiteri 1994; Komoo 1997; Page 1998) 'geotourism' was undefined until the early 1990s. Within Malaysia, 'geotourism' (Komoo & Deas 1993; Komoo 1997) and 'tourism geology' were employed; the latter for a form of applied geology

From: BUREK, C. V. & PROSSER, C. D. (eds) *The History of Geoconservation.*
Geological Society, London, Special Publications, **300**, 37–60.
DOI: 10.1144/SP300.5   0305-8719/08/$15.00 © The Geological Society of London 2008.

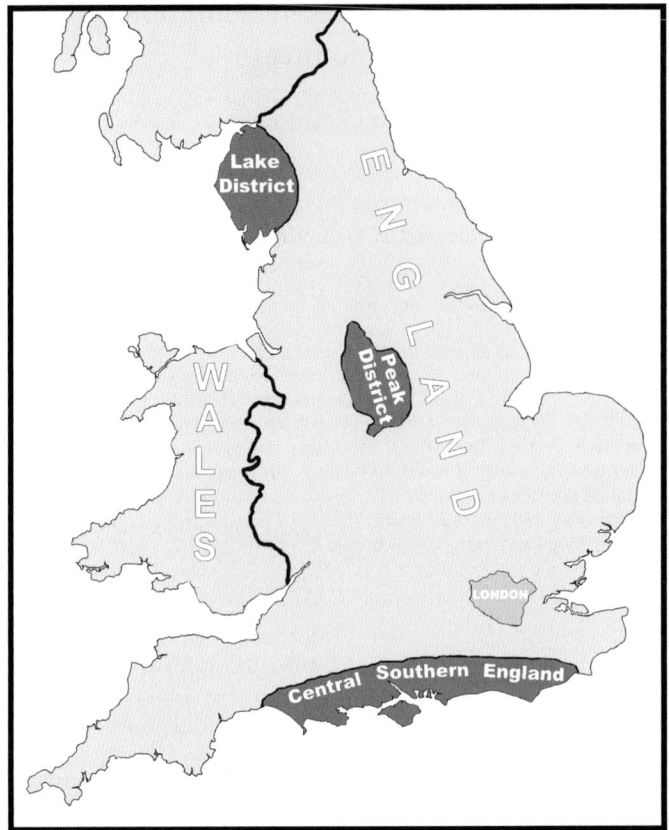

Fig. 1. Map showing the three major geotourism regions.

supporting ecotourism's growth putting: '. . . conservation geology at the same level of importance as . . . conservation biology . . .' (Komoo 1997, p. 2973). Geotourism's first published definition appeared in a 1995 commissioned professional interpretation magazine article: 'The provision of interpretive and service facilities to enable tourists to acquire knowledge and understanding of the geology and geomorphology of a site (including its contribution to the development of the Earth sciences) beyond the level of mere aesthetic appreciation' (Hose 1995a, p. 17). This had evolved from a working definition for evaluation-research informally undertaken for *English Nature* (Hose 1995b) on 'site-specific geological interpretation' 'The promotion and explanation to a non-specialist audience of the geologic features and/or significance of a delimited area by either a fixed facility and/or populist publication' (Hose 1994, p. 2). Following ongoing research it has been refined to: 'The provision of interpretative facilities and services to promote the value and societal benefit of geological and geomorphological sites

and their materials, and to ensure their conservation, for the use of students, tourists and other casual recreationalists' (Hose 2003).

Such definitions encompass an examination and understanding of geosites' physical basis, together with their interpretative media and promotion, as well as geoscientists' lives, work, collections, publications, artworks, field-notes, personal papers, workplaces, residences and even final resting places. Martini (2000) suggested geotourism could fund geoconservation, due to governments' unwillingness to do so. Other European writers have employed broader definitions such as: 'travelling in order to experience, learn from and enjoy our Earth heritage' (Larwood & Prosser 1998, p. 98) but with the apposite assertion that it is partly: 'a consequence of successful Earth heritage conservation' (Larwood & Prosser 1998, p. 98). The chief criticism of such definitions is that, lacking any overt geoconservation message, they might encourage 'geo-exploitation'. Geotourism gained widespread recognition within the UK's geoscience community as a consequence of the first dedicated national conference at the Ulster Museum in 1998.

Geotourism acquired broader usage in the United States of America (USA) where all previous work was ignored alongside *National Geographic*'s claims to have singularly coined the term for: 'a destination's geographic character—the entire combination of natural and human attributes that make one place distinct from another ...' (Stueve *et al.* 2002, p. 1) and is merely sustainable tourism with a holistic approach to landscape. It has been erroneously suggested that their approach has led to the term's rapid acceptance, but since: 'If you travel to see particular scenery or wildlife or experience a particular local culture, climb a particular mountain or kayak a particular river, then in this sense you would be a geotourist' (Buckley 2003, p. 79) it is a superficial re-branding of existing tourism activities; its supporters dismiss geology-focused geotourism as a: 'rather small specialist subsector!' (Buckley 2003, p. 79). Conversely and correctly, the Australian study noted that: 'Fossicking ... comprises one of the world's largest single hobby groups' (Jenkins 1992, p. 129).

Geotourism as special interest tourism

Geotourism *sensu stricto* is a form of 'special interest' tourism: 'when the traveller's motivation and decision-making are primarily determined by a particular special interest ... implies 'active' or 'experiential' travel' (Hall & Weiler 1992, p. 5) that developed from the 1980s as a field of tourism studies when significant changes in the nature of tourism product development and consumption were recognized. It is a growing tourism segment overlapping other emerging tourism forms such as 'eco-tourism', 'sustainable' and 'alternative' tourism and potentially with 'educational travel', 'environmental', 'nature-based' and 'heritage' tourism. The activities and motivations of special interest tourists have been summarized by Weiler & Hall (1992) and similarly for special interest travel which is: 'for people who are going somewhere because they have a particular interest that can be pursued in a particular region or at a particular destination' (Read 1980, p. 195) emphasizing the centrality of interest and place to recreational activity. Two major geotourist groups can be recognized:

1. Educational: from pre-school to postgraduate undertaking geological study as either part of some geographical, environmental or dedicated programme of study; for whom much modern dedicated geotourism provision in the form of visitor centres and trails are intended; and

2. Recreational: amateurs, as individuals or in parties whose idea of an ideal day visit is to look for fossils and pretty rocks and minerals in a pleasant scenic setting, whose expertise ranges from beginner to knowledgeable lifelong enthusiast; for whom geoparks are especially intended.

Some research indicates that tourism involving active components inclined towards conservation, scholarship, science and environmental awareness (Heywood 1990, p. 46), is a restricted market, dependent upon the better-educated and wealthier tourist broadly corresponding to Plog's (1974) 'allocentric' tourist (Hall & Weiler 1992, p. 4). However, this is probably only true of 'dedicated geotourists' (Hose 2000, p. 136), whose interest in geosites is mainly personal education and intellectual improvement. The market for 'casual geotourists' (Hose 2000, p. 136), whose primary motivations are pleasure and social, is considerably larger. Research on the latter group (Hose 1997, pp. 2956–2957; Hose 2000, pp. 137–138; Hose 2005, p. 55) has identified their general characteristics; significantly they visit for informal educational experiences for themselves and accompanying children.

Influences on geotourism's growth

Geotourism's growth is linked to the public's exposure to geology during their school years and to the mass media promotion of geological concepts and attractions. The first national school geology syllabus was published in 1890 by the British Association (Hamilton 1976, pp. 105–107) but failed to influence the late nineteenth and early twentieth century curricula; indeed geology received scant attention in schools, except as an adjunct of physical geography, until the middle of the twentieth century; Cowham's 1900 The School Journey promoting physical geography fieldwork is a classic text. A major constraint on geology's wider promotion in schools was the limited supply of geology graduates; indeed even in the 1930s the UK's universities only produced some thirty geology undergraduates annually (Hamilton 1976, p. 105). In the post-war period of educational innovation geology had a role in schools' scientific education, especially in the newly established secondary moderns in which: 'Experiments in integrated sciences, field sciences, rural and environmental studies developed in the non-examination secondary schools. Its role as an observational integrator with field-based work was accepted with enthusiasm in many areas' (Hamilton 1976, p. 110). There was an underlying assumption that it was unsuitable for grammar schools where it was an adjunct to geography (Hamilton 1976, p. 110), as in today's comprehensive schools, geology was, as earlier in the century, usually taught by geographers: 'in accord with their needs and methods ... any identifiable geology or geomorphology had a largely service role, where information provided by Earth scientists was functional in understanding geographical issues' (Fisher 1994, p. 477).

From the early 1960s (Hamilton 1976) to the late 1980s geology education underwent major growth with university departments expanding: 'to an unprecedented size and with more honours graduates in geology completing their training in any one of the many universities in one year than were produced for the whole country per year before World War II' (Hamilton 1976, p. 105) with many, especially those who had undertaken joint honours degrees, entering the teaching profession. Following those boom years up to a half of the departments were merged with other disciplines or closed. Currently, geology is popular with first year undergraduates, underpinning environmental science programmes, with around a mere 5% then specializing in the discipline. The expansion in university provision was matched by an increased uptake at school examination level; the need to promote and support geology education was recognized by the 1967 founding of the Association of Teachers of Geology, becoming the Earth Science Teachers' Association in 1988, that organizes an annual conference, produces a journal and develops learning support materials; the latter are supplemented by various curriculum initiatives.

Modern demand for discrete geological education within the national curriculum and at school examination level is very limited (King 1993; King & Jones 1999; Jones & King 2005). In the former, its initial contribution (Hawley 1996) was later substantially reduced (Hawley 1998) and geology's annual uptake as an examination subject in England and Wales was only around 2000 GCSE entries and 2500 'A' level entries by the mid-1990s, falling to around 600 for the former with a knock-on effect on the latter yet to be seen (King & Jones 2006) by the early 2000s. In July 2007 the Welsh examining board (WJEC) suggested withdrawal from 2010 (although a re-launched innovative multi-media-based examination and supporting materials might secure its future) of the UK's only geology GCSE would compound this lamentable situation: 'it is vital for our future supply of geologists ... The profile of Geology in the under-18 age-group would dwindle, with serious consequences for the future of the discipline' (Anon 2007). Such changes inevitably impact upon undergraduate uptake.

The burgeoning list of geology field-guides is some measure of the interest in geology in schools and universities, as well as the public. Despite the demonstrable increase in modern field-guides, geology is not as enthusiastically and proportionately pursued by the public as it was in the late nineteenth century when there were numerous gifted amateur field naturalists at work and popular local societies to support them (Allen 1978). Geology was also at the cutting edge of scientific enquiry, following an earlier period of anecdotal observation, that coincided with the opening up of the countryside to new travellers or tourists; the latter were drawn to various geology-based attractions or 'geo-attractions', both natural and artificial, as well as museums.

Geological interpretation

From the late-eighteenth century onwards there were several British innovations in geology's presentation. The Oatlands Park grotto near Weybridge (Barton & Delair 1982), an architectural folly built between 1760 and 1778 (but demolished in 1948), was encrusted with fossils (especially ammonites) from near Bath, modern shells and natural and artificial cave features such as stalagmites. Dorset provided the fossils on which the world's first attempt, De la Beche's 1830 Duria antiquior to visualize the appearance of fossils in life was based. The world's first geological theme park, with laudable educational aims, with three-dimensional reconstructions of prehistoric animals and plants on accurately rendered geological sections was at the Crystal Palace (Hawkins 1854; Doyle 1993, 1995, 2008; Doyle & Robinson 1993; McCarthy & Gilbert 1994). The world's first urban geology trail, of thirty stone pillars with biblical quotes in keeping with the moral educational role ascribed to geology's study, was established by 1881 in a Rochdale churchyard (Baldwin & Alderson 1996, p. 227). The preservation of spectacular fossils in situ was developed in the UK in the late nineteenth century; a 'fossil forest' unearthed in 1873 at the South Yorkshire County Lunatic Asylum (Sorby 1875, p. 458; Cleal & Thomas 1995a, pp. 208–210) was subsequently protected by two small viewing sheds. The second such forest, discovered in 1887, 'Fossil Grove', Glasgow (Cleal & Thomas 1995b, pp. 189–191), is the UK's longest continuously open geo-attraction.

After these promising early innovations little was achieved until the 1960s; for example, the Wren's Nest National Nature Reserve (Prosser & Larwood 2008) after its 1956 designation as the UK's purely geological one was a management challenge to the then Nature Conservancy when: 'there was a danger from over-collecting, and the approach changed to one of look-and-see rather than hammer-and-take' (Robinson 1996*a*, p. 211). Several versions of its trail field-guide were supplemented by an on-site display (Robinson 1996*a*) in 1996. *The Mortimer Forest Geology Trail* (Lawson 1973, 1977) near Ludlow, the first purposely established educational geology trail, was completed in 1973 following co-operation between the Forestry Commission and the Nature Conservancy Council. A late-1980s' development were geology-focused visitor centres commonly providing a range of activities such as talks, identification services and guided walks; for example, the National Stone Centre (Thomas & Hughes 1993) and the Charmouth Heritage Coast Centre (Edmonds 1996).

A major boost to geotourism were the numerous interpretative panels placed from the early 1990s at geosites popular with tourists; for example, by English Nature at Scarborough, Yorkshire and Hunstanton, Norfolk (in 1993); by RIGS groups at Cleeve Common near Cheltenham (in 1998) and Moorfield and Wellfield Quarries near Huddersfield (in 2001); and by Wildlife Trusts, as at Brown End Quarry in Staffordshire (Green 2008) (originally in 1991 and renewed in 2004). The English Nature sponsored Regionally Important Geological and Geomorphological Sites (RIGS) initiative (Harley & Robinson 1991; Harley 1996), and the consequent RIGS Groups (Burek 2008*a*, 2008*b*) and Earth Heritage Trusts, launched in the 1990s resulted in a burgeoning supply of trails, panels and field-guides; for example the North-East Wales (NEWRIGS) RIGS group began their ongoing series of trail leaflets, such as the innovative 1997 'Steaming Through the Past' (Burek & France 1998; NEWRIGS 1997) based on the route of a preserved steam railway at Llangollen and an example that, for sustainable tourism purposes should be more widely adopted. Commemorative plaques, such as that for the Silurian at Ludford Corner near Ludlow, are an uncommon geotourism offering and should be more widely employed again so that tourists are made aware of the significant role that Britain played in geology's development and its present international significance.

A dual approach incorporating interpretative and geoconservation measures, but essentially concerned with the latter, are collecting facilities from clearing rock faces (Anon 1994, p. 16) and providing spoil material, with limited or no on-site interpretation and relying on trail-guides (Duff *et al.* 1985, pp. 61–64); for example, 1980s' work to remove the 'blight' of colliery waste tips near Radstock in Somerset (Robinson 1993, p. 20) revealed an unrecognized Upper Carboniferous land arthropod fauna and the UK's richest insect fauna; subsequently the Writhlington National Nature Reserve was established (Jarzembowski 1989, p. 219). Selling minerals and fossils to tourists was widespread in the nineteenth century; indeed: 'Fossils had been collected at sundry localities for sale to visitors in the latter part of the eighteenth century, especially at Lyme Regis and Charmouth' (Woodward 1907, p. 115) and continues today (Taylor 1992). Presently, there is limited collaboration to provide interpretation during fossil recovery between commercial collectors and public agencies such as museums; the best-publicized involved a 1980s Glasgow housing estate fossil excavation (Wood 1983) that capitalized on the potential for good public relations and income generation through retailing souvenirs and conducting tours. Geological roadshows (Reid 1993), such the Dudley Rock and Fossil Fair, are a recent and sometimes commercial development. However, the chief source for the past 150 years of much popular geological exposure, setting aside the mass media, is museums.

The first public geology museum was the Geological Survey's London-based 1841 Museum of Economic Geology, initially adjacent to Scotland Yard; it displayed useful minerals and stones, manufactured items such as gun barrels and had a laboratory in which the public could have samples of rocks and soils analysed (Bailey 1952). The Natural History Museum, with some geology displays, opened to the public in 1857 and was specifically aimed at attracting a very broad, including working-class, audience (Bennett 1996). In 1851, the Museum of Economic Geology reopened to the public near Piccadilly, together with a Government School of Mines and of Science applied to the Arts, with exhibits supporting lecture programmes. It closed in 1923 and reopened in 1935 in purpose-built South Kensington premises during the Geological Survey's centenary year. Until the late 1960s, its displays were virtually unaltered three-dimensional representations of the Survey's regional guides. Its ground-breaking 1973 multimedia exhibition The Story of the Earth (Tresise 1973; Dunning 1974, 1975; Walter & Hart 1975) was the first permanent exhibition to cover plate tectonics. It was followed by other innovative exhibitions in the late 1970s and 1980s; one of these, British Fossils, adopted a modified traditional specimen-rich approach. One of the most successful of the new galleries was Dinosaurs for which the museum departed from its normal practice of populist soft-backed geological booklets for an A3 hardback publication (Gardom & Milner 1993); undoubtedly aided by *Reader's Digest* sponsorship. A £12 million programme, with sponsorship from mining and power companies, to replace and renew individual displays (Clarke 1991) and whole exhibitions (Dagnall 1995; Robinson 1996*b*; Smith 1996) was instituted in the 1990s; the first two new galleries, The Power Within (geology based) and Restless

Earth (geomorphologically based), opened in 1996 (Sharpe *et al.* 1998).

In the mid-1990s, spectacular multi-media exhibitions opened at the National Museum of Wales (Kelly 1994; Johnston & Sharpe 1997) and the National Galleries of Scotland. Similar approaches to exhibition design were adopted, on much smaller budgets, in the provinces; for example at Liverpool Museum where Earth Before Man, replacing an older geology gallery (Tresise 1966), opened in 1973. However, most provincial museums did not adopt the approach (Knell 1993, p. 20) and concern was expressed about the quality of their displays: 'Poor displays serve only to reinforce negative attitudes to geology ... Textbook in style and jargon-riddled—even the keen amateur would have difficulty in understanding some geological displays.' (Knell & Taylor 1991, p. 24). Most of these interpretative offerings have been supported, with varying levels of competence, by a range of outdoor panels and publications.

Many of the new techniques in museum and visitor centre exhibitions were pioneered by the mass media. The communicative potential of the mass media such as radio was recognized in the USA well before the UK; probably the first example was in 1954 when WGBH(Boston) broadcast a twenty-hour classroom geology course (Lyons *et al.* 1993). Over thirty years later, in 1988, the BBC employed a similar format for its first series of six school's radio programmes on geology. Amongst the first television programmes to explore emerging geological concepts were two one-off BBC programmes, the 1972 The Restless Earth (on plate tectonics) and the 1974 The Weather Machine (on ice ages), both accompanied by hardback books (Calder 1972, 1974). In 1978 the BBC broadcast On the Rocks, a further education production, and in 1988 six magazine-style radio programmes, Rock Solid; both were accompanied by paperback books (Wood 1978; Grayson 1988*a*, *b*). In 1998 the BBC broadcast a spectacular series of programmes (narrated by an eminent biologist—Aubrey Manning) Earth Story, accompanied by a coffee-table style hardback (Lamb & Singleton 1998). The sole major commercial television offering was the 1990 series Landshapes. In recent years there has been considerable cross-over between the mass media and the internet.

The sublime, the picturesque, the romantic and the neo-romantic

'Tourist' first appeared as an English synonym for 'traveller' in the late eighteenth century and was incorporated within the title of probably the first national guidebook, Mavor's (1798–1800) The British Tourists; or Traveller's Pocket Companion, through England, Wales, Scotland, and Ireland. It has been asserted that the: 'The traveller exhibits boldness and gritty endurance under all conditions (being true to the etymology of 'travel' in the word 'travail'); the tourist is the cautious, pampered unit of a leisure industry. Where tourists go, they go *en masse*, remaking whole regions in their homogeneous image.' (Buzzard 1993, p. 2) underscoring the major debate from the earliest appearance of tourism about the potential for tourists to destroy the very thing they go to see; consequently several less pejorative terms have been employed (such as 'lakers', 'excursionists' and 'eco-tourists') over the past two-hundred years to describe those persons who visit scenic places for essentially leisure purposes. Until the mid-eighteenth century for such persons: 'the preferred rural landscape was generally a humanized scene of cultivation, evidence of the successful control of nature' (Towner 1996, p. 138). Subsequently, three major aesthetic movements are considered by literary and art historian scholars to underpin our changing perception of and subsequent relationship with the British landscape.

The first of these, the pursuit of the 'sublime', was overlapped and followed by the 'picturesque', whilst the 'romantic' was an over-arching dominant eighteenth century movement from about 1780 to 1850. These movements reflect three intertwined threads from the late seventeenth through to the mid-nineteenth centuries, the:

1. Nature of the travellers or visitors;
2. Meaning ascribed to, and understanding of, the natural phenomena witnessed by travellers and visitors; and
3. UK's fundamental shift from a rural (land-based economy) to an industrial (manufacturing-based) society and the concomitant rise of the middle-classes in numbers and influence.

They were recorded and promoted by artists, diarists and writers, especially those, now well known to the public as, quintessentially English: John Constable, J.M.W. Turner, John Keats and William Wordsworth; equally there are those generally unknown to the greater populace such as William Gilpin and Samuel Palmer. In the twentieth century their place would be taken by: Paul Nash, Ben Nicholson, John Piper and Stanley Spencer; although their landscapes are generally less idealized and more stylized (see Mullins 1985).

Helpfully, Wordsworth's Guide to the Lakes distinguished two landscape formation phases that contributed separate but linked elements: 'Sublimity is the result of Nature's first great dealings with the superficies of the earth; but the general tendency of her subsequent operations is towards the production of beauty; by a multiplicity of symmetrical parts uniting in a consistent whole' (Wordsworth 1835, p. 35); thus it was occupied with nature's first great dealings, the masses of rock, hill and lake, to solicit from the spectator a feeling of awe and wonder at their wildness and ruggedness as was the case with the earliest of travellers into the British countryside from the late seventeenth century. Conversely, the cult of the 'picturesque' delighted in the softer effects stemming from nature's subsequent operations producing the variegation and harmony expressed, for example, by the curve of river's meander or lake shore, the grouping of the rocks and trees which flank them, the interplay of light and shade over these features and the subtle colour gradations that seemingly meld the scene; an approach adopted from the late eighteenth century by the pioneering travellers into the British countryside. Essentially the landscape is framed as if for a picture and hence the significance of the establishment of scenic 'stations' or viewpoints (see later).

Waterfalls were especially favoured by those interested in the sublime and the picturesque and the term 'cataractist' briefly passed into uncommon usage. That these two aesthetic movements overlapped seems clear from the meaning for 'picturesque' as 'What pleases the eye' from the 1801 Supplement to Dr Johnson's Dictionary; by the opening of the nineteenth century it was in broad enough popular to overlap with the 'sublime'. The 'romantic' was a movement that saw the expression of the feeling of landscape and its evocation in art and literature, especially poetry. From the late twentieth century the emergence of 'neo-romanticism' can be recognized in which the countryside is a place for leisured pursuits and an exercise backdrop for; a re-working of earlier themes within the framework of post-modern consumerism and exemplified by the popularity of The Country Diary of an Edwardian Lady (Holden 1977) (it remained in the Sunday Times bestseller list for over three years, and number one on the same list for longer than any other book), altogether something of a misnomer given its author's residence in the rapidly urbanizing turn-of-the-century West Midlands! The impact of these various aesthetic movements on the traveller, the forerunner of the modern geotourist, can most readily be traced in relation to the landscapes in which they were first perceived and then promoted, largely in the first two regions considered in this chapter, and the events and players involved can be summarized and the significance of the metropolitan influence on modern geotourism can be seen in the last region. Britain's geotourism flourishes best where the landscape is dramatic and readily accessible such as the (not necessarily) mountainous, uplands and especially the cliffed coast.

The Peak District and the earliest geotourists

The Peak District is the birthplace of geotourism because it was the first region to be explored, from the late seventeenth century onwards by travellers visiting caves and mines readily accessible from the industrializing provincial centres such as Derby and Sheffield; they could also break their journey to visit a pottery or a mill, much as modern tourists visit the region's industrial archaeology sites (Harris 1971).

By the time the railways had reached and opened up the Peak District its main contribution to the development of landscape-based tourism had already been achieved. Although the railway reached Derby as early as 1839 it was not until 1863 that passengers could travel to Buxton, via Matlock (Awdry 1990). Before the nineteenth century only the social élite could make their way, at least in any style and that was literally horse-powered and on mainly unmetalled roads, across Britain; consequently there was a limited supply of publications to inform the traveller. One of the earliest such publications, Britannia (Ogilby 1675), was a strip-map of the roads of England and Wales; it was far too bulky to carry on a journey until its 1720 pocket edition (Ogilby 1720).

Celia (Fiennes 1949), an elite traveller in the last two decades of the seventeenth century, obviously managed without it when she rode to Buxton and explored Poole's Hole, afterwards riding to the Ashbourne copper mines; because of the latter visit she is amongst the earliest of recorded geotourists. In the late eighteenth century the farmer Arthur Young travelling via Ashbourne to Dove Dale wrote in 1771:

It is bounded in a very romantic manner by hills, rocks and hanging woods; which are extremely various; and the hills in particular of a very bold and striking character ... The rocks ... forming a wide assemblage of really romantic objects ... (in Trench 1990, p. 158)

By the close of the seventeenth century, the region's main sights had been organized and promoted, by the books of Thomas Hobbes of 1678 (see Fig. 2) and Charles Cotton of 1681, into just seven accessible 'wonders': 'two fonts' (the ebbing and flowing wells at Tideswell and St Ann's Well), 'two caves' (Poole's Hole and Peak Cavern), 'one palace' (Chatsworth House), 'one mount' (Mam Tor) and 'a pit' (Eldon Hole pothole) and appeared

```
D E
Mirabilibus Pecci:
BEING THE
VV O N D E R S
OF THE
P E A K
I N
D A R B Y - S H I R E,
Commonly called
The Devil's Arfe of Peak.

In Englifh and Latine.

The Latine Written by Thomas Hobbes
of Malmsbury.

The Englifh by a Perfon of Quality.

London, Printed for William Crook at the Green
```

Fig. 2. Facsimile of the frontpiece of *De Mirabilis
Pecci: Being the Wonders of the Peak* ... (Hobbes 1678).

in an early nineteenth century traveller's account
(Warner 1802).

Despite Daniel Defoe describing the region as a
'houling wilderness' in the 1720s (Defoe no date,
vol.1, p. 160) and the 'wonders' overrated, contemporary travellers flocked to them if only so they
could agree with his observations and thus appear
educated and superior; their accounts record the
shabbiness and commercialism of tourism provision. The caves of the 'White Peak' (the area
underlain by Lower Carboniferous limestone)
were the most popular attractions. Cotton noted
that an old woman made a living as the keyholder
for Poole's Hole; by the 1750s its entrance was surrounded by modest cottages whose inhabitants
made a similar living. Defoe commented, with
redolence to modern geotourists' expectations
raised by hyperbolic promotion, that:

The wit that has been spent ... had been well enough to raise the
expectation of strangers, and bring fools a great way ... but is ill
bestowed upon all those that come ... with a just curiosity ...

when they go to see it, they generally go away, acknowledging
that they have seen nothing suitable to their great expectation, or
to the fame of the place. (Defoe no date, vol.1, p. 168)

The antiquary William Bray, recollecting his 1773
visit to Peak Cavern, indicating the commercialization of the cave experience by the late eighteenth
century, wrote:

you come to the first stream, the roof gradually sloping to within
two yards of the surface. This water is to be crossed by lying
down in a boat, which is pushed forward by the guide. You then
come to an apartment of great extent, with several apertures
atop. After crossing the water a second time, on the guide's
back, you enter Roger Rain's House, so called from the continual
dropping of the roof. Here you are entertained by a company of
singers, who, having taken a different route, are stationed in a
place, called the Chancel. (Bray 1783, pp. 338–339)

Almost concomitantly the first account of Derbyshire's rocks and minerals appeared, with John
Walcott's (1778) *An inquiry into the original
state and formation of the Earth*. The Hon.
Mrs Murray's 1799 visit to Peak Cavern inspired
practical advice to would-be cave visitors, who:

should provide a change of dress, and they need not fear getting
cold or rheumatism... carry also your night-caps, and a yard of
coarse flannel, to pin on the head, so as to let it hang loose over
the shoulders; it will prevent the dripping from the rocks in the
cave from wetting and spoiling your habits, or gowns; also take
an old pair of gloves, for the tallow candle, necessary to be
carried in the hand, will make an end to all gloves worn in the
cavern. (Murray 1799, p. 8)

There is no evidence that the near contemporaneous
1809 *Petrificata Derbiensia*; or figures and descriptions of petrifactions collected in Derbyshire
by William Martin and the (1811) *General View
of the agriculture and minerals* of Derbyshire by
John Farey had a wide readership amongst these
travellers; likewise for White Watson's ground-
breaking cross-sections of 1811, *A delineation of
the Strata of Derbyshire* and 1813, *A section of
the strata in the vicinity of Matlock Bath*, although
travellers visited his Bakewell museum-shop.

Specimens of Blue John and other coloured
fluorspars and limestones (in their native or
carved forms such as vases) were sold, to adorn
the cabinets of curiosities that marked out the
adventurous traveller, at the mines and caves as
well as shops in Matlock and even Derby cathedral.
In the opening years of the nineteenth century, when
Jane Austen's 1813 Pride and Prejudice had
Elizabeth Bennet including the chance of collecting
'a few petrified spars' as one of the attractions of
a Peak District visit, at least one writer was caution-
ing about the artificial colouring of many fluorspar
specimens! This overt commercialism of the Peak
District eventually prompted travellers to seek
curious natural phenomena elsewhere, initially in
the Pennine Dales and then the Lake District.

Thomas West, in an *Appendix* to his 1778 Lake District guidebook, gave an itinerary of the region's major caves and limestone crags; his subtitle *Some Philosophical Conjectures on the Deluge, and the Alterations on the Surface and Interior Parts of the Earth Occasioned by This Great Revolution of Nature* indicates how observations of natural phenomena inspired speculation on geology. As Cope noted in the 1976 populist *Geology Explained in the Peak District* that the region has lost little of its appeal to the true traveller. For the specialist, Sheffield University's field-guide (Neves & Downie 1967) followed the format of the Geologists' Association (GA) Centennial field-guides in some fifteen (out of twenty-four) detailed day-long excursions. The more widely available, if somewhat technical, *A geological field guide: The Peak District* (Simpson 1982) was published as one of series of such guides to popular tourists areas. Rather surprisingly the GA left it until 1877 (but visited again in 1899 and 1904) to organize a major excursion to the region; this was over five-days with much of the time spent in the 'White Peak'. The National Stone Centre (Thomas & Prentice 1984) near Wirksworth, the UK's first such visitor centre opened in 1989, at the junction of the 'White Peak' and the 'Dark Peak'. The Peak District National Park, the first such established in 1951, is the UK's most visited national park.

The Lake District and early geotourists

Numerous historians and writers of literary and art criticisms have charted the stages in the Lake District's rise to its present popularity with tourists. However, they do not dwell upon the region's extensive industrial, especially mining, interest (Marshall & Davies-Shiel 1977). Indeed this was deliberately neglected in the promotion of a wild rural idyll, especially by the 'Lake School' of poets (Samuel Taylor Coleridge, Robert Southy and William Wordsworth), initially named in the *Edinburgh Review* of August 1817 (Daiches & Flower 1979, p. 115). A definitive account of its geological exploration was recently published by The Geological Society (Oldroyd 2002) and a populist account of the personalities involved is also available (Smith 2001). Celia Fiennes rode through the Lakes in 1698 and recorded potted char and bread recipes rather than the scenery, presumably because it was an unprofitable barren untamed wilderness. Daniel Defoe in the 1720s considered the region wild, barren and frightful. However, from the 1750s travellers to the region, or 'lakers', visited because of the perceived quality of its scenery and its antiquities; the poet

Thomas Gray visited in 1767 and 1769, the farmer Arthur Young in 1768 (1771), the artist William Gilpin in 1772 (1786) and the antiquary William Hutchinson in 1773 (1774). The letters written by Gray describing his second visit were published in William Mason's posthumous edition of his work in 1775. Thomas West organized the chief sights into a guidebook in 1778; the first devoted to the Lake District that set the pattern for later tours and all their associated guidebooks. Illustrated guides to the region's antiquities were also available (e.g. Hutchinson 1794) by the close of the eighteenth century. West established 'stations' at which travellers might best view the scenic wonders whilst reading the accompanying description. All of this was well before the railway finally reached Windermere in 1847.

Prior to this tourist provision, travellers recorded their impressions and these might be published as journals. Numerous amateur and professional artists visited and recorded the sights in various media; these were sometimes published as sets of engravings either on their own or accompanying prose or poetry. For example, the young J. W. M. Turner made his living as a topographical artist executing numerous watercolours and a few oil paintings; two of the latter Lakeland landscapes were exhibited at the Royal Academy in 1798, the most noteworthy being *Morning Amongst the Coniston Fells, Cumberland*. Turner contributed two plates to Samuel Rogers' 1834 Poems. Gray on his journey from the southern end of Derwentwater into Borrowdale wrote that it reminded him of Alpine passes where travellers' caravans were threatened by avalanches! Gilpin writing on the same valley noted:

As we proceeded in our route along the lake, the road grew wilder, and more romantic. There is not an idea more tremendous, than that of riding along the edge of a precipice, unguarded by any parapet, under impending rocks, which threaten above; while the surges of a flood, or the whirlpools of a rapid river, terrify below. (Gilpin 1786, vol.1, p. 187)

Young describing the same locality, from the perspective of somebody concerned with taming nature, thought the lake elegant but the surrounding mountains were too wild with 'dreadful chasms'. Derwentwater was especially popular with early tourists; as West's guidebook noted the view of the lake from Cockshott Hill on its north-eastern shore came close to fulfilling the ideal requirements of the 'picturesque':

On the floor of a spacious amphitheatre, of the most picturesque mountains imaginable, an elegant sheet of water is spread out before you, shining like a mirror, and transparent as chrystal; variegated with islands, that rise in the most pleasing forms above the watery plane, dressed in wood, or clothed with forest verdure, the water shining round them. (West 1778, pp. 89–90)

Whilst today Cockshott Hill would be a 'viewpoint' or 'beauty spot', West and his contemporaries called it a 'station'. His guidebook consists almost entirely of selected, numbered and described stations; the descriptions guided tourists with the textual precision necessary in the days before Ordnance Survey maps and their grid, as for example Coniston's first station:

A little above the village of Nibthwaite the lake opens in full view. From the rock, on the left of the road, you have a general view of the lake upward. This station is found by observing an ash tree on the west side of the road, and passing that till you are in a line with the peninsula, the rock is then at your feet. (West 1778, pp. 50–51)

These stations, along with some of his own invention, were indicated on the maps issued from 1783 by Peter Crosthwaite of Keswick who styled himself as a geographer and hydrographer. He added a couple of his stations for the Derwentwater area; for that halfway up Latrigg he had a flight of steps cut and a marked cross on the ground and the other being most conveniently near to his museum and advertised by beating a gong, that could be heard several miles away. He also published his museum's customers' names in the local weekly newspaper with 1540 persons so noted in 1793 (Ousby 1990, p. 172).

After Arthur Young toured the region in 1768 he asserted that the stations be made readily accessible to a greater range of travellers than thitherto possible because of the precipitous nature of many paths; because woodland obscured some he recommended that trees should be pruned or removed to admit the views and that resting places be provided. The stations often had structures provided for travellers' advantage and the landowners' commercial exploitation, but these were not universally appreciated; James Plumtre in 1799 considered the first Windermere station 'too finished and artificial' (in Ousby 1990, p. 158). Richard Warner's 1802 account of *A Tour through the Northern Counties* includes route maps that could lead modern tourists to still popular places. West and his contemporaries, in reducing the Lake District's scenery to a series of stations with erudite explanations, established a tourist behaviour which can be witnessed almost anywhere today with, for example, well-advertised car-parking where they can take a snapshot.

Likewise the development of natural rock features as tourist attractions. The Bowder Stone (a perilously balanced 2000-ton boulder in Borrowdale) was turned into the region's first developed natural tourist attraction (see Fig. 3) by 1807 with a little mock hermitage, a new druidical stone and a house for a guardian who would show it to travellers; also all the fragments around its base were cleared away and a hole excavated through which visitors could shake hands. When William Green's *Tourist's New Guide* was published in 1819 it had further declined and the guardian when travellers arrived began:

an exordium preparatory to the presentation of a written paper, specifying the weight and dimensions of the stone ... The movement of the hand towards the pocket, is an act John understands as

Fig. 3. The Bowder Stone, Borrowdale from a late nineteenth century postcard; the first developed geo-attraction in the Lake District.

well as any member of the fraternity to which he belongs. (Green 1819, vol. 2, p. 134)

The closure of continental Europe to the British from 1789 to 1815 during the French Revolution and the succeeding Napoleonic wars gave an impetus to the Lake District's exploration, especially insofar as its landscape could be promoted as mimicking that of the Alps, which Wordsworth had already visited. By the opening decade of the nineteenth century the vast bulk of the work to make the landscape readily accessible to visitors had been completed. New paths had been cut to the waterfalls and stations and resting places in the form of summer houses provided; the downside of this was that the visitors were threatening the very landscape they had come to view by this excess of provision. This danger was recognized by probably the best remembered of the early, but by no means the earliest (for which perhaps Richard Warner on Pocklington's Island, Derwentwater in 1802 can lay some claim), landscape conservationist, Wordsworth. This is witnessed by Wordsworth's various remarks in his *Guide to Lakes* (Wordsworth 1820) and in two public letters of 1844 and the sonnet opposing the Kendal and Windermere Railway (Faith 1990, pp. 53–56; Selincourt 1977, pp. 146–166). However, all is not what it seems to the modern eye and Wordsworth, who privately bought shares in the venture, was actually seeking to protect the area from a new form of tourist, not the educated elite for whom West's and his own guidebook had been written, but the mass lower middle and working class tourist from the nearby northern towns of Manchester and Leeds. He had recognized that the steam-powered railway, then rapidly encircling the Lake District would disgorge tourists at Bowness above his beloved Lake Windermere from cheap trains in their hundreds, many of whom would visit tawdry ale-houses of ill repute. The railway reached Windermere as early as 1847, but passengers for Keswick (for Derwentwater) had to wait until 1865, when the expanding north-western rail network linked London with Carlisle station (opened in 1846) and enabled ready access to the northern Lake District. Whilst much of the limited track had been laid to carry away the region's exploited rich mineral resources, it increasingly turned to tourism as those revenues declined; the Kendal to Windermere line had been specifically opened to promote tourism to revive the local economy, although by then shoemaking was beginning to replace the declining woollen industry. Wordworth's spiritual successor Ruskin echoed similar disapproval of the railways thirty years later when an extension to that branch line was proposed.

Such condemnation was not universal and Harriet Martineau's 1855 *A Complete Guide to the English Lakes* was most encouraging to such mass tourists who, she suggested, should spend a day in the mountains with map and compass, although the services of a guide were recommended for mountain ascents of mountains such as Scafell and Helvellyn. The Lake District was especially well promoted, at least to the elite educated and discerning tourist, by William Green who in 1819 published the illustrated *Tourist's New Guide*. He was well acquainted with Wordsworth who fulsomely complimented the book in his *Guide* as a: 'complete Magazine of minute and accurate information.' (Wordsworth 1835, p. 6). If Green popularized the Lake District in pictures, Wordsworth did as much and probably more in prose and poetry. His *Guide* (see Fig. 4) was innovative in that, apart from

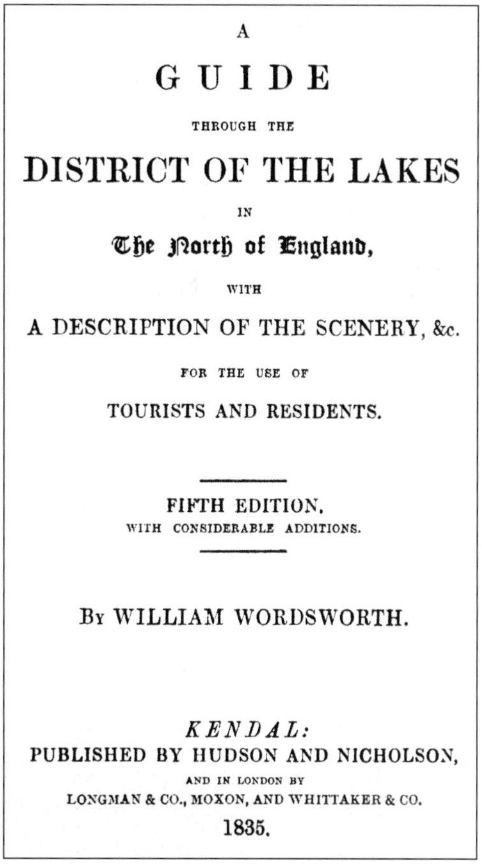

A

GUIDE

THROUGH THE

DISTRICT OF THE LAKES

IN

The North of England,

WITH

A DESCRIPTION OF THE SCENERY, &c.

FOR THE USE OF

TOURISTS AND RESIDENTS.

FIFTH EDITION,
WITH CONSIDERABLE ADDITIONS.

By WILLIAM WORDSWORTH.

KENDAL:
PUBLISHED BY HUDSON AND NICHOLSON,
AND IN LONDON BY
LONGMAN & CO., MOXON, AND WHITTAKER & CO.
1835.

Fig. 4. Facsimile of the frontpiece of The River Duddon, A Series of Sonnets; Vaudracour & Julia: and Other Poems. To which is annexed, A Topographical Description Of the Country of the Lakes, In the North of England (Wordsworth 1835).

merely describing what could be seen, it also sought to link the landscape to natural history (including geology with which he was evidently acquainted as evidenced by his work of 1814), history and people; its opening paragraph indicates the author's desire: 'to furnish a Guide or Companion for the Minds of Persons of taste, and feeling for Landscape, who might be inclined to explore the District of the Lakes with that degree of attention to which its beauty may fairly lay claim' (Wordsworth 1835, p. 1). It was developed from the anonymous text accompanying an 1810 volume of the Rev. Joseph Wilkinson's engravings, for which Wordsworth had little real regard. Gradually, it evolved into an appendix to his River Duddon sonnets in 1820 and then finally into a guidebook in its own right by 1822; continuously revised and expanded, it ran into several more editions up to the 1840s; and following Wordsworth's intercession, the later editions included three letters on geology by Adam Sedgwick.

Subsequent guidebooks moved away from mere descriptions of stations to a more holistic approach that encouraged serious landscape study and some reverence for the works of the Creator. As such they informed visitors from the nearby towns on their day and weekend trips; an example of these being the 1865 visit by a Quaker party to Loweswater (Hodgson & Lunt 1987). These guidebooks were noteworthy for their strong literary content drawing the reader's attention to site-focused poetry and other literary allusions; such an approach is discernible in Charles Mackay's 1846 *The Scenery and Poetry of the English Lakes* that even employed Shelley's brief stay in the Lakes as an excuse to quote the description of the waterfall from 'Alastor' at the otherwise unremarkable Stockghyll Force. There was a crossover with the more conventional guidebooks which began to include similar literary allusions. For tourism the creation of these literary landscapes is probably the most enduring of the achievements of these travel pioneers and to some extent the approach pervades numerous modern guidebooks.

The first populist account of Lakeland geology was Johnathan Otley's (the Lake District's earliest geologist) 1823 *A concise description of the English Lakes... and observations on the mineralogy and geology of the district.* Adam Sedgwick supplemented Hudson's 1842 *Complete Guide to the Lakes with some geological observations.* One of the most popular twentieth century accounts of Lakeland geology was Shackleton's 1967 pocket-sized illustrated *Lakeland Geology: the introduction* to which overtly ignores geoconservation with an emphasis on hammering and collection! The introduction to Prosser's 1977 *Geology Explained* in the Lake District summarizes the

area's appeal: 'The Lake District not only holds tremendous scenic attraction for tourists, but is also one of the most geologically rewarding regions in Britain.' This appeal has been further enhanced by the publication of several field-guides, both technical (Cumberland Geological Society 1982; Dodd 1992) and populist (Lynas 1994) since the 1980s. Surprisingly, given the ready rail access, the GA left it until 1881 and 1900 to visit Lakeland. Today, the Lake District National Park is the UK's second most visited such area.

Central southern England and the birth of the modern geotourist

The development of railways from the 1840s opened up southern Britain to geological enquiry and a particular central southern England (Hampshire, the Isle of Wight, Dorset and Sussex) benefited (Awdry 1990). In the east, Brighton was linked by railway to London in 1841, Portsmouth (for the Isle of Wight) in 1859 and Bournemouth in 1870. Whereas in the west, Weymouth was linked to the expanding rail network in 1865, but Lyme Regis had to wait until 1903 although nearby Bridport had a railway connection by 1857. The western part of the region is much promoted as the writer Thomas Hardy's 'Wessex' (Daiches & Flower 1979, pp. 158–171). Much of the region's popularity has long been from student practitioners' exposure to its classic geosites in their formative years and the publication of numerous field-guides and field excursion reports since the early part of the nineteenth century. The GA published field meeting excursions from the mid-nineteenth century onwards; in the late-1950s it began publication of its Centennial field-guides that influenced two generations of geologists. However, the GA delayed an excursion to the region until 1864, after the opening of the London and Portsmouth Railway, and its subsequent nineteenth century visits can be summarized (see Table 1).

The relationship between these excursions and the expanding railway network is clear; each excursion was announced in the GA's Circular written up in its *Proceedings*, as for example that to Lyme Regis in 1906 (Lang 1906). The region remained popular with the GA in the twentieth century with, for example, excursions to the Isle of Purbeck (Arkell 1934) and Lyme Regis (Barnard *et al.* 1950). A county-focused examination of the region's field-guides highlights key issues in the development and promotion of especially dedicated geotourism provision. The first county to be described in any detail, undoubtedly because of its proximity to London with its active geological

Table 1. *Geologists' Association's Main Nineteenth Century Excursions to Central Southern England*

Date(s)	Main places visited
1864	Isle of Wight
1866	Isle of Wight; Brighton
1879	Weymouth & Portland
1880	Bournemouth
1881	Isle of Wight
1880, 1888, 1894	Barton Cliffs
1882	Isle of Purbeck; Hastings & Battle
1882, 1883, 1885	Worth, Balcombe, East Grinstead, Haywards Heath
1885	Sherbourne & Bridport
1887	Brighton (Newhaven, Seaford)
1889	Lyme Regis & Weymouth
1891	Isle of Wight; Selbourne
1895	Isle of Wight

community centred on The Geological Society, was Sussex.

The first attempts to summarize its geology, *The Fossils of the South Downs* (Mantell 1822) and *Illustrations of the Geology of Sussex* (Mantell 1827) were costly, illustrated, library volumes, containing locality-specific information. Mantell's *Geology of the South-East of England* (1833) is a truly pocket-sized volume; its Sussex information includes some geotourism souvenir information: 'The beach near this place [Rottingdean] contains semi-translucent pebbles of agate, and chalcedony, of a bluish grey colour. These are collected by visitors and when cut and polished are used for bracelets and other ornamental purposes ...' (Mantell 1833, pp. 40–41). In 1846 he published *A Day's Ramble in and around the Ancient Town of Lewes* with one of its eleven chapters on geology. Concomitantly, Dixon's 1850 posthumous publication, *The Geology and Fossils of the Cretaceous and Tertiary Formations of Sussex*, was a worthy supplement, going into a second edition in 1878 with reproductions of the plates from Mantell's *Fossils of the South Downs*. Little followed these nineteenth century Sussex field-guides, although there were several for the London region; the preface to one such highly descriptive text indicated, like that of many modern authors', it aimed to be a handy field-guide:

useful to many who are interested in the natural sciences and desire a field-acquaintance with the geological formations which occur in their district. To those who take pleasure in country walks the route here described will prove attractive, while a knowledge of the structure of the country will add interest to future excursions. (Davies 1914)

Unusually, one of Davies' later field-guides (1939) reversed the usual practice of beginning with the older rocks (instigated by Lyell), so that users

started with rocks containing fossils similar to modern forms before progressing onto the less familiar older forms; it also promoted GA membership and the benefits of using museums and public libraries. From 1958 onwards the GA published several Sussex field-guides as centennial volumes such as *Geology of the Central Weald: The Hastings Beds* (Allen 1960). The descriptive itineraries are well-detailed routes, with some attempt at interpretation. *The Weald* (Gibbons 1981) was a new format of pocket-sized non-specialist field-guides, noteworthy for including a geoconservation message and a cover incorporating *A Geological Code of Conduct*. The introduction to a modern populist mainly descriptive highly illustrated text for Hastings has a commonly adopted approach for non-geologists, beginning:

with a brief geological history of the Wealden district... followed by detailed descriptions of two field trips ... studying the evidence gathered at local sites, an attempt will be made to reconstruct the ancient environments and communities. To achieve this we must try to 'read' and interpret all the available clues, rather like a detective solving a mystery. (Brooks 2001, p. iv)

Until the advent of steam-powered ferries in the late nineteenth century, access to the Isle of Wight was difficult. However it has attracted numerous publications because: 'In the splendid sections exposed along the line of coast we are enabled to examine the strata and trace their relationships to each other ... while the beauty of the scenery ... lends an additional charm to the investigation' (Harrison 1882, p. 103). The aesthetic landscape emphasis underpins many modern populist field-guides. The island's earliest field-guide dates from the first quarter of the nineteenth century, *A Description of the Principal Picturesque Beauties, Antiquities, and Geological Phenomena, of the Isle of Wight* (Englefield 1816). This is a large illustrated volume with much detail on woodlands and trees, containing twelve letters by Thomas Webster (later the Geological Society's first full-time employee as its museum keeper and eventually London University's first Professor of geology (Woodward 1907, pp. 47–48)) with detailed observations upon coastal geology; its significance is its seminal approach and early genesis, as its preface indicated it was:

the result of observations made in the years 1799, 1800, and 1801 ... these materials were ... prepared for publication in the year 1802 ... Circumstances ... occasioned me to lay aside the whole for ten years ... I had expected that my work would be superseded by the labours of later travellers ... I was induced to re-examine my own work, and at length prepare it for publication. (Englefield 1816, pp. i–iii)

The area of this publication is similar to that of Mantell's (1847) later groundbreaking *Geological Excursion Round the Isle of Wight and the Adjacent Coast of Dorsetshire* (see Fig. 5). It represents the

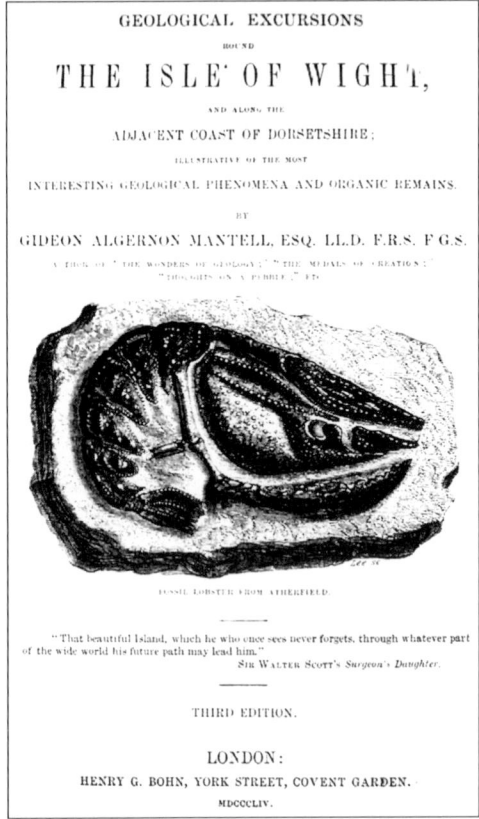

Fig. 5. Facsimile of the frontpiece of the third and last edition of *Geological Excursion Round the Isle of Wight and Along the Adjacent Coast of Dorsetshire* (Mantell 1854).

genesis of the readily portable, pocket-sized illustrated field-guide, inexpensive enough to be taken into the field (which might explain the poor condition of many non-library copies) and selling well enough to warrant two further editions in 1851 and 1854. The first edition's preface justifies the volume in a similar manner to modern field-guides:

... the Geology of the Island is but little known or regarded by the majority of the intelligent persons who every season flock by thousands to its shores, and, rapidly traversing the accustomed routes, visit the picturesque localities noted in the numerous handbooks, and take their departure without suspecting that they have been travelling over a country rich with the spoils of nature ... of the highest interest to the instructed observer.

Even the inhabitants ... manifest an extraordinary degree of apathy in everything relating to the Geology of the Island. (Mantell 1847)

The Geological Survey published an account of the island's geology in the mid-nineteenth century (Bristow 1862). An unlikely attempt to fill the inquiring tourist niche market for casual visitors was a subscription volume, *A Concise Exposition of the Geology, Antiquities, and topography of the Isle of Wight* (Wilkins 1861). The later *A Popular Guide to the Geology of the Isle of Wight* (Norman 1887) had a map and twenty-two plates of fossils and topographic views—many more than its modern equivalents. Its essentially stratigraphic accounts, focused on fossil collecting localities, are accompanied by appendices dealing with Newport Museum and the Ventnor Collection. Its preface noted:

I published, in the Isle of Wight Advertiser, a series of letters upon Geology ... At the solicitation of many friends and readers, and seeing that Dr. Mantell's—the only popular work on the subject —was published nearly forty years ago, I have decided with much diffidence to re-publish these letters in the form of a Popular Guide ... I have endeavoured to give my descriptions in such a manner that the work may be useful to the geologist as well as entertaining to the general reader ... (Norman 1887, pp. iii–iv)

A populist field-guide, with a fossil collecting emphasis, from the first quarter of the twentieth century's was prefaced by the geohistorical note: 'The Isle of Wight is classic ground of Geology. From the early days of the science it has been made famous by the work of great students of Nature ...' (Hughes 1922). The island has long attracted specialist groups' excursions; the Quaternary Research Association's guidebook (Barber 1987), by twenty authors, is noteworthy for its profusion of maps and diagrams and technical writing. The newest GA field-guide (Insole *et al.* 1998) has twenty brief itineraries with numerous diagrams and route maps. The Geological Survey's large-format foldout field-leaflet (Gallois 1999) lacks specific geosite detail. Its much earlier memoir reprinted in 1990 (White 1920), has been the only major work on the island's geology for over eighty years. Late twentieth century geoconservation concerns prompted the late-1980s' first regional geoconservation publication *Guidelines for Collecting Fossils on the Isle of Wight* (Anon, no date) because: 'In recent years, increasing pressure has been imposed on the eroding coastline by growing numbers of collectors and visitors.' Concomitantly, concern was also being expressed in educational quarters (Hawley 1994) about national geoconservation issues (see also Munt 2008).

Dorset has been significant in scientific geology's development since the beginning of the nineteenth century. It had to wait until the mid-nineteenth century for a dedicated field-guide (Damon 1860); towards that century's close it was noted that: 'Some portions of Dorset—more especially the coast—offer such unrivalled opportunities for geological study, that they early attracted the attentions of writers on that science' (Harrison 1882, p. 72). The first populist field-guide, published in 1860 with a second edition almost a quarter of a century later, was for Weymouth; its preface has a remarkably modern theme:

... a guide to the Geology of the district in question, written in a somewhat scientific yet elementary and popular form ...

An endeavour has been made in some measure to supply the above-mentioned want, by pointing out in the following pages where the various formations can be best examined, and by rendering the latter more easily identified by means of views, sections and other illustrations. (Damon 1860)

A steady stream of Dorset field-guides was published in the twentieth century, especially following the designation of the UNESCO World Heritage Coast (Boylan 2008). Davies' influential mid-1930s' field-guide (Davies 1935) went to a second edition from which the introduction noted:

To the student of geology Dorset is well known, at least from books ... a poor substitute for hammering the actual rocks and seeing how they fit into the structural fabric of the country ... where a beginner can get so clear an insight into geological structures and the work of the agents of erosion. (Davies 1956, p. 1)

and that it would save students preliminary library research! Bird's 1995 *Geology and Scenery of Dorset* is an example of those guides forging a link between what the tourist sees and the underlying geology, whilst others merely focus on fossil collecting (Coram 1989; Clarke 1998). The 1990s popular *Lulworth Visitor Centre guidebook* (Pfaff & Simcox 1998) noted: 'all are struck by the beautifully shaped coast and the contrasting rocks that form it. This booklet has been written for both geography students and visitors who would like to understand a little more about this unique landscape.' The GA has published several guides to

the region, especially for the coastal sections (Ager & Smith 1973; Kirkaldy 1976; House 1989; Allison 1992). The western coastal section of this region became England's first natural UNESCO World Heritage Site, popularly promoted as the 'Jurassic Coast' because of its significant contribution to the development of geology and its spectacular coastal sections.

Summations and additions

Several summations follow from the preceding discussion. Geotourism's development required the identification and promotion of geological phenomena concomitant with tourists' willingness to engage with seemingly untamed landscapes to which reliable access was available. The beginnings of tourism, and the antecedents of geotourism, date from the late seventeenth century when long distance passenger transport essentially relied on roads. The first turnpike road was built in 1706 and by 1800 there were some 1600 (Briggs 1983, p. 207) linking the major cities and towns; any journey away from these was still on unmetalled tracks that especially in the winter months, tested both horses and coaching technology to the limit. Journey times by modern standards were slow with the London to Manchester route taking four days in 1754 and by 1790 two days. A swifter means of transport was clearly required. The genesis of mass tourism was Thomas Cook's organized rail excursions that started in 1841, although they were preceded by other excursions from 1836 (Brendon 1991, pp. 5–9). However, from the 1830s onwards the railways provided the major means of transport to field areas for most nineteenth century geotourists. By the 1850s, railways and steamships were opening up the countryside and coast respectively, with some 8000 miles of track (Briggs 1983, p. 210) linking England's cities and major towns, to the newly emerging excursionists:

The excursion fever which seized the working and lower classes in the 1850s was the result of a variety of forces; not the least of which was the desire of the railway companies to fill their trains, and the desire of the public itself to take to the rails. Having multiplied excessively during the years from 1845, the private railway companies fought to gain passengers, much as package holiday companies do today, lowering prices to perilously low levels in order to increase traffic ...

The result of this acute competition was improved conditions of travel for the third-class travel for the third class traveller who had ridden in open trucks in the forties, but by the 1860s was riding in upholstered seats in enclosed carriages. (Swinglehurst 1974, p. 22)

Wealthy travellers still went by, and often in their own, coach and horses despite the resources required—around five acres worth of hay and oats per horse per year and an army of blacksmiths (Briggs 1983, p. 211); indeed:

Travel had been a rarity for all but a minority throughout much of the eighteenth century. Getting around was difficult, and expensive. It was only with the arrival of maps and guidebooks, the creation of the turnpikes and the improvements to the technology of the stagecoach ... that even many of the elite began to travel. (Flanders 2006, p. 211)

During the century even the social elite began to adopt first class railway conveyance (Faith 1990, p. 17). The railway companies pioneered the development of well appointed hotels for the large excursionist parties. The first of these, in 1837, was Bridge House at the southern end of London Bridge and the first at a station was at London Euston that opened in 1839; meanwhile near to the Peak District, Derby's Midland Hotel opened in 1840 (Crawford 1990, p. 78). The railways (Fig. 6) not only offered a swift and reliable service compared with the turnpikes, especially for female field geologists (Burek & Kölbl-Ebert 2007a and b), as published in Bradshaw from 1839 onwards they also provided new exposures during their construction. The GA made considerable use of railways (see Fig. 7 and Burek 2008b); indeed, its first published excursion in 1860 to Folkestone employed the services of the South Eastern Railway, whose line had reached the town in late 1843 before reaching Dover in 1844 (Green 2008).

By the close of the nineteenth century the bicycle was also a popular form of transport for urban visitors to the countryside (Watson & Gray 1978, pp. 122–140; Burek & Kölbl-Ebert 2007a and b; Burek 2008) as promoted by the Cyclists' Touring Club following its 1878 foundation. However, the GA's attempts, following an 1898 proposal by one of its former presidents, at dedicated cycling provision was short-lived; the initial 1899 cycling geology excursion was to the Wokingham area and some seven others were organized before their 1907 withdrawal because of poor attendance, generally fever than ten members (Green 1989, p. 24). This was the same year that the GA employed the motor omnibus following their appearance in the first decade of the twentieth century, for its Tonbridge excursion and this became their common mode of transport after the First World War, as it did for many holidaymakers and day-trippers. During the twentieth century the rise of private motor vehicle ownership opened up the countryside to the elite before the Second World War and to the masses in the 1960s (Lavery 1971, pp. 97–111) when car ownership more than doubled from 5.7 million in the decade's opening year and two-thirds of holiday-makers travelled by car. Motorists' guidebooks and educational posters were published by the major petrol suppliers and Shell in particular promoted nature education, including geology (Shell 1964). Concomitantly the 1960s witnessed not only the decline of the branch line railway network that had began in the 1930s but, the rationalization of the railway network with around one third being closed to passenger traffic (Simmons & Biddle 1997).

By the turn of the twentieth century the guidebook had become an established travellers' tool given: 'in the nineteenth century [it] was, initially, a British and a German invention, people from those countries being the first to have the money and the intellectual curiosity to travel, at least in any numbers.' (Sillitoe 1995, p. 221) and likewise the first English geology field-guide was published in the early nineteenth century; Otley's 1823 *Lake District guide* is perhaps the first to be tourist-focused, whilst Mantell's 1847 *Geological Excursion Round the Isle of Wight* is the first pocket-sized geotourists' guidebook. Bædeker was the first guidebook publisher to employ asterisks (single or double) as marks of commendation for hotels and restaurants, for views and sites of outstanding natural beauty, and for works of architecture and art with the intention of familiarizing readers with the significant sites and sights encountered on their travels; 'starred in Bædeker' became synonymous for high quality. Bædeker's first England guide was published in German in 1862.

The first tourist guidebooks were for a wealthy discerning educated elite readership; however, by the middle of the nineteenth century it was a smaller market than that of the burgeoning middle-classes with more modest means and education, the market into which Bædeker tapped, and consequently quotations were few and brief. His main publishing rival was Murray, whose guidebooks were written for the wealthy educated elite traveller; Bædeker's guidebooks were always cheaper (in price and production quality) and sold in greater numbers. John Murray also produced forty volumes of populist county and cathedral guides. Numerous rival publications, especially in regional series, were produced, such as forty-four volumes published by Black's of Edinburgh from 1826 onwards. Later, Baddeley's nineteen-volume Thorough Guides with maps and plans by Bartholomew were published, followed by Ward Lock's almost ninety-volume 'Red Shilling Guides' and Methuen's fifty-volume 'Little Guides' and twenty-four-volume 'Highways and Byways' series.

Thus, by the last quarter of the nineteenth century good quality and affordable guidebooks existed for much of Britain, many of which had some limited mention of geology. From the middle of the nineteenth century onwards dedicated geotourists could avail themselves of a burgeoning range of geology textbooks and field-guides

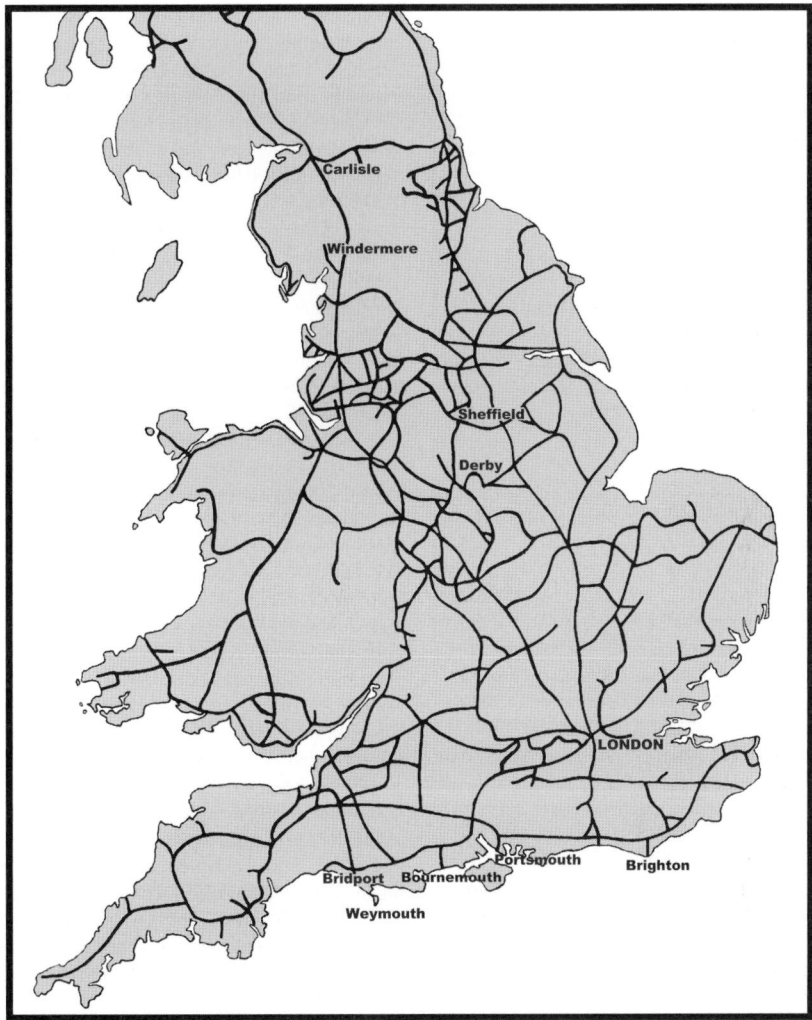

Fig. 6. 1870 railway map of England and Wales showing key localities mentioned within the text.

(finances permitting because of the expense of low print runs). Two major works summarizing southern Britain's geology, Phillips' 1818 *A Selection of Facts from the Best Authorities, Arranged so as to Form an Outline of the Geology of England and Wales* and Conybeare & Phillips's 1822 *Outlines of the Geology of England and Wales* were available. The former was the first to compile all that was known about stratigraphy. The latter became the standard introduction to geology, which through its various revisions represented the evolving compromise between the Bible and geology.

Concomitant with these guides, the first geological textbooks and field manuals appeared; principal amongst the latter were De La Beche's 1831 *A Geological Manual*, 1835 *How to Observe* and 1851 *The Geological Observer*. Some were aimed at the public and children such as Mantell's 1849 *Thoughts on a Pebble* and Hawks' 1914 *The Earth shown to the Children*. In the late nineteenth century county-based geology accounts appeared in trade directories; one set, Harrison's 1882 *Geology of the Counties of England and of North and South Wales*, was bound into a single volume; its modern counterpart is Anderson's 1983 *Field Geology in the British Isles*. The excursions of the GA were published twice, as *Geologists' Association: A Record of Excursions Made Between 1860 and 1890* (Holmes & Sherborn 1891) and *Geology in the Field: The Jubilee Volume of the Geologists' Association (1858–1908)* (Monckton & Herries 1910), as discrete volumes providing invaluable accounts of its excursions and field practices and attitudes of the day to

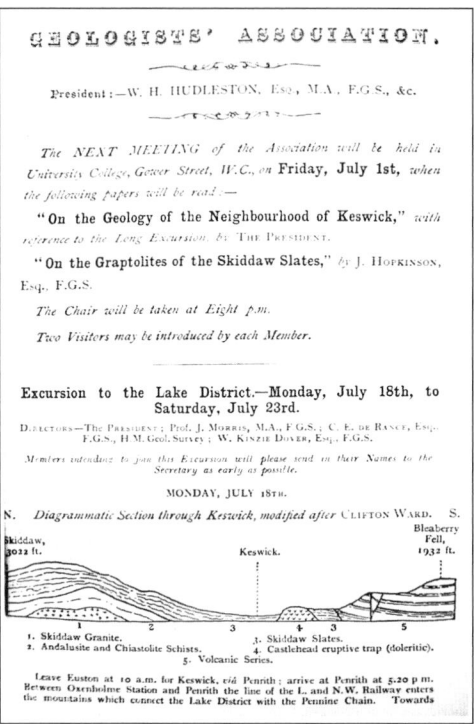

Fig. 7. Geologists' Association Circular announcing 1881 Lake District field excursion; note the use made of the railway for both travel (train departing London Euston 10am and arriving Penrith at 5.20pm) and observation (between Oxenholme and Penrith it enters the mountains connecting the Lake District and the Pennines).

geoconservation. The most influential specialist's field-guides appeared in the mid-twentieth century, the *Centennial Guides of the GA*; published from 1958 onwards, they have evolved into high quality publications in recent years. The non-specialist's *Geology Explained* series, published by David and Charles, appeared in the late-1960s.

Conclusions and the future

The locations and activities that were established by the late nineteenth century are embedded within much modern geotourism provision. The main change has been in the mode of transport and the relationship that geotourists enjoy with the landscape. Roads have replaced mass rail transport as the preferred mode of travel for tourists. There is a greater tourist emphasis on pleasure and leisure than intellectual endeavour and religious awareness. A new impetus was given, from the mid-twentieth century, to scenic tourism by the national

recognition and protection of outstanding scenery as National Parks and Areas of Outstanding Natural Beauty (AONBs) following the 1949 National Parks and Access to the Countryside Act (Prosser 2008). In the National Parks, measures and funding were available to facilitate and promote access for tourists; however, the AONBs lacked such resources and progress was painfully slow (Green 1996, pp. 102–105). The 1968 Countryside Act further promoted scenic tourism, albeit on the urban fringe, with the establishment of Country Parks as areas of open country for informal recreation and to relieve pressure on the National Parks and AONBs. Some of these Country Parks, such as Park Hall in Staffordshire, were established on old mining or quarrying sites, but their geological interest was generally ignored in their presentation.

UNESCO international recognition has been available since 1972; the most significant is citation within the World Heritage List containing around 630 sites, 7% of which are inscribed primarily for geological interest. Because not all significant geosites meet the 'outstanding universal value' criterion required by the World Heritage Convention an alternative was considered (Boylan 2008). In 1999 the Geoparks programme (incorporating the author's geotourism approach) envisaged that it would recognize the relationships between people and geology and the potential for economic development; it would promote landscape elements, rather than small geological outcrops and be managed holistically to protect and enhance the natural characteristics. The concept is modelled on UNESCO's *Man and the Biosphere* programme; this emphasizes the links between conservation and development, additionally linking science, education and sustainable development. It was suggested that the major benefit of the programme would be: '... focusing attention directly on geological and geomorphological conservation and the related issue of sustainable development.' Geosites within Geoparks must be scientifically significant and have educational potential together with some aesthetic appeal. Ideally the geological interest should be allied to some archaeological, historical, cultural or ecological interest. Within Geoparks the sale of geological material, local and imported, is banned and educational provision is required to maintain membership.

The establishment of European Geoparks as: 'A small network of European sites which includes territories with a significant geological heritage and a sustainable development strategy ...' was supported by UNESCO (Jones 2008). Geoparks are the UK's main twenty-first century geotourism development. Focused on economic and social regeneration they continue and develop the trends in interpretative provision resulting from the 1970s recognition of industrial heritage. The two regions i.e. the Peak District and the Lake District which are most significant in geotourism's early development, could be similarly recognized and the relationship between their geology and scenic beauty should be better promoted. The main access to the Geoparks, even when heritage lines are included, is not rail transport (around one-tenth of trips) but road (some three-quarters of trips), creating further environmental pressures on protected and conserved landscapes.

Looking to the future and the development and promotion of Geoparks, greater and better use of the internet to illustrate and promote geosites can be expected; virtual field trips will become a feature of interpretative provision, especially as Geoparks and protected landscapes both vie for custom and seek measures to conserve their geological assets. For actual visits sustainable tourism transport in such landscapes is a poorly addressed area of academic and practitioner study; the use, at least from nodal park and ride schemes, of rail and bus services will need to be encouraged. Geo-interpretation will become more widespread and integrated within other discrete heritage industry offerings, especially those of industrial archaeological and heritage transport nature. Commemorative plaques could be more widely employed so that tourists are better made aware of the significant role that Britain played in geology's development and its present international significance. Geoparks will undoubtedly bring new and large audiences to geology that will generate new demands and pressures on geosites. Naturally, there will be an ongoing debate about the balance between geoconservation and geo-exploitation, including sustainable access measures such as transport. Further detailed geotourism research will be required as an aid to Geopark selection and their sustainable management. Finally, for most casual and many dedicated geotourists given that the setting of geosites is their principal attraction and the development of Geoparks promotes such aesthetic considerations, James Boswell's 18 August 1773 remark to his wife as he prepared to set out with Samuel Johnson on their tour through Scotland and the Hebrides: 'Madame, we do not go there as to paradise. We go to see something different from what we are accustomed to' provide a fitting summation of geotourism's likely continuing appeal into the twenty-first century.

References

ANON. no date. *Guidelines for Collecting Fossils on the Isle of Wight.* Newport, Isle of Wight County Council.

ANON. 1994. *County Durham Geological Conservation Strategy.* Durham County Council, Durham.

ANON. 2007. GCSE Geology Under Threat. *UKRIGS Newsletter,* **5**, 2, 1.

AGER, D. V. & SMITH, W. E. 1973. *Guide No. 23: The Coast of South Devon and Dorset between Branscombe and Burton Bradstock.* Geologists' Association, London.

ALLEN, P. 1960. *Guide No. 24: Geology of the Central Weald: The Hastings Beds.* Geologists' Association, London.

ALLEN, D. E. 1978. *The Naturalist in Britain: A Social History.* Allen Lane/Penguin Books, London.

ALLISON, R. J. 1992. *Guide No. 47: The Coastal Landforms of Dorset.* Geologists' Association, London.

ANDERSON, J. G. C. 1983. *Field Geology in the British Isles: A Guide to Regional Excursions.* Pergamon Press, Oxford.

ARKELL, W. J. 1934. Whitsun field meeting, The Isle of Purbeck. *Proceedings of the Geologists' Association,* **45**, 412–419.

AWDRY, C. 1990. *Encyclopaedia of British Railway Companies,* Wellingborough, Patrick Stephens Limited.

BAILEY, E. 1952. *Geological Survey of Great Britain,* Thomas Murby and Company, London.

BALDWIN, A. & ALDERSON, D. M. 1996. A remarkable survivor: a nineteenth century geological trail in Rochdale, England. *Geological Curator,* **6**, 227–231.

BARBER, K. E. (ed.) 1987. *Wessex And The Isle of Wight Field Guide. (prepared to accompany the Annual Field Meeting held at Southampton and Cowes 21–25 April 1987).* Quaternary Research Association, Cambridge.

BARNARD, T., CAPEWELL, J. G. & LANG, W. D. 1950. Whitsun field meeting at Lyme Regis, 1948. *Proceedings of the Geologists' Association,* **61**, 156–158.

BARTON, M. E. & DELAIR, J. B. 1982. Oatlands Park Grotto And Its Ammonite Fossils. *Geological Curator,* **3**, 375–387.

BENNETT, T. 1996. How the other half visit. *Museums Journal,* **96**, 37.

BIRD, E. 1995. *Geology and Scenery of Dorset.* Ex Libris, Bradford on Avon.

BOYLAN, P. J. 2008. Geological site designation under the 1972 UNESCO World Heritage Convention. *In*: BUREK, C. V. & PROSSER, C. D. (eds) *The History of Geoconservation.* The Geological Society, London, Special Publications, **300**, 279–304.

BRAY, W. 1783. *Tour through some of the Midland Counties, into Derbyshire and Yorkshire by William Bray, F. A. S. Performed in 1777 paraphrased in* MAVOR, W. (1798–1800) *The British Tourists; or Traveller's Pocket Companion, through England, Wales, Scotland, and Ireland. Comprehending the Most Celebrated Tours in the British Islands* (6 volumes) volume 2. B. White, London.

BRENDON, P. 1991. *Thomas Cook: 150 Years of Popular Tourism.* Secker and Wurburg, London.

BRIGGS, A. 1983. *A Social History of England.* Wiedenfeld and Nicholson, London.

BRISTOW, H. W. 1862. *The Geology of the Isle of Wight.* Geological Survey, London.

BROOKS, K. 2001. *Geology and Fossils of the Hastings Area.* Ken Brooks, Bexhill-on-Sea.

BUCKLEY, R. 2003. Environmental Inputs and Outputs in Ecotourism: Geotourism with a Positive Triple Bottom Line? *Journal of Ecotourism*, **2**, 76–82.

BUREK, C. V. 2008a. History of RIGS in Wales: an example of successful cooperation for geoconservation. *In*: BUREK, C. V. & PROSSER, C. D. (eds) *The History of Geoconservation*. The Geological Society, London, Special Publications, **300**, 147–171.

BUREK, C. V. 2008b. The role of the voluntary sector in the evolving geoconservation movement. *In*: BUREK, C. V. & PROSSER, C. D. (eds) *The History of Geoconservation*. The Geological Society, London, Special Publications, **300**, 61–89.

BUREK, C. V. & FRANCE, D. E. 1998. NEWRIGS uses a Steam Train and Town Geological Trail to raise Public Awareness in Llangollen, North Wales. *Geoscientist*, **8**, 8–10.

BUREK, C. V. & KÖLBL-EBERT, M. 2007a. The historical problems of travel for women undertaking geological fieldwork, Women in the History of Geology, V. *Geology Today*, **23**, 30–32.

BUREK, C. V. & KÖLBL-EBERT, M. 2007b. The historical problems of travel for women undertaking geological fieldwork. *In*: BUREK, C. V. & HIGGS, B. (eds) *The Role of Women in the History of Geology*. Geological Society, London, Special Publications, **281**, 115–122.

BUZZARD, J. 1993. *The Beaten Track: European Tourism, Literature, and the Ways to Culture 1800–1918*. Clarendon Press, Oxford.

CALDER, N. 1972. *The Restless Earth*. BBC Books, London.

CALDER, N. 1974. *The Weather Machine*. BBC Books, London.

CLARKE, G. 1991. Geology and the public at the Natural History Museum. *Geology Today*, **7**, 217–220.

CLARKE, N. J. 1998. *Lyme Bay Fossils Beach Guide*. Nigel J. Clarke Publications, Lyme Regis.

CLEAL, C. J. & THOMAS, B. A. 1995a. Wadsley Fossil Forest. *Geological Conservation Review Volume 12: Palaeozoic Palaeobotany of Great Britain*. Chapman and Hall, London, 208–210.

CLEAL, C. J. & THOMAS, B. A. 1995b. Victoria Park. *Geological Conservation Review Volume 12: Palaeozoic Palaeobotany of Great Britain*. Chapman and Hall, London, 188–191.

CONYBEARE, W. D. & PHILLIPS, W. 1822. *Outlines of the Geology of England and Wales*. William Phillips, London.

COPE, F. W. 1976. *Geology Explained in the Peak District*. David and Charles, Newton Abbot.

CORAM, R. 1989. *Finding Fossils in Lyme Bay: A Guide to Lyme Regis, Charmouth and Surrounding Areas*. British Fossils, Wimbourne.

COTTON, C. 1681. *The Wonders of the Peake*. Printed for Joanna Brome, London.

COWHAM, J. H. 1900. *The School Journey: A means of Teaching Geography, Physiography and Elementary Science*. Westminster School Book Depot, London.

CRAWFORD, D. 1990. *British Building Firsts: The First Castle to the First Airport*. David and Charles, Newton Abbot.

Cumberland Geological Society. 1982. *Rocks and Fossils: A geological field guide: The Lake District*. Unwin, London.

DAICHES, D. & FLOWER, J. 1979. *Literary Landscapes of the British Isles: A Narrative Atlas*. Paddington Press, New York.

DAGNALL, P. 1995. A New Era for the Earth Galleries. *Environmental Interpretation*, **10**, 4–5.

DAMON, R. 1860. *Handbook of the Geology of Weymouth and the Isle of Portland*. E. Stanford, London.

DAVIES, G. M. 1914. *Geological Excursions Round London*. Thomas Murby, London.

DAVIES, G. M. 1935. *The Dorset Coast: A Geological Guide*. A and C Black, London.

DAVIES, G. M. 1939. *Geology of London and South-East England*. Thomas Murby, London.

DAVIES, G. M. 1956. *The Dorset Coast: A Geological Guide*. 2nd edn. A and C Black, London.

DE BASTION, R. 1994. The private sector—threat or opportunity? *In*: O'HALLORAN, D., GREEN, C., HARLEY, M., STANLEY, M. & KNILL, J. (eds) *Geological and Landscape Conservation*. The Geological Society, London, 391–395.

DE LA BECHE, H. T. 1831. *A Geological Manual*. London.

DE LA BECHE, H. T. 1835. *Report on the Geology of Cornwall, Devon and Somerset*.

DE LA BECHE, H. T. 1851. *The Geological Observer*, London.

DEFOE, D. no date. *A Tour Through England and Wales Divided into Circuits* (in 2 volumes). J. M. Dent, London.

DIXON, F. 1850. *The Geology and Fossils of the Cretaceous and Tertiary Formations of Sussex*. London.

DIXON, F. (revised by Jones, T. R.) 1878. *The Geology of Susses or The Geology and Fossils of the Cretaceous and Tertiary Formations of Sussex*. (2nd edn) William J. Smith, Brighton.

DODD, M. (ed.) 1992. *Lakeland Rocks and Landscape: A Field Guide*. Cumberland Geological Society, Kendal.

DOYLE, P. 1993. The Lessons of Crystal Palace. *Geology Today*, **9**, 107–109.

DOYLE, P. 1995. *Crystal Palace Park Geological Time Trail: Paxton's Heritage Trail Number Two*. Croydon, London Borough of Bromley.

DOYLE, P. 2008. A vision of 'deep time': the 'Geological Illustrations' of Crystal Palace Park, London. *In*: BUREK, C. V. & PROSSER, C. D. (eds) *The History of Geoconservation*. Geological Society, London, Special Publications, **300**, 197–205.

DOYLE, P. & ROBINSON, J. E. 1993. The Victorian geological illustrations of Crystal Palace Park. *Proceedings of the Geologists' Association*, **104**, 181–194.

DUFF, K. L., McKIRDY, A. P. & HARLEY, M. J. (eds) 1985. *New Sites for Old: A Student's Guide to the Geology of the East Mendips*. Nature Conservancy Council, Peterborough.

DUNNING, F. W. 1974. The Story of the Earth exhibition at the Geological Museum, London, *Museum*, **26**, 99–100.

DUNNING, F. W. 1975. The Story of the Earth in the Geological Museum: A view from the inside. *Geology— Journal of the Association of Teachers of Geology*, **6**, 12–16.

EDMONDS, R. P. H. 1996. The potential of visitor centres: a case history. *In*: PAGE, K. N., KEENE, P., EDMONDS, R. P. H. & HOSE, T. A. *English Nature*

Research Reports No.176—Earth Heritage Site Interpretation in England: A review of Principle Ttechniques with Case Studies. English Nature, Peterborough, 16–23.

ENGLEFIELD, H. C. 1816. *A description of the principal picturesque beauties, antiquities and geological phenomena of the Isle of Wight. With additional observations on the strata of the Island, and their continuation in the adjacent parts of Dorsetshire by Thomas Webster*. T. Webster, London.

FAITH, N. 1990. *The World the Railways Made*. Pimlico, London.

FAREY, J. 1811. *General View of the Agriculture and Minerals of Derbyshire*. 3 vols. McMillan, London.

FIENNES, C. 1949. *The Journeys of Celia Fiennes* (edited by MORRIS, C.). Cresset Press, London.

FISHER, J. A. 1994. The changing nature of Earth science fieldwork within the UK school curriculum and the implications for conservation policy and site development. *In*: O'HALLORAN, D., GREEN, C., HARLEY, M., STANLEY, M. & KNILL, J. (eds) *Geological and Landscape Conservation*. The Geological Society, London, 477–481.

FLANDERS, J. 2006. *Consuming Passions: Leisure and Pleasure in Victorian Britain*. Hodder and Stoughton, London.

GALLOIS, R. 1999. *Holiday Geology Guide: Isle of Wight*. British Geological Survey, Keyworth.

GARDOM, T. & MILNER, A. 1993. *The Natural History Museum Book of Dinosaurs*. The Natural History Museum, London.

GIBBONS, W. 1981. *The Weald: Rocks and Fossils—A Geological Field Guide*. Unwin, London.

GILPIN, W. 1786. *Observations, Relative Chiefly to Picturesque Beauty, Made in the Year 1772, on Several Parts of England: Particularly the Mountains and Lakes of Cumberland, and Westmorland* (in 2 volumes). London.

GRAYSON, A. 1988a. Radio Geology. *Geology Today*, 4, 62–63.

GRAYSON, A. 1988b. *Rock Solid: Britain's Most Ancient Heritage*. Natural History Museum, London.

GREEN, B. 1996. *Countryside Conservation*. (3rd edn). E. and F. N. Spon, London.

GREEN, C. P. 1989. Excursions in the past: a review of the Field Meeting Reports in the first one hundred volumes of the Proceedings. *Proceedings of the Geologists' Association*, 100, 17–29.

GREEN, C. P. 2008. The Geologists' Association and geoconservation: history and achievements. *In*: BUREK, C. V. & PROSSER, C. D. (eds) *The History of Geoconservation*. Geological Society, London, Special Publications, 300, 91–102.

GREEN, W. 1819. *The Tourist's New Guide, Containing a Description of the Lakes, Mountains, and Scenery in Cumberland, Westmorland, and Lancashire, with Some Account of Their Bordering Towns and Villages. Being the Result of Observations Made During a Residence of Eighteen Years in Ambleside and Keswick* (2 volumes). Lough & Co., Kendal.

HALL, C. M. & WEILER, B. 1992. What's Special about Special Interest Tourism? *In*: WEILER, B. & HALL, C. M. (eds) *Special Interest Tourism*. Belhaven, London, 1–14.

HAMILTON, B. M. 1976. The Changing Place of Geology in Science Education in England. *Journal of Geological Education*, 24, 105–110.

HARLEY, M. J. 1996. Involving a wider public in conserving their geological heritage: A major challenge and recipe for success, *Geoscience Education and Training In schools and universities, for industry and public awareness*. Balkema, Rotterdam, 725–729.

HARLEY, M. J. & ROBINSON, E. 1991. 'RIGS—a local Earth-science conservation initiative'. *Geology Today*, 7, 47–50.

HARRIS, H. 1971. *Industrial Archaeology of the Peak District*. David and Charles, Newton Abbot.

HARRISON, W. J. 1882. *Geology of the Counties of England and of North and South Wales*. Kelly and Co/Simpkin, Marshall and Company, London.

HAWKINS, B. W. 1854. On Visual Education As Applied To Geology Illustrated By Diagrams And Models Of The Geological Restorations At The Crystal Palace. *Journal of the Society of Arts* and republished as a leaflet in London by James Tennant, 149 Strand.

HAWLEY, D. 1994. Conservation from the chalkface. *Earth Heritage*, 26–27.

HAWLEY, D. 1996. Urban Geology and the National Curriculum. *In*: BENNETT, M. R., DOYLE, P., LARWOOD, J. G. & PROSSER, C. D. (eds) *Geology On Your Doorstep*. The Geological Society, London, 155–162.

HAWLEY, D. 1998. RIGS for Education. *In*: OLIVER, P. G. (ed.) *Proceedings of the First UK Rigs Conference*. Worcester, Herefordshire and Worcestershire RIGS Group, 65–78.

HAWKS, E. 1914. *The Earth Shown to the Children*. T. C. and E. C. Jack, London.

HEYWOOD, P. 1990. Truth and Beauty in Landscape—trends in landscape and leisure. *Landscape Australia*, 12, 43–47.

HOBBES, T. 1678. *De Mirabilis Pecci: Being the Wonders of the Peak in Darby-shire, Commonly Called the Devil's Arse Peak. In English and Latine. The Latine Written by Thomas Hobbes of Malmsbury. The English by a Person of Quality*. William Crook, London.

HODGSON, M. & LUNT, L. 1987. *Excursion to Loweswater: A Lakeland Visit 1865*. Macdonald Orbis, London.

HOLDEN, E. 1977. *The Country Diary of an Edwardian Lady*. Michael Joseph, London.

HOLMES, T. V. & SHERBORN, C. D. 1891. *Geologists' Association: A Record of Excursions Made Between 1860 and 1890*. Stanford, London.

HOUSE, M. 1989. *Geology of the Dorset Coast*. Geologists' Association, London.

HOSE, T. A. 1994. *Interpreting Geology at Hunstanton Cliffs SSSI Norfolk: a summative evaluation (1994)*. The Buckinghamshire College, High Wycombe.

HOSE, T. A. 1995a. Selling the Story of Britain's Stone. *Environmental Interpretation*, 10, 16–17.

HOSE, T. A. 1995b. Evaluating interpretation at Hunstanton, *Earth Heritage*, 4, 20.

HOSE, T. A. 1996. Geotourism, or can tourists become causal rocklounds? *In*: BENNETT, M. R., DOYLE, P., LARWOOD, J. G. & PROSSER, C. D. (eds) *Geology on your Doorstep: the Role of Urban Geology in*

Earth Heritage Conservation. Geological Society, London, 208–228.

HOSE, T. A. 1997. Geotourism—Selling the earth to Europe. *In*: MARINOS, P. G., KOUKIS, G. C., TSIAMAOS, G. C. & STOURNASS, G. C. (eds) *Engineering Geology and the Environment*. A. A. Balkema, Rotterdam, 2955–2960.

HOSE, T. A. 2000. European Geotourism—Geological Interpretation and Geoconservation Promotion for Tourists. *In*: BARRETINO, D., WIMBLEDON, W. P. & GALLEGO, E. (eds) *Geological Heritage: Its Conservation and Management*. Instituto Tecnologico Geominero de Espana, Madrid, 127–146.

HOSE, T. A. 2003. *Geotourism in England: A Two-Region Case Study Analysis*. Unpublished PhD thesis, University of Birmingham.

HOSE, T. A. 2005. Writ in Stone: A Critique of Geoconservation Panels and Publications in Wales and the Welsh Borders. *In*: COULSON, M. R. (ed.) *Stone in Wales*. Cadw, Cardiff, 54–60.

HOUSE, M. 1989. *Geology of the Dorset Coast*. Geologists' Association, London.

HUDSON, J. 1842. *Complete Guide to the Lakes*. Hudson, Kendal.

HUGHES, J. C. 1922. *The Geological Story of the Isle of Wight*. Stanford, London.

HUTCHINSON, W. 1774. *An Excursion to the Lakes, in Westmoreland and Cumberland, August 1773*. London.

HUTCHINSON, W. 1794. *The History of the County of Cumberland, and Some Places Adjacent from the Earliest Accounts to the Present Time* (in two volumes). Carlisle.

INSOLE, A., DALEY, B. & GALE, A. 1998. *Guide No. 60: The Isle of Wight*. Geologists' Association, London.

JARZEMBOWSKI, E. 1989. Writhlington Geological Nature Reserve. *Proceedings of the Geologists' Association*, **100**, 219–234.

JENKINS, J. M. 1992. Fossickers and Rockhounds in Northern New South Wales. *In*: WEILER, B. & HALL, C. M. (eds) *Special Interest Tourism*. Belhaven, London, 129–140.

JOHNSTON, D. & SHARPE, T. 1997. Visitor Behaviour At The Evolution of Wales Exhibition, National Museum and Gallery, Cardiff, Wales. *The Geological Curator*, **6**, 255–266.

JONES, B. & KING, C. 2005. The Ups and Downs of A Level Geology 1971–1998—A Numerical Picture. *Teaching Earth Sciences*, **18**, 33–34.

JONES, C. 2008. History of Geoparks. *In*: BUREK, C. V. & PROSSER, C. D. (eds) *The History of Geoconservation*. Geological Society, London, Special Publications, **300**, 273–277.

KELLY, J. 1994. Cardiff Arms Its Park. *Museums Journal*, **94**, 25–26.

KNELL, S. 1993. No Stone Unturned. *Museums Journal*, **93**, 20.

KNELL, S. & TAYLOR, M. A. 1991. Museums on the Rocks. *Museums Journal*, **91**, 23–25.

KING, J. H. 1993. Earth Science in the National Curriculum of England and Wales. *Journal of Geological Education*, **41**, 318.

KING, C. & JONES, B. 1999. Geology in the Curriculum: trends in examination entry figures, 1971–1998. Challenge. *Teaching Earth Sciences*, **24**, 187–191.

KING, C. & JONES, B. 2006. The A Level Geology Challenge. *Teaching Earth Sciences*, **31**, 29–31.

KIRKALDY, J. F. (REV. MIDDLEMISS, F. A., ALLCHIN, L. J. & OWEN, H. G.) 1976. *Guide No. 29: The Weald*. Geologists' Association, London.

KOMOO, I. 1997. Conservation geology: A case for the ecotourism industry of Malaysia. *In*: MARINOS, P. G., KOUKIS, G. C., TSIAMBAOS, G. C. & STOURNAS, G. C. (eds) *Engineering Geology and the Environment*. Balkema, Rotterdam, 2969–2973.

KOMOO, I. & DEAS, K. M. 1993. *Geotourism: An Effective Approach Towards Conservation of Geological Heritage*. Symposium of Indonesia—Malaysia Culture, Unpublished paper. Badung, Malaysia.

LAMB, S. & SINGLETON, D. 1998. *Earth Story: The Shaping Of Our World*. BBC Books, London.

LANG, W. D. 1906. Excursion to Lyme Regis, Easter 1906. *In*: WOODWARD, H. B. & YOUNG, G. W. (eds) *Proceedings of the Geologists' Association*, **19**, 323–324, 328–329.

LARWOOD, J. & PROSSER, C. 1998. Geotourism, Conservation and Tourism, *Geologica Balacania*, **28**, 97–100.

LAVERY, P. 1971. *Recreational Geography*. David and Charles, Newton Abbot.

LAWSON, J. D. 1973. New exposures on forestry roads near Ludlow. *Geological Journal*, **8**, 279–284.

LAWSON, J. D. 1977. *Mortimer Forest Geological Trail*. Nature Conservancy Council, London.

LYNAS, B. 1994. *Lakeland Rocky Rambles: Geology Beneath your Feet*. Sigma Leisure, Wilmslow.

LYONS, P. C., ROBERTSON, E. C. & MILTON, L. 1993. C. Wroe Wolfe's Geology Course on Radio-Station WGBH (Boston) in 1954. *Journal of Geological Education*, **41**, 170.

MCCARTHY, S. & GILBERT, M. 1994. *The Crystal Palace Dinosaurs: The Story of the World's First Prehistoric Sculptures*. Crystal Palace Foundation, London.

MACKAY, C. 1846. *The Scenery and Poetry of the English Lakes: A Summer Ramble*. Longman & Co., London.

MAINI, J. S. & CARLISLE, A. 1974. *Conservation in Canada: A Conspectus—Publication No.1340*. Department of the Environment/Canadian Forestry Service, Ottawa.

MANTELL, G. A. 1822. *The Fossils of the South Downs*. Lupton Relfe, London.

MANTELL, G. A. 1827. *Illustrations of the Geology of Sussex*. Lupton Relfe, London.

MANTELL, G. A. 1833. *The Geology of the South-East of England*. Longman, Rees, Orme, & etc., London.

MANTELL, G. A. 1846. *A Day's Ramble in and Around the Ancient Town of Lewes*. Henry Bohn, London.

MANTELL, G. A. 1847. *Geological Excursion Round the Isle of Wight and Along the Adjacent Coast of Dorsetshire*. Henry G. Bohn, London.

MANTELL, G. A. 1849. *Thoughts on a Pebble*. Reeves, Benham and Reeve, London.

MANTELL, G. A. 1851. *Geological Excursion Round the Isle of Wight and Along the Adjacent Coast of Dorsetshire*. (2nd edn) Henry G. Bohn, London.

MANTELL, G. A. 1854. *Geological Excursion Round the Isle of Wight and Along the Adjacent Coast of Dorsetshire*. (3rd edn) Henry G. Bohn, London.

MARSHALL, J. D. & DAVIES-SHIEL, M. 1977. *The Industrial Archaeology of the Lake Counties*. David and Charles, Newton Abbot.

MARTIN, W. 1809. *Petrificata Derbiensia; or figures and descriptions of petrifactions collected in Derbyshire*. D. Lyon, Wigan.

MARTINI, G. 1994. The protection of geological heritage and economic development: the saga of the Digne ammonite slab in Japan. *In*: O'HALLORAN, D., GREEN, C., HARLEY, M., STANLEY, M. & KNILL, J. (eds) *Geological and Landscape Conservation*. The Geological Society, London, 383–386.

MARTINI, G. 2000. Geological Heritage and Geo-tourism. *In*: BARRETINO, D., WIMBLEDON, W. P. & GALLEGO, E. (eds) *Geological Heritage: Its Conservation and Management*. Instituto Tecnologico Geominero de Espana, Madrid, 147–156.

MARTINEAU, H. 1855. *A Complete Guide to the English Lakes*. J. Garnett, Windermere.

MAVOR, W. 1798–1800. *The British Tourists; or Traveller's Pocket Companion, through England, Wales, Scotland, and Ireland. Comprehending the Most Celebrated Tours in the British Islands* (in six volumes). London.

MONCKTON, H. W. & HERRIES, R. S. (eds) 1910. *Geology in the Field: The Jubilee Volume of the Geologists' Association, 1858–1908; 2 vols*. Stanford, London.

MULLINS, E. 1985. *A Love Affair With Nature*. Phaidon, Oxford.

MUNT, M. 2008. A history of geological conservation on the Isle of Wight. *In*: BUREK, C. V. & PROSSER, C. D. (eds) *The History of Geoconservation*. The Geological Society, London, Special Publications, **300**, 173–179.

MURRAY, Hon Mrs. 1799. *A Companion and Useful Guide to the Beauties of Scotland, to the Lakes of Westmorland, Cumberland, and Lancashire; and to the Curiosities in the District of Craven in the West Riding of Yorkshire. To Which Is Added, a More Particular Description of Scotland, Especially That Part of It, Called the Highlands*. G. Nicol, London.

NEVES, R. & DOWNIE, C. 1967. *Geological Excursions in the Sheffield Region and the Peak District National Park*. University of Sheffield, Sheffield.

NEWRIGS. 1997. *Steaming through the Past: A Geological Trail for Llangollen Valley*. NEWRIGS/Chester College, Chester.

NORMAN, M. W. 1887. *A popular guide to the geology of the Isle of Wight, with a note on its relation to that of the Isle of Purbeck*. G. H. Brittain, Ventnor.

OGILBY, J. 1675. *Britannia, Volume the First: or An Illustration of the Kingdom of England and the Dominion of Wales: by a Geographical and Historical Description of the Principal Roads thereof. Actually admeasured and Delineated in a Century of Whole-Sheet Copper-Sculps. Accommodated with the Ichonography of the Several Cities and Capital Towns; and Compleated by an Accurate Account of the More Remarkable of Passages of Antiquity, Together with a Novel Discourse of the Present State*. Ogilby, London.

OGILBY, J. 1720. *Britannia Depicta or Ogilby Impov'd; Being a Correct Coppy of Mr. Ogilby's Actual Survey of all y Direct & Principal Cross Roads in England and Wales: Wherein are exactly Delineated & Engraven, All y cities, Towns, Villages, Churches, Seats & scituate on or near the Roads with their respective Distances in Measured and Computed Miles. And to render this work universally usefull & agreeable, [beyond any of its kind] are added in a clear & most compendious method. A full & particular description & account of all the cities, borough-towns, towns-corporate & their arms, antiquity, charters, privileges, trade, rarities, & with suitable remarks on all places of note drawn from the best historians and antiquaries*. Thomas Bowles, London.

OLDROYD, D. R. 2002. *Earth, Water, Ice and Fire, Two Hundred Years of Geological Research in the English Lake District*. The Geological Society, London, Memoirs, 25.

OTLEY, J. 1823. *A concise description of the English Lakes, and adjacent mountains, with general directions to tourists; and observations on the mineralogy and geology of the district*. L. Otley, Keswick.

OUSBY, I. 1990. *The Englishman's England: Taste, Travel and the rise of tourism*. Cambridge University Press, Cambridge.

PAGE, K. N. 1998. England's Earth Heritage Resource—an asset for everyone. *In*: HOOKE, J. (ed.) *Coastal Defence and Earth Science Conservation*. The Geological Society, London, 196–209.

PFAFF, M. & SIMCOX, D. 1998. *Lulworth Rocks*. Weld Estate, Lulworth.

PHILLIPS, W. 1818. *A Selection of Facts from the Best Authorities, Arranged so as to Form an Outline of the Geology of England and Wales*. Phillips, London.

PLOG, S. C. 1974. Why destination areas rise and fall in popularity. *The Cornell Hotel and Restaurant Administration Quarterly*, **15**, 55–58.

PROSSER, C. D. 2008. The history of geoconservation in England: legislative and policy milestones. *In*: BUREK, C. V. & PROSSER, C. D. (eds) *The History of Geoconservation*. Geological Society, London, Special Publications, **300**, 113–121.

PROSSER, C. D. & LARWOOD, J. G. 2008. Conservation at the cutting-edge: the history of geoconservation on the Wren's Nest National Nature Reserve, Dudley, England. *In*: BUREK, C. V. & PROSSER, C. D. (eds) *The History of Geoconservation*. Geological Society, London, Special Publications, **300**, 217–235.

PROSSER, R. 1977. *Geology Explained in the Lake District*. David and Charles, Newton Abbot.

READ, S. E. 1980. A prime force in the expansion of tourism in the next decade: special interest travel. *In*: HAWKINS, D. E., SHAFER, E. L. & ROVELSTAD, J. M. (eds) *Tourism Marketing and Management Issues*. George Washington University Press, Washington D.C.

REID, C. G. R. 1993. The Local Geologist 12: The road show. *Geology Today*, **9**, 147–153.

ROBINSON, E. 1993. Who would buy a coal tip? *Earth Science Conservation*, **32**, 20–21.

ROBINSON, E. 1996a. Geology reserve: Wren's Nest revised. *Geology Today*, **12**, 211–213.

ROBINSON, E. 1996b. Earth Galleries open: Impression 1. *Geology Today*, **12**, 128–130.

ROGERS, S. 1834. *Poems*. (6th ed.). T. Cadell & E. Moxon, London.

ROSS, M. 1991. *Planning and the Heritage: Policy and Procedures*. Spon, London.

SELINCOURT, E. (ed.) 1977. *Wordsworth's Guide to the Lakes, The Fifth Edition (1835) With an Introduction, Appendices and Notes Textual and Illustrative*. Oxford University Press, Oxford.

SHACKLETON, E. H. 1967. *Lakeland Geology*. Dalesman, Clapham.

SHARPE, T., HOWE, S. & HOWELLS, C. 1998. Gallery Review: Setting The Standard? The Earth Galleries At The Natural History Museum, London. *The Geological Curator*, **6**, 395–403.

SHELL, 1964. *The Shell Book of Nature*. Phoenix House, London.

SILLITOE, A. 1995. *Leading the Blind: A Century of Guidebook Travel 1815–1911*. Macmillan, London.

SIMMONS, J. & BIDDLE, G. (eds) 1997. *The Oxford Companion to British Railway History*. Oxford University Press, Oxford.

SIMPSON, I. M. 1982. *Rocks and Fossils A geological field guide: The Peak District*. Unwin, London.

SMITH, A. 2001. *The Rock Men: Pioneers of Lakeland Geology*. Cumberland Geological Society, Kendal.

SMITH, P. J. 1996. Earth Galleries open: Impression 2. *Geology Today*, **12**, 130–132.

SPITERI, A. 1994. Malta: a model for the conservation of limestone regions. *In*: O'HALLORAN, D., GREEN, C., HARLEY, M., STANLEY, M. & KNILL, J. (eds) *Geological and Landscape Conservation*. The Geological Society, London, pp. 205–208.

STUEVE, A. M., COCK, S. D. & DREW, D. 2002. *The Geotourism Study: Phase 1 Executive Summary*. www.tia.org/pubs/geotourismphasefinal.pdf.

SWINGLEHURST, E. 1974. *The Romantic Journey: The Story of Thomas Cook and Victorian travel*. Pica Editions, London.

TAYLOR, M. A. 1992. The local geologist 1: Exporting your heritage? *Geology Today*, **7**, 32–36.

THOMAS, I. A. & HUGHES, K. 1993. Reconstructing ancient environments. *Teaching Earth Science*, **18**, 17–19.

THOMAS, I. A. & PRENTICE, J. E. 1984. A consensus approach at the National Stone Centre, UK. *In*: O'HALLORAN, D., GREEN, C., HARLEY, M., STANLEY, M. & KNILL, J. (eds) *Geological and Landscape Conservation*. The Geological Society, London, 423–427.

TOWNER, J. 1996. *An Historical Geography of Recreation and Tourism in the Western World 1540–1940*. Wiley, London.

TRENCH, R. 1990. *Travellers in Britain: Three Centuries of Discovery*. Arum Press, London.

TRESISE, G. 1966. *The Earth Before Man: A Guide to the Geology Gallery, Liverpool Museum*, Liverpool Museum, Liverpool.

TRESISE, G. 1973. The story of the Earth at the Geological Museum. *Museums Journal*, **73**, 71–72.

WALTER, H. M. & HART, D. 1975. The Story of the Earth. *Geology—Journal of the Association of Teachers of Geology*, **5**, 92–97.

WARNER, R. 1802. *A Tour through the Northern Counties of England, and the Borders of Scotland*. (2 vols). R. Cruttwell, Bath.

WATSON, R. & GRAY, M. 1978. *The Penguin Book of the Bicycle*. Penguin, Harmondsworth.

WATSON, W. 1811. *A delineation of the strata of Derbyshire, forming the surface from Bolsover in the east to Buxton in the west*. W. Todd, Sheffield.

WATSON, W. 1813. *A section of the strata in the vicinity of Matlock Bath, Derbyshire*. Chesterfield.

WEILER, B. & HALL, C. M. (eds) 1992. *Special Interest Tourism*. Belhaven, London.

WEST, T. 1778. *A Guide to the Lakes: Dedicated to the Lovers of Landscape Studies, and to All Who Have Visited, or Intend to Visit the Lakes in Cumberland, Westmorland and Lancashire*. B. Law, etc, London.

WHITE, H. J. O. 1920. A Short Account of the Geology of the Isle of Wight (1990 reprint with a new bibliography). HMSO/British Geological Survey, London.

WILKINS, E. P. 1861. *A Concise Exposition of the Geology, Antiquities, and topography of the Isle of Wight*, T. Kentfield, Newport.

WILKINSON, J. 1810. *Select Views in Cumberland, Westmoreland and Lancashire by the Rev. Joseph Wilkinson, Rector of East and West Wretham in the County of Norfolk, and Chaplain to the Marquis of Huntley*. R. Ackerman, London.

WOOD, R. M. 1978. *On the Rocks: A Geology of Britain*. BBC Books, London.

WOOD, S. P. 1983. The Bearsden Project or Quarrying for fossils on a Housing Estate, *Geological Curator*, **3**, 423–434.

WOODWARD, H. B. 1907. *The History of the Geological Society of London*. The Geological Society, London.

WOODWARD, H. B. 1907. *The History of the Geological Society*. The Geological Society, London. WORDSWORTH, W. 1814. *The Excursion, being a Portion of the Recluse, a Poem*. Longman, Hurst, Rees, Orme and Brown, London.

WORDSWORTH, W. 1820. *A Guide Through the District of the Lakes of Northern England with A Description of the Scenery, &c. for the use of Tourists and Residents*. (5th edn) Longman, Hurst, Rees, Orme and Brown, London.

WORDSWORTH, W. 1835. *The River Duddon, A Series of Sonnets; Vaudracour & Julia: and Other Poems. To which is annexed, A Topographical Description Of the Country of the Lakes, In the North of England*. Hudson and Nicholson, Kendal.

YOUNG, A. 1771. *The Farmer's Tour through the East of England Being the Register of a Journey through various Counties of this Kingdom, to Enquire into the State of Agriculture, &c. Containing I. The Particular Methods of Cultivating the Soil. II. The Conduct of Live Stock and the Modern System of Breeding. III. The State of the Population, the Poor, Labour, Provisions, &c. IV. The Rental Value of the Soil, and its Division into Farms, with Various Circumstances Attending their Size and State. V. The Minutes of above Five Hundred Original Experiments, Communicated by Several of the Nobility, Gentry, &c. With Other Subjects That Tend to Explain the Present State of English Husbandry*. W. Strahaan and W. Nicoll, London.

The role of the voluntary sector in the evolving geoconservation movement

CYNTHIA V. BUREK

University of Chester, Parkgate Road, Chester CH1 4BJ, UK
(e-mail: c.burek@chester.ac.uk)

Abstract: The role of the voluntary sector in geoconservation has a long history. However, its involvement in biodiversity conservation is even longer. A contrast is made between the biodiversity and geodiversity voluntary sectors through time. With the start of the movement arguably by the National Trust in the late nineteenth century, the baton (or hammer) has been taken up by geological societies locally and nationally, by individuals and more recently by the RIGS initiative. The word voluntary in no way diminishes the work undertaken and achieved by these people. It can be argued that without them geoconservation would not exist. This paper explores their contribution using case studies: National Trust and UKRIGS as national organizations, the RIGS movement as a local initiative, the Chester Society of Natural Science as 'local' interest and the work of individuals through time. The latest Local Geodiversity Action Plans (LGAPs) development as a recent historical phenomenon is explored and the importance of *local* as context for geoconservation illustrated.

The role of the voluntary sector is considered to have been significant in the geoconservation movement from its inception. To establish this, it is necessary first to look at the definition of voluntary as an acceptable term.

Voluntary is defined as 'acting under one's own free will not constrained' and volunteer as 'one of whom of his own free will takes part in any enterprises' (Little *et al.* 1973). As such it is the opposite of compulsory. Today the meaning normally implies 'Acting or done willingly and without constraint or expectation of reward' (Farlex 2007). This is important as, firstly, the question of payment and, secondly, the amateur versus professional status of the individual are not now primary considerations and thus do not apply to the term. The third important fact about voluntary work is that much of it is carried out at the local level. The importance of local action is a strong driver in voluntary conservation but not the only one. A personal moral, ethical or philosophical attitude towards nature can play a part as described later.

The question of defining voluntary groups within the conservation movement has always proved difficult. Voluntary organizations are classified by some, as organizations which are mainly run by volunteers, or those that rely on volunteers in order to carry out significant portions of their work (Marren 2002). Others define them as organizations which are non-statutory and independent of state control, and that do not distribute profits for private gain. They are commonly classed as non-governmental organizations (NGOs) and sometimes as pressure groups such as Friends of the Earth.

These, clearly, are not only local but global. They may act locally but think globally. Hence the difference between individual and collective voluntary action must be recognized.

History of the voluntary conservation movement

Voluntary organizations and volunteers have existed within nature conservation in the UK for a considerable time and have been dubbed 'The voluntary army' by Marren (2002). This voluntary army expanded in the 1970s as shown in Figure 1. The reasons for this are discussed later in this paper. In the UK, the top three voluntary organizations dealing with nature conservation in its widest sense (shown in Table 1) are the National Trust founded in 1895, (The National Trust for Scotland was founded in 1931), the Royal Society for the Protection of Birds in 1891 and the Wildlife Trusts in 1926. Geodiversity conservation is not explicit in their titles and the last two emphasize specifically biological conservation and therein lies the dilemma for geoconservation. Initially conservation was regarded holistically but as time has moved on, conservation has split into its constituent disciplines. Indeed it could be argued that biological conservation has been perceived as more important by the general public and thus legislators, than abiotic conservation.

Consequently, in 1889, a group of women in Didsbury voluntarily formed the 'Fur, Fin and Feather Folk' in order to protest against the massacre

From: Burek, C. V. & Prosser, C. D. (eds) *The History of Geoconservation.*
Geological Society, London, Special Publications, **300**, 61–89.
DOI: 10.1144/SP300.6   0305-8719/08/$15.00 © The Geological Society of London 2008.

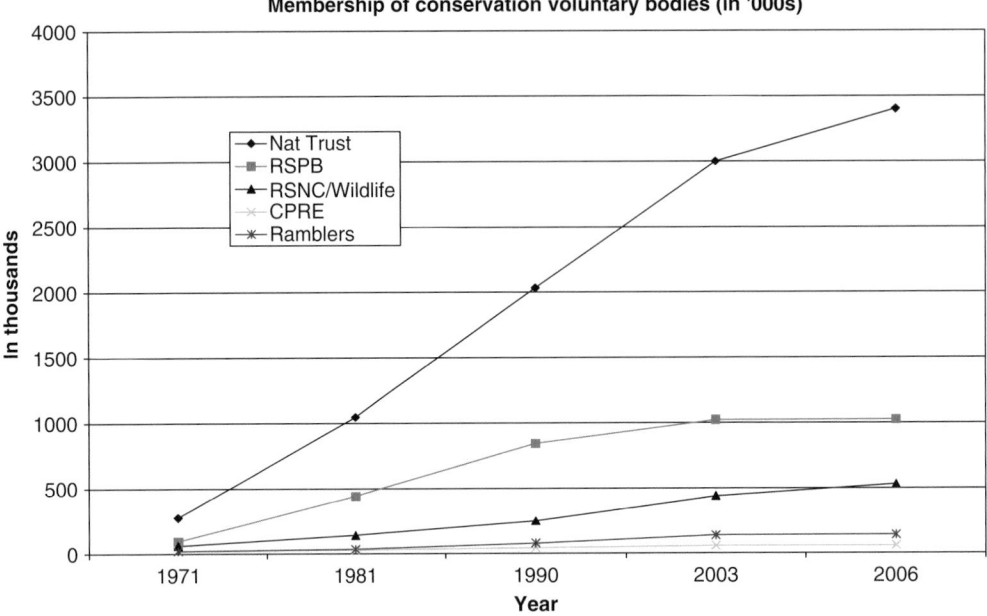

Fig. 1. Voluntary membership of top conservation bodies during last 35 years in UK.

of birds purely for clothing. Within one year the group had more than 5000 members. By 1891, the Didsbury group plus the ladies attending Mrs Phillips' Fur and Feather meetings at her house in Croydon amalgamated to become the *Society for the Protection of Birds* (RSPB 2007). One of the first nature conservation Acts spearheaded by this interest in bird conservation and put before Parliament was the Importation of Plumage (Prohibition) Bill in 1908, but it was not passed until 1921.

In 1912, the *Society for the Promotion of Nature Reserves* (SPNR) was inaugurated, created by Charles Rothschild, who saw the need for an integrated string of reserves across the country. The SPNR aimed to collect and collate information regarding areas of land in the UK which retained 'primitive conditions' and contained rare and local species liable to extinction; to prepare schemes showing which areas should be secured as nature reserves; to obtain such areas; and to preserve for posterity as a national possession some parts of the UK, its floral, fauna and geographical features (Rothschild & Marren 1997).

In 1915, a provisional list of potential reserves was presented to the UK Board of Agriculture. These consisted of some 284 sites covering Britain and Ireland, graded into 3 categories. Many of these sites remain today and are prime places for wildlife (Rothschild & Marren 1997). Following the First World War, and the death of

Rothschild, the conservation movement suffered as funds were scarce and public interest declined greatly (National Archives 2006). However, all but 20 of Rothschild's list in England, which were destroyed before the Nature Conservancy was inaugurated, are now owned or managed by voluntary bodies such as the National Trust, RSPB, the Wildlife Trusts or other conservation organizations such as Natural England. Most sites are Sites of Special Scientific Interest (SSSIs) and thus have legal protection.

These two early voluntary national conservation movements illustrate the importance birds and landscape had to Victorian society. The former has gone on to become one of the most important wildlife organizations for bird protection in the world. The role of the National Trust will be dealt with later. They also encompassed the traditional pre-1970 philosophical stance towards nature, i.e. the Romantic, and Stewardship of Nature beliefs (Sarre & Reddish 1996). Most pre-1970 attitudes are anthropocentric in approach.

Nature conservation moved from an anthropocentric bias towards an ecocentric slant during the 1970s and then oscillated back to a more anthropocentric interest at the end of the twentieth century. This can be illustrated, firstly, by the rise of economic justification for nature conservation based on the health of individuals and societies (English Nature 2002, 2006; Tzoulas & James 2004; Millennium Ecosystems Assessment 2005; Pretty *et al.* 2005)

Table 1. *The Voluntary army*

Organization	Start date	Membership	Aim	Geodiversity/Geoconservation
National Trust	1895	3.4 million (297 000 in National Trust for Scotland)	The protection of places of historic interest or natural beauty	We protect over 700 miles of coastline in England, Wales & Northern Ireland. In total we look after 617 500 acres (250 000 hectares) of countryside, moor land, beaches and coastline
Royal Society for the Protection of Birds	1891 (Royal 1904)	>1 million	To work for a better environment that is rich in birds and other wildlife	182 nature reserves covering 126 846 hectares home to 80% of our rarest or most threatened bird species
Wildlife Trusts	Norfolk 1926, Yorkshire 1946	670 000	Working for an environment rich in wildlife for everyone. 47 local groups	Manage 2200 nature reserves covering more than 80 000 hectares
Conservation Corps/BTCV	1959/1970	130 000	Supporting volunteering opportunities throughout the UK and across the globe	
Friends of the Earth	1971	250 000	Defends the planet and champions a healthy and just world	
Greenpeace	1971	176 000	Greenpeace stands for positive change through action. We defend the natural world and promote peace. We investigate, expose and confront environmental abuse by governments and corporations around the world. We champion environmentally responsible and socially just solutions, including scientific and technical innovation	
WWF World Wildlife Fund for Nature	1961	257 000	The mission of WWF is to stop the degradation of the planet's natural environment, and to build a future in which humans live in harmony with nature, by: conserving the world's biological diversity; reducing pollution & wasteful consumption	Works in over 90 countries ensuring that the use of renewable natural resources is sustainable

(Continued)

Table 1. *Continued*

Organization	Start date	Membership	Aim	Geodiversity/Geoconservation
Plantlife	1989	12 000	Plantlife is the only charity working solely to protect Britain's wild flowers and plants, fungi and lichens, and the habitats in which they are found	23 sites around England, Scotland, Wales and the Isle of Man, covering a total of approximately 4500 acres
Woodland Trust	1972	150 000	The UK's leading conservation charity dedicated to the protection of our native woodland heritage	Own and care for over 1000 woods, covering over 50 000 acres
Wildfowl & Wetland Trust	1946	139 042	The UK's only specialist wetland conservation charity with a national network of wetland visitor centres. WWT is a world leader in the protection of ducks, geese, swans and flamingos and the wetlands they inhabit	2000 hectares in 9 centres across UK
Marine Conservation Society	1978	4000	The UK charity dedicated to caring for our seas, shores and wildlife	
Ramblers	1936	143 000	Britain's biggest walking charity, working for over 70 years to promote walking and to improve conditions for everyone who walks in England, Scotland and Wales	Association for the Protection of Ancient Footpaths in the Vicinity of York, formed in 1824, and the Manchester Association for the Preservation of Ancient Footpaths (1826). It was not until 1865 that a body was created to fight for the open spaces in London
Council for the Protection of Rural England	1926	60 000	CPRE wants a beautiful, tranquil and diverse countryside that everyone can value and enjoy; a working countryside that contributes to national well being by enriching our quality of life, as well as providing us with crucial natural resources, including food; to see the sustainable use of land and other natural resources in town and country. To ensure that change and development respect the character of England's natural and built landscapes, enhancing the environment for the enjoyment and benefit of all	

UKRIGS	1999	46 groups	'The Association will encourage the appreciation, conservation and promotion of Regionally Important Geological and Geomorphological Sites for education and public benefit'	The umbrella organization for the 46 RIGS groups involved in active geoconservation through site notification and maintenance. Over 3000 RIGS notified
AWRG Association of Welsh RIGS Group	1996	3 groups	Safeguarding Welsh geodiversity	The umbrella organization for the Welsh RIGS groups involved in active geoconservation through site notification and maintenance. Over 500 RIGS notified
Warwickshire Geological Conservation Group	1990	45	To raise awareness of geology and landscape through education. To conserve and protect geological sites in the Warwickshire area	Primarily look after the over 90 RIGS
Geology Trusts	2003	7 groups	To work in partnership within a larger and well co-ordinated network, developing good working practice and setting out clear, measurable objectives to ensure the continuation and progression of geoconservation work.	Have attracted over £500 000 in funding for this work
RIGS Groups	1992	530	Regionally Important Geological/geomorphological Sites (RIGS), designated by locally developed criteria, are currently the most important places for geology and geomorphology outside statutorily protected land such as Sites of Special Scientific Interest (SSSI). The designation of RIGS is one way of recognizing and protecting important Earth science and landscape features for future generations to enjoy	53 RIGS groups
Black Country Geological Society	1975	>30		Wide-ranging geoconservation at all levels. Only Voluntary group to be operating a geological recording scheme prior to 1990 and used as the model for the RIGS scheme

and, secondly, in recent decades with the growth of archaeological conservation popularity (possibly due to the impact of the TV programme *Time Team* during the 1990s and early part of the twenty-first century). Geodiversity conservation still lags behind despite the increase in exposure by the mass media especially the BBC with programmes such as *Planet Earth, Earth Story, Coast* and *Landscape Mysteries*, with their accompanying books (Lamb & Sington 1998).

Scale of the British voluntary sector

The scale of voluntary participation in nature conservation in general, is enormous at the local level, medium at the national level, and small at the international level. Geoconservation is a small percentage of this and this begs the question 'Why?' This was further explored by Green (1994), who also established two types of geoconservation being undertaken by voluntary groups in the early 1990s. First, the more popular communication role producing interpretation boards, leaflets and raising awareness among the general public. The second is less popular and involves improving access to the site whether that is by site clearance or negotiation with landowners (Green 1994). In all, Green questioned 36 mainly voluntary geological organizations. Much of the work reported involved the *Geologists' Association* and this is dealt with elsewhere in this volume (Green 2008). The other large group is the *Open University Geological Society* whose main aim is not necessarily geodiversity conservation. This is true of many groups but awareness of geoconservation has increased significantly since 1993. The importance of local and the 'not in my back yard' (NIMBY) attitude come to mind as personal ownership and spatial scale are easier to understand at the local level.

The main nature conservation organizations are shown in Table 1. Together, the top three have more members than the population of Finland. Individually all three have a membership larger than the population of Luxembourg. The National Trust alone has a membership equal to the population of Uruguay or just less than Lithuania or Albania. All the population figures are based on the Central Intelligence Agency (CIA) world fact book (2006). The increase in membership during the move to a more ecocentric perspective during the 1970s is clearly shown in Figure 1. Attitudes to conservation in general changed significantly during the 1960s, once Earth had been viewed from space in 1961 by Yuri Gagarin:

When I orbited the Earth in a spaceship, I saw for the first time how beautiful our planet is. Mankind, let us preserve and increase this beauty, and not destroy it! (Gagarin 1962)

This awareness of vulnerability culminated in 1969 with man stepping onto the Moon. Some would argue that this alone contributed to an interest in and the subsequent rise of a widespread environmental movement:

I think astronauts and cosmonauts the world over have come back from their first mission with a renewed appreciation for how fragile the planet is, and how we have to take care of it (John Fabian 2002)

It suddenly struck me that that tiny pea, pretty and blue, was the Earth. I put up my thumb and shut one eye, and my thumb blotted out the planet Earth. I didn't feel like a giant. I felt very, very small. (Neil Armstrong quoted in Cosgrove 1994)

In order to establish the importance of scale to the voluntary movement, three case studies will be examined in detail in relation to geoconservation. At the national level this will be The National Trust, and at the local level the Chester Society of Natural Science. The third is the RIGS movement which is a recently established voluntary initiative but serves to show the importance of both local and national volunteers.

First steps for voluntary geoconservation

Geoconservation is defined in this paper as the conservation of geodiversity, or maintaining and enhancing the geodiversity of an area or object. Conservation is different to preservation which seeks to keep things as they are or were. Although preservation is applicable to some geodiversity scenarios such as limited mineral or fossil intrinsic sites (e.g. Fossil Grove, Victoria Park, Glasgow front cover), conservation allows nature to evolve and change with time. So rising sea levels, mass movements, moving coastal dunes or natural erosion of cliff faces are all predicted and planned for. Indeed conservation management and action planning should carry this out.

Geodiversity is defined by Prosser (2002) as 'the variety of rocks, fossils, minerals and natural processes' forming our landscapes and soils. However, the term *geodiversity* is a relatively new word with its origins embedded in a Tasmanian forestry document (Sharples 1993) and its growth as a term is discussed by Gray (2004). The intention for geodiversity conservation, not its stated aim, is accepted in this paper. The term *geoconservation* as opposed to Earth science conservation is again more acceptable (Prosser 2002); however Stevens (1994) struggled with the definition using both the terms *geological conservation* and *Earth heritage conservation* exploring their potential and scope. Burek (2007) further defined geoconservation in her inaugural address at the University of Chester in 2006 as 'the stage upon which all forms of life are actors'. This was the first chair of geoconservation to be established in Europe. Although the nineteenth and early twentieth century volunteers were not necessarily safeguarding geology *per se* (see Burek & Prosser 2008, table 1), they were safeguarding landscapes and can therefore be regarded as part of the geoconservation movement if we define geoconservation as geodiversity conservation.

Having explored these definitions, we can now search for the first steps that volunteers took towards carrying out this geoconservation intention by exploring the beginnings of The National Trust.

The founders of the National Trust and their work in open spaces

The National Trust was formed in 1895 by three remarkable individuals who complimented each other in skills, interests and abilities. All three were eminent in their fields. Octavia Hill was internationally known for her work in housing and urban open space protection; Sir Robert Hunter was a solicitor to the Post Office and an authority on the legal status of common land; and Canon Hardwick Rawnsley had successfully campaigned for the protection of the Lake District. Together they were a formidable team. The early history of The National Trust is adequately described elsewhere (Darley 1990; Murphy 2002; Waterson 2003) so only its application to geoconservation will be discussed here. The formation of The National Trust can trace its origins during the nineteenth century, to the mainly middle class public concern about the enclosure of commons.

Commons Preservation Society and Epping Forest

In 1865, the Commons Preservation Society (CPS) was formed originally to protect Epping Forest, to the NE of London, from enclosure and development as it had been Queen Elizabeth I's hunting ground. The importance of open space within an urban area was seen as beneficial to the health and happiness of society. At this point there was no intention of geodiversity conservation but it was inherently implied. It can also be argued that it was purely anthropocentric. Both Robert Hunter (from 1868) and Octavia Hill (from 1875), were involved. The former's success in stopping enclosure of Epping Forest culminated in Queen Victoria declaring on 6 May 1882 at High Beech, Epping Forest:

it gives me the greatest satisfaction to dedicate this beautiful Forest to the enjoyment of my people forever (from Addison 1977)

The Epping Forest Act of 1878 made the Corporation of the City of London the Conservators and

the last part of the Act states: 'as far as possible to preserve the natural aspects of the Forest.'

The difference between conservation and preservation is blurred here. Preservation meaning to keep something as it is, whereas conservation allows evolution to occur by working with nature. Clearly the intention was to inhibit manmade developments but not to inhibit natural progress. Edward North Buxton was one of the first verderers, and served for 44 years. He rose to the challenge and developed principles to be followed to safeguard the natural aspect (Addison 1977). These were:

- the variety of the scenery;
- the preservation of natural features;
- the restoration of the natural aspect where this had been lost; and
- regeneration.

This included protecting the underlying geology by implication if not directly. The importance of soils in the area has also been safeguarded, so it is truly the geodiversity of the area that was and still is being conserved.

In 1958 Qvist (from the London Corporation who has been responsible for the management of Epping Forest since the 1878 Act) under a heading 'Natural Aspect' talked mainly about the habitat and species of the forest not the geodiversity but again it is implied. There is a small section one paragraph long under topography, which directly mentions the geology. It states:

the [Epping] Forest comprises largely London Clay, overlaid in places by Claygate Beds, Bagshot Beds and Pebble Gravel. Essentially it is a long gravel ridge separating the fertile agricultural valleys of the Lea and the Roding and more broken into than much of the surrounding country, by small streams in valleys cut deep down to the clay bottoms, with springs breaking frequently about the 300 feet contour. The gravel top is flat over large areas; much broken by old gravel pits and spoil heaps (Qvist 1958)

This one small piece explicitly recognizes the importance of the underlying geology for the habitat and vegetation and as such raises awareness of geoconservation, albeit in a minimal way. The appointment of the City of London as the conservators of Epping Forest in lieu of the Crown with the wide powers they were granted, has allowed this forest so close to London to remain effectively undeveloped by man forever. The role of volunteers in this process backed by the legislation should not be underestimated.

Robert Hunter

After his achievement with Epping Forest, Robert Hunter went on to become the chief solicitor for the Post Office. He maintained his links with the

CPS and became their vice chairman. He was also chairman of the Kent and Surrey Footpaths Committee and in 1896, published his book, *Open Spaces, Footpaths and Rights of Way*. This book became a handbook and guide for voluntary groups such as the rambling clubs, the CPS and other organizations concerned with preservation of open spaces. It was clearly written and easy to understand (Murphy 2002). Hunter was also President of the first Federation of Rambling Clubs, later to become The Ramblers. In recognition of his services to the open spaces movement especially for the benefit of the lower classes, he was knighted in 1894, an honour never bestowed on another champion of open spaces, Octavia Hill (Fig. 2).

Octavia Hill and the Kyrle Society

In 1878 The Kyrle Society was formed to introduce colour and beauty to the poor. This voluntary body was first postulated by Octavia Hill's sister Miranda Hill following the Public Health Act of 1875. The

OCTAVIA HILL IN 1882.

Fig. 2. Octavia Hill, Founding member of the National Trust.

Act encouraged the purchase of open spaces for the working classes to enjoy through walking and for relaxation and pleasure. It was thought that this would improve the health and well being of the poorer classes. As Octavia Hill stated, these gardens would provide 'Open-air sitting rooms for the poor' (Hill 1877, quoted in Murphy 2002). This society had a subcommittee for open spaces and in 1879, Robert Hunter was the Chair and Octavia Hill the Treasurer. This evolved into The Open Spaces Preservation & Land Development Society (1885). Although geology was not explicit, geodiversity with its coverage of landscape and landforms was.

Canon Rawnsley and the Lake District Defence Society

At the same time, the Lake District Defence Society had been formed in the NW of England supported by people such as Beatrix Potter and Canon Rawnsley. The slow expansion of the railways into the Lake District was seen as an omen by many people including John Ruskin, William Wordsworth and Canon Rawnsley (Sarre & Reddish 1996; Hose 2008). The society itself was a response to the proposed Keswick-to-Buttermere railway line and soon had a membership of 600. They succeeded in halting this development.

The common aim of all these early societies was the promotion of preservation of open space for society at large. The gathering momentum of preserving the landscape led eventually to the setting up of the largest landowner in UK to safeguard the natural heritage of Britain—The National Trust.

National Trust and geoconservation

The name National Trust was suggested by Robert Hunter after Octavia Hill's suggestion of 'The Commons and Gardens Trust' and other suggestions such as a 'National Trust for Historical Sites and Natural Settings' were turned down (Darley 1990). The fact that it should be a trust for the future was the idea of Octavia Hill, as Hunter originally favoured a company. (This was probably due to his background as a solicitor.) The National Trust Act 1907 secured its future so that it could purchase and manage land and accept bequests. Its property rights were inalienable and could not be taken away without an Act of Parliament. It was one of the first organizations to be set up with the conservation of wildlife habitats and geological features in the countryside amongst its express objectives. By 1912 it had 500 members (Rothschild & Marren 1997). The first property, obtained in 1896, was a building, Clergy House in Alfriston in Sussex, followed quickly by the trust succeeding in buying the cliff top Barras Head near Tintagel. The purchase price of £505 was raised publicly. Thus the first natural site was a geodiversity site bought for its romantic associations with King Arthur. In 1899 part of Wicken Fen was purchased for £10 from Charles Rothschild. Further natural sites followed.

The National Trust accepted Derwentwater in 1902 after a public appeal for £6500 was secured. This was the first of its Lake District acquisitions in memory of John Ruskin who had died in 1900. The Trust later acquired Hindhead Common in 1906, Cheddar Gorge in 1910, Blakeney Point in 1912 and Box Hill in 1914. During the Trust's first 25 years, it acquired over 80 properties of which 60 were open spaces including six stretches of coastline. Blakeney Point in Norfolk was over 1100 acres of sand dunes, scrub and shingle. The Norfolk Trust and the botany department of University College, London (UCL), managed it jointly. From 1906, it served as a field station for Professor Oliver and Agnes Arbor (mother of Muriel), (Robinson 2007). This association between UCL and National Trust continued throughout the twentieth century (Robinson, pers. comm.). Charles Rothschild again gave money anonymously for this purchase (Murphy 2002, Rothschild & Marren 1997). This shows the importance of landscape and by implication, geoconservation to the early National Trust.

The National Trust today is the largest voluntary organization of its kind in Britain with a membership of over three million in England and Wales and over 240 000 in Scotland (Bremner 2001) out of a total UK population of just over 60 million. This represents a growth from 0.5% of the total population in 1971 to 5% of the population in 2006. The growth of The National Trust and its importance within British society is shown in Figure 3.

Although not all members are active volunteers in the true sense, it is significant that they have given their membership fee voluntarily even though the cynical would argue this is to save the entrance fee. In Scotland one person in 25 is an active volunteer (Bremner 2001). In 2006, the National Trust began to develop both a draft Geology Policy to safeguard its extensive geodiversity heritage and a second policy for the collection of geological materials. They were published in June 2007 (Cordrey & Ford 2007). The National Trust Geological Policy preamble states that:

The National Trust considers geology to be the foundation of our natural and cultural environment. Geology strongly influences the landscape and wildlife, and is the basis of all landforms, including soils, as well as all life forms .... This policy ... is an attempt to bring together and set out the Trust's policy towards the conservation of its extensive and significant geological resource and in safeguarding the conservation interests and natural processes associated with the wider geological environment.

The Geological Policy has three bullet points which cover the philosophy of the Trust:

- The Trust will care for the natural and cultural geological significance of all our properties;
- The Trust will inform conservation and manage change in the geological environment and its features through learning, identifying, recording,

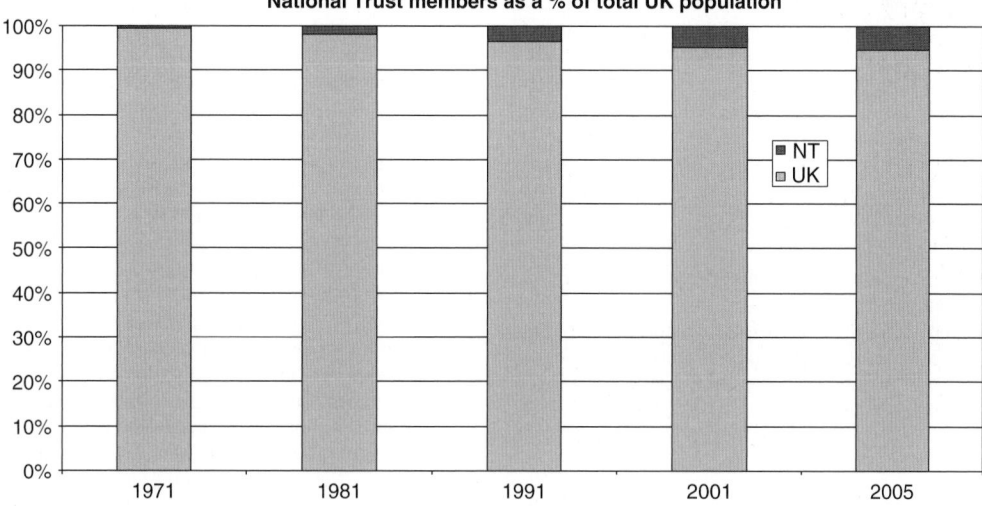

Fig. 3. Growth of the National Trust.

understanding and communicating its significance; and
• The Trust will share the geological significance
  of our properties with members, visitors and
  stakeholders for all to appreciate and enjoy.

The stated approach of the National Trust is to: 'take a
long term, strategic and practical approach to caring
for its geology and one that seeks to work with
natural processes, wherever possible' a true geoconservation stance. The National Trust must be
applauded for this sustainable policy as it carries on
from the National Trust Act 1907 which sought to
conserve wildlife habitats and geological features
amongst its expressed aims at that time.

The specific collecting policy recognizes 'that
geological materials can have scientific, recreational
and aesthetic value' mirroring the criteria for RIGS
in many ways. It aims to 'promote responsible and
safe geological collecting on Trust land with appropriate minimal loss and damage to geological
specimens and sites and lastly seeks to share
the significance and beauty of geological specimens
with local communities, interest groups and individuals for all to appreciate and enjoy'. The National
Trust is such a large land owner in Britain that it
must benefit the geoconservation movement to have
such a responsible policy statement for all to see.

The volunteer and geoconservation—the geovolunteer

As we explore the geoconservation volunteers, we
need to find a new term to describe these individuals
or groups of people. Thus the newly defined word
geovolunteer is offered as a geoconservation volunteer, who gives up his/her time freely without
expectation of reward to forward the cause of
geodiversity conservation. This is used rather than
geological or geomorphological, as it has a wider
application. The geovolunteer needs to understand
fully what is being conserved to be effective.
Thus education across the whole population is
necessary to recruit successful and enthusiastic
volunteers.

(Geo)volunteers therefore fall into three
distinct groups:
• Individuals who collect, both fossils and
  minerals;
• Individuals who display materials in both
  museums and as personal ornaments; and
• All of the above who collectively work through
  voluntary societies.

The local volunteer

In order to illustrate the importance of local volunteers it has been necessary to choose two case

studies, one from the earlier part of the geoconservation movement and one from the latter part
of the twentieth century. The first voluntary
body with its army of volunteers and epitomising
the growth of scientific societies in the Victorian
age is the Chester Society of Natural Science.
The second is the government initiative of the
1990s and the development of the RIGS
movement, which will be used to look at two
scales; firstly at local level involvement and
secondly at UKRIGS as the national organization.
The Association of Welsh RIGS Groups (AWRG)
as a regional body of volunteers is dealt with
elsewhere in this volume (Burek 2008*b*).

However before going into the detail presented
by these case studies some generic points need to
be established.

The role of the individual or local volunteer

The role of the local volunteer has varied with
time. Initially the volunteer was regarded as a collector of fossils, minerals and other geological
memorabilia. The collections were displayed by
the upper classes in their houses as curiosities
and the display cabinets were talking points (Phillips 1990; Knell 1997; Walley 1997; Stott 2003;
Meadows 2004). However they were merely collections and initially little scientific work was
undertaken on them but this was still a form of
ex-situ geoconservation (Burek & Prosser 2008;
Table 1). Later the volunteer moved on to recording details of location, date of collection and sometimes details of a scientific nature along with
sketches and illustrations. The work of Etheldred
Bennett and William Hamilton are good examples
(Burek 2001*a*, *b*, 2002, 2004*a*). Another interesting
example of this is Anna Thynne (neé Beresford,
1806–1866) and her marine aquariums (Stott
2003). She was a great friend of Mary Buckland
and together they voluntarily collected and carried
out experiments on nature. Anna concentrated her
efforts on madrepores in aquariums.

As time moved on, the role of the volunteer
expanded to encompass interpretation and teaching
in a voluntary capacity. The latest roles undertaken,
often but not always by volunteers, are as members
of a pressure group usually seeking funding for a
particular conservation task or to undertake work
for nomination of a particular site for protection.
Thus the role of the volunteer has evolved within
geoconservation from a simple collector to a position of considerable power in the protection of
sites within the planning system. We will explore
the evolution of these roles by examining in
depth one particular scientific society in the NW
of England in the late nineteenth and early
twentieth century.

Voluntary societies

The first voluntary local society with an interest in geology was the Royal Geological Society of Cornwall established in 1814. The second was the Natural History Society of Northumberland, Durham and Newcastle upon Tyne. It was recorded that on 25 April 1846 the Reverend Vicar of Newcastle who was in the chair declared: 'that a society be formed under the name of Tyneside Naturalists Field Club for the practical study of natural history in all its branches' (Coxe 1846).

The society sprang from an earlier one founded in 1829, and had strong connections through mine owners and managers who were early members of the Geological Society, under Lindley and Hutton. Coal Measure flora was accumulated to the extent that the Hancock Museum was founded. Sites were secured through the county extending to the Farne Islands (Robinson, pers. comm.). The Berwickshire Naturalists' Club was instituted on 22 September 1831. The statement of the following year quoted in the presidential address mentions invertebrates, fishes, reptiles and birds as well as geology and plants as interest areas (Johnston 1834).

The third was the Edinburgh Geological Society founded in 1834 by a few individuals who met in 'the classroom of Mr. Alexander Rose, lecturer on Geology and Mineralogy, 1 Drummond St'. The object of the society was 'to promote and extend knowledge of the science of geology, including mineralogy and other collateral branches of science'. The members consisted of 'Gentlemen all of whom were engaged in business' (Lyon 1867).

The Leicester Literature and Philosophical Society was formed in 1835 with Warwickshire Natural History and Archaeological Society following quickly afterwards in 1836 (see Radley 2008). *The Yorkshire Geological Society* came into being in 1837 and then *Cotteswold Naturalists' Field Club* along 'a model of a similar society in Berwickshire' (Baker 1853). The first meeting 'took place on 7 July 1846 at the Black Horse in Birdlip where the original members of the club were proposed and elected' (Baker 1953). In 1849, the *Somerset Archaeological and Natural History Society was* set up after discussions among 'several gentlemen of Taunton and its neighbourhood.' It was one of several such organizations established in the English shires during the 1840s, and reflected the early Victorian flowering of interest in county history, archaeology and the natural environment (Barber 1980). By 1851, the society had 420 members, including large numbers of the Somerset gentry and clergy, and that year published the first volume of its annual Proceedings. In the early years, the members also started to collect items for a society library and museum. The Woolhope Naturalists' Field Club was founded in 1851, Glasgow Geological Society (1858), Liverpool Geological Society (1859) and Bristol Natural History Society (1862). All these societies looked at natural history in its widest context so geology and landscape featured strongly within their remit. As far as can be determined, the Chester Society of Natural Science was the thirteenth or fourteenth natural history society to be set up in this mid-Victorian period and consequently is examined in depth as a representative example of one such society with a strong interest in geology.

Chester Society of Natural Science

The Chester Society of Natural Science (CSNS) was the brainchild of Charles Kingsley and colleagues. Canon Kingsley was one of four canons serving Chester Cathedral for three months each; Kingsley's months being May, June and July. He gave up a Cambridge professorship to take up his appointment in 1869 and took up residence in 1870. During this residency he explored with several other Cestrians the possibility of setting up a Natural Science Society and was prepared to test the waters by offering classes. By May 1870 he had 40 names prepared to go to 12 weekly botanical lectures alternating in and out of the classroom. The lectures had a caveat. 'The Canon stipulated that his class should consist of young men only'. However when the Chester Society was formed, it did not discriminate on grounds of gender and was open to all; 'All persons eligible to become members' (Siddall 1911). The Society was formed on 26 May 1871 after this series of lectures given by Kingsley. At the first general meeting, on 12 June 1871, the society was officially instigated with a committee and Charles Kingsley as President. Honorary members of the Society contained many distinguished scientists including several geologists (Sir Charles Lyell, Sir Philip de Malpas Grey-Egerton, Professor Huxley, Frank Buckland, Boyd Dawkins and Professor McKenny Hughes) (Chester Society of Natural Science 1874). Writing to Charles Lyell, Charles Kingsley explained his reasons:

I have a great favour to ask: I have just started a Natural Science Society—the dream of years. And I believe it will march. But I want a few great scientific names as Honorary Members. That will give my plebs, who are men of all ranks and creeds of course, the feeling that they are initiated into the great freemasonry of science, and that such men as you acknowledge them as pupils. Your most faithful and loyal pupil, C. Kingsley. (Siddall 1911)

Charles Lyell duly obliged, and donated some of his books to the new society. Kingsley's prediction was correct and by the time the first list of members was published three weeks later, it contained 106 names. Membership grew and at the end of the first year stood at 454 members. The Society's initial aim was 'the promotion of the study of natural science by lectures, field meetings, the reading and discussion of papers and other suitable means' (Siddall 1911). Thus the prime purpose of the society was to instruct and with this in mind and at the insistence of A.O. Walker, one of the founding members and later president in October 1871, three sections were formed. 'Geological, Botanical and Zoological. The meetings were held on the first three Thursdays of the month with a general meeting on the last Thursday' (Robinson 1971). Thursday remains the meeting day even today. Walker himself was a meteorologist. As the society grew, so did the interests of the members, with the introduction of natural philosophy and microscopy in 1881, literature and photography in 1891 and astronomy and art in 1901. Each section had its own chair and secretary and held lectures. This growth in diversity is also reflected in its change of name from Chester Society of Natural Science in 1871, to Chester Society of Natural Science and Literature in 1888 to Chester Society of Natural Science, Literature and Art in 1898. By 1880 there were 561 members (Chester Society of Natural Science 1881) and by 1901, 954 members (Stolterfoth 1902).

Geology in the Chester Society

When first initiated, the society had a strong geological presence, but as time went on this became more diffuse. The role of geology within the Society represents a trend which can be followed through several different local voluntary societies at this time, so its history is representative of voluntary societies countrywide.

In 1871, Charles Kingsley encouraged membership of the Chester Natural Science Society (CNSS) through his series of six geology lectures in the King's School in the heart of Chester. This is now occupied by the Queen's School. They were published in 1873 as 'Town Geology' and dedicated to his class members (Kingsley 1873). The subjects covered in this course were:

• The soil of the field;
• The pebbles in the street;
• The stones in the wall;
• The coal in the fire;
• The lime in the mortar; and
• The slates on the roof.

In the preface to his book, he explains why it is necessary to teach about natural history in general and geology specifically:

I know few studies to compare with Natural History; with the search for the most beautiful and curious productions of Nature amid her loveliest scenery. I have known again and again working men who in the midst of smoky cities have kept their bodies, their minds and their hearts healthy and pure by going out into the country at odd hours, and making collections of fossils, plants, insects, birds or some other objects of natural history. (Kingsley 1873)

In his first lecture he starts by explaining the importance of geology.

'The most important facts of geology do not require, to discover them, any knowledge of mathematics or of chemical analysis: they may be studied in every bank, every grot, every quarry, every railway-cutting, by any one who has eyes and common sense, and who chooses to copy the late illustrious Hugh Miller, who made himself a great geologist out of a poor stonemason ... And thus geology is (or ought to be) in popular parlance, the people's science.'

This then is the background with which the first president of the society sought to set up the society.

At its inception in 1871, Dr Henry Stolterforth, a keen geologist himself, became the scientific secretary and he prepared the annual reports for the next 36 years until his death in 1911. This might account for the very full geological records in the reports. In the first year, five lectures on local geology were given to the geology section by Messrs. Shrubsole, Shone, Walker, Macintosh and Cross. The average attendance was about 50 (Siddall 1911). In addition to lectures, excursions were also held and normally led by the president. Thus the fieldtrips had a strong geological/botanical bias. The first trip went to Helsby Crag to look at the Triassic red sandstone and the second to Nannerach in North Wales to look both at the Carboniferous Limestone and its associated flora; both were extremely well attended. The Inaugural Conversazione held at the Chester Town Hall had over 300 people attending and contained fossil exhibits by several members.

By autumn 1872, three lectures on physical geography were supplementing the ordinary geology lectures. In the summer of that same year excursions were being held locally to Llangollen, Delamere and further afield to Dolgellau. 500 people attended the last one and were accompanied up above Llyn-y-Gader, where Canon Kingsley gave a lecture on the geology and topography of the area. It is debateable how many people heard every word, especially as the band of the 14th Regiment was playing music for dancing at the local hotel (Siddall 1911).

The success of the society continued into the following year with over 300 attending the second conversazione and viewing the Provis collection of fossils purchased for the society by the Marquess(sic) of Westminster (Siddall 1911) as well as fossils from the local glacial till (Boulder Clay). By 1873, train excursions were becoming very popular with over 200 members and friends attending the Church Stretton outing. The geologists took charge and led the excursion up Lightspout Valley and Longmynd (Fig. 4). This drawing of the party scrambling up Lightspout Valley was by Mr Alfred Sumner and shows the esker of sand, which choked the Church Stretton Valley (Stolterfoth 1874).

The third conversazione held in 1873 was the last to be attended by Kingsley as he had taken up a position in Westminster Abbey and the following year he spent much of his time in America. In 1874 Mr Mackintosh held a series of lectures in the autumn on geology, fossils were exhibited and excursions held, so there was no shortage of geological expertise within the society.

The charismatic nature of the Canon and his enthusiasm for Natural Science must be held, in part, responsible for the growth of the society but it must also be remembered that several other famous geologists were associated with the society namely Professor McKenny Hughes, Dr Ethel Skeat and Professor Boswell. Ethel Skeat and her work, both at the Queen's School teaching geography and science and as a researcher in the NE Wales area have been described in detail elsewhere (Burek & Malpas 2007). She gained her doctorate as one of the steam boat ladies (Higgs & Wyse-Jackson

2007) while she was a member of the society in 1905. Professor McKenny Hughes, from the geology department at Cambridge, was the Society's second president after Kingsley died in January 1875. McKenny Hughes accepted the presidency in Chester because his father was Bishop of St Asaph and he had family connections in the area. He held that post for 16 years. During this time geology and geomorphology blossomed. After the first president's death, a Kingsley Memorial Fund was set up. Table 2 and Figure 5 show the first six presidents during the nineteenth century.

The roles of the Chester Society Volunteers

The roles undertaken by the members of the CSNS were numerous and all contributed to the eventual understanding of the importance of conservation in a local setting. They collected, donated, ran free lectures, opened a museum, ran a science school, went on excursions and fieldtrips, presented papers, attended lectures and gave away prizes and medals. All contributed to the greater understanding of the role of geodiversity within people's lives and led to recognition of the value of Earth science within the larger nature conservation movement.

Fig. 4. Chester Society of Natural Science outing to Church Stretton, 1873.

Table 2. *Nineteenth century Presidents of the Chester Society of Natural Science with disciplines*

Date	Name	Specialist area
1871–1875	Sir Charles Kingsley	Botanist/Geologist
1875–1891	Professor McKenny Hughes	Geologist
1891–1892	A. O. Walker	Meteorologist
1892–1893	Duke of Westminster	—
1893–1895	W. M. Dobie	Ornithologist
1895–1897	H. Stolterforth	Geologist

Collectors and donators

Collecting and donating is described in detail in the Society's records from its beginning. The turning point for the Society however, came in 1886 with the building of the Grosvenor Museum on land donated by the Duke of Westminster (Robinson 1971) and which is still in existence today. The front entrance of the museum features on the geological guide to *Chester, Walking through the Past*, originally developed by NEWRIGS and funded by Chester City Council in 2000. This is the starting point for the trail (Fig. 6).

Fig. 5. First six presidents of the Chester Society of Natural Science.

Fig. 6. Grosvenor Museum Chester on the front cover of the Chester Geology trail.

Examples of the variety of both donations and donors is given below. In 1873, 30 geology books were donated by the late Miss Potts:

During the past year, through the liberality and kindness of the late Miss Potts, our Society has come in to possession of a most valuable geological and botanical collection. This collection was the result of a long life of scientific pursuits, and when properly arranged and labelled, will be a valuable addition to our present collection. (Stolterfoth 1874)

In 1895, the library book catalogue listed over 18 pages of geology books including Cuvier's *Recherches sur les ossemens fossils* (11 Vols), Lyell's *Principles of Geology* (43 vols) and McKenny Hughes *Geology of Anglesea*. The range of books was astonishing for such a small Society (Chester Society of Natural Science and Literature 1895).

In 1886, the chairman of Halkyn Mines Co. donated 'A fine block of lead ore' and Dr H. Thomas presented Mountain Limestone fossils from the Great Orme Head (Shrubsole 1887). In 1902, Mr F. E. Rooper gave 1 slab of graptolites; 1 specimen of *Orthocerus* (both from the Wenlock Shale) (Newstead 1901). In 1925; Mrs Ethel,

G. Woods (nee Skeat) and Miss Margaret C. Crosfield gave their collection of graptolites and a copy of their reprint *Silurian rocks of the Clwydian range* to the Grosvenor Museum in Chester. It 'is a valuable addition' reported Alfred Newstead FRS (Curator and Librarian) on 27 May 1926 (Newstead 1926). Their importance and the rediscovery of the sites and the samples (Fig. 7) led to conservation status and the RIGS designations in 2005–6 (Malpas 2006). This represented a geoconservation achievement after 80 years (Burek & Malpas 2007).

Each annual report detailed the development of the Society's collections and shows how well the curators cared for them. The four curators up until 1911 were a formidable set; G. W. Shrubsole, A. B. Strahan, Robert Newstead and his younger brother Alfred Newstead. The first two had trained as geologists and understood the importance of obtaining a fine collection while the later two were natural scientists. By 1889 there were 1623 specimens in the museum being curated by Robert (later Professor) Newstead. These were used to illustrate lessons given to schoolchildren and in 1920, over 1620 children visited the museum. In 1903–4 there were

Fig. 7. A graptolite specimen from the Grosvenor Museum Ethel Skeat & Margaret Crosfield 1925 donated collection. Note their initials on the label. Specimen from the Clwydian Mountains.

over 11 032 visitors to the museum who each paid a small sum to enter (Robinson 1971). The scale of interest in natural science was astounding.

Education

The society believed in both education for themselves and for others at all levels. Thus they endorsed the running of free geology lectures for the general public in 1893 and 1896–7 and supported and ran a children's 'school for science' in the museum. This attracted grants and from 1888–1905, the curator's salary was paid for by the Education Committee. To complement this, two further forms of self-education were undertaken, excursions and lectures.

Excursions and fieldtrips

Field excursions carried on throughout the first 50 years of the society. The variety of fieldtrips was not limited by locality, type of geology or distance from Chester. Many excursions were by train (Fig. 8) although evening walks were also popular in the summer. Figure 8 shows the original flier for the field excursion to Rhyd-y-Mwyn and the limestone gorge and cliffs of the Alyn on Friday 18th September 1874, which was to take place by train. In 1882, on 2 August an excursion to the slate mines at Blaenau Ffestiniog was undertaken and in 1886, there was an excursions the Gresford Colliery and to Farndon and Holt, to look at the Triassic sandstones. In 1898, there was a trip to Bull Bay, Anglesey (Fig. 9). Although there were several ladies present, none of them are named in the photo caption. This was often the plight of early lady geologists (Higgs *et al.* 2005; Burek & Higgs 2007). In 1902, excursions were run to Beeston Castle in May; to Rhydymwyn and the Leet in June; to West Kirby and Hilbre Island in the middle of the Dee Estuary in July and to Delamere Forest and the Meres at the end of July.

There was a dip in attendance at field excursions during the mid 1880s, especially during the year 1886–7

> On the whole the excursions were not well attended; on two or three of the days the weather was deplorably wet, and the fact that many of our leading members could not give the time and attention to the excursions which they had done in previous years, was another drawback. This was owing to the extra work entailed in the moving and arranging of our possessions in the new building ... the Museum. (Stolterfoth 1887)

This then seems to have picked up again.

> The increased attendance at the Excursions and Field meetings were even more marked and at these many members displayed a keen interest in Botany, Geology and Entomology. (Miln & Shepheard 1901)

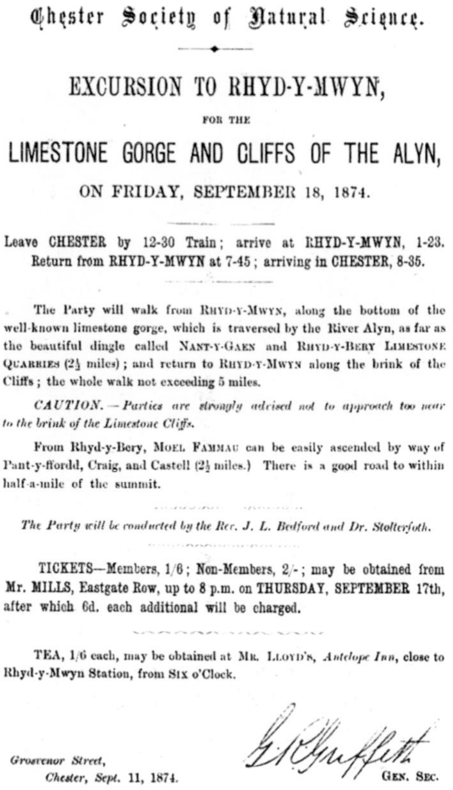

Chester Society of Natural Science.

EXCURSION TO RHYD-Y-MWYN,

FOR THE

LIMESTONE GORGE AND CLIFFS OF THE ALYN,

ON FRIDAY, SEPTEMBER 18, 1874.

Leave CHESTER by 12-30 Train; arrive at RHYD-Y-MWYN, 1-23.
Return from RHYD-Y-MWYN at 7-45; arriving in CHESTER, 8-35.

The Party will walk from RHYD-Y-MWYN, along the bottom of the well-known limestone gorge, which is traversed by the River Alyn, as far as the beautiful dingle called NANT-Y-GAEN and RHYD-Y-BERY LIMESTONE QUARRIES (2½ miles); and return to RHYD-Y-MWYN along the brink of the Cliffs; the whole walk not exceeding 5 miles.

CAUTION.—Parties are strongly advised not to approach too near to the brink of the Limestone Cliffs.

From Rhyd-y-Bery, MOEL FAMMAU can be easily ascended by way of Pant-y-ffordd, Craig, and Castell (2½ miles.) There is a good road to within half-a-mile of the summit.

The Party will be conducted by the Rev. J. L. Bedford and Dr. Stolterfoth.

TICKETS—Members, 1/6; Non-Members, 2/-; may be obtained from Mr. MILLS, Eastgate Row, up to 8 p.m. on THURSDAY, SEPTEMBER 17th, after which 6d. each additional will be charged.

TEA, 1/6 each, may be obtained at MR. LLOYD'S, *Antelope Inn*, close to Rhyd-y-Mwyn Station, from SIX o'Clock.

Grosvenor Street,
Chester, Sept. 11, 1874.

GEN. SEC.

Fig. 8. Flier for the field excursion to Rhyd-y-Mwyn and the limestone gorge and cliffs of the Alyn in September 1874.

The popularity of evening walks did not seem to decrease in the same way. Geological walks are still undertaken during *Science Festival Week* starting at the Grosvenor Museum thereby continuing a tradition of over a century.

Lecture programme

Examples of the numerous geology lectures given are recorded in the annual reports of the Society. A small selection shows the variety of topics covered (Table 3). Only two ladies gave talks in the first 40 years of the life of the society. They were, Ethel Skeat and Mrs Grindon. The high quality of the lecturers is illustrated in 1882 when seven lectures were given by 3 professors, 1 clergyman, 1 QC and 2 esquires. Of these, five were on geology, one on chlorophyll and one on Sir Charles Kingsley.

Messrs. O E. Jones, J. Matthews, W. Nicholls, J. Simon, G. P. Miln, A. H. Fish, F, Thomas, A. Hamilton, J. Nicholson,
F V. Dutton, J. D. Siddall, A. O Walker, J. Bairstow, J. L. Denson, R. Newstead, H. Stolterfoth

Fig. 9. 1898 excursion to Bull Bay, Anglesey, Wales.

Prizes and medals

The society also believed in rewarding outstanding contributions to scientific study, awarding both medals and prizes (Fig. 10). Out of the first seven awards of the Kingsley Memorial Medal, five were given to geologists. For example, the Kingsley Memorial Medal was given to Professor McKenny Hughes (Fig. 11) in 1880, for

The high position he has attained in the scientific world, the real work he has accomplished in the field of geology, together with the kindly zeal and interest he has ever evinced in the welfare of our society, all marked him out as a worthy recipient of a memorial

Table 3. *Lectures given at the Chester Society of Natural Science*

Date	Lecture title	Presenter
1879	On the History of Geology in England during the last forty years	Mr D. Mackintosh FGS
1882	Water considered as a geological agent	Professor McKenny Hughes (President)
1886	Caves and cave deposits	Professor McKenny Hughes (President)
1886	The structure and origin of meteorites	H. Clifton Sorby FRS
1895	Life history of a mountain	Mrs Leo H. Grindon
1900	Notes on the geology of the north coast of Anglesey	Mr J. R. Siddall
1903	Colwyn Bay in the Ice Age	A. O. Walker
1905	Jurassic shorelines: or a fragment of world history	Miss E. G. Skeat D.Sc
1913	Origins and character of limestone	Mr A. W. Lucas FGS

Fig. 10. Past Presidents and Kingsley Memorial Medallists' Society board in the Grosvenor Museum lecture theatre 2007, Chester.

which seeks to keep alive the memory of our Founder. (Chester Society of Natural Science 1881)

and to Mr Shrubsole in 1883

for the good work he has done in the district in geological research, particularly in regard to the Polyzoa of the Mountain Limestone. (Stolterfoth 1884)

In 1880, the junior prizes and certificates were awarded for the best essay on salt. The winners show the range of entrants from near and far: (1) Lewis Fenn (Wallasey Grammar School); (2) John Thomas (Technical Dayschool Chester); and (3) Herbert Hulse (Victoria Road British School, Chester).

In 1882, a prize of £60 was offered for the best collection of the 'corals of mountain limestone' (Chester Society of Natural Science 1883). The generosity of the society towards the local population ensured that interest was kept alive in the natural sciences and with it an understanding of the value of conserving specimens.

The importance of geological activities in the early history of the Chester Society for Natural Science has been demonstrated and the wide range of activities, including conservation, in which the volunteers became involved, remained buoyant until the 1940s. Comment on the war years from 1939–1945 by Robinson (1971) shows the contribution a voluntary society made in enriching the lives of people far from home. Talks,

friendships and education all flourished. Without doubt some of the lectures would have been on the natural history of the area in its widest sense including conservation within the museum.

The War lasted until 1945 and during those six years there were few Section Meetings, though individual members kept in touch as far as their war-time duties would allow and a small amount of research work was done. On the other hand the public demand for lectures grew until by 1945 the membership once again stood at around a thousand and the lectures had to held in the Cathedral Refectory to accommodate the larger audience and much of the credit for those war-time lectures must go to the honorary Secretary D. L Miln. There was a great influx of people during the War particularly those in the older age groups either as private evacuees or serving or working in the district. It was these people who swelled the membership roll and who also regularly attended the lectures. (Robinson 1971)

After the war, many of the sections were no longer sustainable and by 1959 the geological section became part of the Charles Kingsley Naturalists effectively ending a named geological presence in the society in its own right for over 85 years. Membership numbers for the whole society are shown in Table 4. Today, few members realize the rich history of geology associated with their society and the programme, although wide ranging no longer has the scientific or geological bias present at its beginning. Geodiversity tends to be covered by geographical talks. Some geological specimens and displays still remain in the museum while

Fig. 11. Professor McKenny Hughes President of the Chester Society for Natural Science.

some material was transferred to Manchester Museum and Liverpool Museum in the 1970s (Hose; pers. comm.). It is to the credit of the keepers of natural history in the museum that the headquarters of the Cheshire RIGS group and its records are stored safely thus continuing the role

Table 4. *Membership of the Chester Society of Natural Science*

Year	Number of members
1874	502
1884	616
1894	629
1904	1101
1914	879
1924	1260
1954	325
2006	127

of geoconservation at the local level. Rockwatch events are run for children, carrying on the tradition of educating school children in the science of geology. Methods may differ but the outcome is the same—an early interest in dinosaurs and fossils with a wider promotion of geoconservation.

Importance of women in the voluntary movement

The presence of women at the heart of the national conservation movement in the UK began with Octavia Hill. They also made their mark at the local level. An analysis of the membership of the Chester Society of Natural Science shows that between 1873 and 1911, the female membership remained at roughly 25–30% (Fig. 12). Today many voluntary geoconservation groups have a high percentage of women members undertaking geoconservation work in their own time of their own free will. This local interest has not changed although the range and location of the conservation has. The difficulty of travel for women during the nineteenth century may account for this imposed local interest (Burek & Kölbl-Ebert 2007a, b). Today freedom of travel for all will not restrict geoconservation activities, but time and family commitments might. Women also raised other conservation concerns. In the early 1900s, Catherine Raisin was concerned about the conservation of her teaching specimens at Bedford College in London (Burek 2003a, b) and many women donated or sold their collections to museums such as Elizabeth and Alice Gray in Scotland and Elizabeth Carne in Cornwall. Further voluntary roles undertaken by women in the early geoconservation movement are covered by Burek & Higgs (2007).

Voluntary local geological societies

Although the geological activities of the Chester Society have been highlighted to illustrate the importance of the local volunteer in the Chester region, the Society did not explicitly have a geology or geoconservation remit. It dealt with the whole area of natural science.

Members taught the value of conservation through education and by establishing a museum early on; it conserved and preserved specimens such as fossils and minerals in an *ex-situ* way (Fig. 7). On the whole, societies with a strong interest in geology *per se* fall into two groups as shown in Table 5, those formed between 1814–1874 and those that formed after 1960. Outside these dates Manchester Geological Association began in 1925 and in 1880, the Sedgwick Club was the oldest student run geological society in the world.

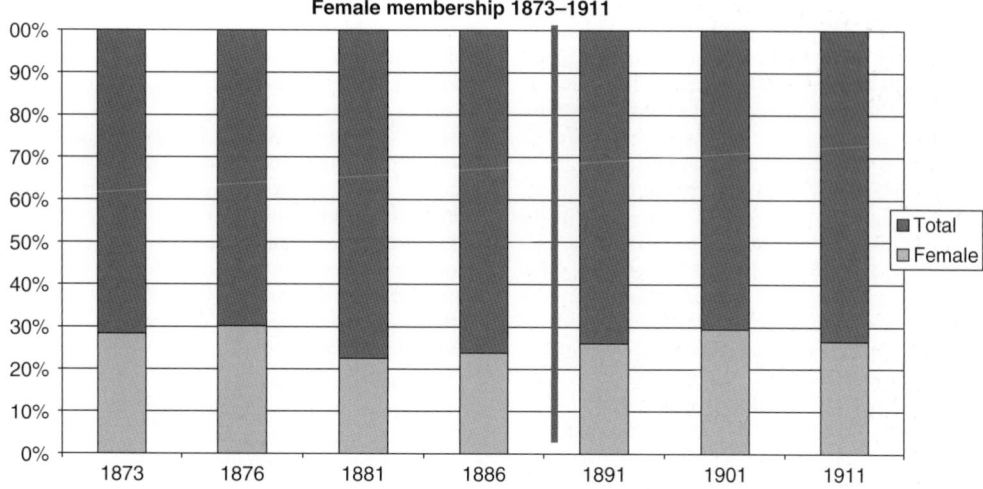

Fig. 12. Female membership of the Chester Society: 1871–1911.

However geoconservation was not always explicitly part of their remit (Green 1994).

Two of the earliest local voluntary societies to address the whole geological site conservation issue were the Black Country Geological Society formed in 1975 and the Russell Society (formed in 1982) which has the conservation of minerals as one of its aims. This is much later than many Biodiversity Conservation Societies (Table 5). However the growth of another voluntary initiative has led to many local geological societies now undertaking geoconservation—the RIGS movement.

Table 5. *Comparison of selected biodiversity and geodiversity conservation voluntary societies*

Biodiversity Conservation Societies	Date
Society for Prevention of Cruelty to Animals	1824
Added Royal	1840
Bristol Naturalists Society	1862
East Riding Association for the Protection of Sea Birds	1867
Chester Society of Natural Science	1871
Geodiversity Conservation Societies	Date
Black Country Geological Society	1975
Russell Society	1982
Warwickshire Geological Conservation Group	1990
RIGS groups	>1990
Gloucestershire RIGS group	1992
Gwynedd & Clwyd RIGS	1993

RIGS movement

The RIGS (Regionally Important Geodiversity (geological/geomorphological) Sites) voluntary movement started as a government agency initiative in 1990. It was a response by the statutory conservation agencies who saw a need for further site protection following the rigorous selection of Geological Conservation Review (GCR) of sites, which had started in 1977 (Ellis *et al.* 1996; Prosser & King 1999; Ellis 2008; Prosser 2008). There were many sites which could not be selected, as they fell outside the GCR criteria but still were worthy of protection. They were not to be regarded as 'understudy' SSSIs but as sites of regional importance in their own right (Harley & Robinson 1991). This led to a discussion between the government agencies and local societies about how to protect them. Eventually the RIGS scheme and its subsequent movement was born. This is embedded in the Nature Conservancy Council (NCC) strategy document entitled *Earth Science Conservation in Great Britain* (Nature Conservancy Council 1990), the second theme of which is to expand the RIGS network. The justification for this was a need for a spread of locally available education sites in addition to the nationally important SSSIs, to meet the increasing demand in 1990 for local geological sites at all levels of education. Many non-SSSIs are widely used by both amateurs and professionals, who study at a local level. An additional factor is that sites with a strong aesthetic appeal are valuable in stimulating public awareness and appreciation of geodiversity and thus their conservation is necessary (Nature Conservancy

Council 1990) The guidelines for expanding the RIGS network suggests that the groups should be informal and voluntary in nature but contain representatives of as many local interest groups touching on Earth science conservation as possible. This voluntary nature initially had two very different consequences. It ensured that, on one hand, RIGS groups could recruit local experts but on the other that they might be perceived as amateurish in the worst possible way and less valued. This is something that the RIGS groups had to fight for some time, as it was perceived by some that this work should be undertaken professionally, i.e. paid for and that the statutory organizations were delegating their own responsibilities.

RIGS are non-statutory geodiversity sites chosen to conserve (not necessarily preserve) a particular interest such as mineralogy or stratigraphy. Burek discussed the difference behind the preservation and conservation philosophy of a site at the second national UKRIGS conference in Worcester (Burek 2000).

The early history of the RIGS movement is covered by Gray (2004), Harley (1994), Harley & Robinson (1991), Prosser & King (1998). The Welsh RIGS movement is covered by Burek (1998, 2000, 2001c, 2008), Rogers (2000) and Tilson (2004) and events in Scotland are adequately covered by Browne (2001, 2002), Browne & McAdam (2003), Butcher (1994), Gordon (2004) and Leys (2000). However a general history of the development of the essentially local RIGS movement is considered necessary to complete the story of the voluntary geoconservation contribution. This culminated in the setting up of the UKRIGS national organization in 1999 following the setting up of a UKRIGS steering group after the first national Worcester conference (Stanley 2000).

Local RIGS groups

Prior to 1990 there were a few geological organizations with conservation schemes for geology, a national one being the Geologists' Association (Green 2008), but they were few and far between. Only six locally based geological conservation schemes existed and few local sites were protected (Harley 1994). The Black Country Geological Society has as one of its two principle aims, the conservation and protection of geological sites (Cutler 1994). The second aim is a full and varied programme of events for its members. The importance of the public face of geology in order to make geoconservation sustainable is recognized and highlighted.

By 1978, over ninety Sites of Importance for Nature Conservation (SINCs) had been identified

and notified locally. These records were held by Stoke City Museum in 1982/3 but moved to Dudley in 1987 when a geological curator was appointed. This latter event is exactly the same set up as happened in Chester 17 years later. A museum location should provide stability, accountability and continuity. Another example was the Hereford & Worcester counties who cooperated closely with their Wildlife Trust partners. The experience gained in these early examples of local protection was used to support the rationale behind the RIGS scheme (Harley & Robinson 1991).

RIGS are nominated for designation based on wider criteria than SSSI:

- The education potential;
- Their research or scientific value;
- Their aesthetic appeal; and
- Their historical associations.

The first geological society to have conservation within its title was the Warwickshire Geological Conservation Group set up in 1990. The Gloucestershire RIGS group set up in 1992, later became the Gloucestershire Geology Trust. Hereford and Worcester Earth Heritage Trust was another early success started in 1996 and one of the first local geological groups to obtain substantial funding from the Heritage Lottery Fund. Cumbria RIGS was also an early success, designating sites and producing booklets and leaflets on important sites. Some RIGS groups such as Derbyshire RIGS and Cheshire RIGS were granted sums of money from their local authority in the early years of the 1990s, to audit and nominate sites for designation but later the groups either declined in membership or were revamped in a different structure.

Each RIGS group is different in its approach and make up. Most of the RIGS groups are entirely voluntary although some may have grants awarded to them for projects or project specific personnel for a limited time period. The RIGS movement is not embedded in legislation but in local expertise, enthusiasm and need. The importance of local in this context cannot be ignored or overstated. After a slow start, the RIGS movement has grown to powerful proportions. It is respected by local planning authorities, government and used extensively in all levels of education (Reynolds 2005).

UKRIGS—a national organization

In 1998, it was recognized generally, and credited to Peter Oliver and Phil Doughty, that there was a need for a national organization to represent RIGS groups (Stanley 2000). This recognition was followed when the first UKRIGS conference was held in Worcester and eventually UKRIGS came into being. The elected members, not the appointed steering group, met for the first time in 2000. Since that date the national committee has met over 60 times and at present has its headquarters at the National Stone Centre in Wirksworth, Derbyshire, UK. Seven national conferences have been held across the country from Edinburgh in Scotland to Llandudno in Wales, from Peterborough in the east to Worcester in the west and from Penrith in the north to Dudley in the Midlands.

A group centred in the SW Welsh borders set up the Geology Trusts in 2003. Although not representative across the whole of the UK they have geoconservation as their remit and have successfully attracted over £1.5 million in grants since 2000 (Herefordshire & Worcestershire Earth Heritage Trust 2005). In 2006, UKRIGS and the Geology Trusts signed a joint statement declaring shared aims thus healing a perceived rift in the geoconservation movement at the national level (Browne & Campbell 2007).

Over the years UKRIGS has represented the face of geoconservation across the whole country and reported on consultation documents from England, Wales and Scotland. It is represented on national committees and stands alongside the statutory agencies, Friends of the Earth, National Trust and Wildlife Trusts. It formulates policy, national strategy and offers guidelines on national issues such as database management, and pricing of information. It produces a national newsletter (first published in May 2000) and maintains a website. One important document produced by the national body in conjunction with NCC in 1999 was the RIGS handbook (Mason 1999). This has become the bible for the movement although by 2007 it was severely out of date and parts were being rewritten.

Membership of UKRIGS has been increasing slowly (Fig. 13). Over the years the progress of the local groups has been monitored periodically by the national committee via questionnaire. These were carried out in 1994, by English Nature, in 1998 and 2006 by UKRIGS and the latter with the Geology Trusts (Table 6). In addition there has been one survey aimed at the UKRIGS newsletter, carried out in 2003 (Burek 2003c). The first 13 issues of the newsletter were evaluated and 24 replies were received from 22 different RIGS groups plus *Historic Scotland* and *ESTA*, two non-RIGS organizations that took the time and trouble to reply. The voluntary bodies

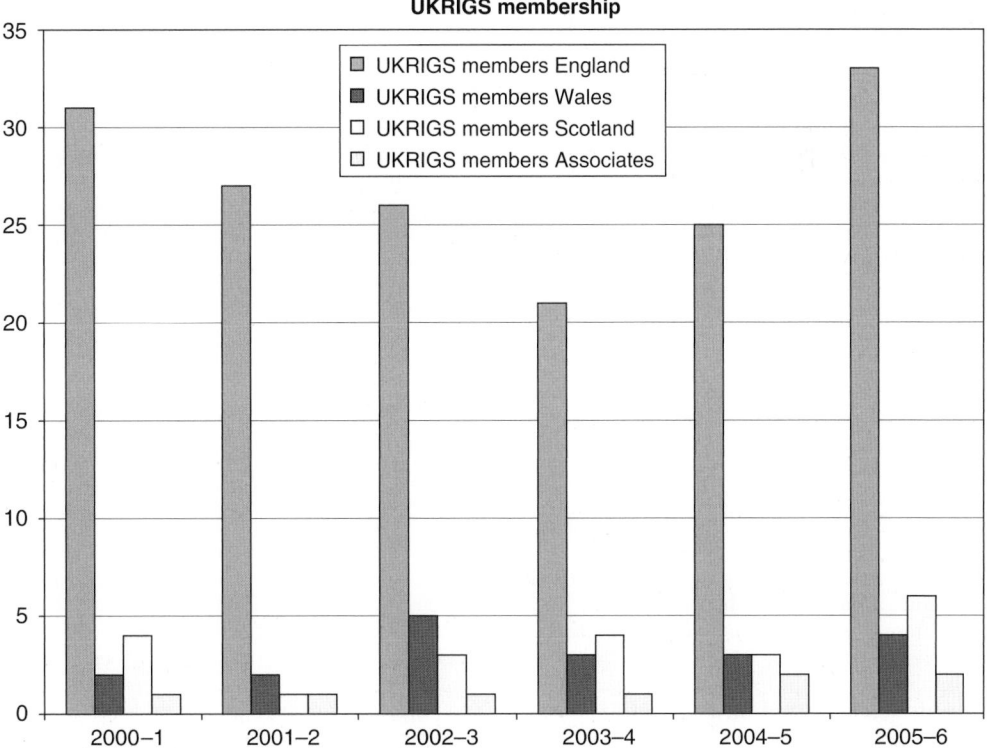

Fig. 13. UKRIGS membership.

Table 6. *Comparison of RIGS groups surveys*

Date	1994	1998	2006
Number of RIGS groups	42	55	56
Number of RIGS groups responding	24	35	38 (68%)
Number of RIGS accepted	829	unavailable	2221
Number of RIGS proposed	1242	unavailable	>3124
Number of people involved	376	unavailable	~1000
UKRIGS Membership	N/A	N/A	43

Data compiled from the RIGS exposure: Couper 1999; Mason 1999; UKRIGS report 2007.

highlighted the importance of local RIGS stories; 'What other groups are doing' and 'getting ideas from other groups' were two representative comments. This still drives the editorial policy of the newsletter today. (The current editor is Cynthia Burek; sub-editor Victoria Page (2008).)

As the RIGS movement matures and the UK Government takes on board the importance of equality between conservation disciplines and recognizes the necessity to safeguard natural resources, it is important to have a strong national face. UKRIGS provides this. The introduction by the Office of the Deputy Prime Minister (ODPM) of the Planning Policy Statement 9 (PPS9) in August 2005 for the first time included geodiversity alongside biodiversity conservation. PPS9 outlines the Government's policy on protection of biodiversity and geological conservation through the planning system but only in England (Prosser 2006, 2008). It recognizes the importance of conserving the best, and RIGS groups, through their role with over 4000 RIGS designations, will help with the protection. This is a clear indication of the importance now attached to the voluntary contribution towards conservation.

Local Geodiversity Action Plans (LGAPs)

Another initiative aimed at the local level for geoconservation was the development of Local Geodiversity Action Plans (LGAPs). These drew on the experience of setting up the Biodiversity Action Plans (BAPs). The BAP system was set up in response to the signing in 1992 of the Treaty on Biological Diversity in Rio de Janeiro, Brazil, by John Major, the Prime Minster of the UK at the time. Thus the process was driven by legislation from above. There is no such legislation for geodiversity and geoconservation so the driver is different. Research was started on the applicability of the biodiversity process to geodiversity in 2001 at the University of Chester, then University College Chester (Burek & Potter 2002, 2006). Conclusions pointed to an acceptability of a modified process tailored to geodiversity with

specific guidelines and suggestions. Three elements were considered of paramount important for successful implementation:

• an audit of all geodiversity related resources;
• boundaries; and
• local voluntary group consultation.

As a consequence of this research, two pilot projects were undertaken based on county administrative areas, Cheshire and Warwickshire. These started in 2002 and by September 2003 the first LGAP had been launched at Chester Museum. There had been some other attempts at LGAP implementation, in Oxfordshire, Devon and Buckinghamshire but these were embedded in the BAP system often under the habitat heading. These two pilot projects were the first to be dedicated to geodiversity outside the biodiversity arena and the Cheshire region LGAP was the first to be publicly available. As these two LGAPs developed, other areas not always on a county basis obtained funding and started the process (Larwood 2002). Often it was perceived as a product not a process and was not sustainable beyond the initial period of funding. Eventually after a year, six LGAPs were evaluated and compared (Burek & Potter 2004). Sustainability was seen as paramount. The Cheshire region LGAP being the oldest was the first one to be evaluated and the percentage of completed actions published (Potter & Burek 2006).

By 2006, over 30 LGAPs had evolved across the UK with examples from Scotland (Burek 2004b) and Wales too (Larwood 2006). Table 7 shows the extent of LGAP development by 2007. The success and transferability of this process is in no small way attributed to the voluntary sector that carries out many of the actions associated with the Action Plan. Many LGAPs are initiated and actively involve the RIGS groups but there are also examples where voluntary bodies, such as the Women's Institute, wildlife trusts, landscape trusts and organizations such as the guides and scouts undertake actions within the process. It is an example of voluntary and governmental organizations operating together for geoconservation and

Table 7. *List of LGAPs either completed or in progress. Date of known publication in brackets*

Abberley and Malverns Geopark (2005)	Doncaster	Lancashire (2004)	Shropshire (2007 draft)	Worcestershire (2006)
Black Country (2006)	Dorset (2005)	Leicestershire and Rutland (2006)	Staffordshire (2004)	Yorkshire Dales & Craven Lowlands (North Yorkshire Geodiversity Partnership) (2007)
Cheshire region (2003, updated 2004)	Durham	North East Yorkshire	Suffolk (2006 draft)	Anglesey (2007 draft)
Cornwall and Isles of Scilly (2005)	Greater Manchester	North Pennines AONB (2004)	Tees Valley (2003)	Clwydian range AONB (2007)
Cotswolds AONB in Gloucester (2005)	Herefordshire (2006)	Northumberland National Park (2007)	Torbay (2006)	Fforest Fawr
Derbyshire and Peak District	Isle of Wight	Nottingham	Warwickshire	Lothian & Borders
Devon (2005)		Oxfordshire (2000 draft)	West Gloucestershire	

sustainability. Only history will decide if one could have performed the task without the other.

By 2006, discussions were underway on the viability of setting up a National Geodiversity Action Plan (GAP) (Burek 2006; Larwood 2006). In October 2006, a small working group met in Chester to debate the need for a National GAP. It was decided to open up the debate to a wider audience. On 19 March 2007 the first ever national workshop was held on the feasibility and implementation of a national GAP. Forty-five participants from all the countries of the UK met representing many different segments of society from industry to RIGS groups; all having been involved in some way with LGAP production. Consensus was not forthcoming on the name except to call it The GAP. This debate

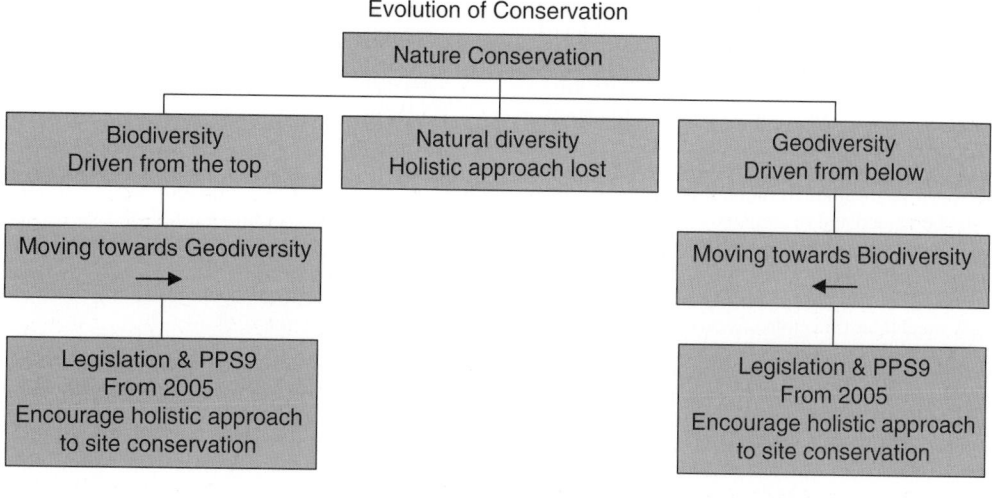

Fig. 14. Relationship of biological and geological conservation.

was appropriately held in Chester where discussions had begun six years earlier. There was agreement over the need for a GAP at the national level and the process had started (Burek *et al.* 2007). Once again Britain was leading the way in geoconservation at the voluntary level.

Conclusions

The role of the geovolunteers has changed with time. Some broad conclusions however can be made. The volunteer army tends to operate more efficiently at the local level and in sheer numbers can make an incredible difference. They can be a driving force for change. Is this sustainable? Opinion varies on this subject as the motives for volunteering may vary. With an ageing population, more people will have leisure time to devote to geoconservation, but the fight is on to overcome the strong public belief that biodiversity is more vulnerable and biological conservation more important. There are many drivers and philosophies for biodiversity and geodiversity conservation but a holistic approach must be adopted in which everyone is a winner. As the geoconservation movement evolves through Geoparks, LGAPs, and geodiversity audits and there is an increasing awareness through the education system, the diverse roles for geovolunteers will increase.

Geodiversity conservation now termed geoconservation for short, seems to oscillate between anthropocentric and ecocentric motives and further work needs to be undertaken on this aspect. However the holistic approach must again be used (Fig. 14) as it was in Victorian times exemplified by the local voluntary natural history societies.

The historical role of women in geoconservation should be highlighted to provide role models for future generations of geovolunteers because their contribution in the early years of conservation is starting to be recognized and their dedication to the cause appreciated.

For geoconservation to succeed in the voluntary sector, it is absolutely crucial that the geovolunteer must be valued and recognized as a driving force so that we can safeguard our sites and landscapes for future generations.

I am grateful for the constructive comments made by the referees E. Robinson and T. Hose. My thanks go to K. Riddington, Keeper of Natural History at the Grosvenor Museum, Chester and secretary of the Cheshire RIGS group, to the archivists at the Geological Society of London and Sedgwick Museum in Cambridge. To all the geovolunteers over the years who have worked often without recognition or praise. Finally to my family for their forbearance in this, the third book I have edited in nearly so many years, I am eternally grateful.

References

ADDISON, W. 1977. *Portrait of Epping Forest.* Robert Hale Ltd, London.

BAKER, T. B. L. B. 1853. 1849 Address read to the Cotteswold Naturalists' Club. *Proceedings of the Cotteswold Naturalists' Field Club*, **1**, 1–8.

BARBER, L. 1980. *The Heyday of Natural History 1820–1870.* Jonathan Cope, London.

BREMNER, D. 2001. *For the Benefit of the Nation—the National Trust for Scotland: The First 70 Years.* National Trust for Scotland, Edinburgh.

BROWNE, M. 2001. *Scottish RIGS Report 2000–2001. In*: ADDISON, K. & REYNOLDS, J. R. (eds) *Upon this Rock—RIGS & and the Planning System.* Proceedings of the Fourth UKRIGS annual conference, Stoke Rochford Hall, 13–15 September 2001, 85–6.

BROWNE, M. 2002. Scottish RIGS Groups annual report October 2001–September 2002: Promoting Geodiversity in Scotland. *In*: TILSON, E, BUREK, C. V. & WOOD, M. (eds) *Promoting Geodiversity.* Proceedings of the Fifth UKRIGS annual conference, Llandudno Clwyd, 3–5 October 2002, 20–23.

BROWNE, M. & CAMPBELL, M. 2007. Geoconservation bodies declare shared aims. *Earth Heritage*, **27**, 3.

BROWNE, M. & MCADAM, D. 2003. RIGS in Scotland 10 years on. *Earth Heritage*, **20**, 8–9.

BUREK, C. V. 1998. *History of RIGS in Wales. In*: OLIVER, P. G. (ed.) Proceedings of the first UKRIGS conference, Worcester, 1998, 147–166.

BUREK, C. V. 2000. The use and abuse of RIG Sites. *In*: ADDISON, K. (ed.) *Geoconservation in Action.* Proceedings of the third UKRIGS annual conference, Newton Rigg, Penrith, Cumbria, 30 August–2 September 2000, 23–31.

BUREK, C. V. 2001*a*. Where are the women in geology? Women in the History of Geology I. *Geology Today*, **17**, 110–114.

BUREK, C. V. 2001*b*. The First Lady Geologist or collector par excellence? Women in the History of Geology II. *Geology Today*, **17**, 192–194.

BUREK, C. V. 2001*c*. Welsh RIGS report 2000–1. *In*: ADDISON, K. & REYNOLDS, J. R. (eds) *Upon this Rock—RIGS & and the Planning System.* Proceedings of the Fourth UKRIGS annual conference Stoke Rochford Hall, 13–15 September 2001, 87–88.

BUREK, C. V. 2002. Where are the women in the history of science, let alone geology? *HOGG Newsletter of the History of Geology Group of the Geological Society of London*, **16**, 12–13.

BUREK, C. V. 2003*a*. Catherine Raisin—a role model professional geologist. Women in the History of Geology III. *Geology Today*, **19**, 107–111.

BUREK, C. V. 2003*b*. Time to take responsibility for collections. *Earth Heritage*, **20**, 22–23.

BUREK, C. V. 2003*c*. Report from UKRIGS newsletter survey. An insert in *UKRIGS Newsletter*, **14**.

BUREK, C. V. 2004*a*. Benett, Etheldred Anna Maria (1776–1845). Dictionary of Nineteenth Century British Scientists. Thoem Press.

BUREK, C. V. 2004*b*. *Application of LGAPs to Scotland. In*: BROWNE, M. & MCADAM, D. (eds) *Good Practice: RIGS Groups Working in Partnership.* Proceedings of the Sixth UKRIGS annual conference,

Ecclesmachen, Broxburn, West Lothis 24–26 October 2003, 32–35.

BUREK, C. V. 2006. Do we now need a National GAP? *Earth Heritage*, **25**, 10.

BUREK, C. V. 2008*a*. *Why Conserve Rocks? Geoconservation Explained*. Inaugural lecture, University of Chester Press.

BUREK, C. V. 2008*b*. History of RIGS in Wales: an example of successful cooperation for geoconservation. *In*: BUREK, C. V. & PROSSER, C. D. (eds) *The History of Geoconservation*. The Geological Society, London, Special Publications, **300**, 147–171.

BUREK, C. V. & HIGGS, B. 2007. The role of women in the history and development of geology: an introduction. *In*: BUREK, C. V. & HIGGS, B. (eds) *The Role of Women in the History of Geology*. Geological Society, London, Special Publications, **281**, 1–8.

BUREK, C. V. & KÖLBL-EBERT, M. 2007*a*. Historical problems of travel for women geologists. *Geology Today*, **23**, 30–32.

BUREK, C. V. & KÖLBL-EBERT, M. 2007*b*. The historical problems of travel for women undertaking geological fieldwork. *In*: BUREK, C. V. & HIGGS, B. (eds) *The Role of Women in the History of Geology*. Geological Society, London, Special Publications, **281**, 115–122.

BUREK, C. V. & MALPAS, J. A. 2007. Rediscovering and conserving the Lower Palaeozoic 'treasures' of Ethel Woods (neé Skeat) and Margaret Crosfield in northeast Wales. *In*: BUREK, C. V. & HIGGS, B. (eds) *The Role of Women in the History of Geology*. Geological Society, London, Special Publications, **281**, 203–226.

BUREK, C. V. & POTTER, J. 2002. Minding the LGAPs— A different approach to the conservation of local geological sites in England? *Geoscientist*, **12**, 16–17.

BUREK, C. V. & POTTER, J. 2004. *Local Geodiversity Action Plans—Sharing Good Practice Workshop*. Peterborough 23 December 2003. English Nature Research Report 601, Peterborough.

BUREK, C. V. & POTTER, J. 2006. *Local Geodiversity Action Plans—Setting the Context for Geological Conservation*. LARWOOD, J. (ed.) English Nature Research Report 560, Peterborough.

BUREK, C. V. & PROSSER, C. D. 2008. The history of geoconservation: an introduction. *In*: BUREK, C. V. & PROSSER, C. D. (eds) *The History of Geoconservation*. The Geological Society, London, Special Publications, **300**, 1–5.

BUREK, C. V., CAMPBELL, S. & LARWOOD, J. 2007. Moving towards a National GAP. *Earth Heritage*, **28**, 18–19.

BUTCHER, N. E. 1994. The work of the Lothian and Borders RIGS group in Scotland. *In*: O'HALLORAN, D., GREEN, C., HARLEY, M., STANLEY, M. & KNILL, J. (eds) 1994. *Geological and Landscape Conservation*. Proceedings of the Malvern International Conference 1993, Geological Society of London, 343–345.

CENTRAL INTELLIGENT AGENCY. 2006. *The World Fact Book*. Directorate of Intelligence, CIA, Washington DC, USA. Website: https://www.cia.gov/cia/publications/factbook/rankorder/2119rank.html (accessed 12.1.2007).

CHESTER SOCIETY OF NATURAL SCIENCE. 1874. The Third annual report of the Chester Society of Natural Science for the year 1873–74 with statement of accounts and list of members, Chester.

CHESTER SOCIETY OF NATURAL SCIENCE. 1881. The tenth annual report of the Chester Society of Natural Science and statement of accounts for the year 1880–81. Chester Society of Natural Science, Chester.

CHESTER SOCIETY OF NATURAL SCIENCE. 1883. The twelfth annual report of the Chester Society of Natural Science and statement of accounts for the year 1882–83. Chester Society of Natural Science, Chester.

CHESTER SOCIETY OF NATURAL SCIENCE AND LITERATURE. 1895. Catalogue of books forming the Library of the Chester Society of Natural Science and Literature. Chester Society of Natural Science, Chester.

CORDREY, L. & FORD, S. 2007. *The National Trust Policy for the Collecting of Geological Materials (Fossils, Rocks and Minerals)*. National Trust Conservation Directorate Swindon.

COSGROVE, D. 1994. Contested global visions: one-world, whole-Earth, and the Apollo space photographs, *Annals of the Association of American Geographers*, **84**, 270–294.

COUPER, P. 1999. Finding what RIGS groups are made of. *Earth Heritage*, **11**, 8.

COXE, R. C. 1846. Formation of the Tyneside Naturalists' Field Club. *Transactions of the Tyneside Naturalists' Field Club*, **1**, 1–5.

CUTLER, A. 1994. Local conservation and the role of the regional geological societies. *In*: O'HALLORAN, D., GREEN, C., HARLEY, M., STANLEY, M. & KNILL, J. (eds) *Geological and Landscape Conservation*. Proceedings of the Malvern International Conference 1993, Geological Society of London, 353–357.

DARLEY, G. 1990. *Octavia Hill—A life*. Constable & Co., London.

ELLIS, N. 2008. A history of the Geological Conservation Review. *In*: BUREK, C. V. & PROSSER, C. D. (eds) *The History of Geoconservation*. The Geological Society, London, Special Publications, **300**, 123–135.

ELLIS, N. V., BOWEN, D. Q., CAMPBELL, S., KNILL, J. L., MCKIRDY, A. P., PROSSER, C. D., VINCENT, M. A. & WILSON, R. C. L. 1996. *An Introduction to the Geological Conservation Review*. GCR Series No. 1 Joint Nature Conservation Committee, Peterborough.

ENGLISH NATURE. 2002. *Revealing the Value of Nature*. Peterborough.

ENGLISH NATURE. 2006. *Natural Remedy: Why the Great Outdoors is Good for Our Health*. English Nature Magazine, **83**, 14–20.

FABIAN, J. 2002. KSC Direct! webcast of the STS-112 launch, Kennedy Space Centre. Website: www.ksc.nasa.gov/nasadirect/archives/KSCDirect/archives/launch/sts112/day3/astro-qa.htm (accessed 16.03.07).

Farlex. 2007. *The FREE Dictionary*. Farlex Inc. Website: www.thefreedictionary.com (accessed 12.1.2007).

GAGARIN, Y. 1962. *Quote for exhibition*. Website: http://all-moscow.ru/culture/museum/astron/astro.en.html (accessed 16.03.07).

GORDON, J. 2004. RIGS in Scotland: Looking ahead. *In*: BROWNE, M. & MCADAM, D. (eds) *Good*

Practice: RIGS Groups Working in Partnership. Proceeding of the Sixth UKRIGS annual conference, Ecclesmachen, Broxburn, West Lothis 24–26 October 2003, 8–12.

GRAY, M. 2004. *Geodiversity: Valuing and Conserving Abiotic Nature.* John Wiley, Chichester.

GREEN, C. P. 1994. The role of voluntary organizations in Earth science conservation in the UK. *In*: O'HALLORAN, D., GREEN, C., HARLEY, M., STANLEY, M. & KNILL, J. (eds) *Geological and Landscape Conservation.* Proceedings of the Malvern International Conference 1993. Geological Society of London, 309–312.

GREEN, C. P. 2008. The Geologists' Association and geoconservation: history and achievements. *In*: BUREK, C. V. & PROSSER, C. D. (eds) *The History of Geoconservation.* The Geological Society, London, Special Publications, **300**, 91–102.

HARLEY, M. 1994. The RIGS (Regionally Important Geological/Geomorphological Sites) challenge— involving local volunteers in conserving England's geological heritage, *In*: O'HALLORAN, D., GREEN, C., HARLEY, M., STANLEY, M. & KNILL, J. (eds) *Geological and Landscape Conservation.* Proceedings of the Malvern International Conference 1993. Geological Society of London, 313–317.

HARLEY, M. & ROBINSON, E. 1991. A local Earth-science conservation initiative. *Geology Today*, **7**, 47–50.

HEREFORDSHIRE & WORCESTERSHIRE EARTH HERITAGE TRUST. 2005. *Working to Record and Protect Geology & Landscape.* Worcester.

HIGGS, B., BUREK, C. V. & WYSE-JACKSON, P. N. 2005. Is there gender bias in the geological sciences? *Irish Journal of Earth Sciences*, **23**, 132–133.

HIGGS, B. & WYSE JACKSON, P. N. 2007. The role of women in the history of geological studies in Ireland. *In*: BUREK, C. V. & HIGGS, B. (eds) *The Role of Women in the History of Geology.* Geological Society, London, Special Publications, **283**, 137–153.

HILL, O. 1877. *Our Common Land.* Macmillan, quoted *In*: MURPHY, G. 2002. *Founders of the National Trust: A Study of Sir Robert Hunter, Octavia Hill and Canon Rawnsley*, National Trust, Enterprises Ltd.

HOSE, T. A. 2008. Towards a history of geotourism: definitions, antecedents and the future. *In*: BUREK, C. V. & PROSSER, C. D. (eds) *The History of Geoconservation.* The Geological Society, London, Special Publications, **300**, 37–60.

HUNTER, R. 1896. *Open Spaces, Footpaths and Rights of Way.* Strahaan & Co, London.

JOHNSTON, G. 1834. Address to the members of the Berwickshire Naturalists' Club September 19 1832. *Proceedings of the Berwickshire Naturalists' Club*, **2**, 4–12.

KINGSLEY, C. 1873. *Town Geology.* Appleton & Co, New York.

KNELL, S. 1997. What's important? *In*: NUDDS, J. R. & PETTITT, C. W. (eds) *The Value and Valuation of Natural Science Collections.* Proceedings of the International conference, Manchester 1995. Geological Society of London, 11–16.

LAMB, S. & SINGTON, D. 1998. *Earth Story.* BBC London.

LARWOOD, J. 2002. Geodiversity and sustainability—the role of Local Geodiversity Action Plans, an overview. *In*: TILSON, E., BUREK, C. V. & WOOD, M. *Promoting Geodiversity.* Proceedings of the fifth UKRIGS annual conference Llandudno, Clwyd, 3–5 October 2002, 5–9.

LARWOOD, J. 2006. LGAPs: Where are we now? *Earth Heritage*, **25**, 9.

LEYS, K. 2000. RIGS in Scotland—a progress report. *In*: OLIVER, P. G. (ed.) *Proceedings of the second UK RIGS conference.* 2–4 September 1999, Herefordshire & Worcestershire RIGS Group, University College Worcester, Worcester, 137–140.

LITTLE, W., FOWLER, H. W. & COULSON, J. 1973. *The Shorter Oxford English Dictionary on Historical Principles. Volume 1.* Clarendon Press, Oxford.

LYON, G. 1867. Memoir of the Society. *Transactions of the Edinburgh Geological Society*, **1**, 1–6.

MALPAS, J. 2006. NEWRIGS Geodiversity audit, 2004–6. NEWRIGS.

MARREN, P. 2002. *Nature Conservation: A Review of the Conservation of Wildlife in Britain 1950–2001.* New Naturalists Series 91. HarperCollins, London.

MASON, V. 1999. RIGS Handbook. *Earth Heritage*, **11**, 5.

MEADOWS, J. 2004. *The Victorian Scientist—The Growth of a Profession.* The British Library, London.

MILLENNIUM ECOSYSTEMS ASSESSMENT. 2005. *Ecosystems and human well being.* Island Press, London.

MILN, G. P. & SHEPHEARD, W. F. J. 1901. *Hon Secretaries report, 30th annual report and proceedings for the Year 1900–1901.* Chester Society of Natural Science Literature & Art, 5–7.

MURPHY, G. 2002. *Founders of the National Trust: A Study of Sir Robert Hunter, Octavia Hill and Canon Rawnsley.* National Trust, Enterprises Ltd.

NATIONAL ARCHIVES. 2006. *Nature Conservation Before 1945.* HMSO. Website: www.ndad.nationalarchives.gov.uk/AH/17/detail.html#n1 (accessed 12.1.2007).

NATURE CONSERVANCY COUNCIL. 1990. *Earth Science Conservation in Great Britain—A Strategy.* NCC, Peterborough.

NATIONAL TRUST. 2006. *Draft Geology Policy.* National Trust.

NEWRIGS. 2000. *Walking Through the Past: A Geological trail Around Chester.* Chester City Council, Chester.

NEWSTEAD, R. 1901. Curator's Report. *31st Annual Report and Proceedings for the year 1901–1902.* Chester Society of Natural Science Literature & Art.

NEWSTEAD, R. 1926. Curator and Librarian's report. *55th Annual Report and Proceedings 1925–1926.* Chester Society of Natural Science Literature & Art.

PHILLIPS, P. 1990. *The Scientific Lady—A Social History of Woman's Scientific Interests 1520–1918.* Weidenfeld & Nicholson, London.

POTTER, J. & BUREK, C. V. 2006. The first Local Geodiversity Action Plan (LGAP): evaluating the Cheshire region LGAP. *Teaching Earth Sciences*, **31**, 19–21.

PRETTY, J., GRIFFIN, M., PEACOCK, J., HINE, R., SELLENS, M. & SOUTH, N. 2005. *A Countryside for Health and Well Being: The Physical and Mental Health Benefits of Green Exercise.* Countryside Recreation Network.

PROSSER, C. 2002. Terms of endearment. *Earth Heritage*, **17**, 12–13.

PROSSER, C. D. 2006. Massive step for geological conservation. *Earth Heritage*, **25**, 8–9.

PROSSER, C. 2008. The history of geoconservation in England: legislative and policy milestones. *In*: BUREK, C. V. & PROSSER, C. D. (eds) *The History of Geoconservation*. The Geological Society, London, Special Publications, **300**, 113–122.

PROSSER, C. D. & KING, A. H. 1998. Regionally Important Geological and Geomorphological Sites: the origin and a forward view. *In*: OLIVER, P. G. (ed.) Proceedings of the first UK RIGS conference, Hereford & Worcestershire RIGS Group, Worcester, 1–8.

PROSSER, C. D. & KING, A. H. 1999, The conservation of historically important geological and geomorphological sites in England. *Geological Curator*, **7**, 27–33.

QVIST, A. 1958. *Epping Forest*. Corporation of London Publication, London.

REYNOLDS, J. 2005. Earth Science education project success. *Earth Heritage*, **24**, 5.

ROBINSON, E. 2007. The influential Muriel Arber: a personal reflection. *In*: BUREK, C. V. & HIGGS, B. (eds) *The Role of Women in the History of Geology*. Geological Society, London, Special Publications, **281**, 287–294.

ROBINSON, H. 1971. *The First Hundred Years 1871–1971*. Chester Society of Natural Science Literature and Art, Chester.

ROGERS, A. 2000. *Report on the activities of the Association of Welsh RIGS groups. In*: OLIVER, P. G. (ed.) Proceedings of the Second UK RIGS Conference 1999. 133–135.

ROTHSCHILD, M. & MARREN, P. 1997. *Rothschild's reserves: Time and Fragile Nature*. Balaban Publishers.

ROYAL SOCIETY FOR THE PROTECTION OF BIRDS. 2007. *Milestones Wildlife Information Factsheet* (3/03). Website: www.rspb.org.uk/about/history/milestones.asp (accessed 17/03/07).

SARRE, P. & REDDISH, A. 1996. *Environment and Society*, 2nd ed. Hodder & Stoughton & Open University.

SHARPLES, C. 1993. *A Methodology for the Identification of Significant Landforms and Geological Sites for Geoconservation Purposes*. Forestry Commission internal report, Tasmania.

SHRUBSOLE, G. 1887. *Curator's Report for 1886–87*. 16th Annual report and statement of accounts for the year 1886–7. Chester Society of Natural Science.

SIDDALL, J. D. 1911. *The Formation of The Chester Society of Natural Science, Literature and Art and an Epitome of its Subsequent History*. Chester Society of Natural Science, Literature and Art, Chester.

STANLEY, M. 2000. Report of the chairman of the UKRIGS steering Group. *In*: OLIVER, P. G. *Proceedings of the Second UK RIGS Conference*. 2–4 September 1999, Herefordshire & Worcestershire RIGS Group, University College Worcester, Worcester 129–131.

STEVENS, C. 1994. *Defining Geological Conservation. In*: O'HALLORAN, D., GREEN, C., HARLEY, M., STANLEY, M. & KNILL, J. (eds) *Geological and Landscape Conservation*. Proceedings of the Malvern International Conference 1993. Geological Society, London, 499–501.

STOLTERFOTH, H. 1874. *Hon. Scientific Secretary's Report*. The Third Annual report of the Chester Society of Natural Science for the year 1873–74 with statement of accounts and list of members. 5–11. Chester Society of Natural Science, Chester.

STOLTERFOTH, H. 1884. *Hon. Scientific Secretary's report*. The Thirteenth Annual report of the Chester Society of Natural Science and statement of accounts for the year 1883–84. Chester, 5–12.

STOLTERFOTH, H. 1887. *Hon. Scientific Secretary's report*. The Sixteenth Annual report of the Chester Society of Natural Science and statement of accounts for the year 1886–87. Chester, 5–10.

STOLTERFOTH, H. 1902. *Hon. Scientific Secretary's report*. 31st annual report and proceedings for the year 1901–1902. Chester Society of Natural Science Literature & Art, 5.

STOTT, R. 2003. *Theatres of Glass, the Woman Who Brought the Sea to the City*. Short Books, London.

TILSON, E. 2004. *RIGS Groups Working Together, UKRIGS/EN Project Report—Oct 2003. In*: BROWNE, M. & MCADAM, D. (eds) *Good Practice: RIGS Groups Working in Partnership*. Proceedings of the Sixth UKRIGS annual conference, Ecclesmachen, Broxburn, West Lothian 24–26 October 2003, 29–31.

TZOULAS, K. & JAMES, P. 2004. *Finding the links between urban biodiversity and human health and well-being*. 4th International Postgraduate Research Conference in the Built and Human Environment, 1–2 April, 2004. University of Salford, Salford.

WALLEY, G. 1997. *The Social History Value of Natural History Collections. In*: NUDDS, J. R. & PETTITT, C. W. (eds) *The Value and Valuation of Natural Science Collections*. Proceedings of the International conference, Manchester 1995, Geological Society, London, 49–60.

WATERSON, M. 2003. *Our Story: For Ever, for Everyone*. The National Trust, Swindon.

The Geologists' Association and geoconservation: history and achievements

C. P. GREEN

Department of Geography, Royal Holloway University of London, Egham, Surrey TW20 0EX, UK (e-mail: greenc@waitrose.com)

Abstract: The vulnerability of the geological record and of the natural environment in general was recognized in the nineteenth century. The Geologists' Association (GA) contributed to the national debate after the First World War and was active in consultations leading to the development of conservation policy and legislation in the 1940s. The geological issues attracted less attention in the 1950s and 1960s, but in the years that followed the GA became increasingly involved in conservation initiatives. The Fieldwork Code was published, the Curry Fund established and the GA became active in supplying the geological input to a wide variety of environmental conservation campaigns.

Concern about the future of the British countryside took shape in the latter part of the nineteenth century. The National Trust for example was founded in 1895 (see Burek 2008a). This concern was coupled with attempts to establish a legal right of access to the countryside for the general public. The GA's involvement in these concerns provides a distinctive insight into their history.

The GA has always been sensitive to the condition and treatment of field localities. From its early years in the 1860s, the GA considered field meetings to be fundamental to its primary object of promoting the study of geology.

The vulnerability of geological sites was drawn to the attention of members on their very first field excursion in 1860 when they went to Folkestone with the somewhat idiosyncratic Samuel Mackie. He circulated a reprint of one of his articles on the Folkestone area which includes the following passage:

It was a treat indeed in my youthful times to see that fossil bank uncovered and display its myriad treasures. Sacks full of beautiful fossils would repay your intertidal toil, but now, daily is that restricted tract most keenly watched by searching eyes; and scantier every day becomes the harvest to be gathered. Mackie (1860)

Mackie was 37 at the time so we can infer the impact of a rapidly widening public interest in geology in the years between say 1830, when Lyell's *Principles of Geology* was published, and the date of the excursion, 1860.

The damage inflicted on field localities by geologists is one of three strands of concern that have shaped conservation awareness within the GA, alongside loss of exposure through development or neglect and the need to retain the goodwill of landowners in the interest of access to geological sites, an issue closely related to their conservation. The following paragraphs outline the history of that growing awareness within the GA and identify the response of the GA to these three concerns.

The early years

In the nineteenth and early twentieth centuries it is unrealistic to argue that the GA had a proactive interest in conservation. Concerns that do appear in GA publications do so as incidental comment in field meeting reports. The tone was generally one of regret, or occasionally irritation. Sometimes the suggested solutions would not now meet with universal approval. For example, the comment of Lamplugh and Cole as they reflected in their Field Meeting Report on a visit to the Yorkshire coast:

. . . a dynamite cartridge or two would soon restore the prolificness of the locality. Lamplugh & Cole (1891)

Despite this apparent lack of sensitivity, the Council Minutes in the first half of the twentieth century reveal the beginning of a more active interest in geoconservation. The first time the Council Minutes refer to a conservation issue is in February 1917 (Mins 3.2.17) when Council agreed to support the efforts of a member 'to secure the preservation of the concretionary Magnesian Limestone on Carley Hill'. As with many other brief reports in the minutes, the outcome is unrecorded.

In the years between the two world wars, popular concern about the preservation of the natural landscape became increasingly widespread and the GA was actively involved in the debate. In 1924, Council appointed a member to attend a conference of the British Association 'on the question of the further protection of sites of

From: BUREK, C. V. & PROSSER, C. D. (eds) *The History of Geoconservation.*
Geological Society, London, Special Publications, **300**, 91–102.
DOI: 10.1144/SP300.7   0305-8719/08/$15.00 © The Geological Society of London 2008.

historic or scientific interest or of natural beauty'
(Mins 7.3.24).

Three years later in 1927, Council was invited to
submit a topic for the BA annual meeting and asked
the BA 'to include in the subject of Nature
Reserves, if not already within its scope, the preservation of historical geological sections' (Mins
20.6.27). Notice that the focus at this time was
rather narrowly on 'historical geological sections',
presumably those associated with major developments of the science. Later in the same year, it
was reported back to Council from the conference
that the BA had agreed 'that preservation of
objects of geological interest should be urged on
the anticipated revision of the National Monuments
Act' (Mins 7.10.27). In the event, no sites of largely
geological interest were recognized as National
Monuments.

Then in 1930, is a record of what can perhaps be
regarded as the first proactive conservation-related
initiative within the GA. Council approved the creation of a 'Preservation of Sections Committee'

(Mins 7.3.30). However, this development does
not seem to have had any practical outcome as
there is no evidence in the GA archives of any
recommendations or activity initiated by this committee. Its apparently brief existence does no more
than illustrate an awareness in the GA of the
conservation issue.

Council was again nudged into action in 1935
(Mins 2.5.35) in response to a letter from an
active member in the Church Stretton area. In this
letter, he dealt first with the discontent of farmers
regarding the behaviour of geologists on their
land, and went on to comment on what he described
as 'the destruction or obliteration of faces of rock by
indiscriminate hacking'. Council published the
letter in the *Circular* for May 1935 under the
heading *An Appeal to Field Parties*. Shortly afterwards, the Council Minutes (Fig. 1) take note of
several letters supporting this appeal and Council
agreed to publish a supplementary notice. This
appeared with the July *Circular* under the heading
A Further Appeal to Field Parties and Geologists

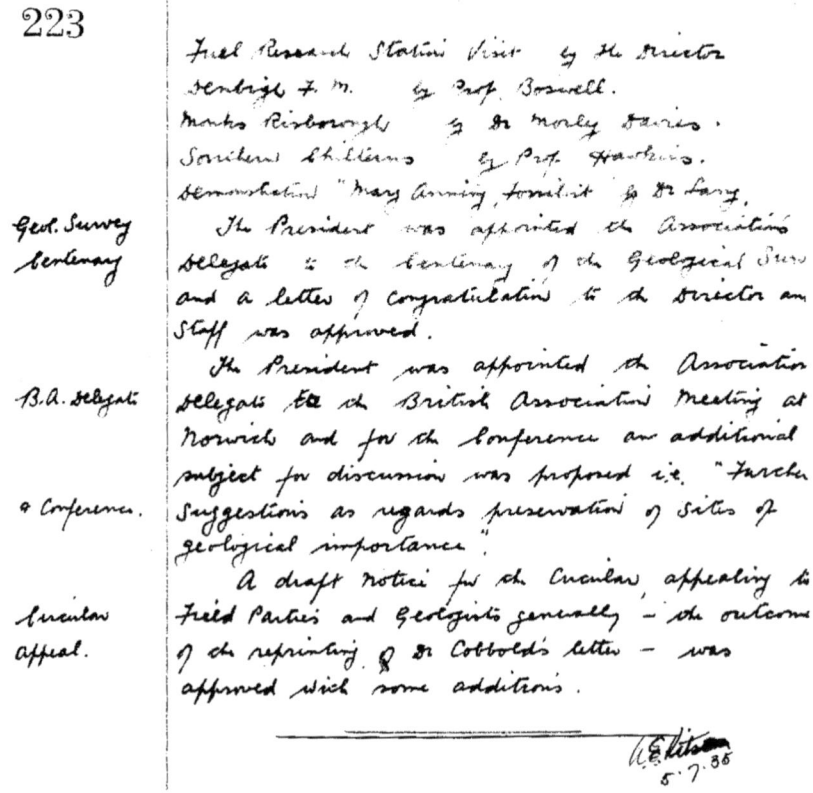

Fig. 1. Council minutes for 5 July 1935 mentioning 'preservation of sites of geological importance' for the British
Association meeting in Norwich and approving a publication on preservation of sites with GA Circular 380.

generally. It dealt mainly with the obligations of geologists towards landowners, but included the following paragraph:

In Quarries - Avoid damage to described and classical sections and the production of needless debris. Preserve special exposures of small outcrops of geological or palaeontological importance. Indiscriminate fossil and implement hunting has caused serious loss to our science at such sites. *Geologists' Association Circular 380, July 1935, Supplementary Notice*

This notice was published as a separate insert and members were advised that 'Copies of this Appeal may be obtained on application to the General Secretary.' As Eric Robinson (1990) has noted, it was a fascinating precursor to the Fieldwork Code published by the GA forty years later.

Council Minutes again record support for geoconservation in 1939 (Mins 2.6.39) when the National Trust was trying to secure access to Harrison's Rocks at Eridge, East Sussex (Fig. 2). Then during the Second World War came the first stirrings that would lead (under the Labour government elected in 1945) to a legislative framework for the conservation of the natural environment.

Hesitation and complacency

From an early stage, the GA was involved in the development of national environmental legislation. In 1941 Council (Mins 9.5.41) appointed a delegate to attend a conference organized by the *Society for*

the Promotion of Nature Reserves (Fig. 3) and two years later when *The Nature Reserves Investigation Committee* was formed by government, the GA was invited by the *SE Union of Scientific Societies* to become involved with the issue of reserving sites of geological interest. The response of Council was interestingly hesitant:

... the practicability of preserving such sites was questioned and Mr Bromehead was asked to convey the view of the Council to the Nature Reserves Committee. (Mins 2.7.43)

However despite this rather frosty response Council had already approved (Mins 6.5.43) a sub-committee to cooperate with the *SE Union of Scientific Societies* on the 'preservation of geological sections and features in the London area and Weald.'

This early interest in the development of national legislation, was not sustained in the years immediately after the Second World War, nor through the 1950s and 1960s when the GA became, once again, largely reactive in its concern for conservation; perhaps it was reassured by the inclusion of sites of geological interest in national legislation to protect the natural environment. It is significant that in the history of the GA, published in 1958 as part of the Association's centenary celebrations, there is no mention at all of conservation issues (Sweeting 1958).

This apparent lack of concern and its gradual replacement by a more positive policy towards

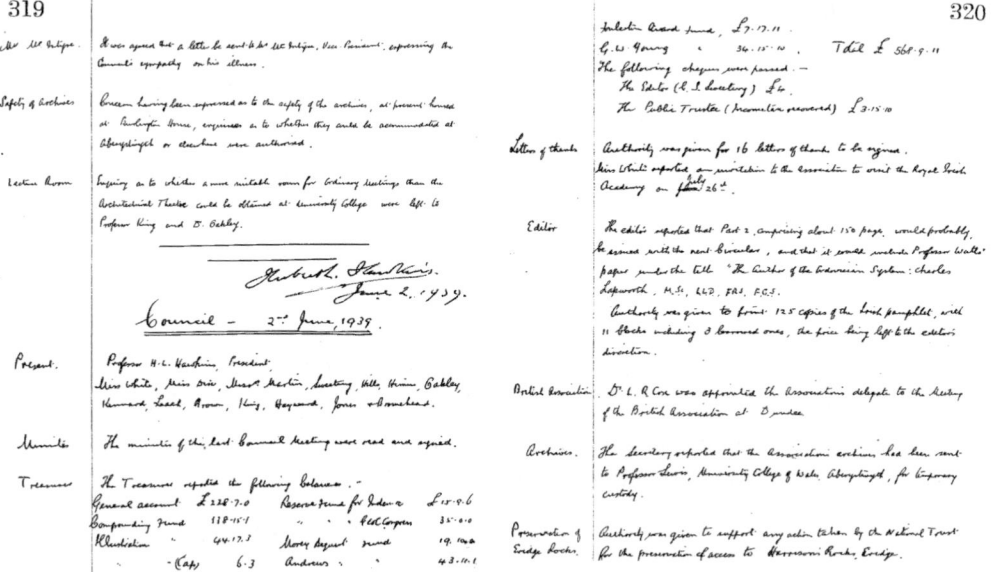

Fig. 2. Council minutes for 2 June 1939 supporting National Trust moves to preserve access to Harrison's Rocks, Eridge, Sussex.

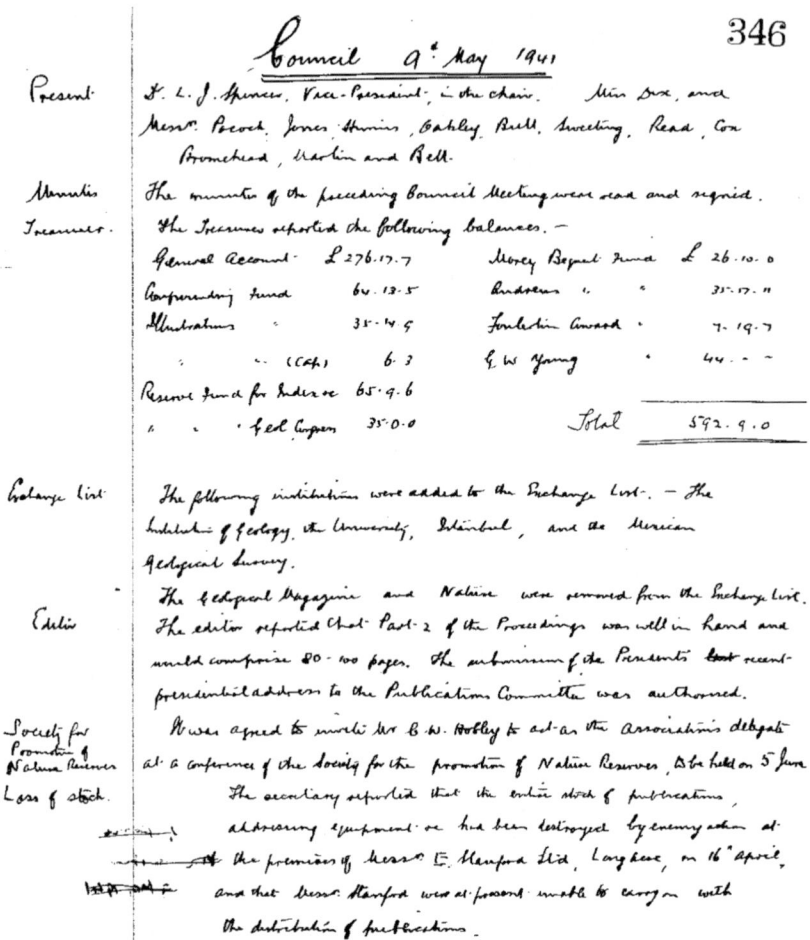

Fig. 3. Council minutes from 9 May 1941 giving active support to the Society for the Promotion of Nature Reserves.

conservation, is well illustrated in the GA's Geological Guides, first published as another element of the centenary celebrations in 1958. When they first appeared, they contained no advice of any kind about conduct in the field. It was not until 1964 that a note was introduced advising users of the guides to seek permission to visit sites. In 1967, the note was extended to encourage users to report to the editor 'any information . . . e.g. infilling of quarries that will help to make the guide more serviceable'. Although this appeal did not express an active interest in conservation, there was at least a recognition that conservation issues existed. At last, in 1971, the note was further extended to include a new paragraph: 'Users of the Guide, in particular those in charge of parties, are ... earnestly requested to avoid over-hammering of exposures'. Finally, in 1973, users of the Guide 'are ... earnestly requested to ensure

that there is no indiscriminate hammering of, or collecting from, exposures'.

Over a period of ten years between 1964 and 1973 this transformation reflected almost exactly the emergence of a new popular awareness of conservation issues; and from the early seventies onward, the GA has been vigorously proactive in its approach to geoconservation.

The years 1972 and 1973 marked a turning point for the GA. Early in 1972 Council received a letter of complaint from a landowner about the mistreatment of property by geologists. In response Council set up a sub-committee 'to consider what action could be taken with quarry owners and others'. At about the same time a notice was published in the *Circular* under the heading *Protection for Geological Sections*, drawing attention to the role of the Nature Conservancy as the national agency responsible for the care of the geological heritage. It is probably no coincidence that the lecture at the July Ordinary Meeting was delivered by Alan Stubbs, an officer of the Nature Conservancy, with the title *The Role of Conservation in Geology and Physiography*. Later in the year Council agreed to organize a half-day meeting at the Geological Sciences Conference in 1973 on *Access to Geological Sites and their Preservation*. The conservation issue had suddenly come alive in the GA.

A new beginning: codes of conduct

Early in 1973, the sub-committee to consider what action could be taken with quarry owners and others presented to Council the first draft of a code of conduct for geological fieldwork. Again, it is no coincidence that Alan Stubbs was elected to Council in March 1973, bringing a Nature Conservancy presence to GA decision-making. He was immediately appointed to convene a committee to redraft the fieldwork code. Since that time, links between the GA and the conservation agencies have remained strong, with NCC and its successors regularly represented on Council and commonly serving as officers. Council gave financial support to a conservation initiative for the first time in 1973. It offered £50 to the Gloucestershire Wildlife Trust towards a total cost of £1500 needed to purchase the Wilderness Cement Works Quarry.

Redrafting the fieldwork code, seeking the support and approval of other societies and obtaining quotations continued through 1973 and 1974 but eventually, early in 1975, the GA took delivery of 100 000 copies of the *Fieldwork Code*. The printing cost of £1317 was met by the GA from the capital of its bequest fund—surely an indication of the importance attached to this initiative. To extend the distribution of the Code as widely as possible, 40 000 copies were delivered to supporting societies and the remaining 60 000 to the GA itself. This was before the GA had an office in Burlington House and the delivery was made to the GA Library at UCL. The GA Library has never occupied a separate space at UCL, so the delivery was no doubt made to the office of Eric Robinson, the GA Librarian at the time and it was he who dealt with the distribution of the Code. He was very active. By January 1976, 45 000 copies of the Code had been distributed, including 10 000 to regional museums, 5500 to the national parks and 3000 to field study centres. In addition, 25 LEAs and a further 100 schools had asked for supplies of the code (Robinson 1976).

Within three years, the first printing of the Code was nearly exhausted and Council agreed in 1978 to print another 100 000 copies. A grant of £750,

the full cost of printing on this occasion, was made by BP (Fig. 4). Between the drafting of the first and second editions of the Code, a new area of concern received legislative recognition through the Health & Safety at Work Act, 1974. The text of the new edition of the Code included wording on safety in the field that had been agreed with the Health & Safety Executive and had the approval and support of the quarrying industry. In the years that followed, the *Fieldwork Code* was adopted in a number of other countries, including Australia and the United States as the model for similar guidance on fieldwork practice.

In 1980, for the first time, a paper on geoconservation appeared in the *Proceedings of the Geologists' Association*. The author was Keith Duff, well known as a distinguished officer of NCC and its successors. His title was *The Conservation of Geological Localities* and he drew attention to what he called 'the geologist-induced problems'. This was at a time when a new source of disfigurement had begun to affect geological exposures. Sampling using portable rock coring equipment became increasingly popular in the 1980s, and increasingly indiscriminate. In 1989 the GA published a separate code *Take Care when you Core* which was widely distributed, in the words of the Annual Report 'to societies, surveys and institutes all over the world'.

When the *Code for Geological Fieldwork* was prepared for a third edition, the substance of the coring code was incorporated in it. The third edition appeared in 1996 with colour and with cartoon illustrations, once again supported by funding from BP.

Making the case for conservation

Although the codes gave the GA a very visible presence in the national and international field of geological conservation, they were not the only GA contribution to the growing awareness of conservation issues. From the 1970s onward the GA regularly made representations in support of conservation initiatives and became involved with other national conservation interests. In the 1970s a particular focus of attention was site documentation. The GA was represented on the Committee for Site Documentation brought together by the Geological Curators' Group and once the National Scheme for Site Documentation was set up, the GA actively promoted it through the Circular (Circular 839, September 1983). Two other issues in which the GA has had a sustained involvement are the protection of limestone pavements and the conservation of peatlands. The GA provided the

Note for Landowners

Landowners may wish to ensure that visiting Geologists are familiar with this Code. In the event of its abuse, they may choose to take the name and address of the offenders and the Institution or Society to which they belong.

Enquiries may be addressed to Dr Eric Robinson, Librarian, c/o Geology Department, University College London, Gower Street, London, WC1E 6BT.

A Code for
Geological
Field Work

The Scope of Geology

Geology embraces the scientific study of the history of the earth, and the past evolution of life upon it. The technical expertise derived from this study also enables our modern civilisation to gain its essential raw materials, for example coal, oil and gas, nuclear fuels, metallic ores, building materials, chemicals for agriculture, and water supplies. It also enables us to develop engineering skills to best advantage, as in the construction of tunnels, dams, roads, and sea defences.

Accessible field sites in good condition are essential for the training of Geologists and for the research needed to make the best use of the natural resources on which we all depend.

ISSUED BY THE
GEOLOGISTS' ASSOCIATION

Printed by courtesy of The British Petroleum Co. Ltd.
by Lindsay & Co. Ltd., Edinburgh

Fig. 4. The Code for Geological Field Work (2nd ed.).

---

only geological representative on the Limestone Pavement Action Group, established in 1992 until 1998 when Cynthia Burek of University of Chester and John Conway of the Royal Agricultural College joined the committee. The GA was the only geological signatory to the charter of the Peat Consortium initiated by the Wildlife Trusts in 1994 and was particularly concerned with land-use conflicts affecting Thorne and Hatfield Moors and the Somerset Lowlands.

The *Circular* and the *Annual Reports* of the GA record the wide range of GA involvement in geological conservation throughout the 1980s and 1990s (see also Burek 2008*a*) and incidentally trace the tireless activity of Eric Robinson, who often took the initiative in conservation matters on behalf of the GA and was formally recognized by Council in 1989 as the GA's Press Officer, a role he had in effect created almost single-handedly over the previous fifteen years. His annual reports in this capacity from 1989 to 1995 reveal just how

widely a GA voice was making itself heard at that time on issues affecting geological conservation. In 1992 alone, representations were made on various site related issues to English Nature, the Department of the Environment, Scottish Natural Heritage, the Royal Horticultural Society and several Local Authorities. In the same year, the GA gave support to the Ramblers' Association in their campaign to achieve wider access to the countryside, 'The Right to Roam'.

Paying for geoconservation: the Curry Fund

For many years, Dennis Curry was a very generous benefactor of the Geologists' Association and when the firm of Curry's changed hands in 1986, the GA holding of Curry shares proved to be worth over £300 000. Council agreed that this large sum should be used to create a grant-giving fund, originally the Geologists' Association Fund but in due course and very properly the Curry Fund of the Geologists' Association. The primary object of the fund is to encourage initiatives within geology which might otherwise not be possible, to encourage innovation, and through far-sighted developments help a wider public to understand and enjoy geology (Fig. 5). The aims of the fund are pursued through two specific avenues: geological publication and geological conservation, with discretion to fund other initiatives that fall outside these two areas.

Table 1 indicates the overall scale of the Curry Fund involvement in the geological arena, disbursing since its formation in 1986, nearly 500 grants and loans totalling nearly £450 000. In that total, geological conservation has attracted the largest share of the grant allocation and Table 2 provides a more detailed insight into this commitment to conservation. As might be expected, acquiring and caring for sites and specimens and presenting them to the public have attracted the largest share of the funding.

BROWN END QUARRY GEOLOGICAL NATURE RESERVE

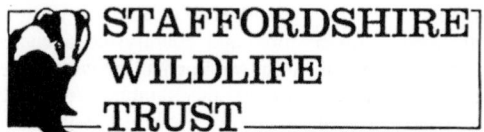

Fig. 5. Leaflet on Brown End Quarry Geological Nature Reserve in Staffordshire published 1991 following funding from Curry Fund in 1987 for site conservation, an interpretation board and literature.

Table 1. *The Curry Fund 1986–2006: grants and loans breakdown*

Grants and loans	£448 959	
Geological conservation 145 awards	£166 009	37.0%
Geological publication 151 awards	£153 578	34.2%
Discretionary* 183 awards	£129 372	28.8%

*Events, educational resources, research, conferences, exhibitions.

Table 2. *The Curry Fund 1986–2006: conservation*

	£	%
RIGS*	**28 626**	**17.2**
Sites	**40 931**	**24.6**
Acquisition	6850	4.1
Maintenance	9498	5.7
Interpretation	24 583	14.8
Specimens	**75 366**	**45.3**
Acquisition	16 562	10.0
Conservation	27 941	16.8
Museum display	13 354	8.0
Curation	17 509	10.5
Other	**21 086**	**12.6**
Publications	5000	3.0
Archive conservation	5688	3.4
Conference	5000	3.0
Miscellaneous	5398	3.2

*Regionally Important Geological/Geomorphological Sites.

Curry Fund grants in support of geoconservation have been distributed within and beyond the British Isles. In 1987, the Curry Fund partly funded site conservation, an interpretative board and a leaflet for Brown End Quarry Geological Nature Reserve (Fig. 5). The largest single grant, of £10 000, was made to the National Museums of Scotland towards the acquisition of 'Lizzie', the earliest fossil reptile in the world. Other Scottish initiatives attracting grants have included the development of the Knockan Crag Centre by Scottish Natural Heritage and the purchase of the Windy Hills site in Aberdeenshire. In Wales grants have been made to support the development and activities of several RIGS groups, including the production of 'Steaming through the past' a geological rail trail through the Llangollen Valley by NEWRIGS (Fig. 6) and including the acquisition by Powys RIGS of the Pales Quarry near Pen y bont (Burek 2008*b*). During 1992–4 the Curry Fund partly

funded the RIGS *Exposure* magazine (Fig. 6). Grants have also been made to groups in Ireland and occasionally to support geoconservation in other countries.

Other highlights of the support given by the Curry Fund to geoconservation include contributions towards the development by the West London Wildlife Group of the Writhlington Geological Nature Reserve; the provision by the Hull Geological Society of a permanent shelter for the famous Rifle Butts section; conservation of the Peterborough Plesiosaur; acquisition by the City of Birmingham of Matthew Boulton's mineral cabinet; funding of overseas delegates to the 1993 Malvern conference; conservation at Ludlow Museum of the great Murchison cross-section and the production of teaching material for the Open University course on Earth heritage conservation. These initiatives give some indication of the broad scope of Curry Fund involvement in geoconservation and the great diversity of the groups that turn to the Geologists' Association for support.

Conclusion

This account of the involvement of the Geologists' Association in geoconservation provides a sense of the gradual development of conservation awareness within the GA and in the wider environmental community in the first half of the twentieth century. It also shows how, in the early 1970s the GA developed a very vigorous proactive commitment to geoconservation, responding and contributing to a growing awareness of conservation issues among the general public. The conservation initiatives taken by the Geologists' Association in the 1970s and 1980s, particularly the publication of the *Fieldwork Code* and the formation of the Curry Fund continue to make an important contribution today to the promotion and management of geoconservation in Britain and beyond.

Alongside the working of these initiatives the GA has developed a strong presence in geological conservation at both national and local levels, representing geology in important debates on the conservation of the natural landscape, notably on peatlands and limestone pavements. A GA representative sits on the Geological Society's Geoconservation Commission and the GA has a long history of positive interaction with the statutory conservation and countryside agencies and with major non-governmental organizations such as the National Trust, the Ramblers Association and the Royal Society for the Protection of Birds. The GA has also regularly made submissions to public enquiries involving local, often site-related, geological issues and has responded to consultative

Fig. 6. NEWRIGS 'Steaming through the past' leaflet, funded by the Curry Fund in 1997.

(a)

SIR DAVID SUPPORTS RIGS

Sir David Attenborough, president of RSNC, The Wildlife Trusts Partnership, has issued a clear statement showing his support for the RIGS movement. By gaining the backing of such an important figure in nature conservation and natural history, RIGS groups will gain greater credibility in many circles including the press and other nature conservation agencies.

The statement also shows the support which RSNC, The Wildlife Trusts Partnership is giving to RIGS.

Sir David Attenborough states,

"I am extremely happy that RSNC, The Wildlife Trusts Partnership is the first voluntary conservation organisation to embrace geology as part of our natural heritage.

Finding fossils was what first sparked off my interest in natural history and I heartily support the conservation of Regionally Important Geological Sites (RIGS), as one of the best ways to inspire more people to explore geology."

The statement can be used by groups in their literature and in press releases.

RIGS LEAFLETS

Two RIGS groups have produced leaflets to publicise their work. Derbyshire's leaflet is a well-produced glossy affair with lots of photographs. Northamptonshire's is much simpler, but has a good outline of the county's geology.

RIGS TAKE OVER

Over 1,380 RIGS have now been registered by more than 20 groups in England alone, narrowly overtaking the number of SSSIs. A flood of applications for the English Nature grant just before the March deadline has meant that the entire budget allocation was claimed. This represents a massive effort by RIGS groups who have documented and submitted sites to local authorities.

The good news is that English Nature are continuing their flat rate grant scheme for a further year, to offset some of the travelling and administration costs of surveying and identifying sites.

For further information about this scheme and how to apply, contact Jo Collinge, Earth Science Branch, English Nature, Northminster House, Peterborough PE1 IUA, or 'phone her on 0733 318322.

Fig. 7. (**a**) Front page; (**b**) Back page of RIGS 'Exposure' Number 2. 'Exposure' was partly funded by Geologists' Association 1992–1994.

(b)

MALVERN CONFERENCE - UPDATE

THE MALVERN CONFERENCE '93

I f you sent off for information on the first international conference on geological conservation, you will no doubt be aware that the programme has now been confirmed and know how much it is going to cost.

If you have generous employers, you will be able to book in for the whole week 18 - 22 July for £350 (residential delegates). However, if your budget does not stretch quite this far, or you only wish to attend part of the conference, you can register on a daily basis for a reasonable £20. The organisers have also planned the sessions so that you don't need to come for the whole week just to cover one theme.

Of the four themes of the conference, two are of particular importance to RIGS. Theme 3 is 'Local Conservation and Community Initiatives'. Chaired by Dr Lars Erikstad (Norway), "Community Goodwill and How It Can Be Tapped and Channelled" will be discussed. Theme 4 is entitled, 'Site Conservation and Public Awareness'. It explores the best ways to

bring a site to life and persuade members of the public of its conservation value.

The other themes of 'Sustainable Development' and 'Landscape Conservation' will be of great value to those involved with site conservation and government.

Optional field trips to the Malvern Hills, Brecon Beacons, the Silurian of the Welsh Borders, Wren's Nest National Nature Reserve and an extra trip to North Wales (22 to 24 July) will give us all a chance to get out and see some rocks.

The field trip on the Tuesday to the Witches Poole RIGS is led by Duncan Hawley of Powys RIGS and is therefore of particular interest to RIGS groups.

Margaret Philips, who is organising the conference, is keen for RIGS members to come along and is trying to gain sponsorship to offer subsidies. She will also provide lists of accommodation to help you plan your stay.

If you have any queries or problems about attending the conference or if you would like to submit a poster presentation please contact:

Margaret Philips
The Company
St John's Innovation Centre
Cowley Road
Cambridge CB4 4WS.
Tel: 0223 421124

EARTH SCIENCE SEMINAR

by Charles Couzens,
Conservation Officer
RSNC, The Wildlife Trusts
Partnership

W hile the leaders of the world were beginning to think about the road to Rio and the Earth Summit, Wildlife Trusts thought about earth science conservation and how to get to Wirksworth. In September 1991 representatives from Trusts throughout England and Wales met to consider the role of Trusts in geological conservation. The two day meeting discussed why geological and geomorphological conservation was important and the role of Wildlife Trusts.

It was clear that geological conservation is a crucial part of nature conservation and important for site protection and environmental education. Full proceedings of the conference have just been published. These cover the theory of RIGS groups, and the experience in one county, ROCKWATCH and the role of RIGS in formal and informal education.

The report is available free from the Lincoln office of The Wildlife Trusts Partnership.

STOP PRESS

New RIGS groups have now formed in Bedfordshire and Clwyd!

RIGS

The RIGS project is sponsored by RSNC, The Wildlife Trusts Partnership, English Nature, Shanks and McEwan, the Clothworkers' Foundation and the Geologists' Association.

RSNC, The Wildlife Trusts Partnership, The Green, Witham

Fig. 7. (*Continued*).

governmental papers dealing with the natural environment and the countryside. The GA obviously has an active commitment to geoconservation. Its distinctive motivation has to do with making geology accessible, so that it can be appreciated and understood by the widest possible audience.

In preparing this paper extensive use has been made of the Council minutes of the Geologists' Association. These are held at the Association's office in the Geological Society apartments in Burlington House. For the early years of the Association the minutes are in bound hand-written

volumes; for more recent years they are filed as loose type-written sheets.

References

BUREK, C. V. 2008*a*. The role of the voluntary sector in the evolving geoconservation movement. *In*: BUREK, C. V. & PROSSER, C. D. (eds) *The History of Geoconservation*. Geological Society, London, Special Publications, **300**, 61–89.

BUREK, C. V. 2008b. History in RIGS in Wales: an example of successful cooperation for geoconservation. *In*: BUREK, C. V. & PROSSER, C. D. (eds) *The History of Geoconservation*. Geological Society, London, Special Publications, **300**, 147–171.

DUFF, K. L. 1980. The conservation of geological localities. *Proceedings of the Geologists' Association*, **91**, 119–124.

LAMPLUGH, G. W. & COLE, E. M. 1891. Excursion to the east coast of Yorkshire. *Proceedings of the Geologists' Association*, **12**, 207–222.

LYELL, C. 1830. *Principles of Geology*. Murray, London.

MACKIE, S. J. 1860. Outline sketch of the geology of Folkestone. *The Geologist*, **3**, 41–45, 81–90, etc.

PROSSER, C. D. 2008. The history of Geoconservation in England: legislative and policy milestones. *In*: BUREK, C. V. & PROSSER, C. D. (eds) *The History of Geoconservation*. Geological Society, London, Special Publications, **300**, 113–121.

ROBINSON, J. E. 1976. Annual Report of the Council of the Geologists' Association for 1975. The Library. *Proceedings of the Geologists' Association*, **87**, 125–126.

ROBINSON, J. E. 1990. 'Clarion o'er the dreaming earth': a personal review of the GA Circular since 1858. *Proceedings of the Geologists' Association*, **101**, 101–118.

SWEETING, G. S. (ed.) 1958. *The Geologists' Association*. The Geologists' Association, London.

The role of the British Geological Survey in the history of geoconservation

ANDREW A. McMILLAN

*British Geological Survey, Murchison House, West Mains Road, Edinburgh
EH9 3LA, UK (e-mail: aamc@bgs.ac.uk)*

Abstract: Over its 172 year history, the British Geological Survey (formerly the Geological Survey of Great Britain) has, through underpinning core activities, its archive and databases and its experienced field staff, provided the geological basis for geoconservation. Evolving activities of the Survey from primary survey and collecting to revision mapping to 3D/4D-modelling reflect changing national needs. In turn, BGS has developed its capability to provide new geological interpretations and a range of publications raising the profile of Earth sciences, both for professionals and for the popular market. Today, BGS's input through networks to geodiversity projects and to newly designated regions such as Geoparks marks a major transition towards a proactive geoconservation agenda in the twenty-first century.

Founded in 1835, the Geological Survey of Great Britain (now the British Geological Survey, BGS) is the world's longest established national geological survey and the United Kingdom's premier centre for Earth science information and expertise. Charles Lyell, in his Presidential Address to the Geological Society in 1836, referred to the need for geological survey to be combined with the geographical survey in progress. He noted the value of obtaining accurate geological information not only for the promotion of geological science 'but also as a work of great practical utility, bearing on agriculture, mining, road-making, the formation of canals and railroads, and other branches of national industry'. It would aid the assessment of the national mineral wealth and formulate what we would call today land utilization policies. Thus the Geological Survey was established by the government expressly for the purpose of producing geological maps of the country based on Ordnance Survey maps as they became available. Accompanying memoirs would be published explaining the geology shown on the maps.

The BGS, its function and history

Geoconservation was not recognized as such in those early years. Indeed the word had not been invented, but although the science of geological conservation or geoconservation has only matured since the 1940s, following the introduction of legislative powers for nature conservation (Gray 2004; Prosser 2008), the functions of the Survey from earliest times have underpinned the objectives of successful geoconservation. These include:

- systematic collection, cataloguing and interpretation of representative materials including fossils, minerals and rocks;
- maintenance of petrological thin section collection;
- the archive of geological photographs started in the 1890s;
- the legacy and collections of the Museum of Practical Geology;
- geological records collections and library;
- the development of the stratigraphical framework of Great Britain;
- the gathering of field data at a range of large scales from 1:10 000 to 1:50 000 and the publication of standards at these scales and at smaller scales (e.g. 1:50 000, 1:250 000 and 1:625 000);
- the undertaking of revision surveys to update and publish geological knowledge; and
- the description of local and regional geology in the form of memoirs, sheet explanations.

Today, BGS (which since 1965 has been a component organization of the Natural Environment Research Council) maintains the UK's National Geoscience Data Centre. This national collection contains the data gathered or generated by BGS and its precursors over more than 170 years of geological survey with records and data that range in age from the early nineteenth century to the present day. Paper archive held by BGS includes maps, notebooks, photographs, memoirs and reports, borehole logs, mine plans and field slips. Many records are in digital format including the 1:50 000 scale geological maps of Great Britain (DiGMapGB-50, see BGS website www.bgs.ac.uk). With the rapid development of new technologies geological data and interpretation are

From: BUREK, C. V. & PROSSER, C. D. (eds) *The History of Geoconservation.*
Geological Society, London, Special Publications, **300**, 103–112.
DOI: 10.1144/SP300.8   0305-8719/08/$15.00 © The Geological Society of London 2008.

provided increasingly in digital format and are integrated into customized Geographical Information Systems (GIS). 3-Dimensional modelling software are enabling, for the first time the development of 3D regional models which have much potential for geoconservation purposes and for geodiversity planning.

This paper outlines aspects of the early history of the Geological Survey and its activities relevant to geoconservation. It shows how, with the post World War II development of geoconservation concept, the BGS is actively contributing to a better public understanding of geodiversity.

Historical backdrop and some early examples of geoconservation

In the latter part of the eighteenth century and early part of the nineteenth century, Earth science took a giant leap forward. Scientific understanding and application was sparked by the twin beacons of intellectual enlightenment and industrial revolution. Clever people, true polymaths, discussed, communicated and debated their findings. James Hutton's observations which supported his *Theory of the Earth* (published in 1795 by the Royal Society of Edinburgh) showed the immense value of geological features and outcrop. These were brilliantly illustrated by his friends John Clerk of Eldin and James Hall (Craig *et al.* 1978).

The necessity of geological maps was recognized in the early years of the nineteenth century. In the making of his seminal geological map of England, Wales and southern Scotland (1815), William Smith was able to interpret the geology and demonstrated how the form of the land was determined by the rocks. Subsequently (1819–24) Smith issued larger scale country maps, coinciding with the issue and distribution of Greenhough's map of England, sponsored by the Geological Society of London (1819). In 1815 Richard Griffith had prepared a geological map of Ireland but it does not seem to have been printed or published. Formal government support of geological survey began the year before in 1814 when John Macculloch was appointed geologist to the Trigonometrical Survey of Great Britain and started his work in Scotland. Although his mission was not to make a geological survey of Scotland, his fine map issued posthumously in 1835 is regarded as the first that was supported, if not commanded, by the British Government. Meanwhile with the establishment of the Ordnance Survey in 1791, opportunity arose to utilize new one inch to one mile base maps of southern England. Under the direction of Colonel Colby several members of the Ordnance Survey staff acquired some geological knowledge and

geological information was recorded on some maps. In 1832 Joseph Portlock was appointed to the survey in Ireland and from then on Henry De la Beche, under the direction of the Board of Ordnance, began the colouring of Ordnance Survey maps of the West of England (Flett 1937). With the political momentum and support of the establishment, the Geological Survey was established in 1835 with De la Beche as its first Director.

Thus began systematic geological surveying, budgets permitting, of the geology of Great Britain, initially in England and Wales but with survey in Scotland (Fig. 1) from the 1850s (Wilson 1977). The true foundation of the British landscape was being discovered. In years to come, the new insights into the understanding of the landscape came with the revelation by Louis Agassiz in the 1840s that much of Britain had been glaciated so that, for example, the 'drifts' defined by Roderick Murchison could be interpreted in the context of a glaciated terrain.

The excitement of these early years galvanized the public to ensure the preservation of fine landscape features. Even before the formation of the national geological survey, there are examples of actions which would now be called geoconservation. These included the saving, by legal action in 1819, of Salisbury Crags below Arthur's Seat in Holyrood Park, Edinburgh from being quarried away (McMillan *et al.* 1999). Here action was directly influenced by concerned citizens for their threatened city landscape. Much later the establishment of the Boulder Committee under the auspices of the Royal Society of Edinburgh initiated a process to identify glacial erratics in Scotland that appeared remarkable in terms of size and superficial markings and to recommend measures for their conservation (Milne Home 1872*a*, *b*). Preservation of the Agassiz Rock in the city also received much attention from the Edinburgh Geological Society from earliest days both to protect the features and to inform the public of the significance of glacial striae (Gordon 1992). In another early example of geoconservation, the preservation of the Carboniferous lycopod (*Stigmaria*) stumps at Fossil Grove, Whiteinch, Glasgow (discovered in 1887) owed much to the efforts of palaeontologists such as R. Kidston and J. Young (formerly a Survey Assistant Geologist and latterly Keeper of the Hunterian Museum) and also to the local authorities for the shelter which was later built to protect the fossil trees (Macgregor & Walton 1948).

The Geological Survey's role in geoconservation

Important early actions of the Geological Survey from its inception in 1835 may be seen to

Fig. 1. Laying the foundations for geoconservation. A group of geologists from the Survey mapping near Braemar, in 1903. From left to right the group comprises John Flett?, Robert Lunn (with camera), E. H. Cunningham-Craig (in kilt) and John Horne. (Reproduced with the permission of the British Geological Survey © NERC. All rights reserved.)

have aided both contemporary and future 'geoconservation' activities.

Museums and collections

Properly maintained and catalogued collections of geological materials form a significant resource. They are in their own right examples of geoconservation and provide an essential reference for field geoconservation. The BGS is custodian of extensive collections of records, materials and data pertaining to the geology and hydrogeology of the UK, its continental shelf and many countries overseas. The collections are of national and international significance. One of the first actions of the newly fledged Survey was the development of a museum at Craig's Court, Whitehall to house De La Beche's collections and the building stone specimens assessed for the then new Houses of Parliament (see below). The collections soon filled

the space available and the case was made for new premises. This resulted in the opening by the Prince Consort in 1851 of the Museum of Practical Geology in Jermyn Street, (221 Piccadilly). Subsequently the museum was relocated to South Kensington where the new building for the Museum of Practical Geology and the Geological Survey of Great Britain was opened by the Duke of York in 1935. It now forms the Earth Galleries of the Natural History Museum. The major reference collections were transferred to Keyworth, Nottingham following the move of BGS there in 1985. The Natural History Museum at South Kensington retained economic specimens, gemstones, minerals and building stones. Successive curators recognized the value of holding and displaying systematic collections of the rocks and fossils and economic minerals for public benefit and instruction.

In Scotland, the close relationship with the Royal Scottish Museum (formerly the Science and

Art Museum and now the National Museum), Chambers Street, Edinburgh resulted in the allocation of gallery space to the Geological Survey of Scotland in 1889 for the display of regional Scottish geology. This was enhanced in 1896 with addition of Professor Heddle's magnificent collection of Scottish minerals (Flett 1937). The resultant collection has been of great value to generations of students of Scottish geology.

Although displays and exhibitions now meet different objectives to engage the modern museum-going public, the underpinning collections still form the basis for geoscientific research. BGS continues to maintain the reference collections at its headquarters in Keyworth, Nottingham and in Edinburgh together with representative onshore and offshore borehole core.

The BGS Palaeontological Collection, numbering 2.5 million specimens, is the most important collection of British fossils in the world. Their integral association with the Survey's 160 year history of mapping British geology means that they are the fundamental biostratigraphical basis for the geological maps which provide the framework for every geological endeavour in Britain. Early palaeontologists of note include Edward Forbes, appointed 1844 followed by John Salter, T. H. Huxley, Robert Etheridge and C. J. Stubblefield. The collections have been enhanced by donations to be held in trust in perpetuity for the nation, some of which predate the Survey. These include specimens from the Geological Society of London collection received in 1911, and specimens figured and described by Murchison in his *Silurian System* published in 1839. The British collection of the Geological Society of London (Moore *et al.* 1991) contains material from such luminaries as Banks, Buckland, Conybeare, Lyell, Murchison, Sedgwick, J. and J. de C. Sowerby and over 600 of the other principal figures of nineteenth century geology. The great bulk of this collection, around 25 000 items, with many type and figured specimens, still remains with the Survey.

Reference collections of rocks, minerals and thin sections are held in a suite of petrological collections held at BGS, Keyworth and Edinburgh. Borehole core has formed part of the Survey collections since its inception. As of 2000, the drill core collection comprised 106 000 one metre boxes holding materials from 2934 boreholes (Hollyer & Wheatley 2000). Indeed the earliest material predates the formation of the Survey and refers to a borehole drilled at Chatham Dockyard, Kent in 1821. Access to minerals and water boreholes by BGS is encompassed by two Acts of Parliament, Section 23 of the Mining Industry Act 1926 and Section 9 of the Petroleum (production) Act 1934. This entitles BGS staff access to log and sample material of any borehole drilled for minerals (including hydrocarbons) over 30 m in depth; Section 205 of the Water Resources Act 1991 allows the same access to boreholes drilled for water greater than 15 m in depth.

Identification of resources

Over the past three centuries with the rapid increase in use of mineral resources to feed firstly the development of the industrial revolution, the requirement to locate, understand and quantify resources became essential. In 1835 the quest for such knowledge was identified as one of the key factors in the establishment of the Geological Survey of Great Britain. One of the most celebrated building stone resource assessments was that conducted by a Select Committee set up to recommend stone for the building of the Houses of Parliament (1839–c. 1852) (Barry *et al.* 1839). Barry's account was the first published detailed survey of the building stone industry of Britain (Lott & Richardson 1997). Although the outcome of the survey was less than satisfactory (the variable quality of the dolomitic limestone recommended resulted in serious differential decay in the building stone), the collection of assessed building stones provided the successful case for the establishment of a Survey Museum in Craig's Court, Whitehall opened to the public in 1841 (Flett 1937).

Stratigraphy

Pioneering stratigraphical studies, resulting from the Primary Survey in Wales during the 1830s, included the development in part of the now familiar Lower Palaeozoic nomenclature and definitions of Sir Roderick Murchison (later to become the second Director-General of the Geological Survey). In the 1870s, revision of the biostratigraphy by Charles Lapworth, a Scottish schoolmaster, in the Southern Uplands of Scotland resulted in the establishment of the Ordovician period. Murchison and Sedgwick also named the Devonian period and discovered fossil fish from the Old Red Sandstone of Scotland. Regional memoirs and sheet explanations accompanied the geological mapping and research. In addition Sir Archibald Geikie, as Director General in his Annual Report for 1883, set in train the ideas for stratigraphical memoirs and the first of these extremely valuable monographs appeared seven years later (Reid 1890). Coinciding with the centenary of the Geological Survey (1935) and the new exhibitions of regional geology at Exhibition Road, the introduction of British Regional Guides proved popular (Department of Scientific and Industrial Research 1936). Aimed at both the interested general reader and as background for the specialist, revised regional guides to 20 onshore regions have served to introduce British stratigraphy to a wider public.

The Geological Survey's contributions to geoconservation in the twentieth century

With the enactment in 1949 of the statutory basis for nature conservation, opportunity arose for geoconservation to be promoted formally. The geological sub-committee of the Nature Reserves Investigation Committee identified some 390 geological localities (including 104 of geomorphological or Quaternary significance) in England and Wales that should be protected as geological reserves. In 1948, J. G. C. Anderson of the Geological Survey identified 60 sites for Scotland. Both lists were very much provisional or *ad hoc* (Gordon 1992). By 1954, with further consideration by the Geological Survey, the official list for Scotland had risen to 169. More systematic surveys were carried out in the 1960s (a precursor to the Geological Conservation Review which began in 1977) and reviewed in an unpublished report by McQuhae & Sargeant in 1978 (see Gordon 1992). During this period as part of the Nature Conservancy Council's process of identifying SSSIs, the Geological Survey was consulted. The exercise benefited from the fact that the contemporary Land Survey Units were geographically defined and staffed by experienced field geologists who 'knew their patch'.

Today, as part of its routine enquiry service BGS is often asked to comment on planning applications whether they are for biodiversity, minerals or for some other development. It is BGS's policy to offer impartial advice, neither condoning nor criticizing a particular development but making factual statements about the geology, hydrogeology, geotechnical information and geomorphology. If certain features are, in the opinion of the geologist, unique or unusual this may be stated.

Safeguarding resources. Both after the First World War and during the Second the Geological Survey undertook systematic strategic surveys of Britain's mineral resources. The value of these surveys was that they not only quantified mineral extent and workability but also provided a background for land-use planning and the safeguarding of resources. Reports published at the time have relevance today as renewed consideration is given to sustainable development of indigenous materials so far as this is possible.

Built heritage. To assist in the identification of materials for built heritage stone repair and conservation work, reference economic memoirs provide details on sources of a wide range of natural resources including slate (Richey & Anderson 1944), mineral resources (including building and road stone) of the Lothians (MacGregor 1945) and limestone (Robertson *et al.* 1949). These provide an invaluable historical record for sourcing of materials. Together with UK-wide information and statistics on current quarrying, materials and products, published regularly by BGS through the *Directory of Mines and Quarries* (Cameron *et al.* 2005, and online via www.mineralsuk.com) these publications inform more recent resource assessment studies such as the recently published BGS *Building Stone Resources Map of Britain* (British Geological Survey 2001) or index volumes such as *Stone in Scotland* (Hyslop *et al.* 2006). In recent decades there has been a resurgence of interest in the use of indigenous natural materials, in particular building stone both for conservation of historic buildings and for new build. The town and city environment in which the majority of the population lives offers many opportunities to develop links between geology and the built heritage (Bennett *et al.* 1996). The opening of new quarries and the re-opening of long-abandoned workings highlight scope for built heritage geoconservation and provides a catalyst to involve the public and professionals in issues such as the use of appropriate indigenous stone in streetscapes and town and urban developments (McMillan *et al.* 2006). BGS has contributed in a variety of ways to promoting this interest through organizations such as the Scottish Stone Liaison Group (Historic Scotland 1997; McMillan & McKinney 2005), the Welsh Stone Forum (Coulson 2005) and the English Stone Forum (Doyle *et al.* 2007 in press).

Geological societies and RIGS. Officers of the Geological Survey have maintained strong links with geological societies and associations across Britain. From the early days, surveyors contributed to the activities of geological societies whose objectives have been the public understanding of geology. Mainly through voluntary effort, but often with the ready support of their managers, survey staff have contributed to the promotion of geoconservation through local activities including public lectures, field excursions and the publication of geological findings in proceedings or transactions.

Development of geodiversity in the twenty-first century: networking links and partnerships

UKRIGS

The 'public good' role of the BGS also translates today into the input BGS geologists have made to the activities of many groups developing Regionally Important Geological and Geomorphological

Sites (RIGS) within the UKRIGS Network. Since the 1980s BGS geologists have offered their professional advice or volunteer their own time to the development of new RIGS sites, preparing posters and leaflets and leading associated tours or excursions. To recognize the mutual benefit of this a close working relationship there is in place a Memorandum of Agreement between UKRIGS and BGS, the objectives of which are shown in Table 1.

Work with statutory bodies and NGOs

In recent years BGS has worked closely with statutory agencies including Scottish Natural Heritage (SNH), Natural England (formerly English Nature), Historic Scotland, CADW and English Heritage. The links have been strengthened as the relevance of Earth science as the underpinning science for biodiversity has been recognized. Another form of link is by representation on external committees such as the Geoconservation Commission under the auspices of the Geological Society (London) (website www.geoconservation.com). This body aims to promote the conservation of our Earth heritage and to ensure that we pass it on in good order to future generations for investigation, education and enjoyment.

BGS has supported the Earth Science Forum for England and Wales (ESEF) and also the Scottish Earth Science Education Forum (SESEF). Both organizations promote Earth science in education and provide resources for teachers, schools and colleges. BGS also collaborates with science centres (e.g. Our Dynamic Earth, website www.dynamicearth.co.uk and the National Stone Centre www.nationalstonecentre.org.uk) and museums. BGS also supports and funds a wide range of collaborative research with universities.

In terms of scientific geoconservation publications, a major input by BGS geologists has been to the Geological Conservation Review (GCR) series. The GCR is led by the Joint Nature Conservation Committee (JNCC), and aims to identify and describe the most important Earth science sites in Britain (Ellis *et al.* 1996). At the heart of the government agencies' objectives has been the challenge to promote the care of Britain's natural environment, its responsible enjoyment and its greater understanding and its sustainable use. An example of joint publications aimed at increasing the general public's understanding of the environment is the SNH/BGS series 'Landscape Fashioned by Geology' Series (McKirdy & Crofts 1999). These books, written with the minimum of jargon, tell the story of the evolving geological and recent history of Scotland through its rocks and landscapes. BGS also supports a range of visitor attractions such as mining visitor centres and heritage

centres. It also participates in 'rock and fossil' events (e.g. the Vogrie Environment Fair) and hosts Open Days for the general public.

Popular events and publications

BGS geologists have contributed to exhibitions, display boards (e.g. at National Trust properties) or in partnership with others (National Museums of Scotland, Our Dynamic Earth, BGS and the Hunterian Museum in Glasgow), involved with raising an awareness of the Earth heritage through geology festivals leading to joint ventures, such as the organization of 'Rock On' the biennial Scottish Geology Festival (formerly Scottish Geology Week). Over the last 15 years the BGS and the Geological Survey of Northern Ireland have produced several popular publications. These include landscape books of Ireland and the Isle of Man (e.g. McKeever 1999; Pickett 2001), popular applied geology 'Whisky on the Rocks' (Cribb & Cribb 1998), holiday geology guides (e.g. Gallois 2001), Falklands Island guides (e.g. Stone & Aldiss 2000), Discovering Geology Fossil Focus Guides (e.g. Wilkinson 2000), building stones leaflets (e.g. Lott & Barclay 2002), and GeoTourism maps (e.g. British Geological Survey 2000).

European Geopark network

A European Geopark is a clearly defined area with important geological heritage in terms of scientific quality, rarity, aesthetic appeal and educational value (Jones 2008). The key functions of a European Geopark are to protect geological heritage, promote geology to the public, and to use geology and other aspects of its natural and cultural heritage to promote sustainable economic development, normally through responsible tourism. All European Geoparks work together in the European Geoparks network, established in 2000. The network consists of some 25 geoparks in 11 countries. The European Geopark network ensures quality of geological heritage, interpretation and education and conservation and also shares expertise. All European Geoparks are also endorsed by UNESCO as a Global Geopark.

BGS has played a valuable advisory role in the establishment of European and UNESCO Geoparks. The first in Britain was the North Pennines AONB in 2003 (Table 2) with the Brecon Beacons National Park being accorded European Geopark status in 2005. The latter's mountains and hidden valleys are the result of nearly 500 million years of Earth history, and the area contains evidence of ancient seas, mountain building and sea level and climate change scattered across a landscape that was shaped by the last ice age. The first

Table 1. *Objectives of the Memorandum of Understanding between UKRIGS and BGS*

Communicating

- Seeking closer understanding of each other's work, and exploring means by which greater collaboration may be established.
- Undertaking regular national level meetings to discuss and share information regarding strategic direction, potential work programmes, joint working, and issues impacting on the objectives, work and running of the respective organizations.
- Providing support to encourage increased contact between BGS programmes and RIGS groups.
- Promoting each other's existence and remit in appropriate literature, relevant presentations, and when meeting influential or key players in the field of geological conservation, both in the United Kingdom and abroad.

Supporting and sharing

- Providing each other with support information, and facilities, as long as this is within reason and is in keeping with the capacities of the respective organizations.
- Working together to facilitate the sharing of information, exchange of ideas, expertise and general learning opportunities. For example through training events, workshops, publications and web-based media and Geodiversity Action Plans.
- BGS will continue to support The Association of UKRIGS Groups throughout the period of this agreement. The Association of UKRIGS Groups will continue to encourage member groups to support BGS in its strategic mapping and scientific programmes.

Working together in partnership

The intention of the MoU is to define the guidelines for the two parties to co-operate in the field of geoscientific data exchange, particularly where mutual interests exist. The intellectual property rights, third party data ownership interests, aspects of commercial sensitivity, societal concerns, and confidentiality, will be paid due regard, protected, and not knowingly infringed.

- BGS and the Association of UK RIGS Groups will meet at least annually to explore and develop areas for joint working. This may result in time and/or financial resources being pooled to develop projects that address issues of mutual concern, in particular Geodiversity Action Plans, geoconservation and related topics.
- Whilst working nationally on strategic issues, BGS and The Association of UKRIGS Groups will encourage cooperation and joint working at a local level between specific BGS projects and individual RIGS groups on more site-specific issues.
- Both parties intend to produce inventories that will cover digital and non-digital data holdings including basic metadata information. When complete, the inventories will be shared between the parties by appropriate electronic means.
- Both will publicise internally what data services are available from the other party, and how to make use of them. They will also publish internally the organizational e-mail addresses of the nominated contacts of each party, to facilitate the effective exchange of data and information by this medium
- It is agreed that neither party to this MoU will seek to profit from the inputs of the other party or undermine their position.
- For Geodiversity Action Plans and related topics that have been funded from external sources, BGS and the relevant RIGS group(s) should work in close cooperation to achieve the agreed output both at the planning and working stages. If the funding levels are insufficient for the full deployment of BGS staff, then BGS will make up any difference or make full funding available to contribute to the success of the project.

European Geopark in Scotland, the North West Highlands Geopark was launched in 2005 at the SNH-owned Knockan Crag National Nature Reserve Visitor Centre (Barron *et al.* 2005). This outstanding area contains some of the most important and diverse geological and geomorphological features and stunning landscapes in Britain and the rocks in the Geopark record the last 3000 million years of history for the landmass that we now know as Scotland. The Geopark also recognizes the diverse natural heritage of the area, local culture and the rich array of historical and archaeological sites. Geologically, the area is dominated by the internationally important Moine Thrust zone, which runs from north to south. In the nineteenth century, this zone puzzled geologists who recognized that packages of rocks were 'thrust' over long distances on top of younger rocks. The Northwest Highlands Memoir (Peach *et al.* 1907) has stood the test of time as a classic groundbreaking publication. Thrusts are now recognized in rocks around the world, including those in the Himalayas and the Alps—but are rarely as accessible as the Moine Thrust at Knockan Crag. As part of

Table 2. *Examples of geodiversity audits and action plans to which BGS has contributed*

Project	Duration	Description
Dundee District Inventory	1995	Inventory of Earth science sites in the Dundee District. BGS contributed to this early audit prepared for SNH which set out methodology and site details for quarries, road cuttings, river sections and for geomorphological features.
County Durham Geodiversity Audit	2003–2004	Conserving geodiversity required a combination of statutory protection for nationally important sites. The protection of both non-statutory sites and geodiversity interests in general impacts the development, and active management of sites and features of importance to geodiversity.
North Pennines AONB Geodiversity Audit and Action Plan	2003–2004	The AONB was awarded the UNESCO-endorsed Global Geopark status in 2003, partly in recognition of the importance of its geology, but also in recognition of local efforts to conserve, interpret and revitalize the area through its Earth heritage. In order to guide understanding and management of the area's unique geological heritage and to support the development of sustainable 'geotourism', the North Pennine AONB Partnership commissioned BGS to advise on the framing of a Geodiversity Action Plan for the AONB. This is the first such study of a protected landscape in this country.
Leicestershire and Rutland LGAP	2003–2007	The principal aims were centred around education, the provision of a detailed audit of all known geological sites in the two counties and the gathering together of data sets of geological information on the counties.
West Lothian Geodiversity Audit	2005–2006	The first to be conducted in Scotland, the audit was undertaken as a means of informing the framing of recommendations and action points designed to guide the sustainable management, planning, conservation and interpretation of all aspects of the Earth heritage of West Lothian. A draft West Lothian Geodiversity Action Plan (WLGAP) was prepared (see Barron & Arkley 2006).
Yorkshire Dales and Craven Lowlands Draft LGAP	2005–2006	The draft Action Plan *Your Dales Rocks* sets out a framework of actions for auditing, recording and monitoring the geodiversity. It is currently a draft and subject to change
Northumberland National Park Sustainable Geodiversity Framework	2005–2007	Evaluating the resource and identifying ways of exploiting it within the co-ordinated framework of the National Park strategy (Lawrence *et al.* 2007). Emphasis is placed on identifying ways by which geodiversity and, in particular, opportunities arising from past and present aggregate extraction might contribute to sustainable tourism and provide opportunities for learning and employment.

recent mapping in the Northwest Highlands, the BGS has published a Walkers' Map (British Geological Survey 2004) which offers an excellent template for future popular publications.

Geodiversity audits and action plans: shaping planning policy

With the expertise and records described previously, the BGS has been well placed to provide advice on geodiversity throughout the UK. Generations of BGS geologists have walked over much of the UK recording information on the geology and, based upon their local knowledge have compiled geodiversity audits and contributed to Geodiversity Action Plans, examples of which are shown in Table 2. One of the earliest examples to which BGS contributed was the Inventory of Earth Science sites in the Dundee District (Hardie 1995). The minerals industry also plays a key role in the conservation of geological heritage. From the creation of geological sections, through to the restoration and management of sites, the industry is uniquely placed to contribute to the conservation, management, scientific, educational and recreational use of geological sites.

There is now an increased emphasis on planning for geodiversity. This is reflected in the recently published PPS9 for England *Biodiversity and geological conservation*, in which Government clearly places geodiversity at the heart of planning policy, and in the number of local and company level geodiversity action plans being produced and implemented. In Scotland, geodiversity has also been recognized in updated Biodiversity Action Plans (e.g. City of Edinburgh Council 2004) and in county natural heritage designations (e.g. Scot 2005).

Concluding remark

The Geological Survey has travelled far in its 172 years. Throughout its history, it has changed course many times to reflect the needs of the nation and to make its geoscience research relevant to the people. It now employs specialists in a wide range of disciplines from physics to chemistry, from information technologists to graphic designers. Yet we can reflect that the Survey's underpinning core activities, its archive and databases and its staff have played and continue to play a significant role in the development of geoconservation to ensure that development in the century ahead is truly sustainable. Charles Lyell would be pleased.

The author, who is the BGS representative on the Geoconservation Commission, acknowledges the assistance of several BGS colleagues in compiling this paper, namely H. Barron, T. Cooper, K. Ambrose, J. Carney, M. Browne, M. Howe, and R. Bowie. He thanks M. Browne and the referees for constructive reviews of the paper. He publishes with the permission of the Executive Director BGS (NERC).

References

BARRON, H. F. & ARKLEY, S. L. B. 2006. West Lothian leads in Scotland. *Earth Heritage*, **25**, 12–13.

BARRON, H. F., KRABBENDAM, M. & TODD, G. 2005. Scotland's first Geopark—The North West Highlands. *Earth Heritage*, **23**, 19–21.

BARRY, C., DE LA BECHE, H. T., SMITH, W. & SMITH, C. H. 1839. *Report on the Selection of Stone for Building the New Houses of Parliament*. Report of the Parliamentary Select Committee.

BENNETT, M. R., DOYLE, P., LARWOOD, J. G. & PROSSER, C. D. (eds) 1996. *Geology on Your Doorstep: the Role of Urban Geology in Earth Heritage Conservation*. The Geological Society, London.

BRITISH GEOLOGICAL SURVEY. 2000. *Tourists' Rock, Fossil and Mineral Map of Great Britain*. Earthwise Publications, British Geological Survey, Keyworth, Nottingham.

BRITISH GEOLOGICAL SURVEY. 2001. *Building Stone Resources Map of Britain (1:1 000 000)*. British Geological Survey, Keyworth, Nottingham.

BRITISH GEOLOGICAL SURVEY. 2004. *Exploring the Landscape of Assynt. A Walkers' Guide to the Rocks and Landscape of Assynt and Inverpolly. Map at*

1:50 000 Scale. British Geological Survey, Keyworth, Nottingham

CAMERON, D. G., BARTLETT, E. L., HIGHLEY, D. E., LOTT, G. K. & HILL, A. J. 2005. *Directory of Mines and Quarries 2005: 7th edn*. British Geological Survey, Keyworth, Nottingham

CITY OF EDINBURGH COUNCIL. 2004. *Edinburgh Biodiversity Action Plan 2004–2009*.

COULSON, M. R. (ed.) 2005. *Stone in Wales: Materials, Heritage and Conservation*. Proceedings of the Welsh Stone Conference, Cardiff 2002. CADW: Welsh Historic Monuments, Cardiff.

CRAIG, G. Y., MCINTYRE, D. B. & WATERSTON, C. D. (ed.) 1978. *James Hutton's Theory of the Earth: The Lost Drawings*. Scottish Academic Press Ltd, Edinburgh.

CRIBB, S. & CRIBB, J. 1998. *Whisky on the Rocks— Origins of the 'Water of Life'*. Earthwise Publications, British Geological Survey, Keyworth, Nottingham.

DEPARTMENT OF SCIENTIFIC AND INDUSTRIAL RESEARCH. 1936. *Summary of Progress of the Geological Survey of Great Britain and the Museum of Practical Geology for the Year 1935*. His Majesty's Stationery Office, London.

DOYLE, P. (ed.) 2008. *England's Heritage in Stone*. Proceedings of a Conference. English Stone Forum, Folkestone, Kent.

ELLIS, N. V., BOWEN, D. Q., CAMPBELL, S., KNILL, J. L., MCKIRDY, A. P., PROSSER, C. D., VINCENT, M. A. & WILSON, R. C. L. (ed.) 1996. *An Introduction to the Geological Conservation Review*. GCR Series No.1 Joint Nature Conservation Committee, Peterborough.

FLETT, J. S. 1937. *The First Hundred Years of the Geological Survey of Great Britain*. His Majesty's Stationery Office, London.

GALLOIS, R. 2001. *Holiday Geology Guide: Stonehenge—the Secrets in the Stones*. Earthwise Publications, British Geological Survey, Keyworth, Nottingham.

GORDON, J. E. 1992. Conservation of geomorphology and Quaternary sites in Great Britain: an overview of site assessment. *In*: STEVENS, C., GORDON, J.E., GREEN, C. P. & MACKLIN, M. G. (eds) *Conserving Our Landscape*. Proceedings of the Conference Conserving our landscape: evolving landforms and Ice-age heritance, Crewe May 1992, 11–21.

GRAY, M. 2004. *Geodiversity—Valuing and Conserving Abiotic Nature*. John Wiley & Sons Ltd, Chichester.

HARDIE, R. A. 1995. *Inventory of Earth Science Sites in the Dundee District*. Report for Scottish Natural Heritage.

HISTORIC SCOTLAND. 1997. *A Future for Stone in Scotland*. Historic Scotland, Edinburgh.

HOLLYER, S. & WHEATLEY, C. 2000. *Users Guide to the British Geological Survey Geological Material Collections*. British Geological Survey Technical Report **WO/00/02R**.

HYSLOP, E.K., MCMILLAN, A. A. & MAXWELL, I. 2006. *Stone in Scotland*. Earth Science Series. UNESCO Publishing, Historic Scotland, IAEG, BGS, Paris.

JONES, C. 2008. A history of Geoparks. *In*: BUREK, C. V. & PROSSER, C. D. (eds) *The History of*

Geoconservation. The Geological Society, London, Special Publications, **300**, 1–5.

LAWRENCE, D. J. D., ARKLEY, S. L. B., EVEREST, J. D., CLARKE, S. M., MILLWARD, D., HYSLOP, E. K., THOMPSON, G. L. & YOUNG, B. 2007. *Northumberland National Park Geodiversity Audit and Action Plan.* British Geological Survey Commissioned Report CR/07/037N.

LOTT, G. H. & BARCLAY, W. 2002. *Geology and Building Stones in Wales (north) and (south).* British Geological Survey, Nottingham.

LOTT, G. K. & RICHARDSON, C. 1997. Yorkshire stone for building the Houses of Parliament (1839–c. 1852). *Proceedings of the Yorkshire Geological Society*, **51**, 265–272.

MACGREGOR, A. G. 1945. *The Mineral Resources of the Lothians.* Wartime Pamphlet, Number 45. Geological Survey of Great Britain.

MACGREGOR, M. & WALTON, J. 1948. *The Story of Fossil Grove.* City of Glasgow Public Parks and Botanic Gardens Department, Glasgow.

MCKEEVER, P. J. 1999. *A Story through Time: The Formation of the Scenic Landscapes of Ireland (North).* Geological Survey of Northern Ireland, Belfast.

MCKIRDY, A. P. & CROFTS, R. 1999. *Scotland: The Creation of its Natural Landscape. A Landscape Fashioned by Geology.* Scottish Natural Heritage, Perth.

MCMILLAN, A. A. & MCKINNEY, A. 2005. A future for stone: a view from the Scottish Stone Liaison Group. *In*: COULSON, M. R. (ed.) *Stone in Wales: Materials, Heritage and Conservation.* Proceedings of the Welsh Stone Conference, Cardiff 2002. CADW: Welsh Historic Monuments, Cardiff, 84–88.

MCMILLAN, A., HYSLOP, E., MAXWELL, I. & MCKINNEY, A. 2006. *Indigenous Stone Resources for Scotland's Built Heritage.* Paper No. **825**. Proceedings of the IAEG 2006 Conference, Nottingham.

MCMILLAN, A.A., GILLANDERS, R.J. & FAIRHURST, J.A. 1999. *Building Stones of Edinburgh.* 2nd edn. Edinburgh Geological Society, Edinburgh.

MILNE HOME, D. 1872a. Scheme for conservation of remarkable boulders in Scotland, and for the indication of their position on maps. *Proceedings of the Royal Society of Edinburgh*, **7**, 475–488.

MILNE HOME, D. 1872b. First report by the Committee on Boulders appointed by the Society. *Proceedings of the Royal Society of Edinburgh*, **7**, 703–775.

MOORE, D. T., THACKRAY, J. C. & MORGAN, D. L. 1991. A short history of the Museum of the Geological Society of London, 1807–1911, with a catalogue of the British and Irish accessions, and notes on surviving collections. *Bulletin of the British Museum, Natural History (Historical Series)*, **19**(1), 51–160.

PEACH, B. N., HORNE, J., GUNN, W., CLOUGH, C. T. & HINXMAN, L. W. 1907. *The Geological Structure of the Northwest Highlands of Scotland.* Memoirs of the Geological Survey of Great Britain.

PICKETT, E. A. 2001. *Isle of Man—Foundations of a Landscape.* British Geological Survey, Keyworth, Nottingham.

PROSSER, C. D. 2008. The history of geoconservation in England: legislative and policy milestones. *In*: BUREK, C. V. & PROSSER, C. D. (eds) *The History of Geoconservation.* The Geological Society, London, Special Publications, **300**, 1–5.

REID, C. 1890. *The Pliocene Deposits of Britain.* Memoirs of the Geological Survey of Great Britain.

RICHEY, J. E. & ANDERSON, J. G. C. 1944. *Scottish Slates.* Wartime Pamphlet, Number 40. Geological Survey of Great Britain.

ROBERTSON, T., SIMPSON, J. B. & ANDERSON, J. G. C. 1949. *The limestones of Scotland.* Special Reports of the Mineral Resources of Great Britain, **35**. Memoirs of the Geological Survey of Great Britain.

SCOT, S. 2005. *Natural Heritage Designations in Fife—a Guide.* Fife Environmental Recording Network and Fife Council, Glenrothes.

STONE, P. & ALDISS, D. 2000. *Falkland Islands: Reading the Rocks—a Geological Travelogue.* Guidecard. British Geological Survey, Keyworth, Nottingham.

WILKINSON, I. P. 2000. *Corals.* Discovering Geology Fossil Focus Guides. British Geological Survey, Keyworth, Nottingham.

WILSON, R. B. 1977. *A History of the Geological Survey in Scotland.* Natural Environment Research Council Institute of Geological Sciences, London.

The history of geoconservation in England: legislative and policy milestones

COLIN D. PROSSER

Natural England, Northminster House, Peterborough PE1 1UA, UK
(e-mail: colin.prosser@naturalengland.org.uk)

Abstract: England, and the UK more widely, have robust and mature statutory and voluntary frameworks for delivering geoconservation. Critical to achieving this advanced position was the inclusion of geoconservation within the first nature conservation legislation enacted in Britain in 1949. The development of this legislation benefited greatly from the wisdom of a number of committees set up to inform government thinking. Many of these committees were advised by the scientific community, including geologists and geomorphologists. The work and influence of these committees in establishing geoconservation as part of statutory nature conservation is explored, and the main statutory and policy milestones which have guided and shaped geoconservation in England since 1949 are described. The rise of the voluntary geoconservation movement in the late 1980s is also explored.

Geoconservation, in terms of conserving, managing, enhancing and promoting scientifically or educationally important geological and geomorphological features, is well established in England. Nationally important geological and geomorphological sites are conserved through conservation legislation and the development planning system as Sites of Special Scientific Interest (SSSIs), whereas locally/regionally important sites are recognized and offered some protection through the planning system as Regionally Important Geological/geomorphological Sites (RIGS). A number of further opportunities to conserve, manage and enhance geologically and geomorphologically important sites exist as a consequence of Government policy.

It was the early recognition of the need for geoconservation in England, and the UK more widely, that resulted in it being included within the robust and mature framework of nature conservation legislation, policy and practice which now operates across the UK. Had events taken a different turn and nature conservation legislation, policy and practice developed without including geoconservation from the outset, experience suggests that securing its inclusion at a later date would have been an extremely difficult challenge. The relatively advanced development of geoconservation in England today is thus largely due to the foresight of those planning for nature conservation in the 1940s.

The first statutory nature conservation legislation to apply in England, The National Parks and Access to the Countryside Act (1949), gives no definitive explanation as to why it includes geoconservation. There are, however, many statements within published reports leading up to this legislation that make clear the thinking at the time, and which would be no surprise today to anyone familiar with the geology or geomorphology of the U.K. These statements are considered in greater detail later in this paper, but in short, they recognize the diversity of the UK's geology and geomorphology, the international importance to science bestowed upon it as a consequence of the pioneering work that took place, the importance of having sites available for educational use in post-war Britain, and very importantly, recognize that 'nature' includes both wildlife and geology/geomorphology as an integrated whole. Although the issue of the resource being subject to threat and damage is not explicitly referred to with regard to geological or geomorphological sites, it was an issue for wildlife sites in what is a relatively populated and developed country. As such, it is also likely to have been in the minds of those considering the safeguard of the geological heritage, as conservation is not usually seen as being worthwhile unless a valued resource is considered to be under threat.

This paper expands upon previous chronological descriptions of the evolution of geological conservation in England (Ellis *et al.* 1996; Thomas & Cleal 2005; Prosser *et al.* 2006). In particular, it explores in some detail the thinking that led to geoconservation being included in the first statutory conservation legislation in England. It goes on to chart 58 years of statutory and policy milestones relating to the national statutory nature conservation effort and then explores the rise of the voluntary geoconservation sector and the policy milestones that have resulted from the activity and campaigning of this movement.

From: BUREK, C. V. & PROSSER, C. D. (eds) *The History of Geoconservation.*
Geological Society, London, Special Publications, **300**, 113–122.
DOI: 10.1144/SP300.9   0305-8719/08/$15.00 © The Geological Society of London 2008.

The lead up to legislation

The earliest engagement in the UK, in what would now be regarded as geoconservation, is described elsewhere in this volume (eg. Doughty 2008; Thomas & Warren 2008). This activity largely revolved around management of collections of geological material but also included some isolated examples of site-based geoconservation, with the fossil forests in Wadsley, Sheffield and Victoria Park, Glasgow being classic examples (Thomas 2005). However, there is no evidence of a structured or national approach to geoconservation being considered in England until the 1940s.

The first mention of geoconservation being considered at a national level appears in the *Report of the Committee on Land Utilisation in Rural Areas* (Scott 1942). This committee, chaired by Lord Justice Scott, and reporting to the Ministry of Works and Planning, was given terms of reference to 'consider the conditions which should govern building and other constructional development in country areas consistently with the maintenance of agriculture, and in particular the factors affecting the location of industry, having regard to economic operation, part-time and seasonal employment, well being of rural communities, and the preservation of rural amenities'.

In terms of geoconservation, paragraph 179, on 'Nature Reservations' is significant. This states that 'We recommend that the Central Planning Authority, in conjunction with the appropriate Scientific Societies, should prepare details of areas desired as nature reserves (including geological parks) and take the necessary steps for their reservation and control...' This is important in that it opens the door for geoconservation, due most likely to the influence of the Committee's vice-chairman, L. Dudley Stamp, who undoubtedly brought his geological/geomorphological background and experience to the deliberations.

The next mention of geology/geomorphology within the context of national conservation thinking occurs within a report entitled *National Parks in England and Wales* (Dower 1945). This was commissioned by the Ministry of Town and Country Planning and set out to 'to study the problems relating to the establishment of National Parks in England and Wales'. It builds on Scott (1942) and makes a number of statements relevant to geoconservation. For example, paragraph 60 refers to '"Nature reserves," covering all natural features—flora, fauna and places of geological interest...', and paragraph 66 notes that most proposed national parks 'are also exceptionally rich in places of special geological interest'. It also proposes a number of potential national parks including geologically rich areas such as the Lake District, Dartmoor and the Peak District.

The first really substantial consideration of geoconservation is included in the report *National Geological Reserves in England and Wales* (Chubb 1945). This report of the Geological Reserves Sub-Committee of the Nature Reserves Investigation Committee built upon the statement by Scott (1942) that nature reserves include geological parks. The sub-committee included G. F. Herbert Smith (Chairman), S. E. Hollingworth, G. H. Mitchell, T. H. Whitehead, D. Williams and K. P. Oakley. Other than the chairman, all members had a geological background, being nominated by either the Geological Survey or the Geological Society and it is recorded that they 'imperturbably held their meetings during the savage attack on the London area by robot planes'. The work of the sub-committee was assisted by fifty 'local' geological advisers including W. J. Arkell, F. W. Cope, Emily Dix, O. T. Jones, W. D. Lang, W. F. Whittard and L. J. Wills.

Early thinking on geoconservation is expressed clearly in Chubb (1945). For example, the foreword to the report makes two very important statements. Firstly, it recognizes that wildlife and geology pose different conservation challenges, stating that 'the necessary conservation measures in the two instances would largely be so different in character'. Secondly, it provides a clear justification for geoconservation, stating that:

The foundations of geological science were laid largely by investigators in this country, and it is important that the evidence upon which their conclusions were based should be preserved for the benefit of students for all time. Although geological features have not the vulnerability of plants or animals, they may, perhaps only from ignorance, easily be damaged or obscured, unless they receive proper care. The country should feel pride in the possession of classic sections or monuments of international fame, and should be anxious for them to be adequately protected. In the United States of America, for instance, many National Monuments have been selected because of their geological interest.

The report proposes a classification of geological reserves listing and classifying 390 sites across England and Wales, splitting them into conservation areas, geological monuments, controlled sections or registered sections. The classification and split of sites in England is shown in Table 1. Locations for these sites on 1 inch maps were deposited in the British Museum (Natural History).

In July 1947, prior to nature conservation legislation being passed in England, two further relevant reports were commissioned and published by the Ministry of Town and Country Planning. The first of these was the *Report of the National Parks Committee (England and Wales)* (Hobhouse 1947). This Committee was set up to develop the thinking of Dower (1945) and to make recommendations about the establishment of national parks and the

Table 1. *The classification of geological reserves as proposed by Chubb (1945)*

Classification type	Definition as given by Chubb (1945)	Number of sites in England identified by Chubb (1945)
Conservation area	Large scale physiographic features and areas containing many items of geological interest.	61
Geological monument	Small-scale geological features and sections of outstanding interest, to be permanently protected and kept in a good state of preservation. Each of these to be provided with a metal notice-plate, briefly explaining its character and origin.	35
Controlled section	Natural sections and artificial sections in a state of disuse, to be subject of control on account of their scientific value, in order to prevent them from being irretrievably obscured by building or dumping of refuse, or otherwise rendered inaccessible	168
Registered section	Sections of exceptional geological importance at present used or worked, to be listed and to be kept under observation by an appointed authority, the owners or lessees being required to give notice to it of their intention to cease operations, in which event the sections in question would be considered for transference to the previous category.	68

conservation of 'wildlife'. The report is significant in that it accepted the need for geoconservation. However, it drew much of its thinking from another 'special committee' which it had set up to provide it with 'specialized scientific knowledge'. The second report was that of this 'special committee', namely, *Conservation in England and Wales: Report of the Wild Life Conservation Special Committee (England and Wales)* (Huxley 1947). The remit of this special committee, chaired by Julian Huxley (Fig. 1), was to consider the proposals set out by Dower (1945) and to advise on any additional measures that would be necessary or desirable for the purposes of 'wildlife' conservation. It included A. E. Trueman and made a number of statements about the importance of geoconservation and the role that geology and geomorphology plays in nature more widely. These included the following:

The reasons for safeguarding geological and physiographical features are not widely appreciated. There are in England and Wales many such features which are of great interest. These are scattered throughout the country, for Great Britain presents in a small area an extremely wide range of geological phenomena. British geologists were pioneers in the creation of scientific geology and have since played an outstanding part in the development of their science. Classical sites are therefore numerous in Great Britain, attracting students from many countries.

But standing level with the biological sciences, though too often neglected in the context of nature preservation, are the geological and physiographical sciences; for it is from the nature and distribution of the rocks and from the configuration

of the Earth's crust that the natural beauty of scenery and its living carpet are derived.

In short, wildlife conservation cannot be separated from nature at large.

The report refers to, and supports, statements in Dower (1945) calling for the conservation of wildlife 'as a broad national objective' with the hope that wildlife policy will be adopted 'as an integral part of a comprehensive programme for conservation and development of our natural resources'. Huxley also recommends that 'the Government should take a general responsibility for the protection and management of certain sites of special, biological, geological, physiographical and other scientific value'.

Further recommendations are that a biological service, 'staffed by scientists with appropriate qualifications should be set up within the Government machine', and that staffing would include specialists in 'relevant subjects, e.g. geology, physiography, soil science'.

With regard to the Geological Sub-Committee's recommendations for 'Conservation areas', six sites, Alderley Edge (Fig. 2), Worm's Head, Avon Gorge, Water End Swallow Holes, Creswell Crags and Wren's Nest (Fig. 3) were accepted as suitable for National Nature Reserve (NNR) status. Of these, only Wren's Nest ever became an NNR (Prosser & Larwood 2008). The report also accepted the concept of protecting 'geological monuments', and 'controlled' and 'registered' sections but this terminology was not carried forward into legislation.

Fig. 1. Sir Julian Huxley (third from the right, with cap and spectacles), Chair of the Wild Life Conservation Special Committee, on a visit to the New Forest in 1960. The report of this committee in 1947 played a vital part in ensuring that geoconservation was included in the first major piece of Nature Conservation legislation. Photo: News Chronicle.

Legislative and policy milestones

National Parks and Access to the Countryside Act (1949)

In 1949, a national statutory approach to nature conservation came into being, with the passing of a nature conservation act including geoconservation, and the establishment, by Royal Charter, of a governmental nature conservation body, the Nature Conservancy. A key part of this act in terms of geoconservation, is Section 23, which states that 'Where the Nature Conservancy is of opinion that any area of land, not being land for the time being managed as a nature reserve, is of special interest by reason of its flora, fauna, or geological or physiographical features, it shall be the duty of the Conservancy to notify that fact to the local planning authority in whose area the land is situated.'

The legislation, built upon the work of Chubb, Dower, Huxley and others, set out a science-led site-based approach to conservation, with features of special scientific interest being identified and protected through a series of discrete sites. As such, only features within designated sites were protected, with anything falling outside being left with no protection at all. This site-based approach, with an emphasis on conserving scientifically important features, has remained the focus of geoconservation at a national level up to the present day (Ellis *et al.* 1996; Prosser *et al.* 2006).

Countryside Act (1968)

This improved the options for managing features within designated sites. For example, Section 15 introduced the facility for the Nature Conservancy to enter into agreements with owner/occupiers where it is expedient in the national interest in order to conserve flora, fauna, or geological or physiographical features.

The Geological Conservation Review (GCR)

The policy decision to undertake this review, initiated in 1977 (Ellis *et al.* 1996; Ellis 2008) was driven by a need to establish a more rigorous and robust, scientifically systematic approach to the identification of nationally important sites for

Fig. 2. Alderley Edge, Cheshire. This Triassic escarpment enriched with copper ores, and a site of pre-historic mining, was one of six geological sites proposed as potential NNRs in 1947, although it never attained NNR status. Photo: Hannah Townley, Natural England.

designation and conservation as geological/geomorphological SSSIs. The fact that sites must first be selected as GCR sites, and thus subject to thorough assessment on a national level by appropriate specialists, before they can be considered as SSSIs, has provided a strong and defendable rationale for the selection and conservation of geological/geomorphological SSSIs.

Fig. 3. Limestone caverns at Wren's Nest, Dudley. This sequence of mined and quarried fossil-rich Silurian limestone was the only one of the six proposed geological NNRs, listed in 1947, to attain NNR status, being declared in 1956. Photo: Geoffrey Prosser.

Wildlife and Countryside Act (1981)

This act improved arrangements for the conservation of SSSIs. Under the 1949 Act, only planning authorities had to be told of the existence of an SSSI, and little information about the conservation interest of the site or what may damage it was provided. Section 28 (1) of the 1981 Act states that:

Where the Nature Conservancy Council are of the opinion that any area of land is of special interest by reason of any of its flora, fauna, or geological or physiographical features, it shall be the duty of the Council to notify that fact–

   (a)   to the local planning authority in whose area the land is situated;
   (b)   to every owner and occupier of any of that land, and
   (c)   to the Secretary of State.

The act also introduced much stronger, and more widely applicable, protection to SSSIs, through requiring owners and occupiers to secure the consent of the Nature Conservancy Council (NCC) (who succeeded the Nature Conservancy in 1973) before they undertook any damaging activities. Prior to this legislation, SSSI protection was relatively weak, and gave no protection against activities outside planning control. To this effect, Section 28 (4) goes on to add that:

A notification under subsection (1) shall specify–

   (a)   the flora, fauna, or geological or physiographical features by reason of which the land is of special interest;
   (b)   any operations appearing to the Council to be likely to damage that flora or fauna or those features.

The act also included provisions to protect limestone pavements, enabling a local planning authority to make a Limestone Pavement Order, prohibiting the removal or disturbance of limestone on or in it.

Environmental Protection Act (1990)

This led to the British conservation agency, the NCC, being split into three country-based agencies: the Countryside Council for Wales, English Nature and Scottish Natural Heritage. The GCR, however, retained a Great Britain-wide coverage through being attached to the Joint Nature Conservation Committee, a fourth body with a remit to lead on certain overarching matters across Great Britain.

Countryside and Rights of Way Act (2000)

This strengthened SSSIs as a conservation tool. It placed increased emphasis on supporting SSSI owners and occupiers in managing, rather than just safeguarding important features within the SSSIs on their land. In particular, the act included new powers that could be used to address the neglect of geological sites, for example degradation of exposures resulting from vegetation growth or burial under weathered material. In addition, the act placed a requirement on public bodies to conserve and enhance SSSIs. It also made it an offence for anyone to knowingly or recklessly damage an SSSI. This change meant that anyone, not just the site owner or occupier, could be prosecuted for damaging an SSSI, allowing more effective action to be taken against visitors damaging an SSSI through undertaking activity such as irresponsible fossil or mineral collecting.

Planning Policy Statement 9: Biodiversity and Geological Conservation (2005)

This described the Government's national policy on the conservation of biodiversity and geological/geomorphological features in England through use of the planning system. It raised the profile of geoconservation amongst local planning authorities and developers and set an expectation that policy and practice relating to development planning would address the needs of geoconservation.

Key points of this policy statement with regard to nationally designated geological/geomorphological sites include the title of the policy statement which explicitly recognizes geoconservation and the 'key principles' for regional planning bodies and local planning authorities which state that:

In taking decisions, local planning authorities should ensure that appropriate weight is attached to designated sites of international, national and local importance; protected species; and to biodiversity and geological interests within the wider environment.

The aim of planning decisions should be to prevent harm to biodiversity and geological conservation interests.

A local/regional approach to conservation: policy milestones

The origin of RIGS

Despite the establishment of a statutory approach to the conservation of nationally important geological and geomorphological features in 1949, it was another forty years before a co-ordinated approach to the conservation of locally and regionally important geological and geomorphological sites came into being. At this time, a number of factors came together which led the NCC and active local geological conservation groups and societies to work together to initiate a nationwide effort to engage in geoconservation on a local and regional scale (Prosser & King 1998). These factors included:

1.   recognition that there were many locally/ regionally important sites needing

conservation, in particular, the GCR site selection exercise was winding down, leaving many locally/ regionally important sites that had not made the grade as SSSIs;

2. recognition that locally/regionally important geological sites need local support if they are to be valued and conserved within local communities;

3. realization that local wildlife sites were better protected than locally/regionally important geological sites because of a strong locally-based voluntary sector; and

4. excellent examples of locally/regionally-based sites with good practice in geoconservation, including the Black Country, Avon and Shropshire, were coming to light.

Consideration of the above and consultation with existing local geological groups resulted in the non-statutory Regionally Important Geological/geomorphological Sites (RIGS) initiative coming into being. The first full description of the RIGS concept is given in the NCC's geoconservation strategy (Nature Conservancy Council 1990a) but early thinking is evident in Harley (1989) which states that:

In an attempt to conserve the most important of these sites, a number of museums, geological societies, wildlife trusts and local authorities have identified the key geological localities in their counties and submitted details of these to the appropriate planning departments for inclusion in their structure plans, thereby affording them an element of informal protection.

It is clear that a network of such sites would usefully complement the SSSI coverage, and form the basis for field education and research at a more local level.

Adopting the term Regionally Important Geological Sites (RIGS) for these sites, and a co-ordinated approach to their selection and conservation, may well be of benefit if a comprehensive nationwide coverage is to be achieved.

Further clarity on the purpose of RIGS is provided in Nature Conservancy Council (1990b) which defines RIGS as 'Any geological or geomorphological sites, excluding SSSIs, in a county (or region in Scotland) that are considered worthy of protection for their educational, research, historical, or aesthetic importance', and goes on to say that they are 'broadly analogous to non-statutory wildlife sites and are often referred to locally by the same name.'

Planning Policy Guidance 9: Nature Conservation (1994)

The first policy recognition of RIGS, reflecting the growth and success of the initiative, appeared in the above government planning guidance note. Paragraph 17 states that 'Regionally important geological/geomorphological sites are being identified by local conservation groups with the involvement in many cases of local authorities. These sites provide valuable educational facilities, and supplement sites notified as SSSIs...'

The Aggregates Levy Sustainability Fund (2002)

The Aggregates Levy, a tax on the extraction of primary aggregate, has generated very significant sums of money. A proportion of this revenue has been made available through the Aggregates Levy Sustainability Fund for environmental projects associated with aggregate extraction sites. The policy decision to include geological/geomorphological site audit, management and promotion amongst those activities eligible for grants, injected significant funding into geoconservation (Fig. 4) on a previously unknown scale (Prosser 2002, 2004).

Planning Policy Statement 9: Biodiversity and Geological Conservation (2005)

This defines the Government's national policy on protection of biodiversity and geoconservation through use of the planning system in England. In terms of local/regional geoconservation, it represents a significant step forwards. In addition to the explicit recognition of geology within its title, key policy principles for regional and local planning bodies include the following statement:

In taking decisions, local planning authorities should ensure that appropriate weight is attached to designated sites of international, national and local importance; protected species; and to biodiversity and geological interests within the wider environment.

In addition, the Guide to Good Practice which accompanies this policy statement endorses another local geoconservation initiative, production of Local Geodiversity Action Plans (Burek & Potter 2004), as a means of delivering the government's policy on geoconservation within a particular government region or local planning authority.

Local sites: guidance on their identification, selection and management (2006)

This guidance from government provides another important policy step, recognizing local/regional geoconservation site systems as equivalent to local wildlife site systems. The guidance fully recognizes the importance of local/regional geoconservation, especially RIGS, and encourages local planning authorities, local/regional geoconservation and wildlife groups to work together to develop a more consistent and integrated approach to the identification and management of 'Local Sites' in England.

Fig. 4. Clee Hill Quarries, Shropshire. An interpretive project developed through a partnership between a RIGS group, a minerals operator and the local community and funded through the Aggregates Levy. Photo: Hanson Aggregates.

Fig. 5. Legislative and policy milestones, from the inclusion of geoconservation in the first nature conservation legislation in 1949 through to a high profile mention in the title of government's planning policy statement in 2005.

The way ahead?

The legislative and policy framework for geoconservation (Fig. 5) continues to evolve, and the latest piece of legislation, the *Natural Environment and Rural Communities Act* (2006), suggests how geoconservation in England may develop in the future. This act resulted in the creation of a new national conservation body, Natural England, bringing together the wildlife and geoconservation functions of English Nature with the landscape functions of the Countryside Agency and the soil and agri-environmental functions of the Rural Development Service. This legislation helps to implement contemporary policy thinking, namely that the natural environment, including geoconservation, is best addressed as an integrated whole. Sixty years on, it appears that the integrated thinking of Huxley (1947), encapsulated by his statement that 'wildlife conservation cannot be separated from nature at large', has once again come to the fore.

The author would like to thank K. Duff (formerly of English Nature and the Nature Conservancy Council) and B. Ing (University of Chester) for reading and commenting on this manuscript.

References

BUREK, C. V. & POTTER, J. 2004. *Local Geodiversity Action Plans: setting the context for geological conservation.* English Nature Research Report, **560**, Peterborough, UK.

CHUBB, L. 1945. *National Geological Reserves in England and Wales.* Report by the Geological Reserves Sub-Committee of the Nature Reserves Investigation Committee. *Conference on Nature Preservation in Post-War Reconstruction. Natural History Survey of Great Britain.* The Society for the Protection of Nature Reserves. British Museum (Natural History), London. September 1945 (Second edition, November 1945), 1–41.

Countryside Act. 1968. HMSO.

Countryside and Rights of Way Act. 2000. Chapter 37. The Stationery Office.

Department for Environment, Food and Rural Affairs. 2006. *Local Sites: guidance on their Identification, Selection and Management.* Department for Food and Rural Affairs.

Department of the Environment. 1994. *Planning Policy Guidance Note 9: Nature conservation.*

DOUGHTY, P. 2008. How things began: the origins of geological conservation. *In:* BUREK, C. V. & PROSSER, C. D. (eds) *The History of Geoconservation.* The Geological Society, London, Special Publications, **300**, 7–16.

DOWER, J. 1945. *National Parks in England and Wales.* HMSO, Cmd. 6628

ELLIS, N. V., BOWEN, D. Q., CAMPBELL, S. *ET AL.* 1996. *An Introduction to the Geological Conservation Review.* GCR Series No 1, Joint Nature Conservation Committee, Peterborough, 1–131.

ELLIS, N. 2008. A history of the Geological Conservation Review. *In:* BUREK, C. V. & PROSSER, C. D. (eds) *The History of Geoconservation.* Geological Society, London, Special Publications, **300**, 123–135.

Environmental Protection Act. 1990. HMSO.

HARLEY, M. 1989. Regionally Important Geological Sites (RIGS). *Earth Science Conservation,* **26**, 13.

HOBHOUSE, A. 1947. *Report of the National Parks Committee.* HMSO, Cmd. 7121.

HUXLEY, J. S. 1947. *Conservation of Nature in England and Wales.* Report of the Wild Life Conservation Special Committee. HMSO, cmd. 7122.

National Parks and Access to the Countryside Act. 1949. HMSO.

Natural Environment and Rural Communities Act. 2006. HMSO.

NATURE CONSERVANCY COUNCIL. 1990a. *Earth science conservation in Great Britain—A strategy.* 1–84. Nature Conservancy Council, Peterborough.

NATURE CONSERVANCY COUNCIL. 1990b. *Regionally Important Geological/geomorphological Sites.* Nature Conservancy Council. A leaflet.

OFFICE OF THE DEPUTY PRIME MINISTER. 2005. *Planning Policy Statement 9: Biodiversity and Geological Conservation.* The Stationery Office.

PROSSER, C. 2002. Aggregates Levy—tax windfall. *Geoscientist,* **12**, Geological Society of London, 16–17.

PROSSER, C. 2004. Geological and landscape conservation—funding at last. *In:* PARKS, M. A. (ed.) *Natural and Cultural Landscapes—The Geological Foundation.* Proceedings of a conference, 9–11 September 2002, Dublin Castle, Ireland. Royal Irish Academy, Dublin, 179–182.

PROSSER, C. D. & KING, A. H. 1998. Regionally Important Geological and Geomorphological Sites: the origin and a forward view. *In:* OLIVER, P. G. (ed.) *Proceedings of the first UK RIGS conference.* Hereford and Worcestershire RIGS Group, Worcester, 1–8.

PROSSER, C. D. & LARWOOD, J. G. 2008. Conservation at the cutting-edge: the history of geoconservation on the Wren's Nest National Nature Reserve, Dudley, England. *In:* BUREK, C. V. & PROSSER, C. D. (eds) *The History of Geoconservation.* The Geological Society, London, Special Publications, **300**, 217–235.

PROSSER, C., MURPHY, M. & LARWOOD, J. 2006. *Geological Conservation: a Guide to Good Practice.* English Nature, Peterborough.

SCOTT, L. 1942. *Report of the Committee on Land Utilisation in Rural Areas.* Ministry of Works and Planning. HMSO, cmd. 6378.

THOMAS, B. A. 2005. The palaeobotanical beginnings of geological conservation: with case studies from the USA, Canada and Great Britain. *In:* BOWDEN, A. J., BUREK, C. V. & WILDING, J. (eds) *History of Palaeobotany: Selected Essays.* The Geological Society, London, Special Publications, **241**, 95–110.

THOMAS, B. A. & WARREN, L. M. 2008. Geological conservation in the nineteenth and early twentieth centuries. *In*: BUREK, C. V. & PROSSER, C. D. (eds) *The History of Geoconservation*. The Geological Society, London, Special Publications, **300**, 17–30.

THOMAS, B. A & CLEAL, C. J. 2005. Geological conservation in the United Kingdom. *Law, Science and Policy*, **2**, 269–284.

Wildlife and Countryside Act. 1981. HMSO, London.

A history of the Geological Conservation Review

NEIL ELLIS

Geological Conservation Review Publications Manager, Joint Nature
Conservation Committee, Monkstone House, City Road, Peterborough PE1 1JY, UK
(e-mail: neil.ellis@jncc.gov.uk)

Abstract: A particularly ambitious programme for overhauling site assessment and
documentation for geoconservation in Great Britain was initiated in the mid-1970s by the
Nature Conservancy Council (NCC) resulting in the formal launch of the 'Geological Conservation Review' (GCR) in 1977. The GCR was a world-first project of its type in the assessment of the
whole geological heritage or 'geodiversity' of a country from first principles. Criteria and assessment methods were developed, with a view to selecting the very best sites to represent the diversity
of British geology and geomorphology. A list of selection categories—GCR 'Blocks'—encompassing British geology and geomorphology was devised. Widespread consultation with geologists and geomorphologists across Great Britain was co-ordinated; and the criteria refined and
interpreted to suit the selection category at hand. Field investigation of proposed sites was a
key component of site selection, although the selection process relied heavily on expert knowledge, literature-review and consensus-building. Almost 3000 sites had been selected for around
100 site selection categories for the GCR 'register' by 1990. Most GCR sites are now conserved
under British law as Sites of Special Scientific Interest. In Great Britain, the GCR has formed the
'benchmark' for attainment of national importance (in a British context), rather than regional
importance; a 'minimum number of sites' criterion is enshrined in the GCR ethos to assist in defining this benchmark. As part of the site-selection process, a considerable archive of information
about sites was amassed. A major publication exercise detailing all of the GCR sites in what
was to become the GCR series of books was devised early on in the GCR programme of work.
An electronic GCR database was also created in the 1980s and has been the subject of ongoing
revision. Although this paper has a historical outlook, the GCR project is not intended to be a finished record of Britain's best sites up to 1990. Instead, the GCR register will keep pace with new
discoveries and developments in geological research. In the future, further re-evaluation and confirmation of the 'conservation value' of each site is envisaged; assessment of additional sites will
also take place, so that the GCR designation can continue to be a hallmark for quality in geoconservation for years to come.

The origins of the GCR

Examples of the *ad hoc* conservation of geological
sites in the UK go back more than a century.
However, the first systematic programme for identifying our best sites for geoconservation in Great
Britain began in the 1950s, using British legislation
to protect them as 'Sites of Special Scientific Interest' (SSSIs), alongside special biological (habitat
and species) features.

A significant innovation in wildlife conservation
in Great Britain was the Nature Conservation
Review (NCR) project, which described areas of
national importance. Its stated aim was to identify
those sites that were of 'greatest value to wildlife
conservation', that is 'most highly concentrated'
in fauna and flora or of 'highest quality'. The
results of the NCR were published in 1977
(Ratcliffe 1977), and the NCR information was
highly valuable in justifying the conservation case
for wildlife and habitat Sites of Special Scientific
Interest (SSSIs), protected under British law.

The NCR was important in helping, by analogy
(Black 1978*a*), to renew efforts in reviewing geological site-selection for conservation purposes in
Britain. Work on revising geoconservation SSSI
lists had already started in the mid-1960s (Black
1969–1977, 1976, 1977*a*, *b*) edited publications
of site reviews on a regional, rather than thematic,
basis), but the geoconservation site lists and information store still needed a significant overhaul.
There were about 1300 geological SSSIs in the
mid-1970s (Wimbledon *et al.* 1995).

A new review programme was envisaged, from
first principles, in the early 1970s, effectively super-
ceding the geo-SSSI revision programme (Black
1978*a*). Perhaps capitalizing on the 'NCR' label
to some extent, a 'GCR', or Geological Conservation Review plan was drawn up by the Nature
Conservancy Council, and approved in December
1974 (Black 1978*a*). However, the NCR and GCR
methodologies, though broadly analogous, were
quite different in detail, particularly in respect
of a 'national' (British) rather than 'regional'

From: BUREK, C. V. & PROSSER, C. D. (eds) *The History of Geoconservation*.
Geological Society, London, Special Publications, **300**, 123–135.
DOI: 10.1144/SP300.10   0305-8719/08/$15.00 © The Geological Society of London 2008.

perspective and in terms of selecting the minimum number of sites, and selecting sites from first principles. The GCR rationale and methods are explored further below.

A working party, established at the planning stage of the GCR project, headed by D. A. Bassett, investigated how the work should be done, and recognized that three main sections of work would be covered in the GCR: a large geological section, a Quaternary science and geomorphology section, and a relatively small speleoglogical section (Black 1978a). A pilot project, beginning 1975, was conducted for the GCR, in which predominantly Middle Jurassic stratigraphy sites were considered (W. A. Wimbledon, pers. comm.).

Although the GCR project began as a selection process to identify the best sites from first principles, it re-considered those sites that were already geological SSSIs, conserved under the provisions of the National Parks and Access to the Countryside Act 1949 (see Prosser 2008), but without the presumption that all, or indeed any, of those sites would be retained for the GCR. Therefore, the GCR was far more than a re-evaluation exercise: it widened the scope of establishing a new conserved geological site 'register', without being prejudiced by a site's inclusion on any previous list. The result was the first, fully comprehensive, review of the geological resource of an entire country. This review was designed to identify, and help conserve, the sites of national and international importance in Britain. No other country attempted such a systematic and comprehensive review of its geological heritage until many years later.

Some two years after the pilot project, the GCR began officially on 3 October 1977 (Wimbledon et al. 1995); the speleological part of the work had actually begun on 1 October 1977 with the letting of a contract to the national caving associations (National Caving Association and BCRA), according to Black (1978a).

It had become clear following the pilot project that significant staffing and funding would be required to assess the geological resource of Britain to achieve the aims of the GCR. Consistent Government funding to support the GCR project began in 1978 (W. A. Wimbledon, pers. comm.).

The aim of the GCR

The aim of the GCR was to assess systematically the scientific part of the geological heritage of Great Britain and to select for conservation those localities that exceed a minimum threshold in their national (British) value to Earth science. It encompassed geology, geomorphology and speleology; fossil, but not present-day soils were included, except where they capped Quaternary deposits.

The rationale was to select sites that would '*represent comprehensively the geological history of Britain and demonstrate the range and diversity of the best Earth science sites*' in the country. Implicit in the project was the gathering of supporting information at the same time as the inventory of the very best sites was being created (Wimbledon et al. 1995).

The GCR was intended to cover the scientific part of our geological heritage only. This is not to say that GCR sites are not also important for other reasons, such as aesthetic appeal or historical significance, but the key concept was that the GCR identified the best sites for scientific research, regardless of aesthetic appeal, ease of access, or value in education. However, most, if not all, of Great Britain's most picturesque geological features are conserved, coincidentally, within the GCR site series, or within National Parks, or are conserved through local protection measures such as nature reserves (see Burek 2008). However, by choosing a scientific basis for the GCR, it has ensured that sites of value to research would be conserved, not just those of 'geotourism' value. The situation could be compared to wildlife conservation: small, aesthetically unattractive plants and animals may have limited appeal to the general public, but are none the less an important part of the natural world and are given conservation protection where necessary.

The Nature Conservancy Council (NCC)—the actual Council itself of that former Government agency—decided to carry out the GCR project in 1974, (with the pilot project beginning in 1975 and a formal launch in 1977). A later decision to make sure that the sites would be conserved as SSSIs was also made by the NCC (W. A. Wimbledon, pers. comm., see Prosser 2008 for a discussion of the SSSI system).

But how are 'best', 'scientifically important' and 'national importance' defined? How does the conservation value of a site with an outcrop of an unusual type of granite compare to a site with relict ice-age shorelines? Clearly, a framework and criteria for GCR site selection were required, and these are outlined below.

Principles of site selection

Framework

GCR blocks. In dividing up the geology and geomorphology of Great Britain into categories against which sites could be selected, around

100 topics were ultimately used. The number was intended originally to be fewer at the outset ('around 40' is cited in Black 1978a, but it is not clear therein whether this number was restricted to geology sites, aside from Quaternary, geomorphology and cave sites). As site-selection work progressed (Black 1978a–i, 1979a–g, 1980a–d, 1982a, b, 1983a–c, 1984, 1985a, b, 1987a, b, 1988a–e), a larger number of categories proved to be more workable, perhaps reflecting division of originally wider categories into smaller ones based more on geological ages than on geological periods.

The GCR categories became known as 'blocks', and they provided an overall structure for site selection and ensured that the different themes of Earth science would receive equal consideration. They can be grouped into seven major themes, listed in Table 1.

Although the relatively common invertebrate fossils do not have a separate selection category in the GCR in their own right, the scientific importance of many stratigraphy sites lies in their fossil content. Therefore, some stratigraphy GCR sites were selected specifically for their fauna that facilitates stratal correlation and enables the interpretation of the environments in which the animals lived.

In contrast to the manner in which most invertebrate fossils are represented in the GCR, fossils of vertebrates, arthropods (except trilobites, which are more common) and plants do have their own dedicated selection categories, owing to the relative rarity of the fossil material.

There was originally an intention to have a 'historical block' (featuring sites considered to be nationally important in the history of geology, see Wimbledon 1991); it was eventually abandoned, since it was believed that the sites that would have been included there would be selected in the geologically themed blocks already.

More detail is given about the derivation of blocks in the GCR in Wimbledon *et al.* 1995 (p. 171–173).

Criteria

GCR site types. Three distinct, but complementary, types of site have been selected for the GCR:

- sites of importance to the international community of Earth scientists;
- sites that are scientifically important because they contain rare or exceptional features; and
- sites that are nationally important because they are representative of an Earth science feature, event or process that is fundamental to understanding Britain's Earth history. 'Nationally important' in the context of GCR site selection refers to importance

to Great Britain as a whole, and means that any site chosen for the GCR has been compared with similar features, where they exist, across the whole of Great Britain.

Therefore, each site selected for the GCR is of at least national (British) importance for geological conservation, and many of the sites are of international importance. The key component is 'representative' sites, ensuring that the coverage of localities selected for the GCR is comprehensive in representing the highlights of British geology and geomorphology.

International importance. This component for GCR site selection ensures that sites of international importance are included to ensure that international responsibilities are met. Five main types of internationally important GCR site can be recognized:

- interval or boundary stratotypes;
- type localities for biozones (rock strata that are characterized by a closely defined fossil content, usually fossil species) and chronozones (rock strata formed during the time-span of the relevant stratotypes);
- internationally significant type localities for particular rock types, mineral or fossil species;
- historically important type localities where rock or time units were first described or characterized, or where great advances in geological theory were first made (e.g. Hutton's unconformity at Siccar Point, Berwickshire, Scotland); and
- important localities where geological or geomorphological phenomena were first recognized and described, or where a principle or concept was first conceived or demonstrated (e.g. Cauldron subsidence at Glencoe, Scotland).

Exceptional features. Many sites have unique, rare or atypical features. For example, at Rhynie in Scotland, a mineralized peat, in the form of chert of Devonian age, preserves a detailed record of an early land ecosystem. The level of microscopic detail preserved in the Rhynie fossils makes this site exceptional. No other sites are known in Britain— or, indeed, the world—to contain such well-preserved fossil material of this age. This makes the Rhynie Chert site irreplaceable. The inclusion of exceptional sites in the GCR ensures that the highlights of British geology and geomorphology are conserved.

Representativeness. Important though international and exceptional sites are, alone they cannot provide the basis for a systematic approach for the selection of sites to cover the essential features of the geological heritage of Great Britain. Therefore the key component of the GCR is provided by the

Table 1. *The site selection categories ('blocks') of the GCR and numbers of sites (as recognized by country conservation agencies at the time of writing). The blocks are grouped into seven themes*

	No. of GCR sites
Stratigraphy (geological time and lithostratigraphy)	
Palaeogene	38
Neogene	26
Cenomanian, Turonian, Senonian, Maastrichtian	37
Aptian–Albian	32
Wealden	43
Berriasian, Valanginian, Hauterivian, Barremian	11
Portlandian–Berriasian	30
Kimmeridgian	13
Oxfordian	31
Callovian	20
Bathonian	48
Aalenian–Bajocian	51
Toarcian	16
Hettangian, Sinemurian and Pliensbachian	30
Rhaetian	13
Permian–Triassic	51
Marine Permian	27
Westphalian	73
Namurian of England and Wales	51
Dinantian of southern England and South Wales	29
Dinantian of Scotland	33
Dinantian of Northern England and North Wales	66
Dinantian of Devon and Cornwall	9
Non-marine Devonian	64
Marine Devonian	55
Prídolí	8
Ludlow	37
Wenlock	47
Llandovery	37
Caradoc–Ashgill	37
Tremadoc	9
Arenig–Tremadoc	5
Arenig–Llanvirn	37
Llandeilo	13
Cambrian–Tremadoc	5
Cambrian	29
Precambrian of England and Wales	36
Palaeontology (based on fossil vertebrates, arthropods excluding trilobites, and plants)	
Tertiary palaeobotany	8
Mesozoic palaeobotany	25
Palaeozoic palaeobotany	44
Pleistocene Vertebrata	42
Tertiary Mammalia	8
Mesozoic Mammalia	10
Aves	8
Tertiary Reptilia	5
Jurassic–Cretaceous Reptilia	29
Permian–Triassic Reptilia	18
Mesozoic–Tertiary Fish/Amphibia	33
Carboniferous–Permian Fish/Amphibia	11
Silurian–Devonian Chordata	56
Palaeoentomology	18
Arthropoda	16
Precambrian palaeontology	7

(Continued)

Table 1. *Continued*

	No. of GCR sites
Quaternary geology ('Ice Age' landforms, stratigraphy, eustasy and isostasy, tufa)	
Quaternary of South-West England	38
Quaternary of Somerset	22
Quaternary of South Central England	22
Quaternary of the Thames	37
Quaternary of South-East England	25
Quaternary of East Anglia	76
Quaternary of Midlands–Avon	20
Quaternary of Wales	72
Quaternary of the Pennines and Adjacent Areas	30
Quaternary of North-East England	30
Quaternary of Cumbria	18
Quaternary of Scotland	140
Geomorphology (the landforms and processes that form the current landscape)	
Caves	49
Karst	42
Coastal geomorphology of England	46
Coastal geomorphology of Scotland	42
Coastal geomorphology of Wales	13
Fluvial geomorphology of England	36
Fluvial geomorphology of Scotland	27
Fluvial geomorphology of Wales	19
Mass movement	35
Igneous petrology (petrology relating to major tectonic events)	
Tertiary igneous	52
Igneous rocks of south-west England	54
Carboniferous–Permian igneous rocks	51
Silurian and Devonian volcanic rocks	16
Silurian and Devonian plutonic rocks	56
Ordovician igneous rocks	61
Structural and metamorphic geology (relating to major orogenic events; also encompasses Precambrian–Cambrian rocks in Scotland that have been significantly deformed)	
Alpine structures of southern England	10
Variscan structures of south Wales and the Mendips	15
Variscan structures of south-west England	31
Caledonian structures of Wales	19
Caledonian structures of the Southern Uplands	9
Caledonian structures of the Lake District	12
Torridonian	11
Moine	76
Lewisian	24
Dalradian	73
Mineralogy (largely based on 'ore provinces')	
Mineralogy of the Lake District	23
Mineralogy of the Mendips	11
Mineralogy of Peak District, Leicestershire, Cheshire and Shropshire	17
Mineralogy of the Pennines	18
Mineralogy of Scotland	55
Mineralogy of south-west England	34
Mineralogy of Wales	44

selection of sites representative of features: events and processes that are fundamental to our understanding of the geological history of Great Britain. With this component, some GCR sites, though less famous than 'classic sites', are no less important since they are a component piece of the 'jigsaw puzzle' of Great Britain's geological history. It is important that representative sites for a GCR block are not selected in isolation from each other: the sites are collectively important in piecing together geological scenarios.

GCR—criteria refinements

How many sites are necessary to represent comprehensively British geology? The GCR aimed to be fully representative of British geology, but this task resembled trying to divide a continuum into component, representative parts.

To help overcome this problem, the GCR is a minimalist scheme.

Minimum number and minimum area of sites. In order to ensure that GCR site status is confined to sites of national importance, the number of sites selected is restricted to the minimum necessary to characterize the GCR block (Black 1978a). Therefore there is a minimum of duplication of features of interest between sites. In this way, the scientific case for conserving a given site is stronger if it is the only one of its kind, or if it is demonstrably the best of a group of similar examples.

The area of a GCR site is always kept to a minimum. For example, in tracing the form of a major structure over a distance of several kilometres, a small number of dispersed, representative 'sample' sites might be selected—the minimum number and size required to describe and interpret the feature adequately. There are, however, exceptions to this general rule: for example, large sites will be required to represent the range of large-scale glacial landforms in the uplands of Wales or Scotland. In contrast, mine spoil heaps, typically of limited size, normally form relatively small sites.

Preferential weightings. All scientific factors being equal, sites that cannot be conserved or that entirely or largely duplicate the interest of another are excluded. Sites that are least vulnerable to potential threat, and are more accessible, are preferred.

Therefore, preference is given to sites that:

- demonstrate an assemblage of geological features or scientific interests;
- show an extended, or relatively complete, record of the feature of interest. In the case of geomorphological sites this often equates to sites that contain features that have been least altered by human activity after formation. For Quaternary subjects, this might relate to sites containing an extended fossil record, including pollen, insects and molluscs, vegetation history or environmental change;
- have been studied in detail and which have a long history of research and reinterpretation;
- have potential for future study; and
- have played a significant part in the development of the Earth sciences, including former

reference sites, sites where particular British geological phenomena were first recognized, and sites that were the focus of studies that led to the development of new theories or concepts.

Application of these criteria ensured that sites chosen for a GCR block have the greatest collective scientific value and can be conserved in a practical sense. Wimbledon *et al.* (1995: appendix 1) provide further explanation of these criteria.

To support the rationale of identifying a minimum number of representative sites a 'network approach' to selection was used, whereby sites were not selected in isolation from each other, but were considered in tandem with other candidate localities (Wimbledon *et al.* 1995, p. 171; Ellis *et al.* 1996, pp. 53–69).

Expertise

The original proposal for GCR site selection was that it was 'to be prepared in-house, largely by using the expertise of temporarily employed staff specially employed on the project, mostly recently-graduated Ph.D. students. The *Review* has been divided into about forty units [proto-GCR blocks] and each is to be entrusted to a single geologist recruited specifically for the relevance of his postgraduate experience' (Black 1978b).

Wimbledon *et al.* (1995) explained this in more detail. Although selection of sites for each block was typically under the stewardship of one individual, external expertise was critically important and always involved: private individuals, universities, research bodies and consultancies were all engaged as necessary. In some cases, the work was achieved by co-ordinating opinions of others, as consortia, for example the British Cave Research Association and National Caving Association. In other cases there was a gathering of small groups of experts, or for some blocks, site lists were produced single handedly (where there were very few experts researching British sites for the subject at hand, e.g. P. Allen, for the Wealden GCR block). Although various approaches were taken to make best advantage of the available expertise, consultation was always a golden thread throughout the GCR site selection process, as reported in serial articles in the *Information Circular* (which transformed into the *Earth Science Conservation* journal from 1978, and which was itself was superceded by *Earth Heritage* in 1994) by Black (1978–1988).

Initially it was hoped that the Quaternary sites may have been selected by issuing a contract in a similar way to the caves part of the work, but Black (1978b) reports that it was decided to take on the work in a similar way to the geology sites through in-house co-ordination.

Methods and working practice of the GCR

Black (1978*a*) emphasized that the criteria for site selection for geological conservation had been in the process of refinement since the 1960s, and therefore well-established in principle even before the outset of the GCR.

Three main methods of working had been employed at the early phases of the GCR: the bulk of the review was to be carried out by post-doctoral workers; purchase of manuscripts to gather together candidate site lists was required; and the gathering of speleological knowledge from the two principal, voluntary caving organizations was engaged. The achievement of a consensus of experts was also emphasized by Black (1978*a*).

An undated internal NCC document indicated that by about 1982 enough work had been done to estimate that there would be up to 2700 GCR sites—many of the localities appearing more than once in the GCR list, owing to localities having two or more 'special features' (i.e. qualifying for more than one GCR block).

As described in the document it was anticipated that site selection would be complete at 31 March 1984; in fact, site selection was largely complete by around six years later. The current status of the GCR—an active site register from which sites can be removed and others added after sufficient ratification—is perhaps different from Wimbledon *et al.* 1995 (in their abstract), which stated that it intended the list to be a 'Domesday Book' of sites (which could be considered to be a once-and-for-all-time record, or alternatively, a 'benchmark' record for comparison later), although they make no further comment about GCR updating in the way that Ellis *et al.* (1996) do.

Work on particular GCR blocks typically followed four stages:

Stage 1: Building and briefing the GCR block team. A co-ordinator for a particular block would advise on site selection criteria and collate the work of a number of consultees. Devising a framework within the block (identifying the geological scope) was undertaken at this stage.

Stage 2: Literature review and site shortlisting. Extensive literature search was undertaken to create a list of sites of potential national or international importance. Each of the sites on the draft list was given standard basic documentation (e.g. site location [a site boundary enclosing the important features of the site, drawn on 1:10 000 Ordnance Survey map], brief summary of scientific interest, possible justification for inclusion—a concise statement, typically between 100 and 200 words in length). Site lists were then circulated among the appropriate experts for critical assessment and comment and a list of potential GCR sites was drawn up. In the case of the Jurassic–Cretaceous Reptilia GCR block, 380 sites were identified from the literature as being potentially special; this number was reduced to about 150 after this initial assessment process (Benton & Spencer 1995, chapter 1).

Stage 3: Field visits and detailed site investigation. At this stage, sites where irreversible damage or deterioration of the features of interest had taken place were dropped from the list.

Stage 4: Final assessment and preparation of GCR site documents. Further scrutiny by consultees, aiming to select the minimum number of representative sites, resulted in a final list of GCR sites for the block. For example, from the list of 150 shortlisted potential Jurassic–Cretaceous Reptilia sites, a final list of 28 confirmed GCR sites was produced.

After over two decades of site evaluation and documentation, the GCR methods and scientific rigour led to the selection of over 3000 sites to form the GCR register. Further details of the GCR criteria and selection processes can be found in Wimbledon *et al.* (1995) and Ellis *et al.* (1996).

GCR site information

Publication of GCR information, in the form of site reports, was always considered to be an integral part of the GCR project (Wimbledon *et al.* 1995, p. 181), effectively producing an encyclopaedic coverage of conserved geological sites in Britain. The intention was that each text would contain a review of the current state of knowledge of the sites in order to provide a definitive scientific reference source on which practical site conservation could be founded and justified. Wimbledon *et al.* (1995) stress that the GCR was intended to be a selection and publication programme exercise carried out in tandem. A standard structured format for the site reports was devised, consisting of an introduction, description, interpretation and conclusion (the conclusion is a justification of conserving the site in plain terms, a statement of the site's key scientific values). A series of around 50 titles for the GCR book series was envisaged at the outset, although some modifications to the intended list subsequently took place such that 44 books are now intended (see Table 2 and bibliography).

Although the very basic GCR information (statement and map for each site) was an essential element of GCR selection practices during consultation about candidate sites, and the foundation of a GCR 'filing system', much more information was needed for the formally selected sites if they were to be conserved using British law, and

Table 2. *List of GCR volumes*

Precambrian and structural geology
Caledonian Structures in Britain (published; volume 3)
Lewisian, Torridonian and Moine Rocks of Scotland (in press)
Dalradian Rocks of Scotland (in prep.)
Precambrian Rocks of England and Wales (published; volume 20)
Fossil Arthropods of Great Britain (in press)
Variscan to Alpine Structures in Britain (in prep.)
Igneous petrology and mineralogy
British Tertiary Volcanic Province (published; volume 4)
Igneous Rocks of South-West England (published; volume 5)
Caledonian Igneous Rocks of Britain (published; volume 17)
Carboniferous and Permian Igneous Rocks of Great Britain (published; volume 27)
Mineralization in England and Wales (in press)
Mineralogy of Scotland (in press)
Palaeozoic stratigraphy
Marine Permian of England (published; volume 9)
British Upper Carboniferous Stratigraphy (published; volume 11)
British Cambrian to Ordovician Stratigraphy (published; volume 18)
British Silurian Stratigraphy (published; volume 19)
British Lower Carboniferous Stratigraphy (published, volume 29)
Old Red Sandstone Rocks of Great Britain (published, volume 31)
British Marine Devonian Stratigraphy (in prep.)
Mesozoic–Cenozoic stratigraphy
British Tertiary Stratigraphy (published; volume 16)
Permian and Triassic Red Beds and the Penarth Group of Great Britain (published; volume 24)
British Upper Jurassic Stratigraphy (published; volume 21)
British Middle Jurassic Stratigraphy (published; volume 26)
British Lower Jurassic Stratigraphy (published, volume 30)
Jurassic–Cretaceous Boundary Rocks in England (in prep.)
British Upper Cretaceous Stratigraphy (published; volume 23)
British Marine Lower Cretaceous Stratigraphy (in prep.)
Palaeontology
Fossil Reptiles of Great Britain (published; volume 10)
Palaeozoic Palaeobotany of Great Britain (published; volume 9)
Fossil Fishes of Great Britain (published; volume 16)
Mesozoic and Tertiary Palaeobotany of Great Britain (published; volume 23)
Mesozoic and Tertiary Fossil Mammals and Birds of Great Britain (published, volume 32)
Pleistocene Vertebrate palaeontology of Great Britain (in prep.)
Geomorphology
Karst and Caves of Great Britain (published; volume 12)
Fluvial Geomorphology of Great Britain (published; volume 13)
Mass Movements in Britain (published; volume 33)
Coastal Geomorphology of Great Britain (published 28)
Quaternary geology and geomorphology
Quaternary of Wales (published; volume 2)
Quaternary of Scotland (published; volume 6)
Quaternary of the Thames (published; volume 7)
Quaternary of South-West England (published; volume 14)
Quaternary of East Anglia and Midlands (in press)
Quaternary of Northern England (published, volume 25)
Quaternary of Southern England (in prep.)

For more information about the books, see the JNCC website www.jncc.gov.uk, or the GCR volume distributor's website www.nhbs.com, or write to GCR Unit, JNCC, Monkstone House, City Road, Peterborough, PE1 1JY.

indeed, would be needed for the planned publication purposes. Therefore, for sites accepted onto the GCR register, longer statements (longer than those described in 'Stage 2' above) describing the scientific importance of the selected sites and citing key references from the literature, were commissioned from relevant experts in the 1980s, typically resulting in two or three pages of A4 typewritten text per site. It was envisaged that the site reports would be '200 to 3000 words in length and would amount to 4 million words for the whole site series' (anonymous unpublished internal NCC document). The intention then was not to include much illustrative material with the text for publication purposes, but that it be supported on the basis of a comprehensive bibliography (Black 1979g). Such information became critically important in the notification of GCR sites as SSSIs, and provided key source material for GCR volumes published in the following years.

Much writing and editorial work with publication in mind was well advanced during the later part of the site-selection programme in the 1980s, under the guidance of the GCR Editor-in-Chief, W. A. Wimbledon.

Dissemination

In the early stages of the GCR project, there was some discussion as to what type of GCR publication should be presented in the pubic domain. It was considered by some that vulnerable localities (such as rare mineral/fossil sites) might come under additional threat if details of location were made too easily available. A letter from distinguished geologists from University of Cambridge to the NCC stated their view that it would be unwise to publish details of GCR sites, fearing that it would lay those vulnerable localities open to irresponsible use or undue pressure from specimen collectors. They preferred that the reports be kept for internal purposes, perhaps made available for public view in 'open files' on request. The matter was discussed with experts, staff and the NCC's Council, but ultimately it was decided to publish the GCR in full. In hindsight it can now be seen that the publication of GCR information has not increased the impact on irresponsible use of GCR sites.

The intended audience for GCR publications consists of those who come into contact with GCR sites in a professional or voluntary capacity, including Earth scientists, for whom the sites are a key resource for geological research, conservation agency staff whose job it is to ensure the sites are well managed, periodically monitored to check their condition, and conserved. However, the introductions chapter and site report conclusions are intended to be understood by non-specialists (through the use of a glossary), whereas individual chapter introductions and site report descriptions and interpretations require use of more technical geological terminology.

With the publication of the first GCR volume in 1989 (Campbell & Bowen 1989) it was firmly established that the GCR information would be placed in the public domain.

There had been some problems with resources (physical and monetary) for the publications

work, and a hiatus as NCC became reformed into three new country conservation agencies and a Joint Nature Conservation Committee created, with ongoing work. However, JNCC took up the GCR publication and information management work in 1990.

The undertaking of the publication programme for the GCR 'Series' of 44 volumes is still ongoing, but 34 volumes have been produced so far (Table 2).

Authorship of GCR publications. In many cases, JNCC commissioned authors to write reports by invitation, in exchange for professional fees. Sometimes, where several individuals or consortia would have been capable of producing a text, Government Treasury requirements were that a 'competitive tender' or 'bidding' system for doing the work was used, although the contract was not always issued to the bidder offering the lowest price— scientific credibility was tantamount. Where possible those people heavily involved with the site-selection process were asked to prepare site reports for publication.

Typically, a book was written by several specialists, under the stewardship of a senior author, under a GCR Editor-in-Chief (W. A. Wimbledon until 1992), and then a GCR Publications Manager (N. V Ellis, the present author) thereafter. JNCC has issued over 100 contracts to write materials for GCR books, combining individual scripts into a single typescript as necesary.

The series of 44 books is based primarily on geological topics, rather than geographical areas, grouping related blocks into manageable subjects suitable for book publication, for example *Caledonian Igneous Rocks* and *Fossil Reptiles of Great Britain* (each covering three GCR blocks).

Problems faced by the GCR publications project

Black (1988e) reviewed some the contemporary problems and opportunities of the GCR project. No project, particularly one on such a grand scale, is ever without its difficulties. Particular problems for the GCR publications/report commissioning work have been as follows.

Funding. Aside from the costs of selecting sites for the GCR, such as the payment of 'block co-ordinators', the cost of GCR writing work is estimated at £1 million. The further cost of publication work is estimated at £500 000 for the whole series, which is offset only partly by book sales. Although this is a not insignificant sum, it does show the high regard that UK nature conservationists at Government level place in geological conservation and the information required to support it. Of course the funding was provided over a long period of time, with publication and writing costs being spread out over nearly 20 years. These figures do not include the costs of site-selection work.

Expertise. Availability of expertise has been a limiting factor, particularly with areas of geology that have attracted fewer workers over the years. In many cases there were only one or two potential individuals with thorough enough knowledge to tackle a particular title of the GCR series. A GCR volume typically takes three to four years to write with someone contracted to do the job who has to balance many other work activities, including research, other contracts and teaching commitments.

Publishing mechanism. At first a commercial publisher was involved with the early publications (Chapman & Hall and successor companies (ITP and Kluwer) for volumes 3 to 14), although the first volume, number 2 of the series, was published by the Nature Conservancy Council directly in 1989 (Campbell & Bowen, 1989). However the commercial publisher eventually pulled out, the titles not being sufficiently profit making. Therefore JNCC was forced either to publish the books itself at cost, or not publish the material at all. Additional funds were secured, and with the advent of digital printing techniques, publication costs were kept to an absolute minimum. Volumes number 15 onwards have been published by JNCC directly.

Furthermore, various programmes for writing and publication work suffered significant delays and setbacks—sometimes as a result of Governmental reviews of conservation priorities, in which the place of the GCR work was re-evaluated. In each of these reviews the value of the GCR hallmark, and indeed the publications themselves, was re-confirmed.

Success in dissemination of GCR information resource. Owing to the costs associated with producing a high quality, but lasting, book product aimed primarily at specialist libraries, the goal of wide public dissemination has not been achieved to the extent that was envisaged. Fewer than 300 copies of each title have been sold, although these books have been sold mainly to libraries for wider use. Some volumes have proved more popular than others, for example the *Fossil Reptiles* volume is out of print.

A further potential drawback is that geological libraries store GCR volumes by topic, so that they will not appear together on the shelf in the way that they were intended; such that the full weight of the GCR 'hallmark' is diluted. Also, owing to the thematic basis, many potential buyers of the books would have been disappointed, because geologists typically require information on their local area, i.e. on a geographical or gazetteer basis. Thus, although the volumes are in the public domain, the target audience has not been reached thoroughly.

Other forms of GCR information management

Further to the commissioning of site reports for publication, a GCR database was begun in the 1980s, to digitize the GCR 'filing system'. Originally this was a simple digital version of the GCR site record sheets. Through subsequent upgrades and additions, and analysis of requirements of conservationists and research scientists, it has been possible to publish this to the World Wide Web.

JNCC has also begun the process of making extracts from published GCR books freely available. PDF files of Chapter 1 of each published volume (2–33, plus most of the introduction volume, No.1 of the GCR Series, (Ellis *et al.* 1996), have been made available on the JNCC website (see e.g. http://www.jncc.gov.uk/earth-heritage/gcr/PDFs/V10/V10chap1.pdf). Work has already begun to convert all of the text and illustrations prepared for publication in book form to a web-based format in due course, which will be published to the JNCC website.

GCR future

The intention has always been that the GCR would be subject to modification, that is, a register subject to change that will be kept up-to-date. Just as a rare plant or animal may enter the 'endangered' list, be discovered in new locations, or recover to levels where it is no longer endangered, new geological discoveries, land-use development, and changes in Earth science theories demand that a site list of the best Earth science sites in a country be subject to review and change if it is to remain up-to-date and to continue to have value as a hallmark for quality. Therefore the GCR project will continue beyond the publications programme. In practice there will be three strands of work: reviewing of GCR site lists; management of GCR information (keeping databases and information stores up-to-date); and further dissemination of the GCR information that have been amassed.

Conclusion

The GCR project has provided the 'backbone' of Earth science conservation in Great Britain since the 1980s. Not only has it resulted in a comprehensive catalogue of Britain's best Earth science sites that will be conserved for future generations, but a large amount of information about the geological history of Britain and geological research has been collected and published. The work has also formed a fraternity of geologists and a group of people with an interest in geoconservation. This GCR information resource is the basis of many geoconservation activities in the UK. The information for 66 GCR sites made a significant contribution to the scientific case for the Dorset–East Devon World Heritage Coast proposal—inscribed for its geology and geomorphology, and has made the scientific case of several Geopark applications.

One of the main aims of the GCR was to make publicly available the reasons for selection of each GCR site, through the publication of site reports, and the management of site data in a database. The information is available in a series of published books: the GCR series.

When complete, the GCR series will be the essential reference and information source about the GCR sites. In total the series provides encyclopaedic coverage of British geology. The publication of the final volume, in 2008, will represent the culmination of this massive body of work. But the completion of the GCR series of books is not the end of the GCR project. It is envisaged that new GCR sites will be identified in the future, and new information about existing GCR sites will be collected. The whole dataset needs to be managed in an accessible central database that is freely available.

The future of the GCR project will be guided by the geological community and those interested in GCR sites, to ensure that such people have the tools and information they need to suit their purposes in geoconservation in Britain.

It is certain that new challenges will face conservation in the future as a result of climate change, land-use changes and pressures, and possibly new, more-holistic approaches to conservation activity that embraces plants, animals, geology/geomorphology, aesthetic and archaeological/historical significance together. The GCR provides a firm foundation from which to tackle such challenges.

References

ALDRIDGE, R. J., SIVETER, D. J., SIVETER, D. J., LANE, P. D., PALMER, D. & WOODCOCK, N. H. 2000. *British Silurian Stratigraphy*. GCR Series No. 19. Joint Nature Conservation Committee, Peterborough.

BARCLAY, W. J., BROWNE, M. A. E., MCMILLAN, A. A., PICKETT, E. A., STONE, P. & WILBY, P. R. 2005. *The Old Red Sandstone of Great Britain*. GCR Series No. 31. Joint Nature Conservation Committee, Peterborough.

BENTON, M. J., COOK, E. & TURNER, P. 2002. *Permian and Triassic Red Beds of Great Britain and the Penarth Group*. GCR Series No. 24. Joint Nature Conservation Committee, Peterborough.

BENTON, M. J. & SPENCER, P. S. 1995. *Fossil Reptiles of Great Britain*. GCR Series No. 10. Chapman and Hall, London.

BLACK, G. P. (ed.) 1969–77. Revision of geological SSSIs. Information Circular of the Nature Consevancy Council, **3**, 4; **7**, 2; **8**, 10; **9**, 1–2, 3–4; **11**, 5–6; **12**, 10–11; **13**, 8.

BLACK, G. P. (ed.) 1976. *Shetland: localities of geological and geomorphological importance*. Nature Conservancy Council, Newbury.

BLACK, G. P. (ed.) 1977a. *Outer Hebrides: localities of geological and geomorphological importance*. Nature Conservancy Council, Newbury.

BLACK, G. P. (ed.) 1977b. *East Anglia: localities of geological and geomorphological importance*. Nature Conservancy Council, Newbury.

BLACK, G. P. 1978a. The Geological Conservation Review. *Information Circular of the Geology and Physiography Section, Nature Conservancy Council*, **14**, 14–15.

BLACK, G. P. (ed.) 1978b. Progress of The Geological Conservation Review. *Earth Science Conservation*, **15**, 1.

BLACK, G. P. (ed.) 1978c. Progress of The Geological Conservation Review – Arenig and Llanvirn Series. *Earth Science Conservation*, **15**, 2.

BLACK, G. P. (ed.) 1978d. Progress of The Geological Conservation Review – Namurian of England and Wales. *Earth Science Conservation*, **15**, 2–4.

BLACK, G. P. (ed.) 1978e. Progress of The Geological Conservation Review – Lower and Middle Lias. *Earth Science Conservation*, **15**, 4–6.

BLACK, G. P. (ed.) 1978f. Progress of The Geological Conservation Review – The Aalenian and Bajocian stages. *Earth Science Conservation*, **15**, 6–8.

BLACK, G. P. (ed.) 1978g. Progress of The Geological Conservation Review – Upper Cretaceous. *Earth Science Conservation*, **15**, 8–9.

BLACK, G. P. (ed.) 1978h. Progress of The Geological Conservation Review – Tertiary Igneous. *Earth Science Conservation*, **15**, 9.

BLACK, G. P. (ed.) 1978i. Progress of The Geological Conservation Review – Caves. *Earth Science Conservation*, **15**, 9–11.

BLACK, G. P. (ed.) 1979a. Progress of The Geological Conservation Review – Wenlock Series *Earth Science Conservation*, **16**, 5–6.

BLACK, G. P. (ed.) 1979b. Progress of The Geological Conservation Review – The Dinantian of Northern and Central England and Wales. *Earth Science Conservation*, **16**, 6–8.

BLACK, G. P. (ed.) 1979c. Progress of The Geological Conservation Review – Pleistocene Deposits of East Anglia. *Earth Science Conservation*, **16**, 8–10.

BLACK, G. P. (ed.) 1979d. Progress of The Geological Conservation Review – Quaternary and Geomorphological Sites in Scotland and Wales. *Earth Science Conservation*, **16**, 10.

BLACK, G. P. (ed.) 1979e. Progress of The Geological Conservation Review – Localities of Palaeobotanical Interest. *Earth Science Conservation*, **16**, 10–12.

BLACK, G. P. (ed.) 1979f. Progress of The Geological Conservation Review – Carboniferous and Permian Igneous Acivity in Central and Southern Scotland. *Earth Science Conservation*, **16**, 12–13.

BLACK, G. P. (ed.) 1979g. Progress of The Geological Conservation Review – Bibliography. *Earth Science Conservation*, **16**, 13.

BLACK, G. P. (ed.) 1980a. Progress of The Geological Conservation Review – Fossil Vertebrate Sites. *Earth Science Conservation*, **17**, 3–5.

BLACK, G. P. (ed.) 1980b. Further Progress of The Geological Conservation Review – Cambrian. *Earth Science Conservation*, **18**, 4–6.

BLACK, G. P. (ed.) 1980c. Further Progress of The Geological Conservation Review – Quaternary of the Thames Valley. *Earth Science Conservation*, **18**, 6–7.

BLACK, G. P. (ed.) 1980d. Further Progress of The Geological Conservation Review – Fossil Insect Localities. *Earth Science Conservation*, **18**, 7–8.

BLACK, G. P. (ed.) 1982a. Further Progress of The Geological Conservation Review – Mass Movement Phenomena. *Earth Science Conservation*, **19**, 27–9.

BLACK, G. P. (ed.) 1982b. Further Progress of The Geological Conservation Review – Fossil Reptiles. *Earth Science Conservation*, **19**, 29–31.

BLACK, G. P. (ed.) 1983a. The Geological Conservation Review. *Earth Science Conservation*, **20**, 15–16.

BLACK, G. P. (ed.) 1983b. The Geological Conservation Review – Bathonian. *Earth Science Conservation*, **20**, 16–17.

BLACK, G. P. (ed.) 1983c. The Geological Conservation Review – Westphalian and Stephanian. *Earth Science Conservation*, **20**, 18.

BLACK, G. P. (ed.) 1984. The Geological Conservation Review reveals new dinosaur remains. *Earth Science Conservation*, **21**, 46–7.

BLACK, G. P. (ed.) 1985a. Progress of The Geological Conservation Review – Fossil Reptile Sites. *Earth Science Conservation*, **22**, 29.

BLACK, G. P. (ed.) 1985b. Progress of The Geological Conservation Review – Pleistocene Sites in the Thames-Avon System. *Earth Science Conservation*, **22**, 36.

BLACK, G. P. (ed.) 1987a. Geological Conservation Review – Progress on GCR Site Selection. *Earth Science Conservation*, **23**, 19–21.

BLACK, G. P. (ed.) 1987b. Geological Conservation Review – Publication Plans for the GCR. *Earth Science Conservation*, **23**, 21–2.

BLACK, G. P. (ed.) 1988a. Geological Conservation Review – Progress on GCR Site Selection. *Earth Science Conservation*, **24**, 24–6.

BLACK, G. P. (ed.) 1988b. Geological Conservation Review – The second GCR contributors meeting. *Earth Science Conservation*, **24**, 27.

BLACK, G. P. (ed.) 1988c. Geological Conservation Review – Progress on GCR site Selection, 31 August 1988. *Earth Science Conservation*, **25**, 40–1.

BLACK, G. P. (ed.) 1988d. Geological Conservation Review – Publication Plans for the GCR. *Earth Science Conservation*, **25**, 41–2.

BLACK, G. P. 1988e. Geological conservation: a review of past problems and future promise. *In*: CROWTHER, P. R. & WIMBLEDON, W. A. (eds) *The Use and Conservation of Geological Sites. Special Papers in Palaeontology*, **40**, 1–200.

BRIDGLAND, D. R. 1994. *Quaternary of the Thames*. GCR Series No. 7. Chapman and Hall, London.

BUREK, C. V. 2008. The role of the voluntary sector in the evolving geoconservation movement. *In*: BUREK, C. V. & PROSSER, C. D. (eds) *The History of Geoconservation*. The Geological Society, Special Publication, London, **300**, 61–89.

CAMPBELL, S. & BOWEN, D. Q. 1989. *Quaternary of Wales*. GCR Series No. 2. Nature Conservancy Council, Peterborough.

CAMPBELL, S., HUNT, C. O., SCOURSE, J. D., KEEN, D. H. & STEPHENS, N. 1998. *Quaternary of South-West England*. GCR Series No.14. Chapman and Hall, London.

CARNEY, J. N., HORAK, J. M., PHARAOH, T. C. *ET AL.* 2000. *Precambrian Rocks of England and Wales*. GCR Series No. 20. Joint Nature Conservation Committee, Peterborough.

CLEAL, C. J. & THOMAS, B. A. 1995. *Palaeozoic Palaeobotany of Great Britain*. GCR Series No. 9. Chapman and Hall, London.

CLEAL, C. J. & THOMAS, B. A. 1996. *British Upper Carboniferous Stratigraphy*. GCR Series No. 11. Chapman and Hall, London.

CLEAL, C. J., THOMAS, B. A., BATTEN, D. J. & COLLINSON, M. E. 2001. *Mesozoic and Tertiary Palaeobotany of Great Britain*. GCR Series No. 22. Joint Nature Conservation Committee, Peterborough.

COSSEY, P. J., ADAMS, A. E., PURNELL, M. A., WHITELEY, M. J., WHYTE, M. A. & WRIGHT, V. P. 2004. *British Lower Carboniferous Stratigraphy*. GCR Series No. 29. Joint Nature Conservation Committee, Peterborough.

COX, B. & SUMBLER, M. 2002. *British Middle Jurassic Stratigraphy*. GCR Series No. 26. Joint Nature Conservation Committee, Peterborough.

DALEY, B. 1985. Progress of The Geological Conservation Review – the Palaeogene Sites of Southern England; their scientific value and conservation. *Earth Science Conservation*, **22**, 30–33.

DALEY, B. & BALSON, P. 1999. *British Tertiary Stratigraphy*. GCR Series No. 15. Joint Nature Conservation Committee, Peterborough.

DINELEY, D. L. & METCALF, S. J. 1999. *Fossil Fishes of Great Britain*. GCR Series No. 16. Joint Nature Conservation Committee, Peterborough.

EMELEUS, C. H. & GYOPARI, M. C. 1992. *British Tertiary Volcanic Province*. GCR Series No. 4. Chapman and Hall, London.

ELLIS, N. V., BOWEN, D. Q., CAMPBELL, S. *ET AL.* 1996. *An Introduction to the GCR*. GCR Series No. 1. Joint Nature Conservation Committee, Peterborough.

FLOYD, P. A., EXLEY, C. S. & STYLES, M. T. 1993. *Igneous Rocks of South-West England*. GCR Series No. 5. Chapman and Hall, London.

GORDON, J. E. & SUTHERLAND, D. G. (eds) 1993. *Quaternary of Scotland*. GCR Series No. 6. Chapman and Hall, London.

GREGORY, K. J. (ed.) 1997. *Fluvial Geomorphology of Great Britain*. GCR Series No. 13. Chapman and Hall, London.

HUDDART, D. & GLASSER, N. F. 2002. *Quaternary of Northern England*. GCR Series No. 25. Joint Nature Conservation Committee, Peterborough.

MAY, V. J. & HANSOM, J. D. 2003. *Coastal Geomorphology of Great Britain*. GCR Series No. 28. Joint Nature Conservation Committee, Peterborough.

MORTIMORE, R., WOODS, C. & GALLOIS, R. 2001. *British Upper Cretaceous Stratigraphy*. GCR Series No. 23. Joint Nature Conservation Committee, Peterborough.

PROSSER, C. D. 2008. The history of geoconservation in England: legislative and policy milestones. *In*: BUREK, C. V. & PROSSER, C. D. (eds) *The History of Geoconservation*. The Geological Society, Special Publication, London, **300**, 113–121.

RATCLIFFE, D. (ed.) 1977. A Nature Conservation Review. Cambridge University Press for Nature Conservancy Council and the Natural Environment Research Council, (2 volumes).

RUSHTON, A. W. A., OWEN, A. W., OWENS, R. M. & PRIGMORE, J. K. 1999. *British Cambrian to Ordovician Stratigraphy*. GCR Series No. 18. Joint Nature Conservation Committee, Peterborough.

SELDEN, P. 1985. Progress of The Geological Conservation Review – Fossil Arthropods: Chelicerates. *Earth Science Conservation*, **22**, 33–4.

SIMMS, M. J., CHIDLAW, N., MORTON, N. & PAGE, K. N. 2004. *British Lower Jurassic Stratigraphy*. GCR Series No. 30. Joint Nature Conservation Committee, Peterborough.

SMITH, D. B. 1995. *Marine Permian of England*. GCR Series No. 8. Chapman and Hall, London.

STEPHENSON, D., BEVINS, R. E., MILLWARD, D., HIGHTON, A. J., PARSONS, I., STONE, P. & WADSWORTH, W. J. 1999. *Caledonian Igneous Rocks of Great Britain*. GCR Series No. 17. Joint Nature Conservation Committee, Peterborough.

STEPHENSON, D., LOUGHLIN, S. C., MILLWARD, D., WATERS, C. N. & WILLIAMSON, I. T. 2003. *Carboniferous and Permian Igneous Rocks of Great Britain North of the Variscan Front*. GCR Series No. 27. Joint Nature Conservation Committee, Peterborough.

TREAGUS, J. E. (ed.) 1992. *Caledonian Structures in Britain: South of the Midland Valley*. GCR Series No. 3. Chapman and Hall, London.

WALTHAM, A. C., SIMMS, M. J., FARRANT, A. J. & GOLDIE, H. S. 1996. *Karst and Caves of Great Britain*. GCR Series No. 12.

WIMBLEDON, W. A. 1991. Historical sites for the Geological Conservation Review. *Geology Today*, **7**, 50–1.

WIMBLEDON, W. A. W., BENTON, M. J., BEVINS, R. E., ET AL. 1995. The development of a methodology for the selection of British geological sites for conservation: Part 1. *Modern Geology*, **20**, 159–202.

WRIGHT, J. R. L. & COX, B. 2001. *British Upper Jurassic Stratigraphy: Oxfordian to Kimmeridgian*. GCR Series No. 21. Joint Nature Conservation Committee, Peterborough.

A historical perspective on local communities and geological conservation

GRAHAM J. WORTON

Dudley Museum and Art Gallery, St James Road, Dudley, West Midlands DY1 1HU, UK
(e-mail: graham.worton@dudley.gov.uk)

Abstract: Like every other human endeavour in the modern world geological conservation has evolved from the changing desires and necessities of evolving communities. Local people in generations past had a need to be 'in tune' with their local landscapes and the environments making use of local materials and landform in order to survive and prosper. These early communities have enriched our modern landscapes with their legacy of geodiversity in altered landscapes, the built environment and venerated geological features. In a modern context we may consider 'geoconservation' to be the 'conservation of geodiversity' and we should acknowledge that local people and historic communities have had a very long involvement in conserving geological heritage although often without consciously doing so. This is particularly the case in terms of landscape and the applied geology of the built environment. Early local communities were usually of low cultural diversity with values of narrow focus. It is commonly local people and local specialist interest societies that have made geological discoveries and brought their importance to the attention of the specialists. In the late 1900s local groups drove the development of what is now the familiar framework for protection of geological heritage and in particular the non-statutory sites and features that rely being locally valued for their sustainability and survival.

This paper cannot be a comprehensive account of the evolution of peoples and their changing engagement with geodiversity. It is, rather, a summary account using selected scattered illustrations of geoconservation through the ages. It examines what we mean by local community and its involvement, either consciously or unconsciously, in protecting geodiversity through stages in human history.

Geoconservation is essentially a modern term for a host of intentions and activities aimed at preserving, conserving and sustaining geological features and processes for the benefit of future generations. People and communities have been conserving geological 'things' throughout human history though only in recent centuries has the intention been directed towards the natural significance of features or landscapes themselves.

When considering the development of peoples, communities and their cultures, and their engagement with the geology around them, it appears that the cultural significance that they placed on their surroundings has often, unintentionally or unconsciously, yielded protection to geological sites and features. Such 'unconscious geoconservation' has often had far-reaching beneficial consequences for geological features in our modern world. Selected examples of this are given below.

At the time of publication there have been many definitions proposed for 'geoconservation' and geodiversity. For example Chris Sharples of Tasmania suggests that 'geodiversity' is geological 'quality that we are trying to conserve' and that 'Geoconservation' is defined by 'the endeavours to conserve it'. Many other definitions echo these principles with varying degrees of additional detail. Perhaps we can simplify what we mean by geoconservation as 'the conservation of geodiversity' in which we encompass the processes of recognizing and assessing the value (or quality) of geological features, collections, sites, monuments, artworks and landscapes and the application of practices for their care, maintenance, management and use for the long term benefit of all.

For the purposes of this paper, geodiversity is taken in its broadest sense to include rocks, fossils, soils, landscapes, biodiversity that is strongly influenced by underlying geology, minerals, mining, geologically significant human history and geologically inspired culture and art.

The development of communities and consequential geoconservation

If we were to summarize currently held beliefs and evidence concerning the path of human evolution then it would appear that disparate nomadic groups of hunter gatherers and traders passing through landscapes eventually settled the land to form more stable communities. They typically did this in places with sources of good tool-making stones and where the landscape and the underlying

From: BUREK, C. V. & PROSSER, C. D. (eds) *The History of Geoconservation.*
Geological Society, London, Special Publications, **300**, 137–146.
DOI: 10.1144/SP300.11   0305-8719/08/$15.00 © The Geological Society of London 2008.

geology also provided a permanent water supply and fertile ground. In effect these settlements became the first industrialized areas. We might consider these early settlements to be fairly independent and sessile which perhaps mainly kept to themselves and developed their own cultures, folklore and characters over time, but were connected often over large distances, by a trade, in workable rocks. People protected the environment in which they lived as it represented their entire livelihood and as such the geological exposures and mines that they worked were vital as sources of stone for tools. They then 'conserved' their geology in the process of industrialization and development of the land.

These ancient human communities had survival and perhaps spirituality at the heart of daily life. It is likely that in such communities understanding of the world was based on a mixture of sound observation and myth. People of such settlements may be expected to have had horizons that were restricted by very limited personal travel and very little personal knowledge of the wider world. There are many good examples of how such communities held geological places within the landscape as special and in consequence venerated them and afforded them cultural protection. A very ancient and well known example (and currently a significant tourism destination) is Australia's Uluru (formerly known as Ayers Rock). Debate continues on when the first Aboriginal peoples moved into the area but the best evidence suggests that it was at least 20 000 years ago. Here they imparted human ritualistic meaning to the landscape which even now draws us to the grandeur of the geological scene that meant so much to them. The geological connection goes still deeper here. In fact the Aboriginal name Uluru, means 'Great Pebble' and we are told that each feature of the rock has a meaning in 'Tjukurpa' or Dreamtime, the traditional Anangu law that explains how the world was created.

There are many similar examples in the ancient world (for example the Egyptian Valley of the Kings, Petra in Jordan and Machu Pichu in Peru) where the veneration of landscape and the development of human structures affords a greater protection than the geological or geomorphological features alone.

In simple terms, the obvious benefits to these early peoples of living together in communities lead to the growth and proliferation of towns and cities and ultimately to the creation of leisured classes, priests, kings and of course the concept of 'wealth'. A surplus of resources enabled power structures to emerge, allowed rulers to travel and leisure time to occupy with the finer attributes of human existence including science, technology and art. Those living 'hand-to-mouth' had no time to innovate. It is only when agricultural practices are well established that there is time available for leisure and 'interesting pursuits' which lead to experimentation and innovation. The uneven distribution of natural resources and, in high latitudes, the dramatic effects of the seasons, necessitated effective communication between occupied areas and created more formalized trade. This in turn spread knowledge and hitherto independent communities assimilated views and ideas from other peoples and cultures. Such communities were better resourced and grew with inevitable expansion of interaction with the land that lead to greater understanding of the Earth, its processes and its useful resources. It is not hard to imagine that in such communities it would have been the people working the land who would have been best placed to make physical discoveries that the privileged classes would be best placed to interpret (within the boundaries of their knowledge at any given time). Arguably in these communities power and status protected geodiversity as curiosities of significance to the influential rather than as a result of recognition or concern about their rarity, fragility or scientific importance.

By the time of the Renaissance period (from about 1500 onwards) there was a general tendency among the aristocracy to be fascinated by natural history and to have their own personal collections of natural objects which they commonly hosted in a 'cabinet of curiosities' in their grand houses. Such collections were the forerunners of modern museums and inspired study of the objects of nature. In Europe a great interest in natural and cultural things from the mid 1550s onwards is evident and also a more systematic study of the natural world emerges in the scientific writing from then on.

Early rumblings of cultural geodiversity in Britain: Dud Dudley 1599–1684

A good example of cultural geodiversity local to the town of Dudley in the West Midlands of England of exceptional importance is the example of the discoveries recorded by Dud Dudley the natural son of the Earl of Dudley. In *Metallum Martis* (1665) he tells of his life and experiments with coal and iron making. In the early 1600s he was given control of one of his fathers iron works and set about finding a new fuel for iron making (as the traditional fuel, charcoal made from timber, was outlawed by Royal Decree at the time. This was due to ongoing threat of war and fear of invasion as well as expanding global import/export trade that required the nations timber for ship building).

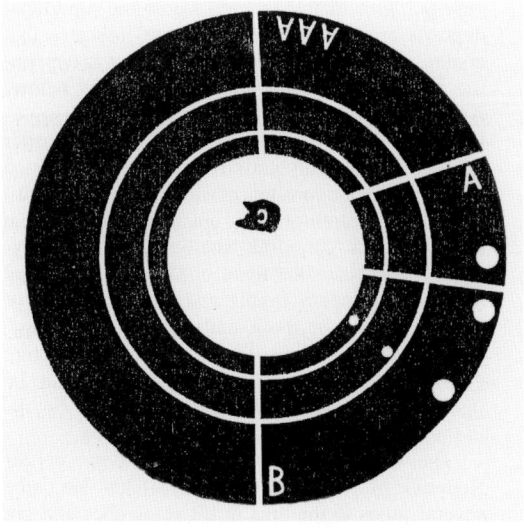

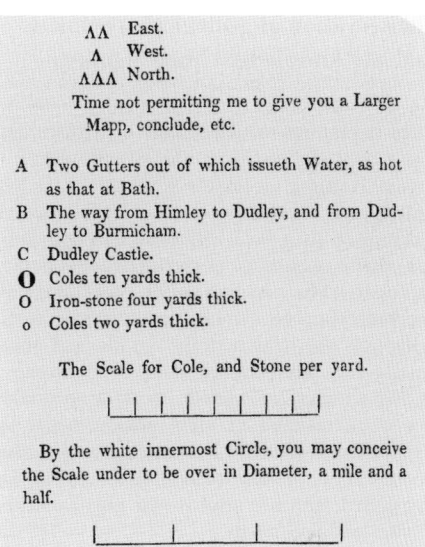

ΛΛ   East.
Λ   West.
ΛΛΛ North.
Time not permitting me to give you a Larger
    Mapp, conclude, etc.

A Two Gutters out of which issueth Water, as hot
   as that at Bath.
B The way from Himley to Dudley, and from Dud-
   ley to Burmicham.
C Dudley Castle.
O Coles ten yards thick.
O Iron-stone four yards thick.
o Coles two yards thick.

    The Scale for Cole, and Stone per yard.

By the white innermost Circle, you may conceive
the Scale under to be over in Diameter, a mile and a
half.

Fig. 1. The first geological map? Mineral seams around Dudley Castle, *Metallum Martis* (1665).

In 1619, he patented a process for the use of coal to make iron and in the wake of this discovery created an industrial frenzy in the Black Country area which turned this part of the Midlands into the world's first industrial area, sometimes referred to as the 'work-shop of the world'.

In terms of geological breakthroughs he had in this work made a fundamental advancement that is generally unacknowledged but enormously sig-nificant in terms of establishing a now universally adopted predictive geological methodology—geological mapping. He published the world's first geological map in *Metallum Martis* in 1665.

This map shown as Figure 1 below is a represen-tation of Castle Hill in Dudley. It shows the out-crops of mineral seams around the hill. This map arguably represents one of the greatest moments in geological reasoning but is clearly still the pre-serve of the wealthy at this time.

Early site recording: Robert Plot 1686

Robert Plot was a scholar at Oxford University who travelled and made an inventory of many natural curiosities in England during the seventeenth century (for example plot 1686). His work rep-resents one of the earliest attempts at systematic recording of the sites and geological features of areas. In his series of English County publications he describes his discoveries during travels around the Midland counties of England. These eclectic works record landscape features, industries and cul-tural identity. They draw attention for the first time

to many geological features and phenomena that are to attract future scholar's attention and begin to make famous a number of local geodiversity sites on the basis of their curiosity alone. He also praises the practical labours of local people. It was a pursuit for the wealthy and privileged at this time and there is no comment or action for the preservation or conservation of the features mentioned in these works.

The advent of site conservation? Pompeii and Herculaneum

Possibly the earliest but certainly one of the best examples of deliberate or conscious conservation of a geologically important site is that of the discov-ery in 1709 of the ash buried communities of Pompeii & Herculaneum in west central Italy. The story of the eruption of Vesuvius was already well known to the European intellectuals from the accounts of Pliny the Younger, however the physi-cal legacy of that eruption had never been found. The discovery of the actual site was made by a local farmer digging a well. The curiosity of this site and its links between a very geologically active landscape and the people who had previously lived in this place was so strong, in this place at this time, that the site was deliberately preserved and excavated for the knowledge and understanding it imparted (and later for the tourism income that it generated). This is perhaps where conscious and deliberate geoconservation of sites begins as an entity in itself.

Conservation of geological specimens and site knowledge: Mary Anning

By the end of the eighteenth century, local people were beginning to make their own mark on this previously exclusive world of privileged geological learning (at least in England where humble individuals were able to rise to geological fame because of their local expertise and dedication in the pursuit of geological knowledge and in particular fossils). Mary Anning (1799–1847) was perhaps the most famous female geological icon of all. Although she searched for fossils and prepared fossils for profit rather than to conserve sites and specimens for science, there is no doubt that her work resulted in great achievements in the expansion of geological knowledge and geological conservation. Today natural heritage in Dorset associated with her affords still greater protection to sites and specimens.

Conscious geoconservation and the impact of art in the eighteenth and nineteenth centuries

During the eighteenth and nineteenth centuries in art and consciousness there was a development of the appreciation of the picturesque which is much promoted by romantic art of the time. This clearly had an effect on the hearts and the minds of those in privileged positions in life. The wealthy created parks with classical landscapes highlighting natural features, sometimes enhancing them, retaining rock faces or 'improving' them but creating access to interesting features. There was a very purposeful intent to preserve the geological features in this process but simply for their aesthetic properties not their intrinsic importance (however the paper by Doyle 2008 relating to the establishment of the landscapes of Crystal Palace Park indicates that sometimes a greater interpretation of the land was intended).

Landowners were at their peak of activity in terms of the wholesale exploitation of land for its mineral wealth. The upper classes became involved in both creating and destroying landscape and revealing the grain of the land in a way and to an extent previously unknown. So the curious descended upon these works and made significant discoveries. The case of Dudley and the exploitation of its Silurian and Carboniferous strata is a classic example of this evolution of science and natural philosophy.

The Black Country experience

Documentary evidence relating to the southern area of the Black Country, where the Coal Measures is very rich in workable mineral seams and superficial deposits are almost entirely absent indicates that coal mining activity has occurred. This was happening from at least 1280 and probably long before. Archaeological evidence in the area demonstrates limestone and ironstone mining much earlier than even this. This activity undoubtedly created masses of rock exposure, however, it was not until the late eighteenth or early nineteenth century that there was any conscious site-based geoconservation. The fame of the town of Dudley had spread as a site of superb fossils and many pioneers of science visited in those early years of geology. One of the local fossils, the Silurian trilobite *Calymene blumenbachii* was among the first to be figured and became known as the 'Dudley Fossil' in the early years of the nineteenth century.

The most famous visits were those of Sir Roderick Murchison in the early 1830s. He came several times to the area in his researches and got to know its miners, gentry and geology. His work here and nearby in the Welsh borders led to his very famous work, *The Silurian System*, which was published in 1839. In that year he also brought a party of scientists from the British Association for the Advancement of Science to Dudley to experience the spectacle of the mines and perhaps this, the greatest of the fossil localities that he had discovered on his journeys.

The local miners had assembled a collection of local fossil finds for the visit which immediately yielded many new species to science. With the encouragement of Murchison the local people decided that these collections should remain together and form a permanent museum in the Town of Dudley. Professor Henry Beckett (1862) wrote 'It will be of interest to add that Dudley Museum originated from the meeting of the British Association in Birmingham in 1839. There were exhibited at Dudley for that meeting a number of Silurian and Carboniferous fossils which belonged to several inhabitants of the town. Consequent on this exhibit of fossils a permanent society was formed and the museum established.' This single event had far-reaching geoconservation consequences.

Dudley and Midland Geological and Scientific Society and Field Club

Murchison's work and influence following his 1839 visit inspired local mine agents and industrialists, together with lay people and patrons including Sir Robert Peel, Lord Ward, Lord Littleton and clergy including the Bishops of Exeter, Lichfield and Worcester, honorary membership of luminaries of the day including professors Buckland and

Sedgwick, Murchison and Phillips to found a geological society and field club. It was in fact founded in 1841 as the earliest Midlands regional geological society with 150 members. Its inaugural address was by given Murchison himself in January 1842 in which he said:

In no part of England are so many geological features brought together in a small compass than in the environs of Dudley, or in which their characters have been successfully developed by the labours of practical men. Where else can then a site be found in which the records of the past can be more successfully preserved, or where we can store up for instruction both the types of primeval life and the evidences of the mighty operations that mark the more ancient conditions of our planet?

The first museum was established in a public house. The generous owner (Mr J. Bennit) offered the first museum this site (the Britannia Inn) in the town centre and permitted alterations to his building to house the museum and society rooms. The museum was open Monday–Saturday between the hours of 10am and 4pm was clearly an entity of some considerable merit as it was later referred to in 1862 by Professor Beckett who alluded to that first society's museum with the words 'the fame of that noble museum extended wherever geology was held in repute' and contained 'unrivalled treasures'.

Murchison and the second British Association visit of 1849

A decade after the publication of the *Silurian System*, Murchison returned to Dudley with a hundred or so scientists of the British Association. They came by canal boat to the limestone mines and were escorted by an estimated 15 000 people lining the huge caverns to hear Murchison speak. As recorded in the *Illustrated London News* (Fig. 2) and depicted in an anonymous 220 line ballad acclaiming Murchison 'King of Siluria', the limestone caverns at this event were illuminated with gas light and coloured flares by Lord Ward, Earl of Dudley and this lead to a new era of geological folklore which echoes in the geotourism of the area today.

These are perhaps among the strangest of stories concerning local geodiversity and geoconservation involving the nuances of local people that have yet come to light and as early examples of museums and important sites we are able, admittedly in an imperfect manner, to track their sustainability through the extant features and literature.

The fortunes of the geological museum have waxed and waned over the years. In 1862 a paper by Hollier (a committee member of the second iteration of the Dudley Geological Society) notes that, although still open, the museum was 'hidden in an old Malt house'. Whether this comment reflects his desire to have the museum in more prestigious accommodation than the public house or whether it reflects an ensuing neglect of years is not clear. There is a clear change in emphasis in Hollier's words, however, which relates to the local people element of the role that this society played in collecting, recording and sustaining the areas geoconservation. He states in the same paper that:

other valuable aids to science originated with the society, which although less generally known, are nevertheless deserving of record. I allude to the field labour of some of the members so honourably acknowledged in the memoirs of the geological survey.

The second Dudley and Midland Geological Society and the fortunes of the Dudley Geological Museum

The early literature of the second society begins in 1862 and its drive was towards the establishment of a new museum of geology for the town. The society had considerable success in this aspiration and later the same year it had raised the money and laid the foundation stone. The following year a curator had been appointed and extramural geology classes were on offer. In 1864 an important exhibition and mining conference was held here. The society membership had swelled to 350 members but the exhaustion of the coalfield and the demise of the mining industry witnessed a similar decline in the welfare of the geological society and by 1901 only 56 members remained.

At some time in the early 1900s the society disappeared and the collections passed into the care of the town council. Objects were for the first time being catalogued and were being redisplayed in 1911 by a Edward Worsey a student of Professor Lapworth of Birmingham University. Letters from him indicate that in the various movements and transfers of the collection, many labels indicating localities had been lost. Nevertheless the new museum (in the old free library now the current Museum and Art Gallery) was opened in Dudley at 3.30pm on 12 December 1912 by Professor Lapworth and a new era of geological conservation and promotion had begun. In the years since this first opening of the new geological museum the collection has enjoyed a home in the town centre. It has also ridden the rise and decline of interest and perceptions of relevance that those early collections and individuals felt, but the ongoing work of local enthusiasts and curatorial staff has nevertheless developed and flourished to be a very considerable geodiversity and geotourism asset of the area.

Fig. 2. London Illustrated News illustration of the British Association to the limestone mines of Dudley (1849).

The third society: The Black Country Geological Society

Many decades after the demise of the second local geological society a new geological society began. The Black Country Geological Society (the BCGS) was established in 1975 following night school classes on geology run as an extramural course from the University of Birmingham. Inspired and thirsty for more, a number of the people attending the course decided to form the new society. One of the establishing principles of the society was that it should be a lobbying organization and practical body for the conservation and development of geological sites and collections. In this single aspect their beneficial impact on the geodiversity of the area has been enormous and makes quite a catalogue some highlights of which are given below.

One of the earliest acts of the society was to lobby the local authority about the state of the reserve geological collections not on display and intervene as museum volunteers in the re-cataloguing and curation of the specimens that had fallen into neglect with the change of emphasis and staff since 1912. By 1984 the collections had been re-housed and a temporary geological curator had been put in place with a manpower

services commission team to take forward the care and development of the collection. This programme funding ceased in 1986 and the collection once again sat dormant.

At this time BCGS also campaigned for the local geological heritage to be included in the planning system as a material consideration in decision-making and for geological heritage eventually to sit as an equal alongside wildlife policy and considerations in local nature conservation strategies and policy documents. Perhaps the society's most important piece of work was their direct role in getting the local authority to create the role of keeper of geology. The BCGS persuaded the GCG to hold their AGM in Dudley in 1986 and at that meeting the GCG made it very clear that the local council was failing in its duties towards the collection and demanded action. This in due course lead to the creation of the new post and in 1987 the first permanent keeper of geology was appointed and a whole new era of geodiversity began.

Since this time members of the society have been very active in site conservation, museum interpretation, promotional events, walks and talks and through hard-won experience have been consultees and advisors on many policy and strategic documents. The BCGS was one of seven founder member regional groups of the 1990 *Earth Science Conservation Strategy*. In 1994 it was the key organization for geological heritage in the *Black Country Nature Conservation Strategy*, and advised on the designation of many SINC's (Sites of Importance for Nature Conservation, local equivalent to RIGS). The work of the BCGS continues and its latest contribution as a key partner in the new Black Country Geodiversity Partnership.

The advent of explicit Nature Conservation Law and its impact in Dudley

Like every other area of classic geology, much of the geodiversity and geoconservation story of Dudley occurred before any formal nature conservation law or policy had been devised and emplaced. Like those other areas Dudley has played an important role in working out what levels of protection and use the geodiversity needs and can sustain. In such an urban area with such a rich industrial heritage it seems that the intimate relationship between the people and the rocks has always been powerful but informal. Wide geodiversity has survived and been sustained here as people valued it fundamentally with cultural ownership rather than because of its superimposed scientific importance.

As the paper by Prosser & Larwood (2008) indicates, the advent of Nature Conservation Law and its initially divisive character ('hands off' or 'keep out' inference) were not welcomed in Dudley. The best known local site of all (the Wren's Nest National Nature Reserve) became the proving ground of how people and internationally important geological heritage might usefully interface and comfortably co-exist. Perhaps the Wren's Nest story is a clear example of how conservation and preservation are often confused. The local experience is that development and use of a geodiversity resource can be compromised by over zealous and misinformed attempts to preserve or 'mothball' an otherwise dynamic geological and cultural resource.

The Black Country Geodiversity Action Plan

The Black Country Geodiversity Partnership was set up in 2004 to rationalize and focus geoconservation efforts in the area. Many organizations representing the local communities here seek consensus and work towards positive change through this process. For the first time geoconservation had a chance of being proactively delivered in a major way rather than the often reactionary system in response to threat that has been the typical situation in the past.

The Black Country Geodiversity Action Plan (BCGAP; Fig. 3) has been produced, listing the agreed priorities for the next year and the mechanism for delivery on these actions. The rationalized and agreed priorities have the advantage of multi-discipline, multi-organizational support at their entry into the tasks and actions lists of the given year and all partners have a interest in its success such that there is genuine commitment to achieving the outcomes set out in the plan. This process is an advent in local geoconservation thinking and delivery in the Black Country. Using a simple format and wide partnership it accesses greater resources and offers wider incentives than the geoconservation alone which is aligned to the various organizational performance targets and issues of the day. It is already transforming the scope of possibilities here and very imaginative projects are being planned for the sustainable beneficial use of the geodiversity in this area.

Looking to the future

The BCGAP has many aims and ambitions. The most far reaching is the establishment of a Black Country Geopark as part of the European/Global Geopark initiative. Pursuit of this accolade will significantly raise levels of general awareness about the geodiversity of the area and its needs. Within this territory exciting large-scale geotourism projects are planned.

At the heart of a 30-year vision of regeneration of the Black Country area, Wren's Nest stands out as a key opportunity, proving ground and innovator for geodiversity in key strategic documents. This will be the greatest of all geoconservation projects attempted to-date. It makes geology and its mass exploitation in the form of historic limestone mines and quarries the centerpiece of radical environmental and economic restructuring of the former industrial area. This will initiate various projects to provide infrastructure and future destinations for 'mass geotourism'. It is also directed at setting up a world class exemplar for geological/geotechnical educational delivery on this site.

By the end of May 2007 two years of design and planning culminated in the submission of the second of two large bids to provide enhanced visitor access and high quality accommodation with additional site enhancement and interpretation on a scale never seen before in the region for geological heritage. The bid was made to funds including the Heritage Lottery Fund and BIG lottery funding streams.

The Black Country
Geodiversity Action Plan

"making a positive contribution to the enrichment of
the Black Country environment and quality of life by
conserving, enhancing and managing the region's
geological heritage and diversity for the benefit of all"

Fig. 3. The Black Country Geodiversity Action Plan.

Split into two phases, the first bid submitted in December 2006 sought to stabilize and re-open to the public (for the first time in 40 years) the Seven Sisters limestone mines complex on the western side of the Wren's Nest hill. This project would include partial support of the most unstable area of the mine with an underground building complex. Designed to be architecturally stunning, the interpretation centre buildings will add a brand new dimension to this already internationally renowned site. It also represents the first ever true visitor facility for the site. The second phase of the bid links with other Black Country partners additional schemes. For Dudley this second phase aims to open up the deep underground mines and canal tunnels beneath the site for people to access for the first time in 200 years. Making the link between the Seven Sisters mine entrances above to the dark labyrinthine underworld of mineworkings, tunnels and canals below it will be a breathtaking insight into the grain of the land and the work of the geologists, miners and engineers of the past.

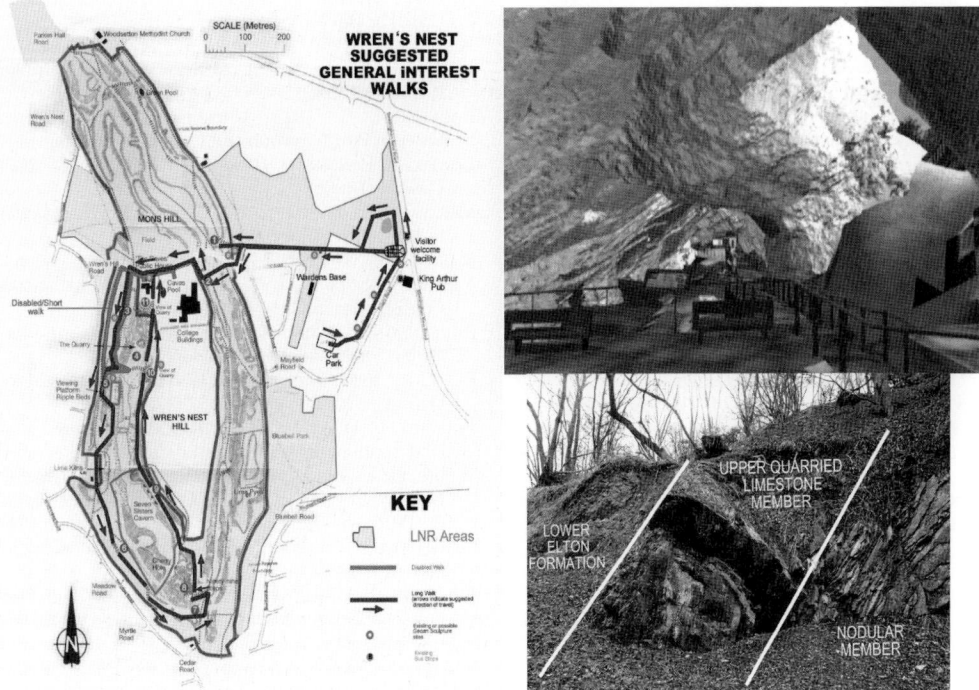

Fig. 4. Engineering a mass geotourism future, the Wren's Nest 'Strata' project.

This will make accessible the entire Much Wenlock Limestone succession in near vertical strata and mined cavities along with canal basins which provide waterway access to the canal networks beyond. This phase will include driving new tunnels and creating a host of underground experiences and exhibitions. It has been designed to have a 100-year life. At the very start of the project sustainability is a core principle for the conservation of this proposed world-class geotourism/geodiversity experience (Fig. 4).

Conclusions

In the course of researching this topic it has become apparent that what we currently understand as 'geoconservation' has a considerable cultural history and will evolve at perhaps an increasing pace as time goes by. The progress of human work through the ages to protect special sources of good rocks and special places of spiritual meaning reflect a direction of travel from unconscious geoconservation to conscious geoconservation. This change has also been from geoconservation being the concern and interest of the privileged and powerful to the domain of the lay person and amateur who cares enough to become involved.

Everyone can be involved now like never before and hopefully more will be inspired and engaged as new initiatives are delivered.

The human lessons that we might pick up from this study are perhaps warnings to the future. Clearly geoconservation as a process needs to move with the times as history evidences that 'Importance' is subjective and temporal. At the moment in modern communities conservation issues and climate change are emotive and increasingly local people feel empowered and are willing to act on matters of concern in the UK today. Within the public sector there is increasing community involvement in decision-making and initiatives and ubiquitous need for justification and performance measurement. Sustainability of our geodiversity will always rely upon communities (whether of local people or knowledge based communities) and their pride in where they live and value they place upon it. It is our greatest challenge perhaps to ensure that geodiversity remains relevant in the longer term changes in the culture of our communities of the future.

I would like particularly to thank J. Hemingway, Archaeological Officer and A. Durkin, Exhibitions Officer at Dudley Metropolitan Borough Council for their expert guidance on matters of human history and the history of art respectively, without which this paper could not have confidently danced across the surface of our human history to look for the connections with our cultural development. I would also like to thank J. Clatworthy of the Lapworth Museum at the University of Birmingham for his access to Lapworth archival materials pertinent to the Dudley case study in this paper.

Select Bibliography

AGRICOLA, G. 1556. *De Re Metallica*. Privately published in London.

BAKER, T. 1848. *Bakers Practical Survey of the Geology, Mineralogy and Historical Events of the District of Dudley*. Privately published in Dudley.

BECKETT, H. 1862. Report of First Annual General Meeting. *Transactions of the Dudley & Midland Geological Society and Field Club*, **1**, December 1862, 3–7.

BOX, J. & CUTLER, A. 1988. Geological Conservation in the West Midlands. *Earth Science Conservation*, **25**, 29–38.

CARVER, M. O. 1987. *Underneath English Towns*. Batsford, London.

CUNLIFFE, B. ET AL. 2001. *The Penguin Atlas of British History*. Penguin, London.

CUTLER, A. 1981. A short history of the Dudley & Midland Geological Societies. *The Black Country Geologist*, **1**, 1–21.

DOYLE, P. 2008. A vision of 'deep time': the 'Geological Illustrations' of Crystal Palace Park, London. *In*: BUREK, C. V. & PROSSER, C. D. (eds) *The History of Geoconservation*. The Geological Society, London, Special Publications, **300**, 197–205.

DUDLEY, D. 1665. *Metallum Martis*. Privately published, printed by T. H., London.

English Nature. 2004. *Geology and Biodiversity—Making the Links*. English Nature, Peterborough.

EVANS, J. G. 1975. *The Environment of Early Man in the British Isles*. BCA, London.

FAUL, H. & FAUL, C. 1983. *It Began with a Stone*. Wiley, New York.

FORTEY, R. A. 2004. *The Earth*. HarperCollins, London.

GALE, W. K. V. 1979. *The Black Country Iron History, a Technical History*. Iron & Steel Institute, London.

GRAY, M. 2004. *Geodiversity, Valuing and Conserving Abiotic Nature*. Wiley, Chichester.

HOLLIER, E. 1862. Reponse to: 'Report of First Annual General Meeting'. *Transactions of the Dudley & Midland Geological Society and Field Club*, **1**, December 1862, 7.

HUTTON, J. 1795. *The Theory of The Earth*. Edinburgh, Cadell & Davies.

JUKES, J. B. 1859. *The South Staffordshire Coalfield*. Memoirs of the Geological Survey of Great Britain, HMSO, London.

LAPWORTH, C. 1912. Unpublished correspondence between Prof Lapworth and Dudley Town Council.

LONES, T. E. 1898. *A History of Mining in the Black Country*. Dudley.

MEGA, J. V. S. & SIMPSON, D. D. A. 1981. *Introduction to British Prehistory*. Leicester.

MURCHISON, R. I. 1839. *The Silurian System*. London.

NATURE CONSERVANCY COUNCIL. 1990. Earth Science Conservation in Great Britain Strategy.

PARKINSON, J. 1804. *Organic Remains of a Former World*. Sherwood, Nealey & Jones, London.

PLOT, R. 1686. *The Natural History of Staffordshire*. Oxford.

PROSSER, C. D. & LARWOOD, J. G. 2008. Conservation at the cutting-edge: the history of geoconservation on the Wren's Nest National Nature Reserve, Dudley, England. *In*: BUREK, C. V. & PROSSER, C. D. (eds) *The History of Geoconservation*. The Geological Society, London, Special Publications, **300**, 217–235.

PROSSER, C., MURPHY, M. & LARWOOD, J. 2006. *Geological Conservation: A Guide to Good Practice*. English Nature, Peterborough.

SARJEANT, W. A. S. & MORRELL, R. W. 1964. The geological societies and geologists of Midland England. *Mercian Geologist*, **1**, 35–48.

STACE, H. & LARWOOD, J. G. 2006. *Natural Foundations; Geodiversity for People Places and Nature*. English Nature, Peterborough.

TAYLOR, C. 1983. *Village and Farmstead: A History of Rural Settlement in England*. George Philip, London.

WOODWARD, J. 1695. *An Essay Toward a Natural History of the Earth*. London.

WORTON, G. J. 1993. A Person on the Inside, Opportunities for Geological Conservation in Local Engineering Projects. *In*: O'HALLORAN ET AL. (eds) *Geological and Landscape Conservation: The Malvern Conference*. Geological Society of London, 359–363.

WORTON, G. J. 1996. Digging up Your Doorstep, Engineers and Their Excavations. *In*: BENNETT, M. R., DOYLE, P., LARWOOD, J. G. & PROSSER, C. D. (eds) *Geology on Your Doorstep: the Role of Urban Geology in Earth Heritage Conservation*. Geological Society, London, 116–127.

WORTON, G. J. 2005. Wren's Nest National Nature Reserve Dudley. *In*: *Involving People in Geodiversity*. JNCC, Peterborough.

History of RIGS in Wales: an example of successful cooperation for geoconservation

NEWRIGS, AWRG & UKRIGS, University of Chester CH1 4BJ, UK
(e-mail: c.burek@chester.ac.uk)

Abstract: The history of RIGS in Wales is an example of a rise, fall and rise again of a voluntary sector organization operating within the field of geoconservation. The Welsh RIGS movement began as small individual groups, which then gave rise to a national organization with specific needs and problems. As RIGS started in Wales before devolution in 1999, it was regarded as a regional grouping but after devolution it was well placed to take the national Welsh agenda of geoconservation forward. This paper explores the development of this co-operation through national conferences, the development of the Association of Welsh RIGS Groups and looks at the unique position that Wales holds within the field of geoconservation. Finally, the achievements of this approach are highlighted.

This paper will detail the history, the unique situation in Wales and the achievements of the Welsh RIGS movement from 1990 to early 2007.

Conservation of Earth science sites is concerned with maintaining rock exposures, landforms and sites where geological processes can be seen in action today. We can conserve our geological history and environment through statutory means but education, research and an awareness of our local natural physical features are becoming increasingly important factors. (Wood 1995)

These opening remarks from the booklet accompanying the first Welsh RIGS Forum in 1994, sum up the philosophy of the Association of Welsh RIGS Groups (AWRG) and the idea of geoconservation. There is a need to both educate geologists about geoconservation and to raise the general level of geodiversity understanding among non-geologists, including the spatial and timescales in which geodiversity operates. This will then facilitate successful conservation measures (Burek & France 1998). The RIGS movement in Wales, through the national AWRG has attempted to educate, research and raise public awareness of geology within Wales, but at a price. Therefore it is deemed necessary to look at the history of the Welsh RIGS Groups and how they have developed.

What is geodiversity?

The term geodiversity was first used in 1993 in a *Tasmanian Forestry Commission* document to describe 'the diversity of Earth features and systems' (Sharples 1993). The second recognized use of the term was at the Malvern International Conference on Geological and Landscape Conservation (1993) in relation to the description of geotope conservation in German-speaking countries as 'management of

geotopes in their natural diversity and characteristics' (Wiedenbein 1994).

By 1996 the Australian Parks and Wildlife Service was using the term as a definition for 'the range or diversity of geological (bedrock), geomorphological (landform) and soil features, assemblages, systems and processes' (Dixon 1996; Australian Heritage Commission 2002). In the UK, geodiversity has been defined by Stanley (2000) in a holistic way as 'the link between people, landscape and their culture: it is the variety of geological environments, phenomena and processes that make those landscapes, rocks, minerals, fossils and soils which provide the framework for life on Earth'. However, a simpler definition has been developed by Prosser (2002) as 'geological diversity or the variety of rocks, fossils, minerals and natural processes' but this omits landforms and soils. The concept that 'geodiversity underpins biodiversity' was first put forward by Burek (2001*a*).

What are RIGS?

Regionally Important Geological/geomorphological Sites (perhaps nowadays better thought of as Geodiversity Sites) are non-statutory sites, which provide an important contribution to the overall network necessary to ensure maintenance of the range and diversity of our Earth heritage. The RIGS scheme was initially set up in 1990 and envisaged as a way of selecting and conserving Earth heritage sites outside the existing GCR/SSSI framework (Ellis *et al.* 1996). It was a government agency initiative under the Nature Conservancy Council, English Nature (now Natural

From: BUREK, C. V. & PROSSER, C. D. (eds) *The History of Geoconservation.*
Geological Society, London, Special Publications, **300**, 147–171.
DOI: 10.1144/SP300.12   0305-8719/08/$15.00 © The Geological Society of London 2008.

England (NE), Countryside Council for Wales (CCW) and Scottish Natural Heritage (SNH). Northern Ireland has its own statutory body, Environment and Heritage Service (Northern Ireland) (Nature Conservancy Council 1990).

The significance of RIGS is stated within, for example, Planning Guidance (Wales), Technical Advice Note (TAN) 5—Nature conservation and Planning. Interesting geological/geomorphological, pedological and/or physiographical features may be notified because of their scientific, educational, historical and/or aesthetic, landscape importance. These can all be incorporated under the concept of geodiversity.

One of the greatest threats to our geodiversity heritage is the lack of awareness of many of our important sites and the misconception that all these sites are robust. Many are extremely vulnerable, particularly those that are geomorphologically related. This raising of public awareness has been at the forefront of the activities of some RIGS groups.

History of the Welsh RIGS movement

Although the RIGS scheme had been initiated in the UK in 1990, within Wales only one RIGS group was successfully founded at this time—Powys—in 1990 by Duncan Hawley. By 1992, they had notified 45 sites to the local authorities. Following this success, it was decided by the CCW that RIGS groups should be initiated throughout Wales. In 1992 CCW, the statutory government conservation agency given the task of safeguarding environmental sites, decided to take action and held the first exploratory meeting in Bangor, North Wales in January 1993. As a result Gwynedd and Clwyd RIGS Group was formed in March 1993. It was soon discovered that this was too unwieldy and covered too large an area. It was subsequently split into two—Gwynedd (now Gwynedd and Môn) and NEWRIGS (North East Wales RIGS). A further meeting was held in Cardiff in December 1993 to set up the southern groups so that between 1993–4, five new RIGS groups emerged over the whole of Wales (Table 1).

The speed with which the groups were set up indicated the importance of solving common problems, which were being identified at the national level, such as difficulties of organization, standardization in choice of sites, methods of recording and site development. This will be discussed under History and development of the AWRG.

Local Welsh RIGS development

By 1995 it had become clear that the number of members of each group was going to be a problem. The North Dyfed (Mid Wales) and

Table 1. *1994 RIGS groups in Wales*

Group name	Year set up	HQ
Glamorgan & Gwent	1994	Cardiff
Gwynedd	1993	Bangor
Mid Wales	1994	Aberystwyth
NEWRIGS	1993	Mold
Pembrokeshire	1994	Haverfordwest
Powys	1990	Brecon

South Dyfed (Pembrokeshire) groups merged to become one—the Dyfed group. In 1998, realignment of boundaries led to the creation of South Wales (Glamorgan and Gwent), Pembrokeshire (Dyfed) and Mid Wales (Powys) groups. The Mid Wales and South Wales groups struggled to keep going in the late twentieth century and had effectively 'disbanded' by 1999. With the increased awareness and funding available from the Welsh Assembly Government they have now become reinvigorated (as discussed below). During this time the situation has remained stable in the north.

The geological expertise available within the different groups was reflected in the different emphasis each gave to their brief. This was normally dictated by three factors:

- the expertise of the individual members of the committee;
- the location of academic institutions within the area; and
- the geology of the area.

Thus in North Wales, the Gwynedd and Môn RIGS group researched, visited sites, assessed and registered RIGS.

Within the NEWRIGS group, initially research and public awareness seemed to take over from site notification (Burek & France 1997, 1998). They produced their bilingual urban trail series entitled *Walking through the past*. The first town described in 1998 was Llangollen; the leaflet was produced in partnership with CCW. Since then various sources of funding such as Geologists' Association's Curry Fund, Awards For All, and Tarmac have been secured to produce the subsequent leaflets (see 2004–May 2007: The Current position).

In South Wales the overriding interest appeared, from afar, to be educational. In Mid Wales there seemed to be an interest in minerals and mining locations. All these activities are laudable in themselves and helped move the RIGS movement forward in many different ways but with a common direction and aim—the conservation of the geodiversity heritage within Wales.

Welsh RIGS in the twenty-first century

2000–2003. The history of Welsh RIGS groups in the early years of the twenty-first century is one of contraction and expansion. The two leading organizations in Wales were the two northern groups Gwynedd and Môn RIGS and NEWRIGS which seemed to go from strength to strength, both concentrating on different areas of activity. The emphasis of each group was steered by local events in their areas as well as different types of expertise within the group. Gwynedd and Môn were influenced significantly by a public enquiry on Anglesey. The Beaumaris Marina public enquiry on 27 July 2000 reached a decision on 29 July 2002 (John Williams, pers. comm.). This site included a RIGS teaching site Gallow's Point, containing 12 'Palaeozoic dykes and their complex contact relationships with the greenschists in the Gwna Group of the Mona Complex' (Wood 2007). Access to these is referred to in the Inspector's report. The success of the outcome to preserve part of the site in front of the marina and to erect a walkway and an information board explaining its importance by the Marina Developers, showed the importance of complete and robust site information. This led the way to establishing strong documentation to support RIGS designations within Wales (Fig. 1).

NEWRIGS, having initially concentrated on raising public awareness through bilingual trail leaflets (Fig. 2), walks and talks because of the importance of the International Eisteddfod at Llangollen, then followed Gwynedd and Môn RIGS group's lead in having strong documentation. The group had also secured funding from CCW to audit and undertake research on the limestone pavements scattered across their area including the tops of the Clwydian hills which contained the Clwydian range Area of Outstanding Natural Beauty (AONB) (Fig. 3).

The variety of activities carried out by RIGS groups is exemplified by the two northern groups throughout 2001 and some highlights of their activities seem appropriate here. The following description is based on the Welsh report given to the fourth UKRIGS annual conference in 2001.

During 2001, activity in all areas varied as the outbreak of foot and mouth took its toll:

Site visits have been severely curtailed, especially in the high-incident areas of Anglesey, Snowdonia and Mid Wales.

A memorandum of agreement was signed in May 2001 between AWRG (Association of Welsh RIGS Groups) and CCW (Countryside Council for Wales), which releases money to the individual RIGS groups through AWRG for site notification. It also gives AWRG recognised [public] support from the national statutory agency. (Burek 2001*b*)

Dennis Wood died in May 2001 while serving as chairman of the Gwynedd and Môn RIGS group. The following note in the Welsh report shows the esteem in which he was held by his peers and the RIGS members.

AWRG has been collecting money to erect a plaque in his memory at Rhoscolyn in Anglesey, his favourite field site ... Dennis at the time of his death was involved in two projects for the Gwynedd and Môn RIGS group. The first was a book on Precambrian geology of Anglesey and the second was a geological interpretation project for Rhoscolyn, based on Dennis's unpublished original maps. Both these projects were carried forward by Dr Margaret Wood and Dr Stewart Campbell of CCW in his memory. (Burek 2001*b*)

Further varied activities and consultations followed, increasingly showing the influence that RIGS groups held.

The Northern Welsh RIGS groups have attended various functions to raise public awareness of their work including the National Eisteddfod in Denbigh. They all now have bilingual posters to advertise their work. They also attended the Wrexham Science Fair and have been consulted on the Compact between the Statutory Environmental Organizations and the Voluntary sector in Wales. (Burek 2001*b*)

Details of the individual North Wales groups followed again showing the differences in approach to activities by the two groups.

Individually Gwynedd and Môn RIGS [group has] notified nine potential sites on Anglesey and the Lleyn Peninsula and visited numerous others. They are involved in cleaning up an erratic site at Llaneilean on Anglesey and have also started on a town trail of Bangor.... Several sites along the newly built A5 have been identified and preserved on Anglesey.

NEWRIGS has:

notified a limited number of sites mostly to Wrexham Maelor Borough Council this year. They have also found the first, easily accessible unconformity between the Ordovician and Carboniferous in the area [at Minera]. Work is progressing at Minera on site clearance with the help of Lafarge. Several other sites including the mining sites identified by the CCW national audit are being processed. Work proceeds on the limestone pavement research with material being produced to put into the national HAP (Habitat Action Plan) database at English Nature headquarters in Cumbria. At present NEWRIGS [is] upgrading [its] statement of interests to fit in with the Gwynedd and Môn forms which CCW requests. (Burek 2001*b*)

During these early years of the twenty-first century, the outbreak of foot and mouth precipitated an increase in the number of town trails produced as they could be used in urban non-foot and mouth areas to explain geodiversity. The Small Grants Fund from Royal Society for Nature Conservation (RSNC) in 2000 supported two of these (Mold and Llandudno), which were completed in April 2001. The group also produced several other town trails the penultimate 'being Denbigh to fit in with the [national] Eisteddfod' location that particular

Instructions underlined for filling in the Site record

Section A

Gwynedd & Môn RIGS Group Site Record

General	Gwynedd & Môn
Site Name: unique to project	**File Number:** your own system
RIGS Number: supplied by GIS later	**Surveyed by:**
Grid Reference: 6 figure	**Date of visit:**
RIGS Category: Scientific, and/or Educational/ Historic/ Aesthetic	**Documentation prepared by:**
Earth Science Category: e.g. Quaternary/ Precambrian / Igneous etc	**Documentation last revised:**
Network: e.g. Precambrian reference section	**Photographic record:**
Sub-network: Fossils	**Date registered:** When document sent **Owner:** **Planning authority:**
Unitary Authority: Named County Council/National Park	
Site Nature: Codes supplied by CCW later	
1:50 000	
1:25 000	
1:10 000	

RIGS Statement of Interest: (this should be short and easily read by a non-geologist— preferably about half a page)   Start with one sentence naming the site and why it is important.   What is its importance in the local and wider context?   Where is it?   Its importance as a scientific resource eg does it have any implications to other features in maybe the landscape, accessibility etc.   Short description of the features and basic interpretation.   Is there a wider implication of the feature e.g. in a British Isles network?

Fig. 1. Welsh RIGS documentation.

year. This was launched by the Chairman of CCW and the Mayor of Denbigh in July 2001.

During 2000, NEWRIGS started to think further afield and 'contacted other groups about co-operating in producing a long-distance trail along Offa's dyke, starting from the north. Work had started but was halted by [the] foot and mouth [outbreak]' (Burek 2001*a*).

Gwynedd & Môn RIGS group produced two bilingual trails during this period *Walking through*

Geological setting/context: This is a general essay which is written for all the sites in a network e.g. all the Precambrian, Carboniferous, or Devonian etc, or all historic sites. It sets the scene for your construction and reasoning behind the representative network you have chosen for your topic. It does not need to be lengthy—one or two pages. Once done it is used as the first part of all your sites in your network. The last paragraph explains what network you have constructed.

Finally, in this section, under the heading **Network** context of the site: write the name of the site stating exactly where the site fits in your network
(the network may have been chosen on stratigraphical grounds or to cover the period/rock types/paragenesis/basin topography etc.)

References: As many as you think are useful in the same format as the example sites.

Section B

PRACTICAL CONSIDERATIONS:

Accessibility: How to find the location.
Availability of parking.
Walking access.

Safety: List any dangerous features e.g. mine shafts, steep slopes, unstable rock, high cliffs, slippery surfaces, state of tides etc. The final sentence should be e.g. 'The normal safety precautions for working in coastal/farming/quarries/etc should be observed and the state of the tide should be monitored.' Omit unwanted words and add other obvious dangers.

Conservation status: These can be found by designations on maps or by contacting the Earth Science Regional Officer, Ray Roberts at CCW 01352 706656. Ray will also produce the final map for you on the GIS once you have sent in a draft boundary on the map supplied by Margaret or Ray. At this stage he also gives the site its Wales RIGS number. Examples of status include: National Park; National Trust; sAONB; Heritage Coast; Local Nature Reserve; Ancient Monument and number; SSSI; GCR; etc.

OWNERSHIP/PLANNING CONTROL:

Owner/tenant: When visiting the site please try to see or find out who owns or tenants the land.

Planning Authority: County Council or NP etc.

Planning status/constraints: Quarries may have planning orders but in general the following is suffice. 'There are no known plans to develop or modify the area covered by this RIGS.

Fig. 1. (*Continued*).

the past: A geological trail for Lower Bangor in 2001 and *Walking through the past: A geological trail for Caernarfon* in 2003 (Fig. 4). These followed the template set by NEWRIGS except they were A3 not A4. By 2003, they had also produced a bilingual booklet *Precambrian Rocks of the Rhoscolyn Anticline.* This was developed from an idea originally conceived by Margaret Wood in 1994

CONDITION, USE & MANAGEMENT:

Present use: What is happening at the moment? Maybe grazing for sheep or cattle, by a public footpath e.g. The Anglesey Coastal Path, a quarry may be operational or disused, a beach used by holidaymakers etc.

Site condition: Are the features of interest in a good condition? Suffering erosion, in danger of becoming obscured etc.

Potential threats: State if any immediate or not. What could cause problems, e.g. engineering works, building on the site, removal of specimens etc?

Site Management: Always state. 'A close liaison should be maintained with the landowner(s) and local planning authority over all aspects of site management. Then, 'The current status and condition of the ..... features are satisfactory/unsatisfactory and (no) modifications are currently required.' (List things which need doing if relevant)

SITE DEVELOPMENT:

Potential use (general): Would it benefit the general public by having some form of information board/interpretation panel/trail/leaflet etc? State if these should be avoided e.g. A sensitive fossil site where collection could damage the interest.

Potential use (educational): Which group of people would get some benefit by visiting or using the site? (primary, secondary schools), interested amateurs, college and university students, researchers, etc.

Other comments: Important notes which do not fit in the categories listed or research details of the site which could be given to potential researchers.

The completed documentation for Trwyn y Parc (weathering site) without the site map and aerial photograph.

Fig. 1. (*Continued*).

---

and was dedicated to Dennis Wood who helped with the research (Wood 2003).

In South Wales the development of RIGS groups proved more difficult to sustain. This may have been due to the presence of an enthusiastic local Geologists' Association group (one of the oldest in the country) or the presence at that time of two universities with strong geology departments. Thus conservation of sites and educational material was already being achieved by a different route. 'The situation in the south of Wales is more problematic. The former chairman [...] has resigned due to travel difficulties and pressure of work ... we await events in that area' (Burek 2001*a*).

2004 – May 2007: The current position. Wide publicity production continued for both northern groups

with trails and posters by NEWRIGS who developed additional guides throughout the period under discussion.

- *Walking through the past: A geological trail around Wrexham* was produced with a small grant from Tarmac in 2004.
- *Walking through the past: A geological guide to Flint* and *Walking through the past: A geological guide to Ruthin* produced with support from CCW in 2005.
- *Walking through the past: A geological guide to St Asaph* and *Walking through the past: Llangollen* second edition were produced with a grant from Awards for All Lottery fund in 2006.

Work continues on producing a town trail for Holywell and a second edition of the Llangollen Steam Train trail. Co-operation between Cheshire

Gwynedd & Môn RIGS Group Site Record

General	Gwynedd & Môn
Site Name: Trwyn y Parc	**File Number:** 0000
Wales RIGS Number: 0000	**Surveyed by:** S. Campbell & M. Wood
Grid Reference: SH 372941	**Date of visit:** 27.10.05
RIGS Category: Scientific	**Documentation prepared by:** S. Campbell
Earth Science Category: Quaternary & Geomorphology	**Documentation last revised:** 21.11.05
Network: Pre-Quaternary landscape evolution	**Photographic record:** CD
Sub-network: Tertiary weathering	**Date registered:** Owner: Planning Authority:
Unitary Authority: Isle of Anglesey County Council	
Site Nature: EC : coastal cliffs & foreshore ED : disused quarry	
1:50 000: Sheet 114, Anglesey/Môn	
1:25 000: Explorer 262, Anglesey West	
1:10 000: SH 39SE	

RIGS Statement of Interest:

Trwyn y Parc is of national (Great Britain) importance for its fossiliferous deposits of Miocene age. Non-marine Miocene sediments containing datable floras are known from only three parts of Great Britain——the southern Pennines, Cornwall and Anglesey. The deposits at Trwyn y Parc have major implications for understanding how the British landscape evolved prior to the Pleistocene ice ages. In particular, they provide critical evidence to show that Anglesey's 'Menaian Surface', namely the broad shape and form of the island's landmass, and the adjacent Snowdonian mountain block were well-established landscape features by the end of the Miocene, that is by 5.3 million years before present.

The geomorphological features at Trwyn y Parc consist of a series of 15 'pipes', mostly circular or ovate in form, located in steeply dipping Precambrian limestone.

Fig. 1. (*Continued*).

RIGS and NEWRIGS hopes to produce a joint trail of Holt and Farndon crossing the English–Welsh boundary.

The total number of trails produced to date is ten, nine of which are bilingual Chester's is not (Fig. 2). Posters included *Castles of North East Wales* (Fig. 5) and the *Geodiversity of North East Wales*, *Wrexham Borras Quarry* and also a general trail showing the location of the town trails. All are bilingual. NEWRIGS also produced a bilingual all-in-one timescale, bookmarker, ruler and photo scale in 2006 funded

The pipes are exposed both in plan on the foreshore and in profile in a series of natural cliff and quarried sections. They occupy an altitudinal range of 12 m. Eight of the pipes contain a karstic infill. In several pipes, the zoned infill comprises multi-coloured clays and breccia truncated by Devensian Irish Sea till. Samples of black clay have yielded pollen and spores of 62 different taxa representing a fossil flora of Miocene age (Walsh *et al.* 1996). Minerals present in the fill include goethite (a product of severe weathering of silicate-bearing rocks subjected to hot and wet climates) and gibbsite (often associated with intense tropical weathering). Mineralogical evidence from one pipe suggests that most of its fill must have been derived from weathering of a large mass of probably acid igneous extrusive rock, possibly a large clast in the Precambrian mélange, for which there is now no other evidence other than its weathering residues.

The pipe deposits are thus interpreted as the remnants of an extensively weathered land surface or saprolite—a deposit of clay and disintegrated rock fragments, formed by the weathering and 'rotting' of rock *in situ*. At Trwyn y Parc, the saprolitic deposits and palaeosols have been selectively preserved because karstic solution allowed them to extend and subside into a series of deep 'pockets' or 'pipes' into the underlying Precambrian limestones. The fossil flora suggests that the climate at the time of formation of the organic clays was temperate/warm temperate. Swamp forests appear to have been juxtaposed with the mixed vegetation of dry uplands. Anglesey's Miocene landscape was clearly dominated by a forest cover of trees and bushes. Because the pipes and their fill at Trwyn y Parc are likely to have been severely truncated by Pleistocene glaciations, it is estimated that northern Anglesey's Miocene land surface lay at *c.* 50 m asl, somewhat higher than today's. It is tempting to conclude that the swamp-forest flora represents the former vegetation cover on this very flat geomorphological feature, whereas the mixed dry upland vegetation represents the cover on the already extant Snowdonian mountain block (Walsh *et al.* 1996). What is clear, however, is that the broad form of the Anglesey and Snowdonian landscape we see today was already well established by the end of the Miocene. This evidence has major implications for understanding the wider geomorphological evolution of the British Isles.

Geological setting/context:

Long-term, pre-Pleistocene landscape development

There is significant evidence that Permo-Triassic erosion effected the primary shaping of the present relief of the Palaeozoic rocks of western Britain. The broad outline and form of Anglesey's coastline and landscape thus appear to have been fashioned by that time. Although the landforms and deposits which adorn the present landscape are the result of processes (especially those of glacial and cold climates) operating in the Pleistocene, the landscapes of western Britain were in fact fashioned over the many millions of years between the Permo-Triassic and the Pleistocene, by a wide range of non-glacial processes operating under diverse, including tropical and sub-tropical, climatic regimes. Depositional evidence for the intervening Jurassic, Cretaceous and Tertiary is, of course, well represented elsewhere in south and eastern Britain, but deposits of these ages were either not deposited or have since been removed from substantial parts of Wales and

Fig. 1. (*Continued*).

by a grant from Awards for All Lottery. Gwynedd and Môn RIGS has continued with its trail series producing *Stone detectives—a geological trail for Conwy town* and *Rhoscolyn:* *Legend in the Rocks—Historical links with Rhoscolyn's geology* in 2004. Gwynedd & Môn RIGS supported by AWRG, Royal Agricultural College and CCW published a bilingual booklet south-west England. At the very best they have been selectively preserved on-land in narrow fault-controlled basins (e.g. the Eocene and Oligocene deposits found in the Bovey-Tracey and Petrockstow basins) and offshore (e.g. the Jurassic and tertiary deposits found in the Cardigan Bay, Central Irish Sea and St George's Channel basins).

The weathering and shaping of the landmass we see today has therefore been a protracted process. Much of the evidence for the geomorphological processes that operated and shaped the landscape during these times was modified or removed by the Pleistocene glaciations and periglacial processes, particularly in Wales, and the vast bulk of the evidence for these pre-Pleistocene events is erosional in nature and lends little precision to the interpretation of events.

The denudation chronology of the British landscape has been detailed in a wide range of publications (see Campbell *et al.* 1998 for a review). Traditionally, geomorphologists have viewed the early to mid-Tertiary as a time when the landmass of Wales and south-west England was subjected to alternating phases of marine inundation and planation and subaerial exposure and weathering: these conditions were used to account for a multiplicity of perceived erosion surfaces—'staircases' of these surfaces were widely invoked throughout Wales (e.g. Brown 1960) and south-west England (e.g. Balchin 1937). Recent evidence, particularly from St Agnes Beacon in west Cornwall and Trwyn y Parc in Anglesey, shows Miocene and mid-Oligocene deposits overlying a prominent erosion surface at between *c.* 75–131m OD. This suggests that surfaces above this general level can be no younger than Miocene and lends much support to the view that large areas of the south-west peninsula and Wales have existed, more or less, in their present forms since perhaps as early as the Eocene. Landscape evolution subsequent to this is likely to have been slow, the sole depositional evidence for a marine incursion onto these ancient landmasses being a minor transgression of late Pliocene age at St Erth, Cornwall.

In recent years, there has been an increasing tendency to invoke only one complex polygenetic erosion surface (e.g. Coque-Delhuille 1982; Battiau-Queney 1984), the constituent landforms having been shaped in tropical and sub-tropical environments (Upper Cretaceous and Palaeocene) and uplifted since the late Miocene (Green 1985). In Wales, this inherited landform has been deeply dissected by Pleistocene glacial activity; in south-west England, periglacial processes appear to have been the chief Pleistocene land-shaping agents.

In view of the above, sites where depositional evidence for pre-Pleistocene land-shaping events is preserved take on a disproportionate importance. In Wales and the Borderlands, evidence for long-term, pre-Pleistocene landscape evolution is represented by a network of RIGS where evidence of Tertiary weathering, the formation of weathered regoliths, saprolites and palaeosols is preserved and has a bearing on the development of specific landforms (e.g. tors) or major landscape features (e.g. erosion surfaces). At some of the sites, evidence for supergene weathering may be superimposed on bedrock previously altered by hydrothermal or pneumatolytic processes associated with intrusive igneous activity. These structurally compromised materials have been susceptible to subsequent exploitation by erosional processes and given rise to distinctive landscape features such as bays and inlets and specific landforms such as tors.

Fig. 1. (*Continued*).

on *Soils in the Welsh landscape* (Conway 2006) (Fig. 6).

However, with the advent of the Welsh Aggregates Levy Sustainability Fund (ALSF) and successful bidding by a joint application in 2004 from both northern groups to perform a geodiversity audit across the whole of North Wales, AWRG was spurred into renewed energy. Two full-time project

Network context of the site

Trwyn y Parc RIGS is one of four sites selected on Anglesey to demonstrate key evidence and concepts concerning the long-term geomorphological evolution of the island. It demonstrates a fossil flora of Miocene age with major implications for understanding the age of Anglesey's land surface (see also Halkyn Mountain RIGS, north-east Wales). Porth Wen RIGS, Porth Swtan RIGS and Porth Padrig RIGS are geomorphologically significant in providing important examples of highly altered bedrock possibly related to non-supergene processes (e.g. hydrothermal activity). Collectively, the sites have major importance for understanding the long-term geomorphological evolution of Anglesey's, and indeed Wales', landscape.

Section B

PRACTICAL CONSIDERATIONS:

Accessibility:
The site lies on the east side of Cemaes Bay in northern Anglesey. It is reached from the A 5025 Amlwch to Valley road, turning north at SH 379938 along a minor road, past the Gadlys Hotel, to a T-junction. Here, a left turn leads south-westwards to a small layby opposite to the entrance to the disused Gadlys (sometimes known as the Penrhynmawr) Quarry. A small number of cars or minibuses can be parked in the layby. The site can be reached on foot along the small lane that leads to Gadlys Farm [check] and through the disused quarry onto the coast.

Safety:

References:

BALCHIN, W. G. V. (1937). The erosion surfaces of north Cornwall. *Geographical Journal*, 90, 52–63.

BATTIAU-QUENEY, Y. (1984). The pre-glacial evolution of Wales. *Earth Surface Processes and Landforms*, 9, 229–252.

BROWN, E. H. (1960). *The Relief and Drainage of Wales.* Cardiff.

CAMPBELL, S. & BOWEN, D. Q. (1989). *Quaternary of Wales*. Geological Conservation Review Series No. 2. Nature Conservancy Council, Peterborough, 237pp.

CAMPBELL, S., HUNT, C. O., SCOURSE, J. D. & KEEN, D. H. (1998). *Quaternary of South-West England*. Chapman & Hall, London, 439pp.

COQUE-DELHUILLE, B. (1982). Importance de l'érosion differentielle et de la tectonique tardi-Hercynienne dans le massif du Dartmoor (Grande-Bretagne). *Hommes et terres du Nord*, 3, 9–26.

Fig. 1. (*Continued*).

officers were appointed for two and three years respectively to carry out this audit. By 2005 a Central Wales area project officer had also been funded for two years by the Welsh Assembly Government (WAG) to produce a geological audit of sites. The NEWRIGS project was successfully completed in March 2006 with over 125 sites notified (Fig. 7; Malpas & Burek 2006). In March

GREEN, C. P. (1985). Pre-Quaternary weathering residues, sediments and landform development: examples from southern Britain. In: *Geomorphology and soils* (eds K. S. Richards, R. R. Arnett, and S. Ellis), George Allen and Unwin, London, 55–77.

GREENLY, E. (1919). *The geology of Anglesey.* Memoirs of the Geological Survey of Great Britain. HMSO, London, 980pp. (2 vols)

GREENLY, E. (1920). 1:50 000 (and 1 inch to 1 mile) Geological Map of Anglesey. Geological Survey of Great Britain, Special Sheet No. 92 and (93 with parts of 94, 105 and 106).

MORAWIECKA, I. M., SLIPPER, I. J. & WALSH, P. T. (1996). A palaeokarst of probable Kainozoic age preserved in Cambrian marble at Cemaes Bay, Anglesey, North Wales. *Zeitschrift für Geomorphologie N.F.* 40, 47–70.

WALSH, P. T., MORAWIECKA, I. M. & SKAWIÑSKA-WIESER, K. (1996). A Miocene palynoflora preserved by karstic subsidense in Anglesey and the origin of the Menaian Surface. *Geological Magazine*, 133(6), 713–719.

WHITTOW, J. B. & BALL, D. F. (1970). North-west Wales. In: Lewis, C. A. (ed.) *The Glaciations of Wales and adjoining regions.* Longman, London, 21–58.

The coastal and quarried cliffs here are not particularly high or precipitous, but the foreshore and cliff areas are especially dissected and uneven. Many of the solution pipes occurring in plan lie in the slippery intertidal zone. The normal safety precautions for working in coastal areas should be observed and the state of the tides should be monitored.

Conservation status:
The site overlaps with Llanbadrig Area GCR site (Precambrian of England and Wales; not yet notified as SSSI) [check] and Gadlys Quarry RIGS (Precambrian fossils). It lies in Anglesey's AONB and Heritage Coast [check]. The adjacent lime-kiln is of historical interest.

OWNERSHIP/PLANNING CONTROL:

Owner/tenant:
The National Trust
Gadlys Farm?

Planning Authority:
Isle of Anglesey County Council.

Planning status/constraints and opportunities:
There are no known plans to develop or modify the area covered by this RIGS.

Fig. 1. (*Continued*).

2007, the Gwynedd & Môn RIGS and in May 2007, the Central Wales RIGS groups successfully finished their projects with 204 and 135 RIGS identified respectively (Fig. 7). Throughout Wales, RIGS groups are being revived or expanded although progress is slower in the South. By mid-2007, South West Wales RIGS group had received financial backing from the WAG to

CONDITION, USE & MANAGEMENT:

Present use:
**The site lies largely on coastal land owned by The National Trust and used for
recreation. A coastal path runs through the site from Cemaes Bay to Porth Padrig
and beyond. The Gadlys Quarry is long disused and the inland parts of the site
are grazed by farm animals.**

Site condition:
**The geomorphological ('pipe') features are in good condition. Those seen in cliff
section are well exposed without suffering excessive coastal erosion, whilst
those seen in plan on the foreshore are exposed to tidal processes, inevitably
including scour and periodic obscuration by shingle and sand.**

Potential threats:
**There are no perceived immediate threats. Engineering works, such as the
construction of retaining walls or slipways, have the potential to remove or
obscure critical deposits.**

Site Management:
**A close liaison should be maintained with the landowner(s) and local planning
authority over all aspects of site management. The current status and condition
of the geomorphological features are satisfactory and no modifications are
currently required.**

SITE DEVELOPMENT:

Potential use (general public):
**Although the site lies in a readily accessible coastal area owned by The National
Trust, the highly specialist nature of the geomorphological feature probably
does not warrant a site-specific interpretation initiative for the public. However,
the site interest could be combined with other local geological features
(including Precambrian fossils) into a more general geological/heritage trail.**

Potential use (educational):
**This RIGS is one of the most important sites in Britain for understanding
long-term landscape development, and is suitable for small parties of students
at A-level and above. It is particularly well suited for demonstrating the complex
concepts of denudation chronology and long-term landscape evolution for
specialist audiences. The site has considerable potential for future research.**

Other comments: **This is one of only three known fossiliferous on-land Miocene
deposits in Britain. As such it is of national (Great Britain) importance.**

Fig. 1. (*Continued*).

produce a geodiversity audit and South Wales was working on setting up a possible RIGS consortium with leadership and partners from National Museum, British Geological Survey, Cardiff University, South Wales GA, Brecon Beacons National Park and the Fforest Fawr Geopark. Fforest Fawr was formed as a Geopark in 2005 (Davies *et al.* 2006) and appointed its first

Fig. 2. Front covers of bilingual 'Walking through the past' town trails—NEWRIGS.

Fig. 3. Taranau Limestone pavement RIGS.

Geodiversity Officer in January 2007. Up until recently local authorities such as Cardiff, Rhondda Cynon Taff and Neath Port Talbot have all undertaken geological audits of their areas generally through the local universities. In Cardiff, the South Wales GA is working with the Local Authority with the hope of identifying potential RIGS by the end of 2007. However WAG and CCW have set up rigorous standards for RIGS documentation which should be adhered to by all RIGS projects with the aim of establishing a common Welsh database.

Although work has progressed at different rates across Wales, the advent of both the Fforest Fawr and Anglesey Geoparks (Anglesey Geopark–GeoMôn is currently seeking European Geopark status) is a boost to the whole geoconservation movement in Wales. The advent of the ALSF in Wales is really making a difference to geoconservation at the local level.

Overall RIGS work features quite highly with local authorities throughout the principality and the place of RIGS within geoconservation is recognized at all levels. With a rigorous and robust set of designation criteria and a database in place, which will eventually be aligned with the UKRIGS Geoconservation Database, RIGS are ensured of recognition in the local authority planning system. Although conflict on RIGS occasionally raises its head as in the case of Holt (Burek 2000; Burek *et al.* 2007) or the Beaumaris Marina (John Williams, pers. comm.), discussion normally leads

to an acceptable compromise. The importance of a geological statement of interest within a RIGS designation has already been tested at a public enquiry on Anglesey at the Beaumaris Marina hearing. This is the ultimate test for RIGS and indeed any geoconservation assessment.

History and development of the AWRG

During the December 1993 RIGS meeting in Cardiff, Mike Brooks suggested that there should be an annual Welsh conference where common geoconservation and RIGS problems could be discussed (Burek 1998). This culminated in the Brecon conference in 1994 (Campbell & Wood 1995). However it was also necessary to have an organization to facilitate these and to provide consultation at the national level. The six chairmen (quickly becoming five) formed a steering committee to explore the national identity of the mainly regional activities of the groups and they finally proposed the establishment of the AWRG, which was ratified by the constituent regional groups and the annual residential forum. The steering committee was renamed the Association of Welsh RIGS Groups (AWRG) in order to distinguish the organization from the event. This met for the first time in 1995 (Burek 1998).

At this point it is beneficial to look at the broad objectives that were set for this exchange of expertise. Eight points were deemed important and are considered the main objectives of AWRG:

- co-ordination of non-statutory Welsh geological conservation through the area RIGS groups;
- exchange of information, ideas and expertise between the area RIGS groups;
- creation of a national profile for RIGS activities in Wales where this may be more effective;
- liaison with CCW and other relevant national bodies, both in Wales and beyond, on geological conservation and related aspects of wildlife, heritage and countryside conservation;
- funding of management and conservation projects from sources identified at a national or regional level;
- consolidation of area group databases into a national scheme;
- development of a forum for all voluntary interests in geology within Wales; and
- organization of an annual, residential forum, rotating its location, administration and chairmanship of the AWRG through each of the area groups.

History of the Welsh Annual Forum

The divergence of ideas of the different groups had been held in check and the overall aim and direction had been achieved by holding an annual forum. It

Fig. 4. Front covers of bilingual 'Walking through the past' town trails—Gwynedd & Môn.

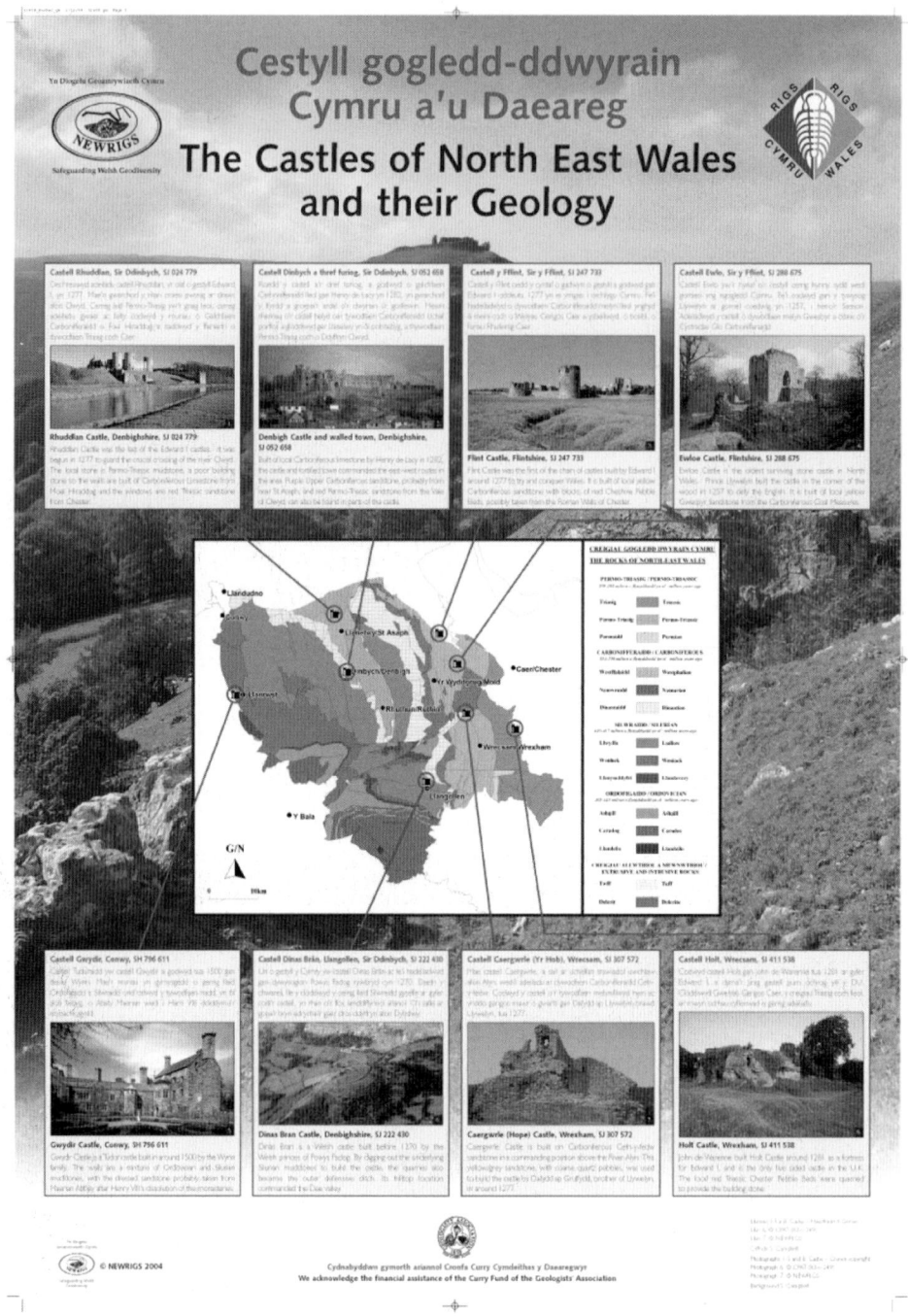

Fig. 5. Poster of Castles of North East Wales.

was decided early on that each group should host a forum (Table 2) alternating between North and South Wales. Each one had concentrated on a different targeted audience and theme. Each annual forum was residential and run along similar lines with strong support from the statutory agency CCW. The first day consisted of a number of invited speakers, a review of what the groups had achieved, and a series of workshops. The second day was devoted to field visits—either to potential

Soils in the Welsh Landscape

(A field-based approach to the study of soil
in the landscapes of North Wales)

Priddoedd yn Nhirwedd Cymru

(Dull ar raddfa maes o astudio pridd yn
nhirweddau Gogledd Cymru)

J. S. Conway

Fig. 6. Soils of Wales booklet.

RIGS or to locations of interest along the theme of the conference. Hence, at Llandudno 'we visited the Great Orme Copper mine (a tourist attraction) in 1995 and in 1997 we visited two potential RIGS proposed for their historical interest and associations with Darwin and Sedgwick' (Burek 1998). At Machynlleth in 1998 there was a visit to a silver mine. The 1999 forum at the National

NEWRIGS

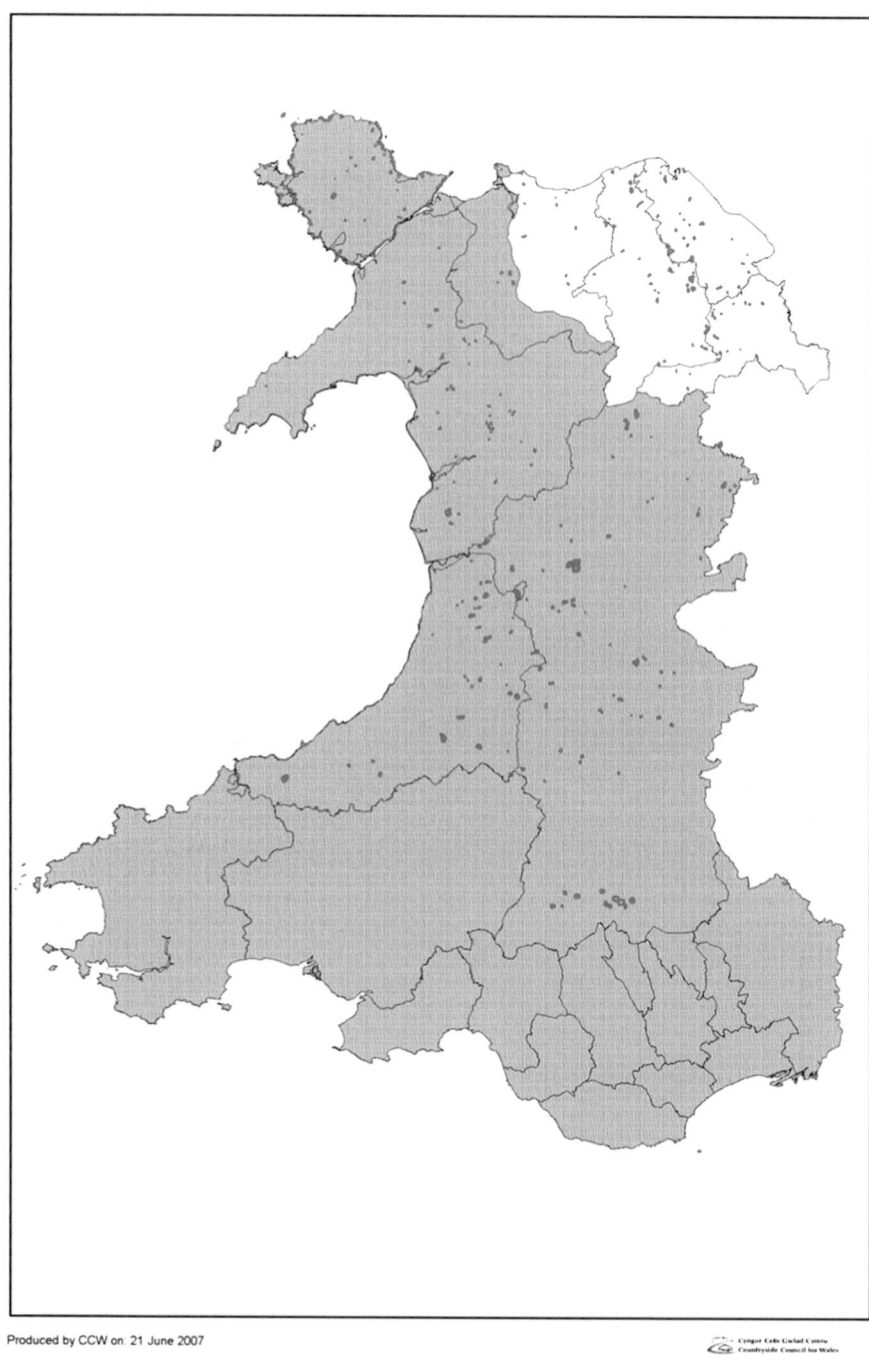

Fig. 7. Map of NEWRIGS area (in white) showing both the boundary with England and the other RIGS groups; Gwynedd & Môn RIGS and Central Wales RIGS groups. Gwynedd & Môn RIGS and NEWRIGS are the two northern groups who carried out the first geodiversity audit funded by the Welsh Assembly Government. RIGS are marked.

Museum of Wales in Cardiff was the second forum aimed at the RIGS members themselves. This took the form of a workshop and practical conference that looked at the different forms of publicity that are available to RIGS members to promote their work to a wider audience.

This completed the circle of national AWRG annual fora as originally conceived (Table 2). All five RIGS groups had held their conference concentrating on an important issue of concern to them (Burek 1998). As a result of the heightened awareness generated through the annual fora, the RIGS process has expanded and many planning authorities have now had to incorporate the designation within their structure or strategy plans as a statutory obligation. Local geological and geomorphological sites now have an even greater chance of surviving for the next generation.

Welsh RIGS context

As a result of a growing national awareness throughout the UK which has emerged from the developing local RIGS groups, a UK-wide national conference was called in Worcester in September 1998 to which all the RIGS groups were invited. From this conference, a steering group was set up to develop a UKRIGS structure and committee. A second national conference, again sited in Worcester, was held in September 1999 to which the steering group reported and an elected 8-person committee was set up to represent all the RIGS groups at a national level as a national body. This full committee met for the first time in Chester in November 1999. The further development of UKRIGS is discussed in detail in Burek (2004). The National UKRIGS logo is based on the Welsh trilobite logo and the form of the annual conferences have been very similar to those pioneered by Wales. AWRG is now a member of the UKRIGS organization and plays a strong part in its development. Within UKRIGS it is regarded as an example of good regional practice although devolution has clouded this spatial terminology.

Unique situation in Wales

Wales is important to geology not least because the names of three periods within the Lower Palaeozoic are of Welsh derivation. Cambrian comes from the Latin name for Wales (Wilson 1994) while Ordovician and Silurian are derived from the names of ancient Welsh tribes. Wales is a nation split in two by its landscape. Large parts of the centre are sparsely inhabited. Much of the Welsh population and infrastructure is polarized either along the south or north coasts. However, it does have a common culture and two common languages. These facts provide Welsh RIGS and indeed all geoconservation with its unique features:

- the location of the population and academic institutions; and
- the bilingual nature of official publications and the funding and expertise needed to supply this.

In many respects Wales is similar to Scotland but there the population is heavily concentrated in the centre and the north and south are sparsely populated. Increasingly a second language is seen in official documents too but it is not as widespread as the Welsh language is in Wales.

Data from the 2001 census showed that the majority of the 2 903 085 population in Wales live either along the south coast and valleys or the north coast and down the English border. Very few live in central Wales. With the exception of Lampeter and Aberystwyth the location of the major academic institutions is confined to either the north or south. Most of those with any interest in geodiversity are clustered in the south with the

Table 2. *National Welsh Forum locations, date, themes and principle audience*

Location	Date	Main theme	Target audience
Brecon	1994	Getting Welsh RIGS together—a Welsh strategy	Welsh RIGS themselves
Llandudno	1995	Public awareness & planning	Planning officers & tourist industry
Swansea	1996	Education	Teachers
Llangollen	1997	RIGS in the community: where are they?	Local community, tourist industry & media
Machynlleth	1998	Minerals & mines in Wales	Mining industry & mineral officers
Cardiff National Museums & Galleries of Wales	1999	Interpretation	Welsh RIGS themselves

exception of Bangor University and NEWI (North East Wales Institute of Higher Education) in Wrexham. This of course mirrors the population (Table 3).

Outside South Wales, NEWI and Bangor have limited courses in aspects of geoconservation at undergraduate level and therefore employ few qualified or professional staff. The British Geological Survey no longer has a permanent office in Aberystwyth as it closed in September 1994 but one opened in Cardiff in 2007. Thus the professional academic base is largely missing. This is born out by a survey carried out at the Llangollen annual forum in 1997. 45 questionnaires were completed by participants at the conference, who were assumed to have an interest in Welsh geoconservation by their attendance. The questionnaire sought to gain an insight into sources of help for RIGS conservation both inside and outside Wales. It also wanted to find out more about the location of Welsh speakers with a geological background for possible translation help. Welsh geological speakers and translators are at a premium in Wales. The questionnaires were then analysed by NEWRIGS (Table 4). This conference was chosen as it represented people prepared to travel to Llangollen and who had an interest in RIGS and geoconservation. Figure 8 shows the distances people had to travel from outside Wales to get to the border. Thus it must be concluded that although many people outside Wales would willingly undertake conservation, most do not live close enough in either time or distance to participate easily in the monitoring of RIGS.

These results show that the majority of participants lived outside Wales, but most have attended Welsh geological excursions and 25% have written papers on Welsh geodiversity. This means that there are few people to undertake

geoconservation work within Wales. Few people however, spoke Welsh (<10%), which mirrors the population of Wales. However, although official documents, often including funding ones from outside sources, are required to be bilingual, there are few Earth scientists employed in the field of geological conservation capable of undertaking this task. This has a knock on effect when transmitting the conservation message across to the general public if notice boards and literature are required to be bilingual. Misinterpretation and wrong translations can easily occur when one word has many meanings.

Thus there is no real focus for geology in north Wales and until recently there was no geological society. The Geologists' Association North Wales Group (Cymdeithas Daeareg Gogledd Cymru), a branch of the Geologists' Association founded in 1994, and a branch of the Open University Geological Society (the North Wales branch) have now rectified this. South Wales, in contrast, has a thriving and very well established Geologists' Association local group indeed one of the oldest in Wales (founded in October 1959 but holding its first meeting on 23 January 1960).

Geological museums are few and far between. There is the National Museum of Wales in Cardiff and limited displays attached to some university departments. Besides these there are small museums in North Wales such as Stone Science on Anglesey and in the north-east, Wrexham Museum which has a limited exhibit mainly applied to coal and tile/brick making.

Welsh geoconservation benefits

As far as geoconservation goes, there are three positive attributes that Wales has over neighbouring England or Scotland:

- a large number of quarrying and mining companies;
- outdoor activities often based on walking, natural history or climbing/potholing, in other words a long-established, healthy tourist industry and a large number of keen landscape amateurs associated with this; and
- an extensive and accessible coastline, much of it already protected by National Park status or SSSI designation.

Consequently there are large numbers of planners either directly concerned with geoconservation issues or indirectly through the National Parks, Areas of Outstanding Natural Beauty (AONBs) and The National Trust along with a geopark established in 2005 and a prospective geopark on Anglesey.

The role of the RIGS movement in Wales is more than about notifying sites, it is popularizing

Table 3. *Welsh Higher Education academic institutions dealing with any subject likely to be linked to RIGS (i.e., Geology, Geography, Oceanography, Conservation, Planning) (UCAS 2003)*

University name
University of Wales, Aberystwyth
University of Wales, Bangor
University of Cardiff
University of Glamorgan
University of Wales, Lampeter
University of Wales College, Newport
North East Wales Institute of HE, Wrexham
Swansea Institute of HE
Trinity College, Carmarthen
University of Wales, Swansea

Table 4. *Summary of questionnaire results from AWRG 1997 Forum attendees (44 respondents)*

Question

	Yes		No		Little
1. Do you live in Wales? (If no please answer questions 5 & 6)	4		41		
2. Have you been on any Earth science fieldtrips to Wales?	38		7		n/a
3. Have you written any papers or done research on Welsh geology/geomorphology?	11		33		n/a
4. Do you/can you speak Welsh?	3		3		39

	<5	**5–15**	**16–50**	**51–100**	**>100**
5. How far in miles do you live from the Welsh border?	1	2	13	12	13

	<0.5	**0.5–1**	**>1–2**	**>2–3**	**>3**
6. How far in time (hours) do you live from the Welsh border?	1	15	11	9	5

geology to a receptive audience—environmentally concerned and outdoor tourists.

The second significant feature of Wales is its language and culture which are also a tourist attraction. However, this is a two-edged sword as many publications must be bilingual English/Welsh. There are two difficulties with this. Firstly bilingualism cuts down the space available on any publication and increases the cost of publication. The importance of visual aids is paramount in these situations. Thus the obvious method of getting round this is to use pictures, maps and diagrams in RIGS publications. 'A picture is worth a thousand words'. This can then appeal and educate across the cultural divide. The second difficulty is adequate translation—bed and cleavage show the incredible difference in meaning a wrong translation can produce by someone not familiar with geological terminology. A specialist science translator has to be used (like Dyfed Elis Grufydd) to make sure that the translations are correct and as has been stated before such people are few and far between.

A third point is that Wales is an important source of all types of stone for building especially limestone, slate and aggregate much of which is exported across the border. This is especially true in North Wales where it finds its way to the urban expansion of Manchester and Liverpool. Therefore, there is a constant change in geodiversity, which needs regular monitoring. This has led to a recent increase in funding for both geological and biological conservation needs through the ASLF.

To summarize, the unusual features of RIGS in Wales all add up to a unique situation:

- a polarized population and infrastructure which produce a geographical imbalance between population centres and academic institutions;
- geological information is primarily located in Cardiff or Bangor with CCW but much geological knowledge and expertise is outside the principality;
- important geological sections of international importance to stratigraphic nomenclature;
- no natural focus for geological expertise in north Wales (except of course Countryside Council for Wales in Bangor);
- a major source of raw building materials and therefore constantly expanding quarries needing monitoring;

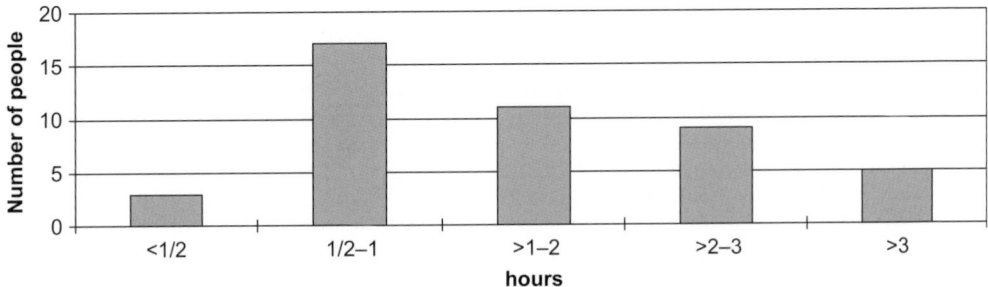

Fig. 8. Time taken to reach the Welsh border by conference attendees.

- a thriving outdoor, well-established tourist industry; and
- a culture that is keen to produce bilingual products.

Welsh achievements in the last 16 years

With a small population and an even smaller resident geological expertise, the establishment of a national RIGS scheme within Wales must rank as a major achievement of the past 16 years. The establishment of the AWRG highlights commitment on this front. In 2006, the Welsh RIGS groups, following an initiative by NEWRIGS, united to produce a bilingual leaflet explaining the importance of RIGS to a wider audience. This was published by AWRG (Campbell 2006). It is probably significant that over the past 16 years the chairman of some of the different RIGS groups have either lived outside Wales or close to the border and most do not speak Welsh. The notification of over 460 RIGS throughout Wales (Fig. 9) is also an achievement not because of the lack of suitable sites but because of the lack of manpower. Until very recently all work was carried out voluntarily by the individual groups with limited financial assistance from CCW. There are at least 100 sites currently being assessed and processed and the mines and mineral sites report, the Minescan report, undertaken by the National Museum of Wales and Countryside Council for Wales has already indicated the need for many new mineral RIGS. The establishment, running and completion of the annual forum is also an outstanding achievement. Every year, for 6 years, interested parties travelled far and wide to discuss geoconservation. Public awareness has been raised through all the media available (Table 5).

Five further achievements are worthy of mention. Firstly, the introduction of a RIGS research programme funded by CCW to undertake various research projects based on the scientific or management aspects of Welsh RIGS. The international importance of limestone pavements as identified by the European Union Habitat Directive in 1992, encouraged three separate contracts to undertake research on the sensitive habitat and its geodiversity firstly in Wales and then later more detailed research in North Wales. To date these have included: identification of limestone pavements in Wales and their flora; limestone pavements, insoluble residue and soil formation in north Wales; grazing management over limestone pavements in Wales; and microclimate and biodiversity on North Welsh limestone pavements. Some of these projects led to publication in scientific research journals such as The relationship between Carboniferous Limestone, insoluble residue and smelting on limestone pavements in

North Wales (Burek & Conway 2000). Many sites were then declared RIGS to give them a measure of protection. More recently projects have been undertaken on: soils of Wales; and the Precambrian of Anglesey.

Secondly, the support of students in their second year of study from the University of Chester to undertake a work placement within the RIGS movement.

Thirdly, RIGS is represented on the Limestone Pavement Biodiversity Habitat Action Plan Steering Group and has input information, which has influenced forthcoming legislation. This means that RIGS has raised public awareness of conservation issues from the local through to the highest levels of government, not a bad achievement for a small, dedicated group of conservation-minded geologists.

Fourthly, the two northern RIGS groups have been jointly granted funds from the WAG administered ALSF to complete a geodiversity audit of the whole of the old counties of Clwyd and Gwynedd and to come up with a strategy and template for the rest of Wales. In 2005 a further grant was given to the revitalized Central Wales group. This audit was completed in May 2007 and a comprehensive database of RIGS is now available for central and northern Wales with guidelines for other groups to follow.

Fifthly, the development of three Local Geodiversity Action Plans, one for Anglesey, one for Fforest Fawr and the other for the Clwydian Range AONB is worthy of mention. These will provide plans to bring geoconservation issues to the attention of the local population and increase their understanding of the issues involved. These holistic action plans, first developed for biodiversity and initially transferred to geodiversity as a pilot in Cheshire in 2002, are being developed across the UK and provide another means of raising the profile of geoconservation across all sectors of the community (Burek & Potter 2002, 2006). Table 6 summarizes the Welsh achievements in geoconservation over the last 16 years.

Lessons learnt

It is not easy to raise public awareness of geological conservation but this *must* be achieved if sites are to be notified and subsequently safeguarded and monitored. Pretty flowers and fluffy animals are emotive; rocks for the most part are not! Enthusiasm and determination to educate and then conserve do pay off in the end but at a price. That price is time. All the members of Welsh RIGS groups are geovolunteers (Burek 2008) and are generally in part or full time employment. Therefore their 'free' time is at a premium. It takes

RIGS in Wales

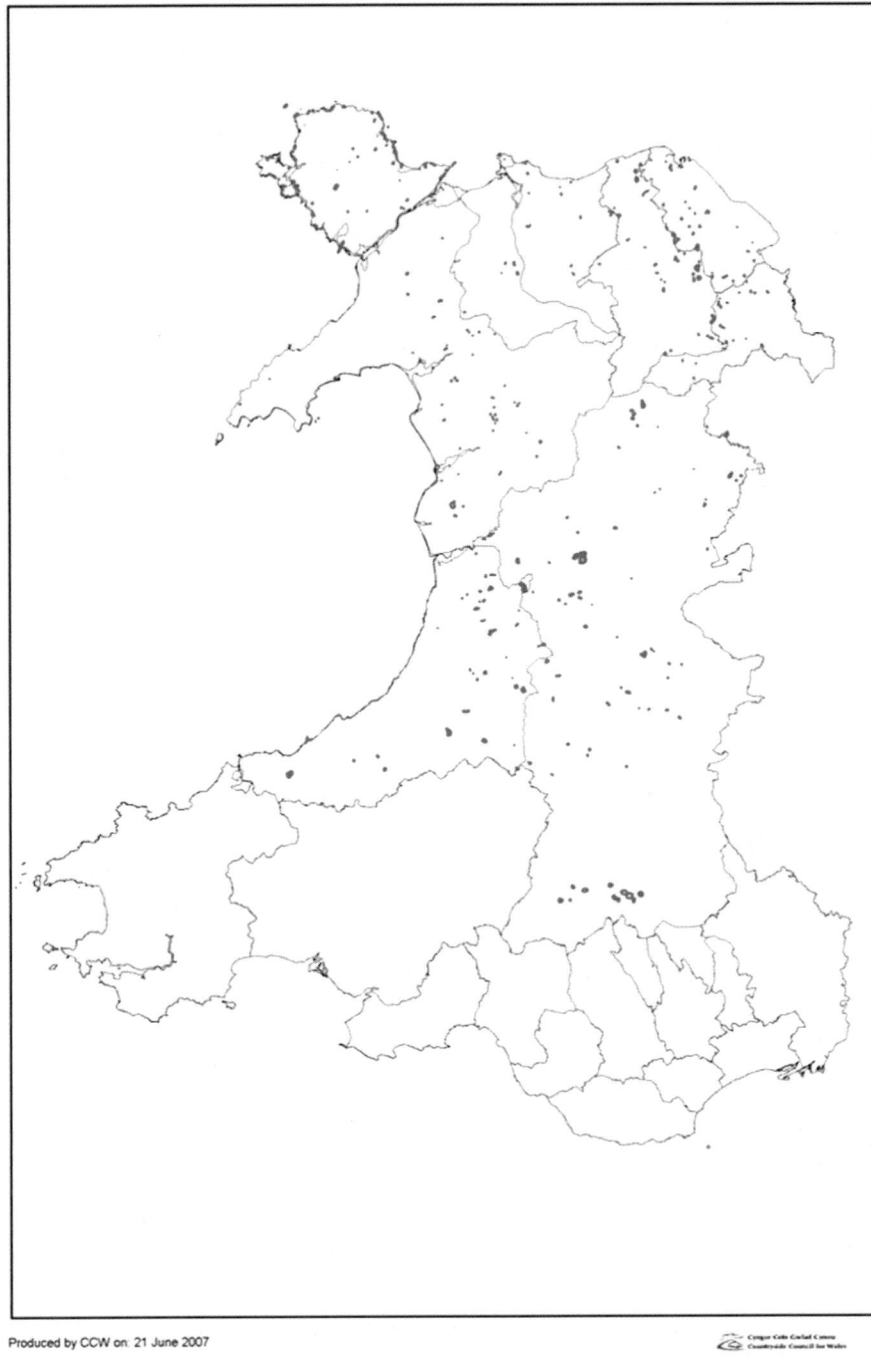

Produced by CCW on: 21 June 2007

Scale 1:1117079   OS base maps reproduced with permission of HMSO. Crown copyright reserved. CCW licence No. 100018813

Fig. 9. Map of Wales showing the wide extent of designated RIGS up to 2007. The large numbers of coastal RIGS on Anglesey shows up clearly. (It should be noted that sites are currently being audited in both South West and South Wales.)

Table 5. *Selected examples of types of media used*

PRINTED MATTER
Popular Newspapers   News items and feature articles
Journals (academic and popular)   Earth Heritage, Breconshire Wildlife & Naturalist
Trail guides, trail boards   Holt, Llangollen
Conference posters   Geoscience 98, GA reunion Cardiff 2004

TELEVISION & RADIO
SC4   News item (Welsh and English)
BBC2   Feature programme
Marcher Sound   interview

PUBLIC INTERACTION: WALKS AND TALKS
Rockwatch children's meetings
WI talk
Eisteddfford tables to raise public awareness of RIGS activities
Classroom visits to local primary schools
Open days at Quarry
Science Festivals at NEWI

Table 6. *Welsh achievements in geoconservation in the last 16 years*

Establishment of a national RIGS scheme
Establishment of AWRG
Notification of over 250 RIGS
Annual residential forum
Research programme
Establishment of work placement for students
Influencing conservation legislation
Raising public awareness of Earth Heritage conservation through many types of media
Funds received from Welsh Assembly for a geodiversity audit for North & Central Wales
Establishment of national database template for RIGS data collection
Establishment of national template for RIGS dissemination of data as reports
Completion of geodiversity audit for North and Central Wales
Establishment of cooperation with Cofnod for wildlife and nature data collection & storage
Funds received from Welsh Assembly for geodiversity audit for South West Wales

---

commitment to the cause and perseverance to succeed. In Wales, the RIGS movement has started out on that never-ending road. Geoconservation must take into account change and therefore, like sustainable development, it will never finish. It is a process and is continuous. Change will also occur in the membership of groups as people move on either literally or metaphorically. This must lead to flexibility in both organization and attitude. Bringing people together, annually at first, was invaluable for discussion of difficulties and problems in this respect. AWRG still meets regularly to discuss problems.

If any lessons are to be learnt from the Welsh experience, it is that a few dedicated people can make a difference and for the sake of the next generation we must do our best to safeguard our Welsh geodiversity heritage.

I would like to thank my two reviewers, S. Howe and S. Campbell, for their constructive comments and additional information, which led to a much better paper. I would also like to thank my fellow chair of Gwynedd & Môn RIGS M. Wood for her support, encouragement and help over the years. Thanks are due to J. Williams of the Ynys Môn/Isle of Anglesey Planning Service for information on the Beaumaris Public enquiry and to R. Roberts of CCW for the RIGS maps. Lastly, the members of AWRG and NEWRIGS for their participation in the process over the last 15 years without which this paper would not have been possible.

References

AUSTRALIAN HERITAGE COMMISSION. 2002. *Australian Natural Heritage Charter*. 2nd edn. Australian Heritage Commission, Canberra.

BUREK, C. V. 1998. *History of RIGS in Wales. In*: OLIVER, P. G. (ed.) Proceedings of the first UKRIGS Conference. UKRIGS Worcester 1998, 147–166.

BUREK, C. V. 2000. The use and abuse of RIGS sites. *In*: ADDISON, K. (ed.) *Geoconservation in Action*.

Proceedings of the third UKRIGS annual conference Newton Rigg, Penrith Cumbria 30 August–2 September 2000. UKRIGS, 23–32.

BUREK, C. V. 2001a. Non-geologists now dig Geodiversity. *Earth Heritage*, **16**, 21.

BUREK, C. V. 2001b. Welsh RIGS Report 2000–1. *In*: ADDISON, K. & REYNOLDS, J. R. (eds) *Upon This Rock*. Proceedings of the fourth UKRIGS annual conference. Stoke Rochford Hall, Peterborough 13–15 September 2001, 87–88.

BUREK, C. V. 2004. What is UKRIGS? *GA Magazine of the Geologists' Association*, **3**, 16–18.

BUREK, C. V. 2008. The role of the voluntary sector in the evolving geoconservation movement. *In*: BUREK, C. V. & PROSSER, C. D. (eds) *The History of Geoconservation*. The Geological Society, London, Special Publications, **300**, 61–89.

BUREK, C. V. & CONWAY, J. 2000. The relationship between Carboniferous Limestone insoluble residues and soils on limestone pavements in North Wales. *Cave & Karst Science*, **27**, 53–59.

BUREK, C. V. & FRANCE, D. 1997. Walking and steaming through the past. *Earth Heritage*, **8**, 8–9.

BUREK, C. V. & FRANCE, D. 1998. NEWRIGS uses a steam train and town geological trail to raise public awareness in Llangollen, North Wales. *Geoscientist*, **8**, 8–10.

BUREK, C. V. & POTTER, J. A. 2002. Minding the LGAPs — A different approach to the conservation of local geological sites in England? *Geoscientist*, **12**, 16–17.

BUREK, C. V. & POTTER, J. 2006. *Local Geodiversity Action Plans—Setting the context for geological conservation*. English Nature Research Report 560, Peterborough, UK.

BUREK, C. V., CONWAY, J., CUBITT, V. & MALPAS, J. 2007. Calling a Holt to conservation conflicts. *Earth Heritage*, **27**, 24–25.

CAMPBELL, S. 2006. Welsh RIGS cooperate on leaflet, *Earth Heritage*, **26**, 3.

CAMPBELL, S. & WOOD, M. 1995. First Welsh RIGS forum. *Earth Heritage*, **3**, 27–28.

Cardiff Computer service website: http://www.cs.cf.ac.uk (accessed 12/8/98).

CONWAY, J. 2006. *Soils in the Welsh Landscape*. Association of Welsh RIGS Groups, CCW, Bangor.

DAVIES, J., HUMPAGE, A. & RAMSAY, T. 2006. Forest fawr: A Welsh First. *Earth Heritage*, **26**, 10–11.

DIXON, G. 1996. *Geoconservation: An International review and strategy for Tasmania*. Occasional Paper 35. Parks & Wildlife Service, Tasmania.

ELLIS, N. V., BOWEN, D. Q., CAMPBELL, S., KNILL, J. L., MCKIRDY, A. P., PROSSER, C. D., VINCENT, M. A. & WILSON, R. C. L. 1996. *An introduction to the Geological Conservation Review*. GCR Series No. 1. Joint Nature Conservation Committee, Peterborough.

MALPAS, J. & BUREK, C. V. 2006. A ton of new RIGS. *Earth Heritage*, **26**, 12–13.

NATURE CONSERVANCY COUNCIL. 1990. *Earth Science Conservation in Britain*. NCC, Peterborough.

PROSSER, C. 2002. Terms of endearment. *Earth Heritage*, **17**, 13.

SHARPLES, C. 1993. *A methodology for the identifications of significant landforms and geological sites for geoconservation purposes*. Forestry Commission, Tasmania.

STANLEY, M. 2000. Geodiversity. *Earth Heritage*, **14**, 15–18.

UCAS website: http://www.ucas.ac.uk?search/index.html (accessed 1.12.06).

WIEDENBEIN, F. W. 1994. *Evaluating volcanological landscapes in Germany and Greece. In*: Brochure of Abstracts. The Malvern International Conference on geological and landscape conservation, Great Malvern 18–24 July 1993. Joint Nature Conservation Committee, 27.

WILSON, R. C. L. 1994. *Earth heritage conservation*. Geological Society of London and Open University, Milton Keynes.

WOOD, M. 1995. *First Welsh RIGS Forum*. Brecon 1994, CCW/RIGS internal publication, 1–2.

WOOD, M. 2003. *Precambrian rocks of the Rhoscolyn Anticline*. Association of Welsh RIGS Groups, Countryside Council for Wales, Bangor, Wales.

WOOD, M. 2007. *Developing a methodology for selecting Regionally Important Geodiversity Sites (RIGS) in Wales & A RIGS survey of Anglesey and Gwynedd*. Vol 1. Gwynedd & Môn RIGS, Countryside Council for Wales, Bangor, Wales.

A history of geological conservation on the Isle of Wight

MARTIN MUNT

*Department of Palaeontology, The Natural History Museum, Cromwell Road, London
SW7 5BD, UK (e-mail: m.munt@nhm.ac.uk)*

Abstract: The Isle of Wight on the south coast of England has near continuous exposures of Early
Cretaceous to Early Oligocene and Quaternary deposits and has long been regarded as a classic
area of British geology. It has a long history of study dating back to the start of the nineteenth
century. The identified threats to geoconservation are coastal erosion, coastal protection
schemes aimed at preventing erosion and fossil collecting. Of these, however, erosion and collecting can also be seen as opportunities. Geology has influenced tourism since the eighteenth century
which subsequently promoted both interpretation and conservation. Collecting of geological
specimens for museum collections is documented as early as 1825. Site-based conservation
began in 1951 with the notification of geological Sites of Special Scientific Interest (SSSIs) and
Regionally Important Geological/geomorphological Sites (RIGS) have been notified since the
mid 1990s. There has been a voluntary set of guidelines for fossil collecting and more recently
the preparation of a Local Geodiversity Action Plan (LGAP) brought together diverse groups
and individuals to begin developing a strategic approach to interpreting and conserving the
island's geological heritage. The main outcome of the LGAP process has been to develop a
partnership with a view to applying for membership of the European Geoparks Network.

The Isle of Wight, situated off the south coast of
England, has near continuous cliff and foreshore
exposures of rocks dating from the early Cretaceous
through to the early Oligocene (see Fig. 1). The
island has long been regarded as a classic area of
British geology and has a history of research
going back to the very beginnings of British geological study. In recent years, the long standing
interest in dinosaurs has gathered pace with the
island becoming a focus for interest in dinosaurs,
not just in the UK, but throughout Europe.
Despite the long history of research, the island
still has much to offer as we refine our techniques
of understanding the past. The island is also a
great resource for promoting the public understanding of the Earth sciences. This has become a key
element of the island's economy, as illustrated by
the building of Dinosaur Isle Museum in Sandown
to replace the old Museum of Isle of Wight
Geology. The conservation challenges on the
island arise from two main threats; firstly coastal
erosion, though more specifically the mechanisms
employed to try to protect against it, and secondly
the high profile issue of fossil collecting.

The geology of the Isle of Wight

The Isle of Wight forms part of the Wessex Basin,
the onshore component of the Wessex–Channel
Basin (Chadwick & Evans 2005), a fault defined
structure developed by the extension of east–west
trending Variscan faulting. However, much of the
existing surface structure is attributable to the
post-Cretaceous Alpine Cycle. The island's
Lower Cretaceous comprises the Wealden Group,
Lower Greensand Group, Carstone Formation,
Gault Clay Formation and Upper Greensand
Formation; the Wealden Group, being the richest
source of dinosaur remains in Europe. The Lower
Cretaceous sequence is widely recognized as a
classic marine transgression sequence. The Upper
Cretaceous comprises the Chalk, most notably the
thickest development of Campanian chalk in
Britain. The youngest parts were, however, eroded
away following Late Cretaceous to early Cenozoic
uplift. The Chalk forms spectacular cliffs at the
eastern and western extremes of the island.

The Cenozoic rocks outcrop on the northern half
of the island as defined by the central east–west
outcrop of the Chalk, referred to locally as the
island's backbone. The Cenozoic rocks comprise
Paleocene, Eocene and Oligocene sequences;
however the oldest parts of the Paleocene are
missing. The island provides excellent exposures
of Paleocene Reading Clay (Lambeth Group), the
Eocene Thames Group, Bracklesham Group and
Barton Group and the largely unique exposures of
the late Eocene to early Oligocene Solent Group.
There is cover of localized Quaternary deposits,
most notably a superb raised beach at Bembridge.

The island is often described as a microcosm of
southern England, with landscape elements which
resemble those of Dorset, Hampshire and Sussex,
attributable to the laterally variable underlying
strata. The SW coast of the island is dissected by
the gorge-like chines. Chines (a local word also

From: BUREK, C. V. & PROSSER, C. D. (eds) *The History of Geoconservation.*
Geological Society, London, Special Publications, **300**, 173–179.
DOI: 10.1144/SP300.13   0305-8719/08/$15.00 © The Geological Society of London 2008.

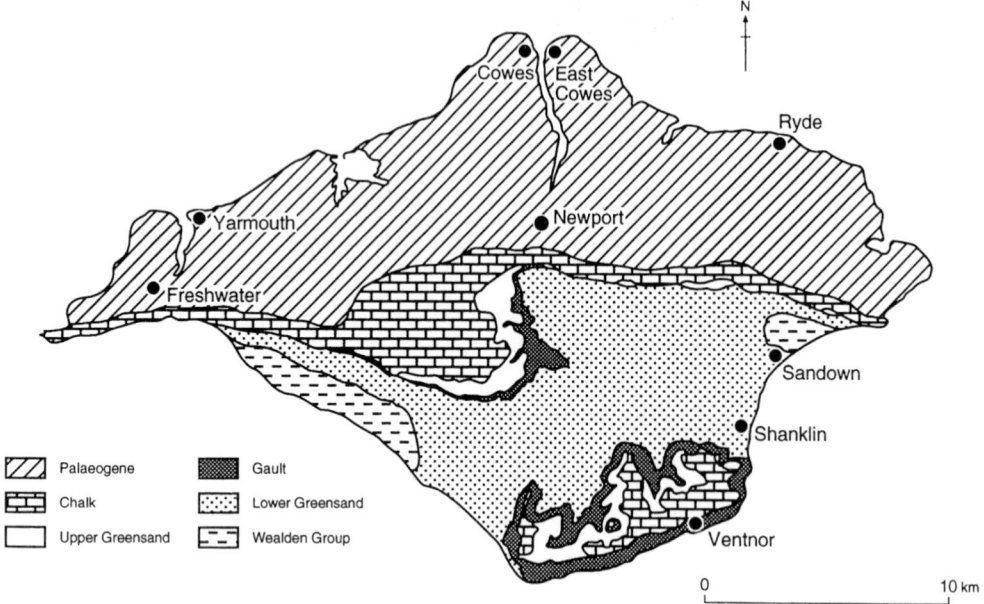

Fig. 1. Outline geological map of the Isle of Wight after Insole *et al.* 1998.

used in east Dorset) are rapidly eroding stream channels, the consequence of the shortening of channel lengths as the cliff line has retreated as a response to post-glacial sea level rise.

The coastal landscape of the island varies considerably. The north coast is dominated by slumped cliff lines in the poorly consolidated Solent Group. These are truncated by estuaries including the River Medina and Newtown Harbour. High chalk cliffs form the eastern and western extremes, there are the sea stacks known as the Needles in the west and sea caves at Culver Cliff in the east. Much of the SE and SW coasts comprise cliffs, commonly terraced by mass-movement; the coastal town of Ventnor is the largest urbanized mass-movement complex in Europe. With rates of erosion in the order of one metre per annum, the erosive effects of the chines, and the mass movement, it is the coastline that presents the setting, challenges and opportunities that define geological conservation on the Isle of Wight.

Discovering the island's geology

In 1799, the topographer Sir Henry Englefield initiated an interest in the Isle of Wight: to lay the foundation of the first scientific study of the island's geology. Englefield employed Thomas Webster to describe the geology, which was later published by Englefield (1816). Webster's work provided the basis for the first memoir of the

island's geology by Forbes (1856). These two works concentrated on what we now call the Solent Group. In the meantime Fitton (1836, 1846) unravelled the Cretaceous sequence using the island's SW coast. Mantell (1847) provided the first overview of the island's geology. Bristow *et al.* (1889) produced the only full geological memoir of the island. This was abridged and modified by Osborne White (1921). Rowe (1908) described the Chalk as part of his series of publications on the Chalk of the English coast. Many fossils collected from the island were first described by J. Sowerby and later J. de C. Sowerby (1822–1829). Excellent overviews of the importance of the island's geology in the development of the geological sciences are provided by Freeman (2004) and Rudwick (2005).

Fossil collecting and conserving the island's geology

On the Isle of Wight, the collecting of geological objects and, therefore, their preservation, seems to have begun in the early nineteenth century, and the realization of the importance of the island's geology had an important influence on the Isle of Wight Philosophical Society, founded in Newport in 1810 (Jackson 1941). Curation of geological specimens on the island began as early as 1825, with the Reverend E. Kell followed by the geologist

E. P. Wilkins. For the next seventy years fossils were collected and put on display in Newport. This collection formed the basis of the current Isle of Wight Council geological collection. We know very little about the early collecting activity on the island. There are however, some records, for example Wilkins (1859) recorded that 'in 1854 I obtained, from near Brighstone, a wagonload of bones, of gigantic size, for the Museum at Newport'.

As with the abundant fossil finds of Lyme Regis in Dorset, much of what was found left the island for collections, many of which have subsequently found their way into collections such as the Natural History Museum and the Sedgwick Museum. The Reverend William Fox (1813–1881) of Brighstone made many important finds, all of which he passed on to Sir Richard Owen at the Natural History Museum (Martill & Naish 2001). Another prolific nineteenth century collector, Mark William Norman, supplemented his mason's income by selling fossils to tourists. However, much has remained on the island in the Isle of Wight Council collection and unlike its more famous rival fossil mecca at Lyme Regis, there remains a dedicated palaeontological museum on the island.

Collecting from the island has continued, and seems no less an important activity today. The local geological collection was relocated to Sandown in 1914 with the opening of the Museum of Isle of Wight Geology. The collections expanded greatly during the 1920s and 30s, and there has been rapid and sustained expansion since the mid-1970s. Full local authority ownership of the collections was secured after the Second World War, and with the foundation of a unitary authority in 1995, the collection has been expanded and is now displayed in Britain's first purpose built dinosaur museum, Dinosaur Isle Museum, in Sandown. The local geological collection comprises approximately 40 000 specimens including over 200 type, figured and cited specimens.

Fossil collecting as a threat: guidelines for fossil collecting

The rapid increase in interest in the island's fossils during the mid-1970s led to great discoveries, such as the skeleton of the carnivorous dinosaur *Neovenator*. Such finds have kept the focus on UK dinosaurs centred on the Isle of Wight, but at a price.

The rediscovery of the island's fossil dinosaurs in the 1970s fuelled competition between fossil collectors; initially between amateur collectors and the museum, joined later by commercial collectors too. The value placed on the island's fossil heritage

increased when the then Museum of Isle of Wight Geology became part of the Isle of Wight County Council and when its first professional staff were appointed in 1976 to redevelop the museum and its collections. This change heightened the desire to retain fossil finds on the island and in public ownership. However, from the mid 1980s there was an increased interest in fossil collecting from commercial collectors and by the early 1990s strong polarized views between staff at the Museum of Isle of Wight Geology and local fossil shop proprietors had developed (see Munt 2001; Simpson 2001). High profile dinosaur finds were increasingly generating arguments over ownership, especially as landowners such as the National Trust developed an increased interest in what was being collected from their land. In an environment where the implications of landownership was not being fully considered, the debate about collecting became one of collecting for commercial gain against taking the 'moral highground' and arguing that fossils belong to society.

With feelings running high a solution needed to be found. In order to attempt to achieve this, different interest groups on the island were brought together with the help of English Nature and with the involvement of other interested parties such as the University of Portsmouth and the Geologists' Association. This resulted in the development of guidelines for fossil collectors from across the spectrum, advice to landowners and notes on responsible collecting practice and the management of fossil collections (Radley 1995). However, these were only guidelines rather than a strict and enforced code of conduct.

This coming together and the associated consultation process between interested parties, took the heat out of the fossil collecting situation at the time but twelve years on, differing views remain; there are still groups and individuals that refuse to speak to one another. One tangible outcome has been the engagement of landowners such as the National Trust which has now developed its own fossil collecting policy. This allows for the collection of small loose fossil specimens from their land but the removal of large objects and the leading of guided walks on their land now requires their permission. Although there have been advances in the management and understanding of fossil collecting on the Isle of Wight, it remains without practical regulation or a serious code of contact. This is regarded by many as a good thing, but not so by others (concerned about damage to the natural heritage and loss of specimens from the island in the same way that the nineteenth century collections described above were lost and dispersed).

With its high profile and publicity it is easy to think of fossil collecting as the major threat to the island's geological heritage. There is much negativity surrounding the subject and there is little doubt that some very important fossils have been lost from the island, both historically and recently. The rapidly eroding cliffs provide a regular and sustainable supply of the commoner fossils and the author takes the view that if they are not collected then they will be lost. What is important is that the rare and important finds are recognized and filtered out into public collections. This can be achieved by effective use of local expertise.

Tourism

Interest in the island's geology at the end of the eighteenth century followed its discovery by the gentry, as it became part of the itinerary during 'the season' (August and September). Brinton (2006) attributes much of the interest to the 'natural beauty and charm of the coast, in particular the chines'. Indeed, Shanklin Chine remains a popular tourist attraction today some two hundred years after its 'discovery'.

Today, tourism is the key element of the island's economy. Webber et al. (2006) estimated that interest in geodiversity generated between £2.6 million and £4.9 million of the island's tourism industry between 2004/2005, when the total was £352 million. With this income generation an impact on geological sites and features as a result of site visits and associated usage is to be expected. Site visits include use by led field trips to interpret geology and collect fossils, with three localities (Yaverland, Compton Bay and Fort Victoria) being the focus of these activities. All three are actively eroding and provide a regular supply of small finds, and as such can be judged to be sustainable for collecting. Trampling, particularly of foreshore exposures has a minor affect compared to the average rates of erosion of one metre per year.

One really positive impact of tourism is in sustaining Dinosaur Isle Museum, which is essentially self-funding through admissions and retail. It thereby helps to secure the future of geological conservation and interpretation on the island. Webber et al. (2006) estimated that the millions generated by geology-related tourism to the island support between 324 and 441 jobs in the tourist sector, a measurable impact of the value of geological conservation.

The threat of erosion

Arguably, the biggest single factor affecting the island's geological heritage is coastal erosion, the effects of which can be quite dramatic with erosion rates around the island of approximately one metre per year. However, erosion provides both threats and opportunities to the island's geology. The loss of exposure of strata of limited spatial extent, such as individual plant debris beds in the Wessex Formation, does occur. Some strata, due to tectonic setting, are only seen in foreshore exposures, notably the King's Quay fish bed of the Fishbourne Member (Upper Eocene Solent Group) and would be affected by sea level rise. Geomorphological features are more vulnerable, the Needles Rock (known as Lot's Wife) collapsed into the sea in 1772 and more recently in 1992 Arch rock in Freshwater Bay collapsed. Some chines have also become dry due to river capture and erosion back to source, the former is seen at Walpen Chine near Blackgang, the latter has occurred in Totland Bay.

The loss of exposure directly to erosion is small compared to the loss resulting from attempts to stop erosion. The larger coastal towns such as Cowes, Sandown, Shanklin and Ventnor all have sea walls. These sea walls probably developed with the towns during rapid urban growth in Victorian times. Other less developed areas such as Totland Bay also have sea walls. Breakwaters have been used at sites such as Colwell Bay to the east of the Totland Bay sea wall and within Sandown Bay. Collapsed sea walls at Priory Bay (near Seaview) and at Gurnard (west of Cowes) record earlier attempts to halt erosion which have since been abandoned. Seawalls have obscured the geology for generations, preventing access to it for study but even where they collapse, such as at Gurnard, they have left the beaches and cliffs both inaccessible and ugly.

Flood defences have been built at a number of locations. Within Sandown Bay at Yaverland, they are integrated with the sea wall built to reduce erosion. Flood defences tend not to impact on geological exposures as they are in low lying areas but do interfere with coastal processes in the estuaries and lagoons they protect, and stop the movement of foreshore sediments. Beach replenishment has been used at Bembridge Forelands to reduce erosion of low cliffs of Quaternary deposits.

Coastal land slipping and associated inland phenomena dominate the Undercliff, running from Shanklin to Blackgang, taking in Bonchurch and Ventnor. Numerous projects include individual slope realignments, hard sea defences and drainage alterations. The import of non-local rocks for both armouring and replenishment has altered beach compositions. Mass movement typifies much of the under-consolidated sediments which comprise both the Cretaceous and Palaeogene sequences. This had resulted in many small schemes around the island, including the use of gabions and netting and often poorly constructed hard solutions.

Such local schemes impact on geology often in SSSI sites such as Whitecliff Bay, the European Palaeogene stratotype.

Shoreline management plans

Coastal management is an important issue for the Isle of Wight Council and individual landowners. Shoreline Management Plans (SMPs) have been developed and are led by the island's own coastal management team. The team at the Isle of Wight Centre for the Coastal Environment are involved in international programmes such as 'Response', a European LIFE Environment Project (see Fairbank & Jakeways 2006), developing strategies for coastal management. SMPs identify the conservation status of coastline areas and set them in context with other priorities such as potential impact of erosion on human habitation. The Department for the Environment and Rural Affairs (Defra) promote four generic approaches (see Table 1) which are implemented locally.

Of the island's coast, all of the SW coast and much of the NW coast fall under no intervention. The remainder, comprising the Sandown Bay area, Ventnor and the NE coast fall largely into the holding the existing line category. Areas around Seaview on the NE corner of the island are examples of where the existing defence line is to be advanced. SMPs can serve to promote the conservation of geological interest, but where they are incompatible with the management required to conserve a geological interest, they may provide a planning framework that will seal the fate of a site.

Conserving the island's geology

By the early 1950s the development of coastal protection schemes such as sea walls was beginning to swallow-up exposures. Notably within Sandown Bay, where sea walls that had been in place for some time had begun to fail and new concrete structures were being built to replace them. It is at this point that local schemes began to record exposures before they were lost. J. F. Jackson was the first professional geological curator of the Museum of Isle of Wight Geology. Five minutes walk from the museum was the failing Yaverland sea wall exposing dinosaur bearing Wealden Marls (Wessex Formation). Jackson photographed this section, along with many others, most spectacularly recording a massive cliff fall near Niton.

In 1951, the first notification of geological sites as SSSIs took place. These included geological interests at Bembridge Down, Bouldnor and Hamstead Cliffs, Compton Down and Hanover Point to St Catherine's Point. Some were revised in 1959 with further revisions and additions right up to the present day. Being a county defined by the sea, most of its important geology is on the coast and most of the island's coast is now contained within geological SSSIs; to a point where few coastal sites of geological significance lie outside SSSIs (see Table 2).

The island is not rich in inland geological sites. Scars in the landscape provide evidence that the island was once dotted with quarries of all sizes in a wide range of rock types. The chalk quarries were recorded by Griffin & Brydone (1911) and Rowe (1908). Large clay extraction sites existed on the banks of the River Medina, at Gunville near Newport and Afton, near Freshwater but none are notified and conserved. Of the chalk quarries, Arreton Down SSSI used to include the biggest chalk quarry on the island, although this is now excluded from the notification. The only other inland geological SSSI is Prospect Quarry, near Shalcombe. This is an exposure of the Bembridge Limestone Formation (Solent Group) in what is now an intermittently active quarry. Shalcombe was the source of some specimens illustrated and described by Sowerby (1822).

Because of the wealth of SSSIs covering the island's coast, leaving few localities outside of conserved areas, the identification of Regionally Important Geological Sites (RIGS) progressed slowly on the island. A RIGS group, based at the former Museum of Isle of Wight Geology was set up in 1990. The first two sites were notified in 1996, to accommodate sites within Sandown Bay, one on the margin of an existing SSSI, the second comprising foreshore geology between existing SSSIs. At the time of writing, a further five sites

Table 1. *Generic approaches for SMPs as defined by the Department for the Environment and Rural Affairs (Defra)*

Category	Description
Hold the existing Line	Maintain and improve existing defences
Advance the existing defence line	Construction of new defences
Managed realignment	Identifying a new line of defence landward of existing lines
No active intervention	No investment in sea defences

Table 2. *A list of geological and geomorphological SSSIs on the Isle of Wight*

Site	Notification year	Interest
Bembridge Down	1951	Cretaceous
Bembridge School and Cliffs	1999	Quaternary
Bonchurch Landslips	1951	Geomorphology
Bouldnor and Hamstead Cliffs	1951	Palaeogene
Brading Marshes to St Helen's Ledges	1951	Palaeogene
Colwell Bay	1959	Palaeogene
Compton Chine to Steephill Cove	2003	Cretaceous
Compton Down	1951	Cretaceous
Gurnard Ledges	1966	Palaeogene
Headon Warren	1951	Palaeogene and geomorphology
King's Quay Shore	1951	Palaeogene
Lacey's Farm Quarry	1993	Palaeogene
Priory Woods	1998	Quaternary
Prospect Quarry	1971	Palaeogene
Watcombe Bay	1971	Cretaceous
Whitecliff Bay & Bembridge Ledges	1955	Palaeogene

are in the process of notification. These comprise sites within existing biological SSSIs and sites not previously notified. The philosophy has been not to notify sites with existing geological designations. In order to integrate conservation notifications within the local planning framework, the designation of RIGS on the island has become integrated with the Local Biodoversity Action Plan.

Besides the notification of RIGS, there have been a number of other major conservation measures in recent years. In 1995, with funding from English Nature, guidelines for fossil collecting were produced (Radley 1995) these are outlined above.

Further funding from English Nature in 2003–2004 allowed the production of a Local Geodiversity Action Plan (LGAP) for the island. This encouraged diverse local groups, landowners and agencies to work together and consider how, for example, geology fits into the aims of the area of the Isle of Wight designated as an Area of Outstanding Natural Beauty. It also allowed cooperative working between Coastal Management and the Dinosaur Isle Museum and enabled the linking up with the Local Biodiversity Action Plan.

The Isle of Wight LGAP took a thematic approach, looking at: the role of the Isle of Wight in the history of the Earth sciences; dinosaurs; the role of geology in the development of the Isle of Wight tourism industry; coastal mass movement phenomena; the Isle of Wight's geological record; fossils and fossil collecting; and geology and landscape. The partnership project was led by staff at Dinosaur Isle Museum with eight other organizations including the National Trust, the Country Landowners Association, the Geological Society of the Isle of Wight and the Isle of Wight LBAP involved.

A very useful product of the Isle of Wight LGAP was the production of a geological audit. This has become the first electronically available audit of the geology of the island and is planned to be available through Isle of Wight Council websites. The LGAP also raised awareness of geology with organizations such as the Isle of Wight AONB and partnerships such as the West Wight Partnership and a lottery funded landscape project. The LGAP has since been used to secure funding for a smaller project auditing dinosaur footprints, funded through the Sustainability Fund of the AONB, and a project raising awareness of geology in Brighstone, funded by the Isle of Wight Economic Partnership.

The future of the conservation of the island's geology

With much of the island's coastal geological heritage within SSSIs and with RIGS filling-in the gaps, the basic geological conservation framework is sound. Interpretation is provided at the Isle of Wight Council funded Dinosaur Isle Museum and the Isle of Wight Centre for the Coastal Environment. Fossil hunting trips are also run from the fossil shops at Blackgang Chine and 'Dinosaur Farm'. The Isle of Wight LGAP has helped to recognize important objectives which will serve to work towards future conservation needs of the island. Currently, a steering group is developing a case for the Isle of Wight to become a member of the European Geoparks Network. With infrastructure such as Dinosaur Isle Museum along with established conservation designations, the Isle of Wight Council anticipates an application for membership to the Geoparks Network by the end of 2007.

Concluding remarks

The early recognition of the importance of the geology of the Isle of Wight provided an impetus to begin the process of collecting fossils which have over two centuries illustrated the value of the island's geology for both conservation of individual objects and the rocks and sites from which they have come. Collecting and conservation of individual finds has ultimately led to the conservation of sites. Interpretation of the geology and landscape is supported by the local authority through Dinosaur Isle Museum and The Isle of Wight Centre for the Coastal Environment. The seemingly unending public interest in dinosaurs has benefited the island as Europe's richest dinosaur locality. A holistic approach of both object collecting and site conservation has developed and is promoted through the LGAP, itself providing a new impetus for international recognition of the importance of the island's geological heritage.

References

BRINTON, R. 2006. *Isle of Wight: The Complete Guide.* Dovecote Press, Stanbridge.

BRISTOW, H. W., REID, C. & STRAHAN, A. 1889. *The Geology of the Isle of Wight.* Memoirs of the Geological Survey England and Wales. Eyre and Spottiswoode, London.

CHADWICK, R. A. & EVANS, D. J. 2005. *A Seismic Atlas of Southern Britain—Images of Subsurface Structure.* British Geological Survey Occasional Paper No. 7. British Geological Survey, Keyworth.

ENGLEFIELD, H. C. 1816. *A Description of the Principal Picturesque Beauties, Antiquities, and Geological Phenomena of the Isle of Wight.* Payne & Foss, London.

FAIRBANK, H. & JAKEWAYS, J. 2006. *Mapping Coastal Evolution and Risks in a Changing Climate.* Isle of Wight Council, Newport.

FITTON, W. H. 1836. Observations on some of the strata between the Chalk and the Oxford Oolite in south-east of England. *Transactions of the Geological Society,* **4**, 103–390.

FITTON, W. H. 1846. A stratigraphical account of the section from Atherfield to Rockenend in the Isle of Wight. *Quarterly Journal of the Geological Society of London,* **3**, 289–327.

FORBES, E. 1856. *On the Tertiary Fluvio-Marine Formations of the Isle of Wight.* Memoirs of the Geological Survey of Great Britain and of the Museum of Practical Geology. Longman, Brown, Green, and Longmans, London.

FREEMAN, M. 2004. *Victorians and the Prehistoric— Tracks to a Lost World.* New Haven, Yale University Press.

GRIFFIN, C. & BRYDONE, R. M. 1911. *The Zones of the Chalk in Hampshire.* Dulau, London.

INSOLE, A., DALEY, B. & GALE, A. 1998. *The Isle of Wight.* Geologists' Association Guide No. 60. Geologists' Association, London.

JACKSON, J. F. 1941. *The Museum of Isle of Wight Geology—History of Collection.* Unpublished MS.

MANTELL, G. A. 1847. *Geological excursions around the Isle of Wight and Along the Adjacent Coast of Dorsetshire; Illustrative of the Mot Interesting Geological Phenomena and Organic Remains.* S. H. G. Bohn, London.

MARTILL, D. M. & NAISH, D. 2001. *Dinosaurs of the Isle of Wight.* Palaeontological Association Field guides to fossils: No. 10. The Palaeontological Association, London.

MUNT, M. C. 2001. Fossil collecting on the Isle of Wight: past, present and future. *In:* BASSETT, M. G., KING, A. H., LARWOOD, J. G., PARKINSON, N. A. & DEISLER, V. K. (eds) *A Future for Fossils.* National Museums & Galleries of Wales, Cardiff, 51–54.

RADLEY, J. 1995. *Guidelines for Collecting Fossils on the Isle of Wight.* Isle of Wight County Press, Newport.

ROWE, A. W. 1908. The zones of the White Chalk of the English coast. *Proceedings of the Geologists' Association,* **20**, 209–339.

RUDWICK, M. J. S. 2005. *Bursting the Limits of Time.* The University of Chicago Press, Chicago.

SIMPSON, M. I. 2001. Pirates or palaeontologists? An alternative view of the Isle of Wight geological experience. *In:* BASSETT, M. G., KING, A. H., LARWOOD, J. G., PARKINSON, N. A. & DEISLER, V. K. (eds) *A Future for Fossils.* Cardiff, National Museums & Galleries of Wales, 54–59.

SOWERBY, J. 1822. *The Mineral Conchology of Great Britain.* Meredith, London.

WEBBER, M., CHRISTIE, M. & GLASSER, N. 2006. *The Social and Economic Value of the UK's Geodiversity.* English Nature Research Report No. 709. English Nature, Peterborough.

WHITE, H. J. O. 1921. *A Short Account of the Geology of the Isle of Wight.* Memoirs of the Geological Survey of Great Britain. HMSO, London.

WILKINS, E. P. 1859. *A Concise Exposition of the Antiquities and Topography of the Isle of Wight.* Kentfield, London.

Context and history of geological conservation in Warwickshire, central England

JONATHAN D. RADLEY

Warwickshire Museum, Market Place, Warwick CV34 4SA, UK

(e-mail: jonradley@warwickshire.gov.uk)

Abstract: The geology of Warwickshire, central England, is diverse but generally poorly exposed. Geological conservation initiatives can be traced back to the mid to late nineteenth century when the Warwickshire Natural History and Archaeological Society amassed locally collected geological specimens and documented local geological sites. The society declined during the late nineteenth century. Following the Second World War, local geological conservation activity was invigorated by national initiatives, leading to establishment of geological Sites of Special Scientific Interest within the county and a site recording programme at the Warwickshire Museum. The Warwickshire Geological Conservation Group was established in 1990. Subsequently, a partnership between that group and the Warwickshire Museum, with support from the Nature Conservancy Council, resulted in establishment of a Regionally Important Geological/geomorphological Sites network. These sites are presently the focus of funded conservation and interpretation projects.

The central English county of Warwickshire (Fig. 1) is dominated by an intensely farmed landscape of rolling hills and valleys, mainly less than 150 m above sea level. Higher ground occurs in the north of the county and along the Oxfordshire and Gloucestershire borders in the south. The Nebsworth Downs on the Gloucestershire border reach over 250 m above sea level. The county is further characterized by a small number of main towns (e.g. Nuneaton, Warwick and Stratford-upon-Avon) interspersed with a much larger number of small market towns. Part of southern Warwickshire's hilly ironstone and Cotswold fringe (Warwickshire County Council 1993*a*) falls within the Cotswolds Area of Outstanding Natural Beauty and the recently established Cotswold Hills Geopark (Robinson 2007). The county's topography, soils, habitats, land use patterns and to a degree, the architecture and industrial history, reflect the underlying geology (Warwickshire County Council 1993*a–c*).

Warwickshire lies on the broad outcrop of continental Triassic and richly fossiliferous marine Jurassic strata that runs across England from the Dorset–Devon border in the SW to the Yorkshire coast in the NE (British Geological Survey 2001). These strata 'young' in the stratigraphic sense towards the SE, culminating in the hilly Jurassic ironstone and Cotswold limestone fringe of southern Warwickshire (Hains & Horton 1969; Fig. 1). The Warwickshire Coalfield, occupying a partly fault-bound, north–south oriented block between the Triassic Knowle and Hinckley basins (Bridge *et al.* 1998; Powell *et al.* 2000; Fig. 1),

diversifies this regional pattern. There, the surface geology is dominated by the continental Upper Carboniferous–Permian Warwickshire Group. The coalfield is bordered in the NE (Nuneaton inlier) by outcropping Upper Carboniferous Westphalian Coal Measures and a narrow, steeply-dipping outcrop of Neoproterozoic (Charnian) volcanic rocks (Caldecote Volcanic Formation) and Cambrian to Ordovician (Tremadoc) marine sandstones and mudstones (Hartshill Sandstone and Stockingford Shale groups), intruded by Ordovician sills (Bridge *et al.* 1998). A smaller inlier of Lower Palaeozoic rocks borders the western side of the coalfield (Dosthill Inlier; Geological Survey of Great Britain (England and Wales) 1922). Quaternary sands, gravels and clays are widespread, including the palaeontologically and archaeologically important Middle Pleistocene deposits of eastern Warwickshire (Shotton *et al.* 1993; Keen *et al.* 2006).

Rocks are still quarried for aggregate, building stone, brick-making and as raw materials for the Rugby cement industry. A single deep mine remains: Daw Mill Colliery, near Arley, NW of Coventry, extracting the Warwickshire Thick Coal (Bridge *et al.* 1998). Several large hard-rock quarries in the Nuneaton inlier are currently dormant, for example Jee's Quarry, Hartshill. Disused quarries are frequently overgrown, slumped and/or flooded, or in some cases landfilled. Shallow road cuttings and sunken lanes tend to be extensively vegetated. Natural exposures, though widespread, are normally deeply weathered and poorly accessible in streams and overgrown ravines.

From: BUREK, C. V. & PROSSER, C. D. (eds) *The History of Geoconservation.*
Geological Society, London, Special Publications, **300**, 181–195.
DOI: 10.1144/SP300.14   0305-8719/08/$15.00 © The Geological Society of London 2008.

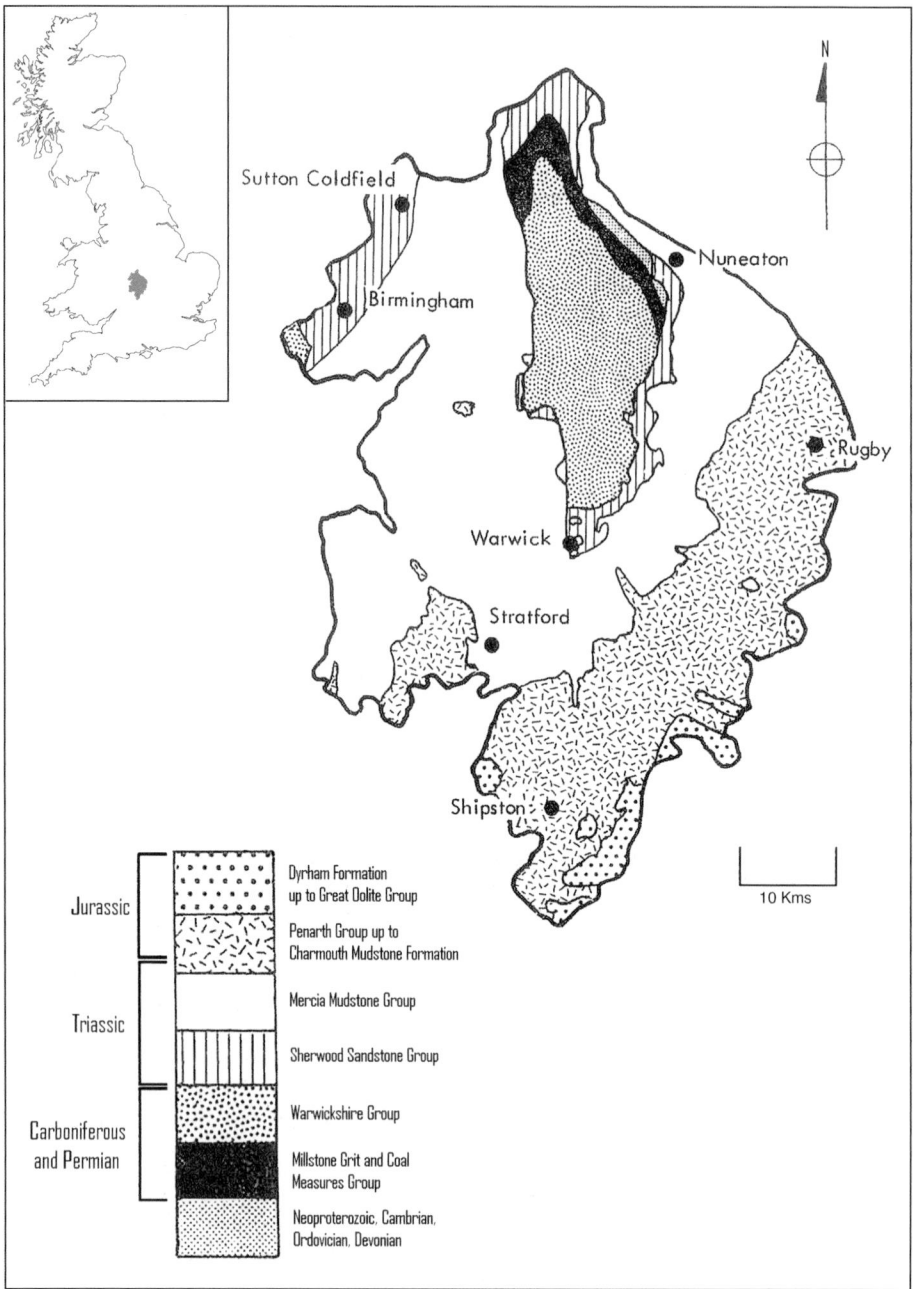

Fig. 1. Outline solid geological map of Watsonian Vice-County 38 (Warwickshire), central England.

History of local geological conservation

Nineteenth century to the 1940s

For the purposes of this paper and present-day local conservation initiatives, Warwickshire is taken as the 'greater Warwickshire' area defined as Watsonian Vice-County 38 (Fig. 1). Geological site conservation within the county essentially postdates the Second World War. During the early nineteenth century, the early days of local geological conservation were characterized by collecting and site recording activity. The Warwickshire Natural History and Archaeological Society (WNHAS) was founded in 1836 and remained active until the latter part of the nineteenth century. One of the principal aims was to initiate a geological collection for display at their museum within the Market Hall in the county town of Warwick. The museum collections soon grew, incorporating locally discovered material and acquisitions from further afield (Green 1986; Crossling 1994; Radley 2006). Establishment of the Warwickshire Naturalists' and Archaeologists' Field Club in the mid-1850s added further impetus.

The Natural History and Archaeological Society's published annual reports (1837–1892) confirm acquisition of many specimens; a significant number of which are documented amongst the Warwickshire Museum's present-day collection (Crossling 1994). With the construction of railways, opportunities arose for recording the geology of important temporary exposures. This was the heyday of local fossil collecting and members of the society, notably the Reverend Peter Bellinger Brodie (founder of the Field Club and long-term geological curator for the WNHAS), documented many exposures and added many specimens to the museum collections (Fig. 2). As a consequence, numerous reports of temporary exposures and quarry sections were published in the society's proceedings and other journals in the latter part of the

nineteenth century (e.g. Brodie 1874; Beesley 1877). The geological interest of the county was also well-known nationally, and groups such as the Geologists' Association visited local sites and the Market Hall museum (Brodie & Kirshaw 1872). The WNHAS declined through the late nineteenth century. In 1932 the collections were transferred to the Warwickshire County Council. A public museum was re-opened at the Warwick Market Hall in 1951 where the council-run Warwickshire Museum has been located to the present day.

Post-war period

National sites. Reflecting a need to conserve Britain's most valuable sites, the National Parks and Access to the Countryside Act of 1949 placed a requirement on the newly formed Nature Conservancy to notify local authorities of the existence of Sites of Special Scientific Interest (SSSIs) (Nature Conservancy Council 1990; Prosser *et al.* 2006). Many of Warwickshire's earliest geological SSSIs were recommended by a committee convened during the Second World War (Anon. 1945) and inspected during the 1950s by the Nature Conservancy (J. A. Irving, pers. comm. 2003). Warwickshire's first eight geological SSSIs were established by the early 1950s and selection continued intermittently until the mid-1970s.

Fig. 2. Fossil fish (*Dapedium* sp.); collected during the mid-nineteenth century from the Early Jurassic Blue Lias Formation at Stockton, near Rugby, Warwickshire. Warwickshire Museum specimen G559; John William Kirshaw collection. John William Kirshaw FGS was an Honorary Curator of Geology for the Warwickshire Natural History and Archaeological Society from 1857 to 1878. Specimen measures 305 mm in length.

The Geological Conservation Review (GCR) was launched by the Nature Conservancy Council (NCC) in 1977, to provide a more systematic audit of British Earth Heritage sites (Ellis 1996). Coinciding with the GCR, the Wildlife and Countryside Act of 1981 required the re-notification or de-notification of previously established SSSIs (see above). By 1990, the NCC had completed GCR site selection, involving trial re-excavation of at least one disused Warwickshire quarry in 1982 as part of the NCC site clearances (Upper Triassic–basal Jurassic Wilmcote Limestone at Temple Grafton; Anon. 1983). The GCR list formed the basis of Warwickshire's present-day

SSSI coverage (Table 1), necessitating de-notification of nine previously selected sites and establishment of fifteen new ones (Nature Conservancy Council 1988a, b; J. A. Irving, pers. comm. 2007).

The statutory protection afforded to Warwickshire's geological SSSIs has proved extremely robust (e.g. Anon. 1987; Glasser 1993). Within the broader geographic context of Vice-County 38 there has been one well-documented failure; that of the former Webster's Clay Pit SSSI, Coventry (fossiliferous Keresley Member of the Upper Carboniferous Meriden Formation; Bridge et al. 1998). Consent for landfill had been established

Table 1. *Geological Sites of Special Scientific Interest in Warwickshire, England (2007)*

Geological Site of Special Scientific Interest	National grid reference	Interest
Boon's Quarry	SP 330 947	Unconformity between Neoproterozoic Caldecote Volcanic Formation and Boon's Member of Cambrian Hartshill Sandstone Formation
Woodlands Quarry	SP 325 947	Home Farm Member of the Cambrian Hartshill Sandstone Formation
Illing's Trenches	SP 324 943	Cambrian Abbey Shale Formation
Griff Hill Quarry	SP 362 888	Ordovician differentiated sill
Kingsbury Brickworks	SP 222 995	Upper Carboniferous Etruria and Halesowen formations
Coten End Quarry	SP 290 655	Fossiliferous Triassic Bromsgrove Sandstone Formation
Guy's Cliffe	SP 293 668	Fossiliferous Triassic Bromsgrove Sandstone Formation
Shrewley Canal Cutting	SP 212 764	Triassic Arden Sandstone Formation
Wilmcote Quarry	SP 151 594	Uppermost Triassic – basal Jurassic Wilmcote Limestone Member of the Blue Lias Formation
Napton Hill Quarry	SP 457 613	Lower Jurassic Dyrham and Marlstone Rock formations
Cross Hands Quarry	SP 269 290	Clypeus Grit Member of the Middle Jurassic Salperton Limestone Formation and Middle Jurassic Chipping Norton Limestone Formation
Harbury Quarries	SP 390 583, also SP 383 589 and SP 385 590	Middle Pleistocene fluvial sands and glacigenic clays
High Close Farm	SP 232 596	Middle Pleistocene Baginton Formation
Ryton and Brandon Gravel Pits	SP 383 763, also SP 386 760 and SP 378 749	Middle Pleistocene Baginton Formation overlain by Fourth Terrace of the River Avon
Stretton-on-Fosse Pit	SP 220 381	Middle Pleistocene fluvial gravels and sands, and glacigenic clays
Waverley Wood Farm	SP 366 714	Waverley Wood Member of the Middle Pleistocene Baginton Formation
Wolston Gravel Pit	SP 325 947	Middle Pleistocene Baginton and Wolston formations
Ailstone Old Gravel Pit	SP 211 512	Fourth Terrace of the River Stour
Broom Railway Cutting	SP 081 529	Second Terrace of the River Avon
River Itchen	SP 404 558	Sinuous fluvial channels and erosional features

through the planning system prior to SSSI notification, resulting in ultimate loss of the site's geological interest (Prosser 2003; Prosser *et al.* 2006). On the positive side, the local geological SSSI network has attracted the attention of researchers, and specialist groups such as the Quaternary Research Association (Keen 1989) and Geologists' Association (Swann 2005). Scientific accounts of the sites have been published within the JNCC's Geological Conservation Review Series (e.g. Rushton *et al.* 1999; Benton *et al.* 2002). Several geological SSSIs have benefited from enhancement through English Nature's Face Lift programme (Murphy 2004). Amongst these, Cross Hands Quarry SSSI, near Long Compton (Middle Jurassic Clypeus Grit and Chipping Norton Limestone; Table 1) has provided a sustainable palaeontological resource for educational groups over several decades (Besterman 1988; Nature Conservancy Council 1989; Radley 2005).

Local sites. Warwickshire Museum's Geological Localities Record Centre (GLRC) was established in 1978 as part of the National Scheme for Geological Site Documentation (NSGSD; Nature Conservancy Council 1990). Intensive data capture was undertaken for the Vice-County 38 area during 1978–1981, principally from field records, maps and published literature held within the museum, largely by volunteers, and employees of the Manpower Services Commission Job Creation Programme. This remains as the geographical basis of present-day county geological conservation provision. From the outset, the GLRC was in the form of paper records (Museum Documentation Association summary sheets) linked to card indexes, map files, a record centre library, and a site pin-map based upon Ordnance Survey 1:50 000 scale topographic sheets. This system is still in use today (2007), and was augmented by the computerized Geological Sites Database GD2 (Copp 1994) in the early 1990s.

The Warwickshire Wildlife Trust, NCC, and the Warwickshire Museum established the Warwickshire Geological Conservation Group (WGCG) in 1990 (Harley 1990), now an affiliated group of the Geologists' Association. This was a response to the NCC's Regionally Important Geological/ geomorphological Sites (RIGS) initiative (Nature Conservancy Council 1990; Harley 1994) and the growing threats to local sites (Besterman 1981; Prosser 2003). From the outset, the WGCG aimed to fulfil the traditional educational and social roles of a local geology group as well as site documentation and conservation, and has worked in close partnership with the Warwickshire Museum.

The WGCG established a RIGS sub-committee during the early 1990s, comprising the museum's Keeper of Geology, the NCC's county officer and a range of amateur and professional geologists and Earth science teachers. Principal aims were RIGS selection and advancement of the RIGS agenda. The selection policy, reflecting NCC's guidelines (Nature Conservancy Council 1990) is still used today and assesses sites for their educational, scientific, historic and aesthetic value. Initially, site selection was governed by an essentially stratigraphical framework, to afford representative, robust coverage of the major lithostratigraphical divisions making up Warwickshire's geological succession. By mid-1994, details of approximately forty RIGS had been lodged within the museum's GLRC and with planning authorities and landowners, establishing a mechanism for non-statutory protection through the planning process (Harley 1994). The list includes several de-notified SSSIs (Table 2).

Other notable achievements for the WGCG during the 1990s included sponsored clearance of the Cambrian–Upper Carboniferous (Millstone Grit) unconformity at Moor Wood railway cutting RIGS near Nuneaton (Anon. 1996), relocation of large rock specimens from Judkins Quarry, Nuneaton and Griff No. 4 Quarry, Bedworth (Tables 1–3) to public access areas (Fig. 3), and construction of an interpreted 'rock garden' at the Warwickshire Wildlife Trust's Brandon Marsh visitor centre.

In 2001 the WGCG gained funding from the Department of the Environment, Transport and Regions (DETR) via the Western Association of RIGS Groups (now The Geology Trusts; Prosser *et al.* 2006), to establish a further forty RIGS (Campbell & Oliver 2002). Given the representative stratigraphical coverage afforded by previously established RIGS, the DETR project necessitated a more flexible assessment and selection process, including sites demonstrating geomorphological features, active geological and geomorphological processes, finer lithostratigraphical divisions and intraformational variation (Table 3). Significantly, this approach echoed the holistic, geodiversity concept that was growing in popularity at that time (Gray 2003; Stace & Larwood 2006).

The WGCG obtained funding from the Aggregates Levy Sustainability Fund (ALSF) during 2002 to provide geodiversity management plans for four county aggregate-producing quarries. At that time, English Nature was starting to promote Local Geodiversity Action Plans (LGAPs) as a mechanism for planned, timetabled delivery of local (principally county-based) geoconservation (Burek & Potter 2002). Also during 2002, the Warwickshire Museum gained funding from English Nature to investigate the feasibility of a county LGAP. This pilot study established a number of themes that indicated the viability of

Table 2. *Warwickshire Regionally Important Geological/geomorphological Sites selected before 2001–2002 RIGS project*

Regionally Important Geological/ geomorphological Site	National grid reference	Interest
Judkins Quarry	SP 343 932	Neoproterozoic Caldecote Volcanic Formation and Triassic basal Mercia Mudstone (de-notified SSSI)
Jee's Quarry	SP 333 940	Cambrian Hartshill Sandstone and Purley Shale formations
Oldbury (Mancetter) Quarry	SP 312 952	Cambrian Outwoods Shale Formation and Ordovician Oldbury Sill
Midland Quarry	SP 350 925	Cambrian Hartshill Sandstone Formation, Ordovician sill and Triassic Bromsgrove Sandstone Formation
Steppy Lane Cutting	SP 3098 9601– 3118 9610	Cambrian Stockingford Shale Group (de-notified SSSI)
Stockingford Railway Cutting	SP 342 921–346 921	Cambrian Stockingford Shale Group
Moor Wood Railway Cutting	SP 321 937	Cambrian Outwoods Shale Formation
Purley (Mancetter) Quarry	SP 305 963	Cambrian Outwoods Shale Formation
Moor Wood Quarry	SP 3170 9395	Cambrian Outwoods Shale Formation intruded by Ordovician sill
Mancetter Hill Quarry	SP 3013 9572	Devonian Oldbury Farm Sandstone Formation
Dosthill Church Quarry	SP 2113 9981	Namurian Millstone Grit overlying Ordovician sill
Baggeridge Brickworks	SP 220 990	Upper Carboniferous Etruria and Halesowen formations
Whateley Quarry	SP 2284 9930	Upper Carboniferous Halesowen Formation
Arley Tunnel Cutting	SP 300 911	Whitacre Member of the Upper Carboniferous Meriden Formation
Hill Cottage Quarry	SP 276 883	Whitacre Member of the Upper Carboniferous Meriden Formation (de-notified SSSI)
Newdigate Railway Cutting	SP 341 868	Whitacre Member of the Upper Carboniferous Meriden Formation
Corley Cutting and Corley Rocks	SP 304 853	Keresley Member of the Upper Carboniferous Meriden Formation
Wickes Store Cutting, Coventry Ring Road	SP 3316 7955	Keresley Member of the Upper Carboniferous Meriden Formation
Gibbett Hill Quarry	SP 3045 7521	Gibbett Hill Conglomerate within the Permian Kenilworth Sandstone Formation
Kenilworth Castle Quarry	SP 278 719	Permian Kenilworth Sandstone Formation
Motslow Hill Cutting	SP 3323 7241	Permian Kenilworth Sandstone Formation overlain by Middle Pleistocene Baginton Formation
Ashow Site	SP 3110 7040	Permian Ashow Formation
Roundberry Quarry	SK 277 038	Triassic Polesworth Formation
Baginton Garden Centre Quarry	SP 339 750	Triassic Bromsgrove Sandstone Formation
River Avon section, Milverton	SP 301 665	Triassic Bromsgrove Sandstone Formation
Rock Mill Quarry	SP 3015 6635	Triassic Bromsgrove Sandstone Formation
Quarryfield House Quarry	SP 359 723	Triassic Bromsgrove Sandstone Formation

(Continued)

Table 2. *Continued*

Regionally Important Geological/ geomorphological Site	National grid reference	Interest
Nursery Cottage (Arden) Brickworks	SP 2055 8290	Triassic Mercia Mudstone Group
Rowington Canal Cutting	SP 200 692	Triassic Arden Sandstone Formation (de-notified SSSI)
Dark Lane Copse Quarry, Lighthorne	SP 325 558	Langport Member of the Upper Triassic Lilstock Formation
Temple Grafton Quarry	SP 121 539	Uppermost Triassic – basal Jurassic Wilmcote Limestone Member of the Blue Lias Formation and Holocene tufa
Southam Cement Works Quarries	SP 420 631	Upper Triassic Lilstock Formation; Saltford Shale and Rugby Limestone members of the Lower Jurassic Blue Lias Formation
Parkfield Road Quarry	SP 493 759	Rugby Limestone Member of the Lower Jurassic Blue Lias Formation (de-notified SSSI)
Meon Hill Barn Landslip	SP 1830 4514	Lower Jurassic Dyrham Formation
Edge Hill Quarries	SP 371 469	Lower Jurassic Dyrham and Marlstone Rock formations
Burton Dassett Hills	SP 395 522	Lower Jurassic Dyrham, Marlstone Rock and Whitby Mudstone formations, Middle Jurassic Northampton Sand Formation
Winderton Road Cutting	SP 341 408	Middle Jurassic Northampton Sand Formation
King's Hill Boulder	SP 326 744	Glacial erratic (de-notified SSSI)
Rawn Hill	SP 312 967	Ordovician intrusion (geomorphological)

Table 3. *Warwickshire Regionally Important Geological/geomorphological Sites selected during and after 2001–2002 RIGS project*

Regionally Important Geological/ geomorphological Site	National grid reference	Interest
Moor Wood Farm Quarry	SP 3135 9415	Cambrian Outwoods Shale Formation
Merevale Lane Cutting	SP 2890 9765	Ordovician Merevale Shale Formation
Polesworth Railway Cutting	SK 2653 0310–2680 0270	Upper Carboniferous Lower and Middle Coal Measures formations
Claybrookes Marsh Spoil Tip	SP 379 770	Upper Carboniferous Coal Measures Group
Griff No. 4 Quarry	SP 362 885	Upper Carboniferous Coal Measures Group
Baxterley Quarry	SP 2825 9710	Upper Carboniferous Halesowen Sandstone Formation
Hill Farm Quarry, Maxstoke	SP 2415 8810	Whitacre Member of the Upper Carboniferous Meriden Formation
Chapel Green, Fillongley	SP 271 854	Keresley Member of the Upper Carboniferous Meriden Formation
Meriden Hill Cutting	SP 2535 8198–2560 8196	Allesley Member of the Upper Carboniferous Meriden Formation
Canley Brook	SP 2900 7775–3088 7735	Upper Carboniferous Tile Hill Mudstone Formation
Cherry Orchard Brickpit	SP 295 721	Permian Ashow Formation

(*Continued*)

Table 3. *Continued*

Regionally Important Geological/ geomorphological Site	National grid reference	Interest
Stiper's Hill Plantation	SK 2715 0248	Upper Permian or basal Triassic Hopwas Breccia, and Triassic Polesworth Formation
North Woodloes Quarry	SP 276 679	Triassic Bromsgrove Sandstone Formation
Paul's Land Quarry	SP 3806 8998	Triassic basal Mercia Mudstone Group
River Avon, Marlcliff	SP 0925 5048	Triassic Mercia Mudstone Group, Blue Anchor Formation and Westbury Formation
Ufton Hill Farm Landfill Site	SP 3932 6165	Cotham and Langport members of the Triassic Lilstock Formation
Round Hill Road Cutting	SP 1430 6178	Upper Triassic – basal Jurassic Lilstock Formation and Wilmcote Limestone Member of the Blue Lias Formation
Southam Bypass Cutting	SP 419 627	Rugby Limestone Member of the Lower Jurassic Blue Lias Formation
Ettington Road Cutting	SP 2644 4915	Lower Jurassic Blue Lias and Charmouth Mudstone formations
Napton Industrial Estate	SP 455 616	Lower Jurassic Charmouth Mudstone Formation
Napton 'Doggers'	SP 4563 6140	Lower Jurassic Dyrham Formation
A422 Quarry, Hornton	SP 377 453	Lower Jurassic Marlstone Rock Formation
Edge Hill Farm Quarry	SP 3650 4660	Lower Jurassic Marlstone Rock Formation
Avonhill Quarry	SP 4172 5075	Lower Jurassic Dyrham, Marlstone Rock and Whitby Mudstone formations
Humpty Dumpty Field, Ilmington	SP 2070 4286	Lower Jurassic Marlstone Rock and Whitby Mudstone formations
Windmill Hill Quarry, Tysoe	SP 3323 4263	Astarte elegans Bed of Middle Jurassic Northampton Sand Formation
Brailes Hill No. 1 – geological	SP 2943 3916	Middle Jurassic Northampton Sand ('Scissum Beds') and Chipping Norton Limestone formations
Weston Park Lodge Quarry	SP 285 340	Middle Jurassic Chipping Norton Limestone Formation
Traitor's Ford Quarry	SP 3355 3621	Middle Jurassic Great Oolite Limestone
Paget's Lane Pit, Bubbenhall	SP 3735 7203	Thurmaston and Brandon members of the Middle Pleistocene Baginton Formation; Thrussington Member of the Middle Pleistocene Wolston Formation
Wood Farm Quarry	SP 373 719	Thurmaston and Brandon members of the Middle Pleistocene Baginton Formation; Thrussington Member of the Middle Pleistocene Wolston Formation
Royal Oak Gravel Pit and Cutting	SP 5468 7342	Hillmorton Member of the Middle Pleistocene Wolston Formation
Marsh Farm Quarry, Salford Priors	SP 0800 5252	Second Terrace of the River Avon
River Avon, Stratford Racecourse	SP 1848 5332	Holocene alluvium of the River Avon
Griff Hollows	SP 3617 8956	Ordovician sill showing spheroidal weathering features (geomorphological)
Oldbury Grange Sills	SP 3158 9425	Hill ridges on outcrop of Ordovician sills (geomorphological)

(Continued)

Table 3. *Continued*

Regionally Important Geological/ geomorphological Site	National grid reference	Interest
Mows Hill Dingle	SP 1351 6957	Possible meltwater-cut gorges and waterfall exposing Triassic Arden Sandstone Formation (geomorphological)
Brailes Hill No. 2 – geomorphological	SP 290 390	Benched hillside profile developed on Lower and Middle Jurassic strata (geomorphological)
Edge Hill Landslip	SP 381 487	Landslips in Lower Jurassic Charmouth Mudstone and Dyrham formations (geomorphological)
Warmington Church Exposure	SP 4097 4746	Post-glacial cambering affecting the Lower Jurassic Marlstone Rock Formation (geomorphological)
Old Milverton River Terraces	SP 2966–2967	First, Second and Fourth Terraces of the River Avon (geomorphological)
Paul's Ford, Attleborough	SP 389 911	River confluence, active fluvial erosion and deposition (geomorphological)
River Blythe Oxbow	SP 1330 7550	Oxbow lake (geomorphological)
River Arrow, Studley	SP 0863	Active shingle bars, palaeochannels and terraces (geomorphological)
Southam Salt Spring	SP 446 605	Saline spring (geomorphological; de-notified biological SSSI)

such a scheme for future geoconservation effort (Burek & Potter 2004). A consultation group was established, drawing representatives from local conservation and Earth science groups, planning authorities, quarry operators and landowners.

The LGAP pilot study was followed by a second project in 2003–2004, similarly funded by English Nature, to establish an LGAP for non-marine Permian–Triassic fossil sites (Burek & Potter 2004). This included an audit of sites and existing

Fig. 3. The Warwickshire Geological Conservation Group's rock display adjacent to Judkins Quarry Regionally Important Geological Site, Nuneaton, Warwickshire. Large boulders retrieved from the quarry comprise pyroclastic and basaltic rocks from the Charnian Caldecote Volcanic Formation, quartz arenites from the Cambrian Hartshill Sandstone Formation, intrusive rocks of Ordovician age and Triassic sandstones and breccias. Numbers painted on the rock specimens refer to information on the interpretation panel.

collections, and identified opportunities for interpretation, education and further research. The resulting action plans addressed site maintenance, monitoring, interpretation and protection, as well as the palaeontological potential of unexposed 'reserves' of Permian–Triassic rock. Subsequent talks with English Nature resulted in a draft framework for a county LGAP, and the adoption of the Warwickshire LGAP project by the WGCG. In the absence of further funding the project was suspended in 2004. At present the Permian–Triassic plan awaits implementation.

The WGCG was awarded further ALSF funding towards the end of 2004 to conserve Middle Pleistocene deposits at Wood Farm Quarry RIGS, near Bubbenhall (Radley & Friend 2006; Table 3, Fig. 4), well-known for recent discoveries of Palaeolithic stone tools (Keen *et al.* 2006). Following excavation, local sections in the unconsolidated, fluvial Thurmaston and Brandon members of the Baginton Formation (Shotton *et al.* 1993; Keen *et al.* 2006) degrade rapidly (Shelton 2004); obscuring well-preserved sedimentary structures (Fig. 4). In an attempt to preserve these, and the junction with the overlying glacigenic Thrussington Member of the Wolston Formation, a small face at Wood Farm Quarry was cleaned and graded, and protected with a weather-proof canvas cover (Fig. 5). Ongoing monitoring of the protected face during quarry restoration has proven the general efficacy of this scheme (Radley & Friend 2006). The project was also a successful public relations and educational exercise, involving production of leaflets, delivery of talks on local Pleistocene

Fig. 4. Partly slumped face in unconsolidated Middle Pleistocene sediments at Wood Farm Quarry Regionally Important Geological Site, near Bubbenhall, Warwickshire. Fluvial gravels and sands in the lower part of the section represent the Thurmaston and Brandon members of the Baginton Formation. These are overlain by clayey till—the Thrussington Member of the Wolston Formation. Face is approximately 6 m high.

Fig. 5. Protective screen covering unconsolidated Middle Pleistocene sediments at Wood Farm Quarry Regionally Important Geological Site, near Bubbenhall, Warwickshire (2006). The planar junction between the arenaceous Brandon Member (Baginton Formation) and the argillaceous Thrussington Member (Wolston Formation) is clearly visible in the upper part of the face.

geology and environments, construction of artificial peels from the sands and gravels, publication of a website, and installation of outdoor and indoor interpretation panels at the nearby Ryton Pools Country Park (Fig. 6). Since 2005, the ALSF has funded a broader scheme to conserve and interpret a network of Quaternary sites in eastern Warwickshire. This has allowed excavation and conservation of sections at the Brandon Marsh Nature Centre and within the Ryton Pools Country Park (Fig. 6), and installation of interpretation boards in nearby villages.

Ongoing geological conservation

Local sites

Working and disused quarries are coming under increasing pressure from landfill and/or development. Since the 1990s, the Warwickshire Museum and WGCG have recommended planning conditions to the Minerals Planning Authority for a range of quarries and landfill sites represented within the RIGS network. Recent planning casework has included applications for Midland Quarry, Southam Cement Works Quarry and Edge Hill Quarry (Table 2), as well as consultation for County Minerals and Waste Local Development Plans and the Office of the Deputy Prime Minister's Planning Policy Statement 9 (Prosser 2006). During 2006, four Warwickshire sites were assessed as part of the national GeoValue project (Purley Quarry RIGS; Griff No. 4 Quarry RIGS/Griff Hill Quarry SSSI; Edge Hill Quarries RIGS; Wood Farm Quarry RIGS; P. W. Scott, pers. comm. 2007; Tables 1–3), aimed at quantifying the conservation value of geological exposures (Scott *et al.* 2007). Warwickshire Museum's collecting activities continue to safeguard material evidence for geological sites under threat.

Fig. 6. Geological interpretation and conservation of unconsolidated Middle Pleistocene sediments at Ryton Pools Country Park, Warwickshire (2007). Interpretation panel in the foreground provides an outline of Middle Pleistocene environments and a context for locally discovered mammalian fossils and stone tools. In the background, dug into a degraded face, the roofed structure (approximately 1 m high) protects a section exposing the junction of the arenaceous Brandon Member (Baginton Formation) and the argillaceous Thrussington Member (Wolston Formation). The nearby steps are cut into clays of the Thrussington Member.

Site clearance work is also undertaken periodically, for example in January 2007 at Hill Farm Quarry RIGS, Maxstoke, re-exposing sections through alluvial conglomerates and sandstones of the Whitacre Member (Upper Carboniferous Meriden Formation; Powell *et al.* 2000; Fig. 7). The WGCG's conservation (formerly RIGS) sub-committee continues to identify and establish geological and geomorphological RIGS. As administrator of the GLRC, Warwickshire Museum presently receives notification of temporary exposures through archaeological field projects, county planners, the WGCG and the general public. Engineering works such as pipeline excavations are generally unviable for long-term conservation agreements but provide opportunities for recording and collecting (Radley 2005). The Warwickshire LGAP project has been re-started following further funding by Natural England (formerly English Nature). A county LGAP is presently being drafted with the assistance of British Geological Survey expertise and resources.

Museum-based collecting

Warwickshire Museum's current Acquisition and Disposal Policy (adopted 2004) states that 'Future collecting will concentrate upon well-documented rock, fossil and mineral specimens from the county that are not represented in the present collections, of better quality than existing holdings, or preserve hitherto unrepresented features of geological significance.' Echoing the RIGS selection philosophy (see above), the policy promotes acquisition of geological specimens that afford balanced representation of the geological components of county geodiversity (*sensu* Gray 2003). Arguably therefore, collecting and site-based geological conservation is more closely integrated than at any other time in the county's conservation history.

Pressures on museum resources and shifting curatorial roles, philosophies, policies and priorities are increasingly constraining and reordering collecting activities in British museums (Knell 2004). Additionally, health and safety legislation and insurance requirements are rendering access to quarries, temporary exposures and other inland geological collecting sites increasingly difficult. Despite this, regular visits to larger quarries (many designated as SSSIs and RIGS), and temporary exposures continue to provide important geological specimens and GLRC records (e.g. Martill 2005; Radley 2007).

Fig. 7. Clearance work being undertaken by members of the Warwickshire Geological Conservation Group at Hill Farm Quarry Regionally Important Geological Site, near Maxstoke, Warwickshire (2007). The quarry exposes ESE-dipping cross-bedded fluvial sandstones of the Whitacre Member (Upper Carboniferous Meriden Formation), 300 m east of the Warwickshire Coalfield's Western Boundary Fault.

Discussion and conclusions

Geological conservation, in the form of collecting and site recording, was initiated in Warwickshire during the 1830s by the WNHAS. This was followed by a general decline in local geological investigation and documentation during the late nineteenth and early twentieth century. The post-World War Two period has witnessed the growth of the national geological conservation movement (Knell 2002; Prosser *et al.* 2006), and eventually, establishment of the WGCG in 1990, strongly influenced by the national scene. The WGCG has maintained a close partnership with the Warwickshire Museum, the latter practising a holistic, integrated approach to geological collection, site conservation and education activity (Besterman 1988). This

partnership has additionally benefited from close working relationships with Natural England and its predecessors, the Geologists' Association, British Geological Survey, landowners, quarry operators and developers, and most recently, the Association of UK RIGS Groups (UKRIGS) and Geology Trusts. In recent years, greater funding availability and the increasing popularity of the geodiversity concept has necessitated significant expansion of the WGCG conservation sub-committee's agenda. This has allowed development of the LGAP programme and completion of several projects focusing on site identification, conservation and interpretation.

Reflecting the national situation, the county geological conservation movement still operates on a small scale in relation to the range of wildlife conservation schemes and projects that are currently in place. However, Warwickshire's wide-ranging geological heritage remains accessible in quarries, cuttings, natural exposures and the collections of the Warwickshire Museum, British Geological Survey and other collecting organizations. Local quarries in particular are under continuing threat from landfill and development, necessitating increasing consultation between owners, developers, the local planning authority, Warwickshire Museum and the WGCG. Geodiversity management plans and action plans should allow a planned, strategic approach to county geological conservation.

Museum and university-based research is realizing the potential of many disused sites (some preserved as SSSIs or RIGS) as sources of specimens for recognized collections, as well as new data for the GLRC and for academic research (Radley 2005), strengthening the case for their long-term conservation (Prosser *et al.* 2006). The WGCG, Warwickshire Museum and Natural England are encouraging active use of local geological sites by geologists and other interested parties, reflecting current policies and philosophy of the geological conservation 'establishment' that promotes a broad 'ownership' of geological science (Knell 2002). The benefits are increasingly evident throughout Warwickshire in terms of the profile of geology, its collections, sites and their conservation.

I would like to thank A. Irving (Natural England) and I. Fenwick (Warwickshire Geological Conservation Group) for commenting on an early version of the text, as well as J. Larwood (Natural England) and M. Bradley (Warwickshire Geological Conservation Group) for further constructive suggestions. The membership of the Warwickshire Geological Conservation Group is acknowledged for ongoing support and enthusiasm for local geological conservation. A. Isham (Warwickshire Museum) assisted with the drafting of Fig. 1.

References

ANON, 1945. National Geological Reserves in England and Wales. *Report by the Geological Reserves Sub-Committee of the Nature Reserves Investigation Committee*. The Society for the Promotion of Nature Reserves, British Museum (Natural History, London).

ANON, 1983. Temple Grafton, Warwickshire. *Earth Science Conservation*, **20**, 35.

ANON, 1987. Boons Quarry saved from tipping. *Earth Science Conservation*, **23**, 28.

ANON, 1996. Railway cutting into history. *Earth Heritage*, **6**, 6.

BEESLEY, T. 1877. The Lias of Fenny Compton, Warwickshire. *Proceedings of the Warwickshire Naturalists' and Archaeologists' Field Club*, **1877**, 1–22.

BENTON, M. J., COOK, E. & TURNER, P. 2002. *Permian and Triassic Red Beds and the Penarth Group of Great Britain*.

Joint Nature Conservation Committee, Peterborough, Geological Conservation Review Series, **24**.

BESTERMAN, T. P. 1981. Geological conservation in the Warwickshire County Council Structure Plan. *The Geological Curator*, **3**, 51–52.

BESTERMAN, T. P. 1988. The meaning and purpose of palaeontological site conservation. *In*: CROWTHER, P. R. & WIMBLEDON, W. A. (eds) *The Use and Conservation of Palaeontological Sites*. Palaeontological Association, London, Special Papers in Palaeontology, **40**, 9–19.

BRIDGE, D.McC., CARNEY, J. N., LAWLEY, R. S. & RUSHTON, A. W. A. 1998. *Geology of the Country around Coventry and Nuneaton*. Memoir of the British Geological Survey. The Stationery Office, London.

BRITISH GEOLOGICAL SURVEY. 2001. *Solid Geology Map, UK South Sheet, 1:625 000 Scale, 4th edn*. Natural Environment Research Council.

BRODIE, P. B. 1874. Notes on a railway-section of the Lower Lias and Rhaetics between Stratford-on-Avon and Fenny Compton, on the occurrence of the Rhaetics near Kineton, and the Insect-Beds near Knowle, in Warwickshire, and on the recent discovery of the Rhaetics near Leicester. *Quarterly Journal of the Geological Society*, **30**, 746–749.

BRODIE, P. B. & KIRSHAW, J. W. 1872. Excursion to Warwickshire, July 10th and 11th, 1871. *Proceedings of the Geologists' Association*, **2**, 284–287.

BUREK, C. & POTTER, J. 2002. Minding the LGAPs. *Geoscientist*, **12**, 16–17.

BUREK, C. & POTTER, J. 2004. *Local Geodiversity Action Plans—Sharing Good Practice Workshop, Peterborough, 3 December 2003*. English Nature Research Report **601**.

CAMPBELL, M. & OLIVER, P. 2002. Western pioneers. *Earth Heritage*, **18**, 6–7.

COPP, C. J. T. 1994. Developing geological site recording software for local conservation groups. *In*: O'HALLORAN, D., GREEN, C., HARLEY, M., STANLEY, M. & KNILL, J. (eds) *Geological and Landscape Conservation*. The Geological Society, London, 371–380.

CROSSLING, J. 1994. Warwickshire Museum. *In*: NUDDS, J. (ed.) *Directory of British Geological Museums*. Geological Society, London, Miscellaneous Paper, **18**, 81.

ELLIS, N. V. (ed.) 1996. *Geological Conservation Review Series: An Introduction to the Geological Conservation Review*. Joint Nature Conservation Committee, Peterborough, Geological Conservation Review Series, **1**.

GEOLOGICAL SURVEY OF GREAT BRITAIN (ENGLAND AND WALES) 1922. *Lichfield (Drift): Sheet 154*. Ordnance Survey, Chessington, Surrey.

GLASSER, N. 1993. Wolston clean-up. *Earth Science Conservation*, **33**, 17–18.

GRAY, M. 2003. *Geodiversity: Valuing and Conserving Abiotic Nature*. John Wiley, Chichester.

GREEN, M. 1986. *Warwickshire Museum 1836–1986*. Warwickshire Museum Publication.

HAINS, B. A. & HORTON, A. 1969. *British Regional Geology, Central England* (3rd edn). HMSO, London.

HARLEY, M. 1990. RIGS initiative takes off! *Earth Science Conservation*, **28**, 3–5.

HARLEY, M. 1994. The RIGS (Regionally Important Geological/geomorphological Sites) challenge—involving local volunteers in conserving England's geological heritage. *In*: O'HALLORAN, D., GREEN, C., HARLEY, M., STANLEY, M. & KNILL, J. (eds) *Geological and Landscape Conservation*. The Geological Society, London, 313–317.

KEEN, D. H. (ed.) 1989. *West Midlands Field Guide*. Quaternary Research Association, Cambridge.

KEEN, D. H., HARDAKER, T. & LANG, A. T. O. 2006. A Lower Palaeolithic industry from the Cromerian (MIS 13) Baginton Formation of Waverley Wood and Wood Farm Pits, Bubbenhall, Warwickshire. *Journal of Quaternary Science*, **21**, 457–470.

KNELL, S. J. 2002. Collecting, conservation and conservatism: late twentieth century developments in the culture of British geology. *In*: OLDROYD, D. R. (ed.) *The Earth Inside Out: Some Major Contributions to Geology in the Twentieth Century*. Geological Society, London, Special Publications, **192**, 329–351.

KNELL, S. J. 2004. Altered values: searching for a new collecting. *In*: KNELL, S. J. (ed.) *Museums and the Future of Collecting, Second Edition*. Ashgate, Aldershot, 1–46.

MARTILL, D. M. 2005. Tubular epibionts with possible symbiotic bacterial biofilm on ammonites from the Lower Jurassic of Central England. *Neues Jahrbuch für Geologie und Paläontologie, Monatshefte*, **2005(6)**, 359–372.

MURPHY, M. 2004. Face Lift improves science research. *Earth Heritage*, **22**, 12–13.

NATURE CONSERVANCY COUNCIL. 1988a. SSSI notification and denotification. *Earth Science Conservation*, **24**, 28–33.

NATURE CONSERVANCY COUNCIL. 1988b. SSSI notification and denotification. *Earth Science Conservation*, **25**, 43–46.

NATURE CONSERVANCY COUNCIL. 1989. *Cross Hands Quarry: Geological Site Description*. Nature Conservancy Council, Peterborough.

NATURE CONSERVANCY COUNCIL. 1990. *Earth Science Conservation in Great Britain: A Strategy*. Nature Conservancy Council, Peterborough.

POWELL, J. H., GLOVER, B. W. & WATERS, C. N. 2000. *Geology of the Birmingham Area*. Memoir of the British Geological Survey. The Stationery Office, London.

PROSSER, C. 2003. Webster's Clay Pit SSSI. Going, going, gone but not forgotten. *Earth Heritage*, **19**, 12.

PROSSER, C. 2006. Massive step for geological conservation. *Earth Heritage*, **25**, 8.

PROSSER, C., MURPHY, M. & LARWOOD, J. 2006. *Geological Conservation; a Guide to Good Practice*. English Nature, Peterborough.

RADLEY, J. D. 2005. The Jurassic of Warwickshire: perspectives on collecting. *The Geological Curator*, **8**, 181–187.

RADLEY, J. D. 2006. Trace fossils: a smaller museum's perspective. *The Geological Curator*, **8**, 247–254.

RADLEY, J. D. 2007. New information on the Lias Group (Late Triassic—Early Jurassic) in the Avon Valley, Warwickshire. *Proceedings of the Cotteswold Naturalists' Field Club*, **44**, 102–107.

RADLEY, J. D. & FRIEND, C. 2006. Soft-sediment challenge. *Earth Heritage*, **26**, 21.

ROBINSON, E. 2007. Those geoparks. *Geology Today*, **23**, 8–9.

RUSHTON, A. W. A., OWEN, A. W., OWENS, R. M. & PRIGMORE, J. K. 1999. *British Cambrian to Ordovician Stratigraphy*. Joint Nature Conservation Committee, Peterborough, Geological Conservation Review Series, **18**.

SCOTT, P. W., SHAIL, R., NICHOLAS, C. & ROCHE, D. 2007. *The Geodiversity Profile Handbook*. David Roche GeoConsulting, Exeter, UK.

SHELTON, P. D. 2004. *The Conservation and Management of Unconsolidated Geological Sections*. English Nature Research Report **563**.

SHOTTON, F. W., KEEN, D. H., COOPE, G. R. ET AL. 1993. Pleistocene deposits of Waverley Wood Farm Pit, Warwickshire, England. *Journal of Quaternary Science*, **8**, 293–325.

STACE, H. & LARWOOD, J. G. 2006. *Natural Foundations: Geodiversity for People, Places and Nature*. English Nature, Peterborough.

SWANN, G. 2005. Fire and ice: GA Excursion to Warwickshire. *Magazine of the Geologists' Association*, **4**, 20–21.

WARWICKSHIRE COUNTY COUNCIL. 1993a. *Warwickshire Landscape Guidelines: Avon Valley, Feldon, Cotswolds*. Warwickshire County Council Planning and Transport Department, Warwick.

WARWICKSHIRE COUNTY COUNCIL. 1993b. *Warwickshire Landscape Guidelines: Dunsmore, High Cross Plateau, Mease Lowlands*. Warwickshire County Council Planning and Transport Department, Warwick.

WARWICKSHIRE COUNTY COUNCIL. 1993c. *Warwickshire Landscape Guidelines: Arden*. Warwickshire County Council Planning and Transport Department, Warwick.

A vision of 'deep time': the 'Geological Illustrations' of Crystal Palace Park, London

PETER DOYLE

Department of Earth Sciences, University College London, Gower Street,
London WC1E 6BT, UK (e-mail: doyle268@btinternet.com)

Abstract: Crystal Palace Park in the London Borough of Bromley is a masterpiece of park design by the visionary Sir Joseph Paxton. Created to house the iron and glass 'Crystal Palace' (the temporary structure built for the 1851 Great Exhibition in Hyde Park), the park was developed on a series of themed terraces, with the palace itself at the top of Sydenham Hill. The terraces were linked by a grand central walkway, and massive fountains played in gigantic fountain bowls. Today, the palace is gone, destroyed by fire in 1936; the fountains are quiet and their bowls occupied by the stadia of the National Sports Centre; and the central walk is interrupted by intrusive twentieth century concrete architecture. But one jewel of the original remains. In the SE corner lies a remnant of Paxton's original English landscape garden, a fragment populated with 'antediluvian monsters' and geological cliffs. This remnant is arguably the world's first attempt at recreating, in a systematic, scientific and ordered way, the geology of the United Kingdom, and its survival and subsequent restoration in 2001 is a remarkable testimony to its constructors and originators.
    This paper examines the background and achievement of this first accurate recreation of geology in a public park, a Victorian monument to the relevance of promoting awareness of the science as a foundation to effective geoconservation.

Geology: the new science of the masses

The birth of geology—the science of the composition, structure and history of the Earth—can be traced back to at least the seventeenth century, but the explosion of popular and professional interest in the science can be placed at around the turn of the eighteenth century (Rudwick 1985, 1992). At this time, geologists and palaeontologists in France, Germany and Britain were shaping the new science, with men such as Georges Cuvier in Paris, catastrophist and identifier of extinction; James Hutton, a Scot, the originator of the concept of uniformitarianism, and discoverer of the vastness of geological time (since dubbed 'deep time'), and the Englishman William Smith, creator of the geological map and interpreter of 'strata' identified by fossils (e.g. see Gould 1990; Rudwick 1985, 1992; Winchester 2001). From the turn of the century through to the 1860s scientific advances came thick and fast in Britain, most published by the Geological Society of London, the world's first geological society, set up in (1807). In the *Transactions of the Geological Society of London* came the first notices of dinosaurs and other extinct organisms (in Buckland (1824) was the first formal description of what became known as a dinosaur, *Megalosaurus*); and the establishment of the formal stratigraphy of the British Isles (in Buckland (1835), the stratigraphy of the

Portland and Purbeck was established, a model of which was later made in Crystal Palace Park).
    By the 1860s geology was so popular that it was included as one of the eight 'greater sciences' on the Albert Memorial in London (the other seven being agriculture, geometry, physiology, astronomy, rhetoric, chemistry and medicine; Brooks 1995). The subject found favour with the masses, and people turned out in their hundreds to hear notables such as Sir Roderick Murchison speaking on his 'Silurian system' underground in Dudley (Barber 1980). Barber's thesis is that geology and palaeontology were part of the popular Victorian obsession with the natural sciences. Geology was available to anyone with access to the countryside, and even, from the mid-nineteenth century onwards, within the confines of the urban park.

Geology in public parks

The birth of the public park has been adequately described elsewhere, and it is sufficient to say that from the early 1840s onwards, formally laid out urban parks were a feature of urban planning (Conway 1991). From the beginning, geology was included in such parks, usually as aspects of 'hard landscape', but also increasingly representative of 'scientific specimens' (Conway 1991; Taylor 1995; Doyle *et al.* 1996). In fact, the history of

From: BUREK, C. V. & PROSSER, C. D. (eds) *The History of Geoconservation.*
Geological Society, London, Special Publications, **300**, 197–205.
DOI: 10.1144/SP300.15   0305-8719/08/$15.00 © The Geological Society of London 2008.

the representation of geological artefacts and of geology *per se* in private gardens and parks can, arguably, be traced back to the 'Grand European Tour' and the development of grottoes on private estates. The eighteenth century Goldney Garden grotto in Bristol is an excellent example (Savage 1989). In public parks, many representations of geology are associated with Joseph Paxton, who had earlier used the concept to good effect in the private estate of Chatsworth. In Birkenhead Park (1847), rockworks represented The Strid in Yorkshire, a rocky gorge cut through Millstone Grit (Taylor 1995). Interestingly, Paxton imported this stone into Birkenhead Park, a park actually founded on other, younger and more brightly coloured red sandstones. This desire for accuracy was to become a focal point for Paxton in developing his gardens, and also, for other designers who used accurate representations of geology in later parks. Good examples include the Khyber Pass in East Park, Hull (1887), the 'Pulhamite' cliffs and crags in Battersea Park (1866–70), and the representation of Thornton Force in Lister Park, Bradford (1903) (see Festing 1984; Conway 1991; Robinson 1994; Taylor 1995; Doyle *et al.* 1996). It was against this backdrop that Paxton was to create his masterpiece; Crystal Palace Park, at the time on the outskirts of SE London.

The embodiment of a Victorian ideal

In June 1854, Crystal Palace Park was opened to the public for the first time. Paxton intended it to be a complex of pleasure grounds to rival those of the Palace of Versailles, housing the reconstructed Crystal Palace—the innovative glass and steel structure built by Paxton for the Great Exhibition of 1851 in Hyde Park (Beaver 1986). Crystal Palace Park was laid out on Sydenham Hill, within the 200-acre grounds previously occupied by Penge Place. The Palace itself was rebuilt on the crest, and a series of terraces were constructed on its slopes, including immense fountains and a large boating lake (Fig. 1). The vision was a complex one, as it led from the delights and peculiarities of the immense interior of the palace itself, to a variety of experiences laid out for the visitor in the surrounding parkland; a perfect embodiment of the Victorian ideal of the continuity of knowledge.

The central grand walkway was the axis of the park, bisecting the formal italianate gardens of the top terrace, the large twin fountains and finally the English landscape garden at the base of the slope. The division of the park into terraces created a subset of gardens that encompassed several of the categories discussed by the influential

Fig. 1. Late Victorian photograph of the 'Crystal Palace' with the 'Geological Illustrations' just on the other side of the lake: to the left, the Secondary Island (with its dinosaurs and other reptiles), to the right, the Tertiary Island. Between them may be seen the Coal Measures.

park designer John Loudon twenty years before (Loudon 1835) in which gardens were classified according to the intention of the designer into scientific, landscape, recreation and burial categories.

Perhaps most important of these was the English landscape garden, which was found to the east of the formal gardens and encompassed the lower part of the park, including the lakes, part of the waterworks and ultimately linked to the great fountain systems. Within it, and associated with the lakes, the manifestations of geology were to be constructed, and they remain there today, a unique component of this extraordinary public park.

What is not clear, however, is who originally had the idea to reconstruct geological environments within the park. The issue is still much debated, and requires further research (McCarthy & Gilbert 1994). Suggested authors of the scheme include Sir Richard Owen, Prince Albert and Sir Joseph Paxton himself. Paxton is the most likely. Whatever the origin of the idea, the Board of the Crystal Palace Company sanctioned the construction of what might now be termed a complex geological 'theme park' in the SW quadrant of the park within the English landscape garden:

It is here that one of the most original features of the Crystal Palace Company's grand plan of visual education has been carried out. There, all the leading features of Geology are found displayed, in so practical and popular a manner, that a child may discern the characteristic points of that truly useful branch of the history of nature. (Anon. 1893, p. 29)

The Crystal Palace Company directors were no strangers to geology (H. Torrens, pers comm.), as several of them were involved in the large civil engineering schemes of the day, and at least one was also the director of a mine company in Clay Cross in Derbyshire. As such it is not unreasonable to expect that they would wish to see representation of what was termed 'that most useful branch of the history of nature', and this is underlined by the fact that many of the geological features were modelled on the geology of Derbyshire. Another contemporary guide to the park points to the author of the scheme as a whole, David Thomas Ansted, late Professor of Geology at Kings College, London, and subsequently at the College of Civil Engineers at Putney:

The original plan of the whole was suggested by Professor Ansted, and arranged with Sir Joseph Paxton at an early period of the laying out of the grounds; and as soon as the state of affairs permitted and the actual earthworks of the Plateau were in progress, a model of the intended structure was completed and coloured geologically by Professor Ansted. The works have been ably constructed from this model by Mr James Campbell, who also procured the stone and other minerals from different parts of the country. (Phillips 1855, pp. 191–192)

There was evidently a close working relationship between Ansted and Paxton. 'The series was carefully tabulated by Professor Ansted, to ensure its geological accuracy, according to Sir Joseph Paxton's designs for the picturesque arrangement of this interesting portion of the grounds' (Anon. 1893, p. 29).

The whereabouts of Ansted's model is not known, and as far as can be ascertained, no plans of Ansted's own vision exist, but it is clear that the landscape was to include a representation of the successive ages of the geology of Britain from the Primary (Precambrian–Palaeozoic today) rocks through to the Secondary (Mesozoic) and Tertiary (Cenozoic–Quaternary today). This is also apparent from Ansted's own writings, which echo the existing materials in the park, and from recent 'geological' mapping of the remaining structures in the park (Ansted 1858; Doyle & Robinson 1993, 1995). Ansted's book displays many similarities to the Crystal Palace tableaux.

Given the industrial links of the company directors, it is not surprising that this vision of the geology of Britain was to include economic rocks and geological structures, together with the remains of relatively newly discovered fossil organisms constructed in a full-size and three-dimensional form. The task of constructing these was the duty and vision of Benjamin Waterhouse Hawkins, who had illustrated the published work and treatises of some of the leading palaeontologists of the day. As 'Director of the Fossil Department of the Crystal Palace', Hawkins, advised by Sir Richard Owen, constructed his full-sized extinct mammals and reptiles arranged stratigraphically in Ansted's geological landscape; a vivid recreation

of the most recent discoveries in a new and exciting science (McCarthy & Gilbert 1994).

In fact, it was Gideon Mantell (to many the discoverer of the dinosaurs) who was originally asked to assist. The Alexander Turnbull Library, Wellington, New Zealand, contains a manuscript extract from the minutes of a meeting of the Board of Directors of the Crystal Palace Company, held on 10 August 1852 at which it was resolved:

that a geological court be constructed containing a collection of full-sized models of the animals and plants of certain geological periods, and that Dr Mantell be requested to superintend the formation of that collection ... (Alexander Turnbull Library MS papers 83, folder 32)

The contemporary guides were well aware of its significance as the most extensive educational endeavour ever in a public park: 'the spectator standing on the upper terrace of the Plateau has before him the largest educational model ever attempted in any part of the world' (Anon. 1893, p. 29).

Constructing a vision of 'deep time'

Ansted's geological framework was completed with the exception of the older Cambrian and Silurian 'greywacke' rocks; a framework intended to illustrate the geological development of Britain on its journey through the 'deep time' of geological history. The younger 'Primary' rocks had a special place adjacent to a vibrant water-course, and here a Mountain Limestone cliff overlain by Millstone Grit and faulted against Coal Measures was constructed, modelled on the Derbyshire Peaks (Fig. 2). All of these geological units were founded on the Devonian rocks of the Old Red Sandstone forming the framework for the main

Fig. 2. 'Primary' rocks in Crystal Palace Park: the restored Mountain Limestone Cliff and cave.

water feature and 'rustic' bridge (McDermott 1854; Anon. 1893).

Adding realism to the Mountain Limestone was the construction of a three-quarters scale lead mine and cave, complete with stalactites. As noted in the guidebooks of the day, this element of Derbyshire realism was created by James Campbell, a mining engineer and member of the Crystal Palace Company board (Phillips 1855). What is surprising about this construction is the sheer technical complexity of the structure. The limestone cliff and mine was completely destroyed in the 1960s during remodelling of the watercourse as a 'water garden', but enough remained to be able to carry out archaeological investigations in 2001. These showed the presence of a stepped limestone cliff constructed over a brick-arched tunnel. The tunnel contained extensive modelling of stalactites and other features associated with the karst landscape of Derbyshire. Importantly, these investigations also revealed that complex mineral veins had been built into the scheme to enhance realism, together with large crystals of typical minerals found in the lead mines that flourished in Derbyshire in the mid-nineteenth century; an echo of the crystal grottoes of the previous century. The coal face also demonstrated considerable complexity, with coal cut in blocks and reconstructed in accurate relations with sandstones and ironstones typical of the Clay Cross Pit (Fig. 3; Doyle & Robinson 1993, pp. 184–7).

Overlying gently tilted Carboniferous rocks was the New Red Sandstone, deliberately placed in unconformity: these sandstones provide continuity with the first of two islands intended to carry the reconstructions of extinct animals, constructed by Hawkins (Doyle & Robinson 1993, pp. 188–9). Downslope from this tableau was the 'Secondary island' itself commencing with tilted New Red Sandstone. Conformable with these were representations of the other major 'Secondary' (Mesozoic) geological units of southern Britain. In turn Lias, Oolite and Wealden rocks succeeded each other, surmounted by Chalk at the head of the island. In place on top of these rocks were reconstructions of animals that had been recovered from the Mesozoic formations during the previous fifty years, and beyond, on a separate island, the mammals of the Cenozoic and Quaternary, on rather more uncertain footings consequent upon the weaker sands, clays and gravels that typify these geological units in Europe (Doyle & Robinson 1993, 1995).

As has been argued by Martin Rudwick, Hawkin's representation of extinct animals in a stratigraphical arrangement was not unique. In fact, it was part of a developing tradition of pictorial representation of geological time in the mid-nineteenth century. But what was unique, and remains so to

(a)

(b)

Fig. 3. 'Primary' rocks in Crystal Palace Park: (**a**) the Coal Measures cliff as it appears today; (**b**) the Coal Measures cliff from a Victorian photograph (Courtesy Mick Gilbert).

this day, is the accurate portrayal of those animals in three dimensions, and set within a framework of rocks that once contained their fossil bones (Rudwick 1992). However, since the restoration, the observer can see the stratigraphy as a backdrop and stage to the recreated animals, with perspective providing continuity to that stratigraphy.

Populating a geological landscape

Hawkins set out his method of working in a lecture delivered to the Society of Arts in 1854, subsequently reprinted for separate distribution by James Tennant, who was later to supply small-scale models of the dinosaurs to educational establishments and museums. Working closely with Sir Richard Owen, Hawkins first created a clay model that was then altered in line with the scientist's vision (Hawkins 1854). This was particularly significant in relation to the reptiles, where active debate raged, a debate well rehearsed by Deborah Cadbury (2001), whose book deals with the antagonism between Mantell and Owen. Much of the battle was fought over the relative size and form of dinosaurs such as *Iguanodon*, Mantell's own discovery. With Mantell's withdrawal from Crystal Palace, it was Owen's vision that was completed.

Having gained agreement, most of the larger animals were constructed as buildings with strong brick piers to support their massive bodies. A range of commonly available building materials were then used to create the overall framework, onto which were attached the carefully moulded outer layers of the dinosaurs, for example. Smaller reptiles were built up carefully *in situ* (Fig. 4) and several of the mammals were built around iron armatures, with delicately moulded lead heads and limbs. One of the largest of the mammals, the *Megatherium*, a gigantic ground sloth, illustrates a fourth technique used by Hawkins, that of careful sculpture from limestone blocks, rather than moulding of cements.

The retinue of animals represented the brightest and best discoveries by mostly British scientists: New Red Sandstone dicynodonts and labyrinthodons from the Cape Province of South Africa and the Midlands of England, respectively (Fig. 4); Liassic ichthyosaurs and plesiosaurs from Lyme Regis, based almost exclusively upon the discoveries of the collector Mary Anning (Fig. 5); the alligator-like *Teleosaurus* from the Lias of Whitby and pterodactyls of the Oolite and Chalk (Fig. 6); the Stonesfield Slate dinosaur *Megalosaurus*, the first of the 'terrible lizards' described by William Buckland in 1824 (Fig. 7), and its prey from the

Fig. 4. The labyrinthodons under construction in 1854 (Courtesy Mick Gilbert).

Fig. 6. *Teleosaurus* (foreground) with Jurassic pterodactyls in the background, after recent restoration.

Fig. 5. The Secondary Island. Ichthyosaurs and plesiosaurs under construction in 1854 (Courtesy Mick Gilbert).

Fig. 7. Close up of the *Megalosaurus*, as it appears today.

Fig. 8. *Iguanodon* standing on Wealden sandstones, as it appears today.

Weald *Iguanodon* (Fig. 8) and *Hyaelosaurus*, both described by Gideon Mantell (e.g. Mantell 1825); and the Chalk marine reptile *Mosasaurus* (Fig. 9). All were constructed on representations of the very rocks that yielded their bones, and described by Owen himself in his own guide to the display, published by the park authorities in 1854 (Owen 1854).

Separated from the Secondary island by a weir was the 'Tertiary' island. A symbolic end to the 'Age of Reptiles', the weir marked the beginnings of the 'Age of the Mammals'. The 'Tertiary'

island was to be populated with mammals from the Cenozoic (and early Quaternary), but surviving records show that just a fraction of the mammals originally intended were built, due to financial difficulties (Doyle & Robinson 1993, 1995; McCarthy & Gilbert 1994). A letter from Hawkins to Sir Richard Owen dated 24 October 1855 illustrates what was completed and what was intended (Owen papers 14/534, Natural History Museum; Fig. 10). Notable mammals constructed were: Cuvier's Paris Basin *Palaeotherium* and *Anoplotherium*; *Megatherium*, a giant ground sloth

Fig. 9. *Mosasaurus* as it appears today.

from South America (Fig. 11); and *Megaceros*, the 'Irish Elk' (Fig. 12). Amongst the other mammals intended were the mammoth, mastodon, *Dinotherium*, and the giant armadillo *Glyptodon*. Birds driven to extinction by human activity, the Dodo the Moa (*Dinornis*)—upon which Richard Owen had built his reputation—were also to be built. All were to be placed upon a geological backdrop of worked aggregates intended to represent the relatively unconsolidated rocks of this interval of geological time. They were never completed.

Changing fortunes, changing fashions

From the 1860s onwards, the park had mixed fortunes, and one-by-one the visionary nature of the park's landscape began to fail or become obscure. The changing nature of park activities—including the garrisoning of troops in two world wars—and changes in local government priorities also

Fig. 11. *Megatherium* as it appears today.

took their toll. By the 1970s the continuity and integrity of the grand idea had become broken and fragmented, and some major features, such as the impressive cliff of Mountain Limestone, had been completely destroyed.

Today, the reconstructed animals are 'buildings' protected by law, and have mostly survived, despite neglect, scorn and derision. Most general accounts of dinosaurs have as a start point the discoveries of Mantell and Buckland, the work of Owen and

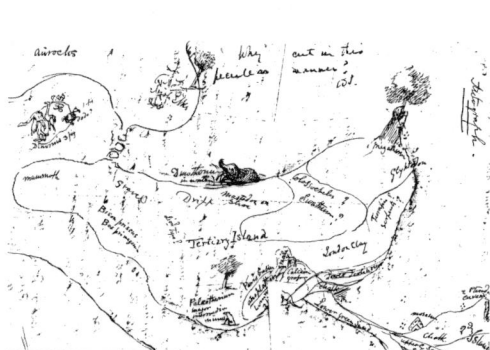

Fig. 10. Manuscript map of the Tertiary Island, reproduced from a letter from Hawkins to Owen preserved in the Owen papers at the Natural History Museum, London. The maps show the extent of what was planned for this island, a fraction of which was constructed.

Fig. 12. The antlers of *Megaceros*, the Irish Elk. The original had fossil antlers, then a common find.

the models in Crystal Palace. The majority make some passing reference to 'inaccuracies' and commonly refer to the quadrupedal stance of Iguanodon and *Megalosaurus*, and the 'mistake' of placing the *Iguanodon*'s thumb-spike on its nose. This trend started in the late nineteenth century, with Henry Woodward of the British Museum (Natural History) being especially scornful:

the late Mr B. Waterhouse Hawkins (formerly a lithographic artist) was for years occupied in unauthorised restorations ... discoveries of later years have shown that *Dicynodon* and *Labyrinthodon* ... were salamander-like reptiles ... that *Iguanodon* did not usually stand on 'all fours' [and] that the horn on its snout was really on its wrist. (Preface in Hutchinson, H.N. 1892, p. iii)

These views, although strictly accurate, are starkly unimaginative, and contributed to the decline of what is in reality a magnificent adventure in science, a bold step of creating a three-dimensional geological textbook in the heart of a London suburb.

Fortunately, with Heritage Lottery backing, the Borough of Bromley has restored and reconstructed the 'Geological Illustrations', dubbed the 'Time trail', for a more modern audience (Doyle 1994). Conservation of the 'buildings' (dinosaurs and mammals), and reconstruction of the geological features included 110 tonnes of replacement Carboniferous Limestone delivered from a source close to the original in Derbyshire (Fig. 2). This was constructed on geological principles as a replica cliff, complete with the original mineral mine and cave. Cotswold Oolite replaced that destroyed by neglect and natural process, to form the perch for two missing Jurassic pterodactyls (Fig. 6). The pterodactyls themselves were constructed to the highest standards, replicas of the 1854 originals. A cliff of chalk complete with flint lines was constructed for the other pterodactyls. Contemporary photographs and fragments close to the original site demonstrated the original form of this cliff, a cliff that had, in common with the rest of the geological display, literally mouldered away into the undergrowth.

The dinosaurs have been carefully restored, the swellings associated with iron bar and cracks associated with age and settlement repaired with high-specification materials intended to last (Doyle 2001a, b). The original paint scheme was assessed: a layer of startling pink in the strata of paints forms the base, but the final coat includes muted greens and greys with appropriate glazes (Figs 6–9). Finally, sensitive planting, reflecting the succession of plant life through time as known in the 1850s, complements the animals and their geological setting. Piece by piece the 'geological illustrations' have risen once again from their municipal park setting, almost 150 years after

they were first conceived, a striking reminder of Victorian ingenuity and scholarship.

The lesson of Crystal Palace Park

The Victorian ideal of Crystal Palace Park was a means of introducing to an urban public, in a practical sense, a science that could otherwise only be seen in popular books and monthly magazines—as new discoveries were being made, and as geologists became public figures. Using the latest science, published in the best journals of the day, the public were entertained by the juxtaposition of solid geology in correct stratigraphical relationships and reconstructed dinosaurs and large mammals. In this way, the Crystal Palace Park experience transcended the mundane, and became more than a gaudy theme park. It served instead as both visual spectacle and outdoor teaching laboratory, a lesson in the promotion of awareness of geology that is rarely attained, with any success, today (Doyle 1993; Doyle *et al.* 1996). This record of innovation, repeated in other Victorian parks serves as a lesson for today; only through increasing awareness of geology will the public be sufficiently engaged to conserve it (Doyle & Bennett 1998). As such, Crystal Palace Park deserves its place in the annals of geoconservation as an outstanding example of the relevance and importance of enhancing awareness of geology in an urban setting.

References

ANON. 1893. *Illustrated Guide to the Palace and Park.* Charles Dickens & Evans, London.

ANSTED, D. 1858. *Geological Science: Including the Practice of Geology and the Elements of Physical Geography.* Houlston & Stoneman, London.

BARBER, L. 1980. *The Heyday of Natural History, 1820–1870.* Jonathon Cape, London.

BEAVER, P. 1986. *The Crystal Palace.* Phillimore, Chichester.

BROOKS, C. 1995. *The Albert Memorial.* English Heritage, London.

BUCKLAND, W. 1824. Notice on the *Megalosaurus* or the great fossil lizard of Stonesfield. *Transactions of the Geological Society of London (Second Series),* **1**, 390–396.

BUCKLAND, W. 1835. On the geology of the neighbourhood of Weymouth and the adjacent parts of the County of Dorset. *Transactions of the Geological Society of London (Second Series),* **4**, 1–46.

CADBURY, D. 2001. *The Dinosaur Hunters.* Fouth Estate, London.

CONWAY, H. 1991. *People's Parks. The Design and Development of Victorian Parks in Britain.* Cambridge University Press, Cambridge.

DOYLE, P. 1993. The lessons of Crystal Palace. *Geology Today,* **9**, 107–109.

DOYLE, P. 1994. *The Geological Time Trail*. London Borough of Bromley.

DOYLE, P. 2001a. Restoring a Victorian vision of 'deep time'. *Earth Heritage*, **16**, 16–18.

DOYLE, P. 2001b. Restoring a Victorian vision of 'deep time': the geological illustrations of Crystal Palace Park. *American Paleontologist*, **9**, 2–4.

DOYLE, P. & BENNETT, M. R. 1998. Earth heritage conservation: the past, present and future agenda. *In*: BENNETT, M. R. & DOYLE, P. (eds) *Issues in Environmental Geology. A British Perspective*. Geological Society, London, 41–67.

DOYLE, P. & ROBINSON, E. 1993. The Victorian 'Geological Illustrations' of Crystal Palace Park. *Proceedings of the Geologists' Association*, **104**, 181–194.

DOYLE, P. & ROBINSON, E. 1995. Report of a field trip to Crystal Palace Park and West Norwood Cemetery, 11 December 1993. *Proceedings of the Geologists' Association*, **106**, 71–78.

DOYLE, P., BENNETT, M. R. & ROBINSON, E. 1996. Creating urban geology: a record of Victorian innovation in park design. *In*: BENNETT, M. R. L., DOYLE, P., LARWOOD, J. G. & PROSSER, C. D. (eds) *Geology on Your Doorstep*. Geological Society, London, 74–84.

FESTING, S. 1984. 'Pulham has done his work well'. *Garden History*, **12**, 138–158.

HAWKINS, B. W. 1854. On visual education as applied to geology. *Journal of the Society of Arts*, **2**, 444–449.

GOULD, S. J. 1990. *Times Arrow, Times Cycle*. Penguin, London.

HUTCHINSON, H. N. 1892. *Extinct Monsters*. Chapman & Hall, London.

LOUDON, J. C. 1835. Remarks on laying out Public Gardens and Promenades. *Gardeners Magazine*, 644–669.

MANTELL, G. 1825. Notice on the *Iguanodon*, a newly discovered fossil reptile, from the sandstone of Tilgate Forest in Sussex. *Philosophical Transactions of the Royal Society of London*, **14**, 179–186.

MCCARTHY, S. & GILBERT, M. 1994. *The Crystal Palace Dinosaurs*. Crystal Palace Foundation, London.

MCDERMOTT, E. 1854. *Routledge's Guide to the Crystal Palace and Park at Sydenham*. Routledge, London.

OWEN, R. 1854. *Geology and the Inhabitants of the Ancient World*. Crystal Palace Library and Bradbury & Evans, London.

PHILLIPS, S. 1855. *Guide to the Crystal Palace and Park*. 5th edn. Bradbury & Evans, London.

ROBINSON, E. 1994. The mystery of Pulhamite and an outcrop in Battersea Park. *Proceedings of the Geologists' Association*, **105**, 141–143.

RUDWICK, M. J. S. 1985. *The Meaning of Fossils. Episodes in the History of Palaeontology*. University of Chicago Press, Chicago.

RUDWICK, M. J. S. 1992. *Scenes from Deep Time*. University of Chicago Press, Chicago and London.

SAVAGE, R. J. G. 1989. Natural History of the Goldney garden grotto, Clifton, Bristol. *Garden History*, **17**, 3–40.

TAYLOR, H. A. 1995. Urban Public Parks, 1840–1900: design and meaning. *Garden History*, **23**, 204–208.

WINCHESTER, S. 2001. *The Map that Changed the World*. Viking, London.

Cavers and geoconservation: the history of cave exploration and its contribution to speleology in the Yorkshire Dales

PHILLIP J. MURPHY[1] & ANDREW T. CHAMBERLAIN[2]

[1]*School of Earth and Environment, University of Leeds, Leeds LS2 9JT, UK*
(e-mail: p.murphy@see.leeds.ac.uk)

[2]*Department of Archaeology, University of Sheffield, Sheffield S1 4ET, UK*

Abstract: Caves are important as they preserve archaeological and palaeoenvironmental data otherwise lost from the land surface. The fragile nature and limited extent of cave deposits is often not appreciated by non-specialists and the activities of the main group of cave visitors (sporting cavers) are viewed as damaging to the cave interior deposits. Potential threats to the cave interior deposits of the Yorkshire Dales National Park including caver activity are reviewed. It is concluded that sporting cavers have added greatly to our knowledge of the archaeological record contained in the caves. They appreciate the value of the underground environment and take steps to preserve the cave interior deposits. Any geoconservation strategy that deals with caves must involve the caving community.

Caves are defined as accessible natural cavities within rock formations. They are found in a variety of rock types and result from a range of different geomorphological processes. Most caves are initially formed by hydrological processes acting on slightly soluble minerals within the rock, giving rise to karst landforms characterized by solution caves, sink holes, lack of active surface water courses and efficient underground drainage. Karst landforms are limited to a few geographical areas in Britain, mainly those where the soils are underlain by the massive limestones of the Devonian, Carboniferous and Permian periods: the Yorkshire Dales has the greatest extent of cave development in England (Waltham *et al.* 1997) (Fig. 1). Caves are also formed by mechanical processes, including the slip-rift fissuring, which can affect well-bedded and jointed limestones and sandstones. Sea caves that are formed in high-energy erosional environments at coastal exposures of rocks can become relict landforms after isostatic uplift.

Caves act as nature's archives. By isolating material from the forces of erosion within the rock mass they prevent, or at least delay, the destruction of otherwise rarely preserved material. Cave interior deposits provide a source of palaeoclimatic, palaeontological and archaeological data from times when surface evidence has been removed, or at least greatly reduced. The processes which lead to emplacement of deposits within a cave are selective, so when studying such deposits it should be remembered that they are not a fully representative subsample of what was present on the surface when the deposits were emplaced.

Within the caves themselves it cannot be presumed that everything will be perfectly preserved as deposits are subject to the ongoing subterranean geomorphological processes such as collapse of sediment stacks and redistribution by cave streams. The spatial distribution of caves is not uniform; the dissolution processes that form the karst landscapes containing caves tend to be confined to carbonate lithologies and some carbonate units contain more caves than others (for instance compare the small number of caves known in the English Chalk with that of the Carboniferous Limestone areas). When studying cave deposits one is not dealing with a geographically uniform spread of sample locations. In spite of these potential biases, the study of cave interior deposits has become central to palaeoclimatic reconstructions and has provided crucially important archaeological evidence. It is no accident that many of the most important sites where evidence for human evolution has been recovered are caves.

Caves pose particular problems from a conservation perspective. Cave interior deposits are by definition not laterally extensive so they must be considered a finite resource when sampling or undertaking an excavation. Most people who visit caves are recreational or 'sporting' cavers. The passage of cavers through a system can lead to the removal of sediment from the cave floor and damage to speleothem deposits. A significant proportion of the caver population have an interest in exploring new cave passages. It is one of the few remaining opportunities to discover something truly new and one of the main ways to do this is to try and dig through sediment blockages. Many

From: BUREK, C. V. & PROSSER, C. D. (eds) *The History of Geoconservation.*
Geological Society, London, Special Publications, **300**, 207–215.
DOI: 10.1144/SP300.16   0305-8719/08/$15.00 © The Geological Society of London 2008.

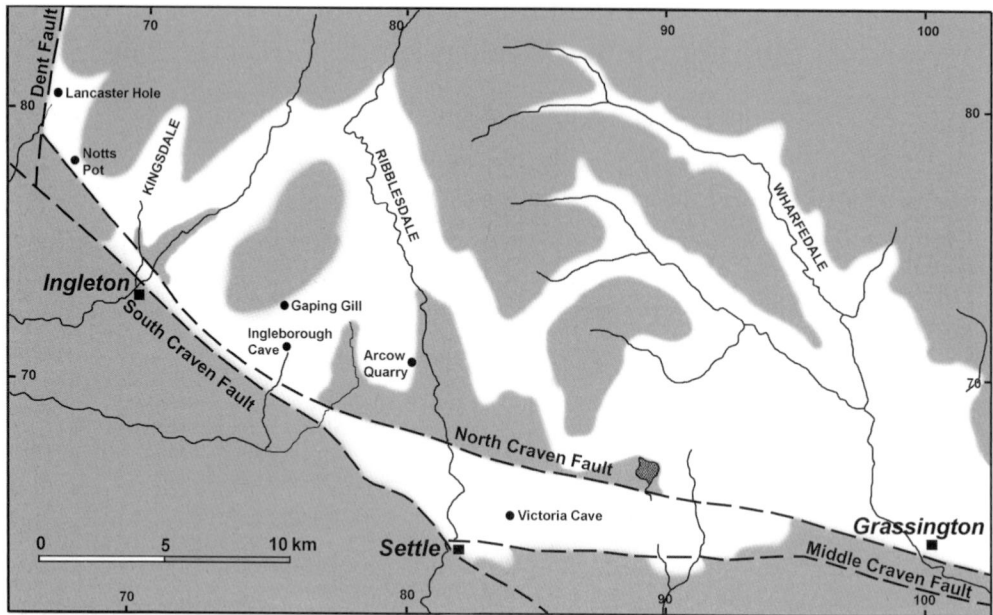

Fig. 1. Map of the study area (From Waltham *et al.* 1997).

of the caving areas of Britain are National Parks or areas of outstanding natural beauty (AONB) and often include National Nature Reserves and Sites of Special Scientific Interest. Caves therefore fall within the conservation frameworks of these areas. The activities of cavers may be seen as a potential threat to the deposits within the caves. The purpose of this paper is to explore some of the geoconservation issues and to show how caver activity has greatly increased our knowledge and understanding of the sediment record preserved within caves. We aim to demonstrate that the caving community values the fragility of the underground environment and is proactive in conserving, protecting and enhancing the caves visited by cavers, illustrated with examples from the Yorkshire Dales.

Bone caves

Both recreational and scientific exploration of the caves of the Yorkshire Dales have proceeded since 1837 when Mr J. Anson Farrer ordered excavations to begin at Ingleborough Cave (Craven 2004; McFarlane *et al.* 2005). During the course of 160 years of exploration the bones of extant and extinct animal species have been recovered from many of the region's caves, although in only half of these instances have the finds been followed up by scientific study. Much of the scientific interest in the Dales caves began in the nineteenth century,

when eight caves were researched and their contents were described in local and national scientific journals. This interest waned during the twentieth century despite the upsurge of discoveries by recreational cavers from the 1940s onwards. During recent decades cavers have continued to discover deposits of animal bones in caves, but the finds have been only sporadically reported in the recreational caving literature and in most cases the material has not been studied by archaeozoologists or palaeontologists.

Twenty-six speleological sites are listed in Murphy (2002) where the occurrences of vertebrate remains were recorded in the sporting caving literature. This exceeds the number of cave sites with vertebrate remains recorded in the scientific literature (Chamberlain 2002). Of the caver-recorded sites only twelve have had formal species identifications made by a competent authority and only three sites have had material carbon dated. The situation is not much better in those sites described in the scientific literature where only four sites have published carbon dates, although two of those sites are also partly constrained by uranium-series speleothem dates. Where identification has been made at caver-recorded sites the most common remains are those of red deer, closely followed by cow, dog and horse. The bias of the caver records towards large animal bones has been discussed by Murphy (2003) who suggested that it is likely to reflect the probability with which finds are noticed and reported by sporting cavers. Although this is a valid explanation, the limitations posed by the use of head-mounted lighting systems by cavers are not generally appreciated by non-cavers. Even in regularly visited sites bones have been passed by cavers for many years before being reported. For example, although bones had already been recovered from the River Junction area of Kingsdale Master Cave (an intensively explored cave and a very popular caving excursion for novice cavers), a horse skull jammed beneath a rock ledge next to the main cavers' path was only noticed in 2002. Another contrast is in the age of the material recovered. Most of the caver-recovered fauna is of domestic or agricultural origin dating to the last few thousand years. However the records of reindeer, auroch and woolly rhinoceros remains from caves in the Yorkshire Dales indicate that cavers do stumble across much older faunal remains.

In part, the remoteness of many of the Dales caves may have militated against archaeological and palaeontological investigations being conducted at the less accessible sites; the caves described in the scientific literature are concentrated in the southern part of the Dales, close to the centres of population and the principal routes of communication along the Wharfe and Ribble valleys (see Fig. 2). As with other caving regions of Britain, the arrival of the railway brought the Victorian educated elite to within an hour's horse or coach ride of some of the more accessible cave systems, and proximity to road and rail networks has been a significant factor in determining which sites have received scientific attention (Craven 2002). Sporting cavers, on the other hand, have not been deterred by surface accessibility and it is geological rather than geographical factors that have influenced the pace and pattern of recent discoveries of new cave systems.

Most of the bone caves from the scientific literature that are listed in Chamberlain (2002) were explored prior to J. W. Jackson's synthesis of British cave archaeology and palaeontology (Jackson 1962) whereas nearly all of the finds in gazetteer of cavers' finds (Murphy 2002) were first described from the late 1960s onwards. This has significant implications for cave conservation because the finds made at some of the sites described in the recreational caving literature are not routinely reported to the statutory authorities who are responsible for nature and heritage conservation in the areas concerned.

Threats to cave conservation

The potential fragility of underground resources in karst environments lends extra importance to the conservation and management of caves and their sediments. Threats to caves and their contents can be divided broadly into human activities, and the actions of physical and biotic processes in the natural environment. Although limestone quarrying is sometimes perceived as an important threat to caves, many quarries within protected areas including the National Parks are approaching the ends of

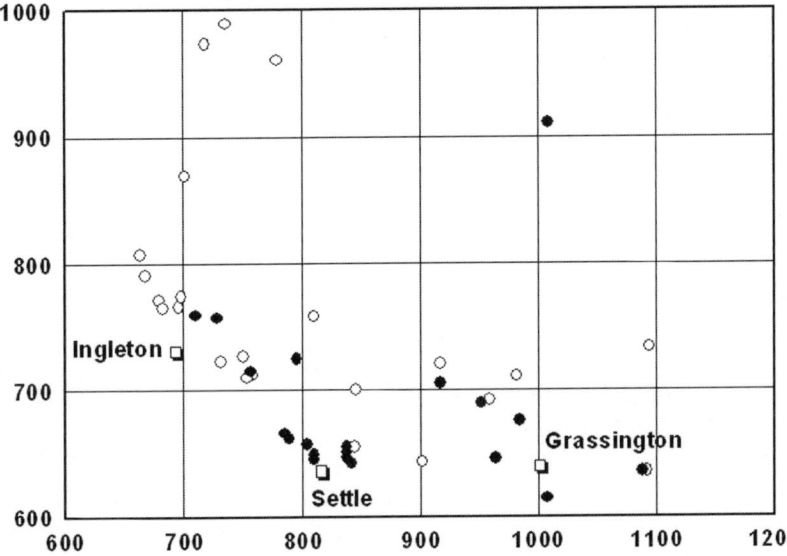

Fig. 2. Locations of bone caves in the Yorkshire Dales. Solid symbols are caves reported in the scientific literature, open symbols are caves reported in the caving literature.

their licences and there are few proposals for new quarries or extensions to existing extraction areas within the Yorkshire Dales. An exception was the recent extension granted to the Arcow Quarry near Horton-in-Ribblesdale, where extraction of Silurian greywacke for use as roadstone threatened caves in the overlying Carboniferous limestone. In this instance conditions on the planning approval provided a limited opportunity for archaeologists to examine the caves and their sediments before these sites were quarried away. Some areas of karst have been severely damaged during recent decades through the removal of pavement formations for decorative rockery stone, and this damage has continued (albeit at a reduced level) despite the implementation of Limestone Pavement Orders under the Wildlife and Countryside Act (Webb 1995).

Farming practices have a relatively benign impact on caves, but there have been some intermittent problems in karst landscapes from pollution of underground autogenic and allogenic water supplies that have been linked to industrial and agricultural effluent disposal. The complexity of hydrological systems in limestone karst areas means that it can be difficult or impossible to trace polluted underground water to a specific surface source. There are also recognizable problems posed by cattle and sheep using cave entrances as shelters. Although some farmers have walled or fenced off cave entrances to prevent animals becoming endangered the animals still congregate near the entrances and can cause trampling and erosional damage to talus deposits in front of the cave: these are often the areas of caves that have heightened archaeological value. In karst landscapes dolines (called shakeholes in the Yorkshire Dales) provide convenient dumping sites for rubbish, building debris, abandoned machinery and other unwanted materials (Fig. 3). In instances where such sites serve as sinkholes in wet conditions this is likely to have adverse effects on local ground water quality. The infilling of such sites also reduces landscape diversity and closes off access routes for fauna into small crevices and fissures in the limestone.

Species of burrowing mammals such as rabbits and foxes pose a slight threat to unconsolidated cave sediments, but badgers can cause more extensive damage to deposits. Badgers prefer smaller, dry, sediment-filled caves for the construction of their setts and in the Yorkshire Dales they appear to be less frequent occupiers of caves than in

Fig. 3. Domestic and agricultural refuse tipped into a closed depression.

more heavily wooded areas of karst such as the Derbyshire and Staffordshire areas of the White Peak. Although badgers cause visible disturbance to cave sediments, regular and careful monitoring of their spoil can be used to detect new instances of archaeological or palaeontological remains in caves.

The management of caves within the Yorkshire Dales has generally been one of benign neglect. Caves tend to be located in less accessible areas of the limestone uplands, and three quarters of Yorkshire Dales caves are located in land that is allocated to woodland or rough grazing. Livestock movements in unenclosed areas of rough grazing are not controlled, and only in a few instances are caves fenced off from surrounding land, so trampling and erosion of cave sediments by livestock can be observed at several cave entrances. In many cases the woodland adjacent to cave sites is unmanaged, and at some sites there is evidence of tree root penetration of rock fissures and cave sediments. Tree fall at cave entrances can also be a significant cause of damage to cave roofs. Overall, the locations of caves in sparsely managed or unmanaged land units implies that when human or natural processes have an adverse impact on caves and their contents this may not be readily observed by land managers, and this reinforces the case for monitoring of cave sites by other visitors such as cavers.

Human visitors are potentially the most destructive threat to caves and their sediments, and adverse impacts due to individual activities within caves include trampling, disturbance and removal of sediments, collection of fossils and speleothems, vandalism, graffiti and littering. Most recreational cavers view caving as a challenging sport, and their primary motivation is to descend to the furthest and/or lowest point in a cave system and then return (Mycroft et al. 1997). A preference is often expressed for 'through trips' that link separate cave entrances, adding to the interest and navigational challenge of the underground route. Many cavers have a desire to explore new caves, but the effort required to excavate in choked or sediment-filled passages without a guaranteed prospect of success deters all but the most motivated explorers and thus digging tends to be an intermittent activity amongst cavers. The focus of interest for discovering new cave passages also tends to be in the deeper areas of existing cave systems, rather than in cave entrances at the ground surface where archaeological and palaeontological remains are more frequently found.

Digging

Digging as a means of discovering new caves is a well-established practice in the karst of the Yorkshire Dales and other areas. This can be either digging from surface karst features in the hope of entering passages below or trying to extend known cave passages by digging through blockages. The scale of some of the digging operations is often not appreciated by non-cavers (Fig. 4). Some sites may be dug regularly for years and involve serious engineering work to gain new passage. For example on Leck Fell in Lancashire it took four years for a group of cavers to excavate nearly 80 m vertically to gain access to a series of passages already discovered by divers (Walsh 2001), and in Gaping Gill cavers spent nearly every weekend for four years digging a new entrance to the system (Haigh 2003). The amount of passageway explored by cavers by digging can be astounding. In the Gaping Gill system by the beginning of the Second World War two and a half miles of passage had been explored (Beck 1984). With the post war rise in popularity of caving coupled with a determination to explore new passage via what ever means was necessary (digging or cave diving) the system had been extended to 7 miles by the mid 1970s (Brook et al. 1975). More recent exploration has now increased the length of explored passage to more than 8 miles.

When undertaking an exploratory dig cavers try to minimize the amount of material removed in order to gain access to open passage. This is not only to minimize the physical effort involved but because opportunities to dispose of debris from the dig site are often limited. In a filled horizontal passage this means digging against the roof as this will probably be where the blockage is the shortest, the fill least consolidated and the probability of intersecting any unfilled roof voids the highest. When digging a filled shaft the cavers will follow a solid wall rather than trying to go down the centre. This allows bracing of the dug shaft and will hopefully reach the top of any ongoing passage with the least volume of material needing to be moved. The strategies employed by cavers mean that in a horizontal passage the cavers concentrate on removing the youngest sediment deposited in the passage and in a vertical shaft fill the oldest deposits at the base of the filled shaft will hopefully not need to be disturbed. This contrasts with the activities of archaeologists whose aim is to unravel as much of the history of the site as possible. This contrast between the nature of a cave exploration dig and an archaeological excavation may provide an explanation for the dominance of recent fauna in the caver reported bone finds. Caver activity tends to be confined to the most recent deposits and underlying older deposits are left undisturbed. As a result, the recovery of animal bones by cave diggers must be taken as an

Fig. 4. Cavers excavating a filled shaft looking for new cave passage.

indication of the potential of a site to contain older remains, and should not be judged as archaeologically unimportant solely on the basis of the age and origins of the bones submitted for identification.

The two cave sites in the Craven area which have produced Ipswichian (OIS 5e) faunas, are both known as a result of major excavations that have penetrated the full depth of cave sediments. In the case of Raygill Fissure, Lothersdale, the cave was exposed as a result of limestone quarrying and the fauna were recovered as the quarry face moved through a filled off-vertical shaft. The Ipswichian fauna were at the base of the completely sediment filled shaft from the base of which a horizontal passage led off (Mial 1880). The site appears to have been a pitfall trap in pre-last glacial times. In the case of Victoria Cave the Ipswichian strata were only discovered as a result of the deliberate sinking of a number of shafts in the floor of the cave as part of a large scale archaeological dig (Tiddeman 1872). In neither case was there any evidence of the presence of such old deposits before quarrying or excavation took place. This shows the possibility of there being more such sites in the Craven area and any sites identified as containing animal remains by cave exploration activity must be considered as potential repositories of older deposits.

Digging as a means of discovering new caves by sporting cavers has made a significant contribution to our knowledge of the archaeological resource of caves in the Yorkshire Dales. The strategies employed by cavers in order to dig through material blocking both horizontal passages and vertical shafts tends to confine the disturbance to the youngest layers in the deposit and has contributed to the bias apparent in the caver records towards more recent bone assemblages. This suggests caver digging activities should not necessarily be seen as a problem by the archaeological community or the statutory bodies responsible for conservation as the damage caused, on the whole, may be relatively limited. It should instead be seen as an opportunity to increase our knowledge of this often neglected field of archaeology.

Speleothem conservation

A speleothem is a secondary mineral deposit found within a cave. By far the most common mineral is calcite. The calcite is deposited as a result of the loss of carbon dioxide from ground water resulting in it becoming oversaturated with calcium. The calcium is then precipitated as calcium carbonate. Though calcite is by far the most common mineral many others do occur in speleothems (Hill & Forti 1997). As well as having an aesthetic value speleothems are important scientifically as they can be dated radiometrically (e.g. Gascoyne *et al.* 1983) and may contain records of isotopic variation and other information which can be studied as climate proxies for palaeoclimatic research (Lauritzen & Lundberg 1999).

In the earliest days of cave exploration speleothems were removed from caves to adorn the gardens of the wealthy (Shaw 1998) and in the nineteenth and early part of the twentieth century many caves suffered serious losses of speleothems. In the well-decorated cave of Eglin's Hole in Nidderdale for instance, discovered in 1715, all the calcite formations had been removed by the

Fig. 5. The Colonnades, Lancaster Hole. Forty years after the repair.

latter part of the nineteenth century (Craven 2006). Since the 1940s there has been an increased appreciation of the importance of speleothems by the caving community. On finding a new cave passage it is now routine to mark either a particular route through a passage to avoid damage or to mark out the boundaries of areas not to be entered by using plastic tape. The cleaning of speleothems that have become covered in mud has also become common place. In the Gaping Gill system the first piece of passage explored after the initial descent of the Main Shaft called Old East Passage was originally beautifully decorated. Over the years many formations had been damaged and become covered in the mud characteristic of much of the cave system. A route through the passage avoiding the surviving formations has been marked out and some formations have been cleaned back to their original state. An extension to Old East Passage called Glover's Chamber was explored in 1995 and in order to preserve the pristine state of the speleothems in this section the way into it is kept hidden except at certain times of the year.

In January 1965 one of a group of stalagmites called the colonnades in Lancaster Hole had been broken into five pieces. With sponsorship from CIBA Ltd the speleothem was repaired using Araldite Resin (Pickup 1966). This repair has now lasted over forty years and other such repairs have been undertaken (Fig. 5). This shows what lengths cavers are prepared to go in order to protect and preserve speleothems. Since the 1970s cavers have cooperated closely with conservation organization, for instance in the designation of SSSIs within karst landscapes (Hardwick & Gunn 1996). In the 1990s the interest in conservation became more formalized with the publication by the National Caving Association (now superceded by the British Caving Association) of its cave conservation policy (NCA 1995) and cave conservation handbook (NCA 1997). Out of the four long-established caving clubs based in the Yorkshire Dales, two have a conservation officer on their committee (Craven Pothole Club and the Red Rose Cave and Pothole Club), the earliest such appointment having been made in 1995.

Conclusions

Although at first glance some of the activities of the caving community may appear to be at odds with the conservation of the very limited resource of cave interior deposits, without the digging activities our knowledge base would be greatly diminished. Cavers who use caves for leisure purposes are generally aware of the need to protect fragile speleothem formations, and caving clubs take an active role in cave conservation including controlling access to vulnerable caves and taping off sensitive or non-access parts of caves. They also serve as an 'early warning system' by alerting conservation organizations to potential threats to underground deposits and formations. The caving community has embraced the importance of a conservation minded approach to visiting caves for a long time even though this is rarely appreciated by those who are not familiar with the caving world. Any attempt to formulate a geoconservation strategy for an area which contains caves must as a first step involve the active caving community in the area and must appreciate and support the conservation activities undertaken by cavers, often out of their precious leisure time and frequently at their own expense.

The authors thank J. N. Cordingley for his critical comments and helpful suggestions on the manuscript. A. Chamberlain acknowledges the support of English Heritage who have funded some of his research on cave archaeology. P. (Footleg) Fretwell provided the photograph of the Collonades.

References

BECK, H. M. 1984. *Gaping Gill: 150 Years of Exploration.* Robert Hale Publishing, London.

BROOK, A., BROOK, D., DAVIES, G. M. & LONG, M. H. 1975. *Northern Caves Volume Three Ingleborough.* Dalesman, Clapham.

CHAMBERLAIN, A. T. 2002. A gazetteer of non-human vertebrate remains from caves in the Yorkshire Dales described in the scientific literature. *Capra* 4 available at: http://capra.group.shef.ac.uk//4/bonecavechamberlain.html/

CRAVEN, S. A. 2002. A history of cave exploration in the Northern Pennines, United Kingdom, from 1838 to 1895. *Cave and Karst Science*, **29**, 21–32.

CRAVEN, S. A. 2004. Ingleborough cave, Clapham, north Yorkshire, England. *Cave and Karst Science*, **31**, 15–34.

CRAVEN, S. A. 2006. An introduction to the speleohistory of upper Nidderdale, Yorkshire, UK, to the early nineteen-sixties. *Cave and Karst Science*, **33**, 33–40.

GASCOYNE, M., FORD, D. C. & SCHWARZ, H. C. 1983. Rates of cave and landform development in the Yorkshire Dales from speleothem age data. *Earth Surface Processes and Landforms*, **8**, 557–568.

HAIGH, D. 2003. In Hensler's footsteps. *Descent*, **175**, 20–23.

HARDWICK, P. & GUNN, J. 1996. The conservation of Britain's limestone cave resource. *Environmental Geology*, **28**, 121–127.

HILL, C. A. & FORTI, P. 1997. *Cave Minerals of the World.* 2nd edn, Huntsville, Alabama: National Speleological Society.

JACKSON, J. W. 1962. Archaeology and palaeontology. *In*: CULLINGFORD, C. H. D. (ed.) *British Caving.* Routledge, London, 252–253.

LAURITZEN, S.-E. & LUNDBERG, J. 1999. Speleothems and climate: a special issue of *The Holocene*. *The Holocene*, **9**, 643–647.

McFARLANE, A., LUNDBERG, J. & CORDINGLEY, J. 2005. A brief history of stalagmite growth measurements at Ingleborough Cave, Yorkshire, United Kingdom. *Cave and Karst Science*, **31**, 113–118.

MIAL, L. C. 1880. Raygill Fissure the cave, its contents. *Proceedings of the Yorkshire Geological and Polytechnic Society*, **7**, 207–208.

MURPHY, P. J. 2002. A gazetteer of non-human vertebrate remains from caves in the Yorkshire Dales referenced in caving club journals and allied literature. *Capra* 4 available at: http://capra.group.shef.ac.uk//4/bonecavemurphy.html/

MURPHY, P. J. 2003. A gazetteer of non-human vertebrate remains from caves in the Yorkshire Dales for which there is no record. *Capra* 5 available at: http://capra.group.shef.ac.uk//5/bonecave2.html/

MURPHY, P. J. & CHAMBERLAIN, A. T. 2002. The bone caves of the Yorkshire Dales—an introduction. *Capra* 4 available at: http://capra.group.shef.ac.uk/4/bonecaveintro.html/

MYCROFT, D., BENTHAM, K. & PAGE, K. N. 1997. *A Management Plan for the Caves of Lathkill Dale, Monyash, Derbyshire*. English Nature Research Report 223. English Nature, Peterborough.

NCA. 1995. *Cave Conservation Policy*. National Caving Association London.

NCA. 1997. *Cave Conservation Handbook*. National Caving Association London.

PICKUP, D. 1966. The truth about the Collonades. *Happy Wanderers Cave and Pothole Club Journal*. Privately published.

SHAW, T. R. 1998. Wookey Hole in Somerset and Pope's Grotto at Twickenham, England. *Cave and Karst Science*, **25**, 29–36.

TIDDEMAN, R. H. 1872. Discovery of extinct mammals in the Victoria Cave, Settle. *Nature*, **VII**, 127–128.

WALSH, A. 2001. The quest for a dry way. *Descent*, **159**, 20–22.

WALTHAM, A., SIMMS, M., FARRANT, A. & GOLDIE, H. 1997. *Karst and Caves of Great Britain*. Geological Conservation Review, **12**. Chapman and Hall, London.

WEBB, S. 1995. Conservation of limestone pavement. *Cave and Karst Science*, **21**, 97–100.

Conservation at the cutting-edge: the history of geoconservation on the Wren's Nest National Nature Reserve, Dudley, England

COLIN D. PROSSER & JONATHAN G. LARWOOD

Natural England, Northminster House, Peterborough PE1 1UA, UK

(e-mail: colin.prosser@naturalengland.org.uk)

Abstract: In 1949, nature conservation legislation was passed in Great Britain which enabled areas of land to be declared as National Nature Reserves (NNRs). In 1956, the Wren's Nest, Dudley, a Silurian (Wenlock) limestone hill, internationally famous for its geology and fossil reef faunas, was declared a geological NNR. The combination of internationally important geology, abandoned, unstable and dangerous quarries and mines, and a large adjacent urban population have provided continual conservation challenges. This paper uses contemporary correspondence to describe the deliberations that led to the Wren's Nest being declared as one of the first NNRs in England. It goes on to describe the major management challenges which have arisen. These include instability and collapse of mine workings, fly-tipping, vandalism and heavy recreational use by the local community. It highlights the conservation solutions that have been developed during the 50 year history of the reserve. These have included management of unstable and dangerous ground, cutting of new geological sections, establishment of geology trails, use of volunteers and the strengthening of local community links. The Wren's Nest has also played an important role in raising awareness of the geological heritage within the local planning authority. This has led to the adoption of geoconservation policies and to the development of projects using the area's geological heritage to attract visitors. Today the Wren's Nest remains important for its geology and is also one of the most significant geological reserves in the world for demonstrating the challenges of geoconservation and how they may be overcome. This historical perspective on 50 years in the life of a reserve provides an insight into the innovative geoconservation solutions developed at the Wren's Nest that can be applied elsewhere. Although the Wren's Nest NNR is internationally known for its geology, and has a very high profile in geoconservation, this paper is the first to attempt to explore the thinking and process that led to this abandoned industrial site, in an urban setting, being declared as one of the first NNRs in the UK.

Wren's Nest Hill is the best known of three related limestone hills, Castle Hill and Hurst Hill being the others, which rise out of the coal-bearing rocks of the South Staffordshire Coalfield, in Dudley, West Midlands. The folded layers of the Silurian Much Wenlock Limestone Formation which form Wren's Nest, have been quarried or mined for hundreds of years. Initially this was to support local agriculture but was later, along with local coal mining, to fuel the development and expansion of the area's iron industry (Warwick *et al.* 1967). Over one hundred years of limestone extraction at the Wren's Nest resulted in a prolific number of extremely diverse and well preserved Silurian (Wenlock) aged fossils coming to light. These magnificent fossils have found their way into museums across the world, and through their cultural association with quarrying and prosperity in the area, they soon established a degree of local fame. One such fossil, the trilobite *Calymene blumenbachii*, a favourite with quarrymen, was given the name the 'Dudley Locust', or 'Dudley Bug' and was included within Dudley's coat of arms (see Mikulic & Kluessendorf 2007). Early observations on the

geology of the area include those of Sir Roderick Murchison, who visited Dudley several times and in 1839 published the 'Silurian System' which established the Silurian Period and included illustrations based on material from Dudley and undoubtedly the Wren's Nest, and Jukes (1859) who provided the first descriptions of the limestones of Dudley.

When mineral working finally stopped on the Wren's Nest in 1924, the hill was left in a somewhat 'awkward' state. Extensive, quarrying and mining had left a scarred landscape of deep trenches, open mine entrances and dangerously undermined and unstable ground. As the rate of working had diminished from hundreds of men in 1833, to just 20 in 1906 (Warwick *et al.* 1967), trees had become established on parts of the hill creating a relatively attractive landscape in places. This returning 'naturalness', the dramatic scars of extraction, the close proximity of an urban population and the spectacular views to be enjoyed from the hill meant that by 1900, the Wren's Nest had changed from being a largely industrial site into an inspiring combination of a local beauty-spot and curiosity. This dual appeal is well illustrated by a range of

From: BUREK, C. V. & PROSSER, C. D. (eds) *The History of Geoconservation.*
Geological Society, London, Special Publications, **300**, 217–235.
DOI: 10.1144/SP300.17   0305-8719/08/$15.00 © The Geological Society of London 2008.

postcards of the area (Fig. 1a–c) dating from the period 1900–1918 and featuring the woodlands, trenches and caverns. These clearly demonstrate local interest in visiting Wren's Nest despite the fact that some mineral extraction was still taking place.

The 20 years following the cessation of mineral extraction were relatively uneventful. Around 1935, council housing estates were built abutting and surrounding the hill, and there was a period of organized tipping, presumably to infill some of the abandoned trenches and mines. The late 1930s also saw an upsurge in geological interest in the site, with a Geologists' Association visit taking place (Butler & Oakley 1936) and publication of a major geological survey (Butler 1939). By the 1940s, the Wren's Nest had become internationally important for its geology and palaeontology.

Fig. 1. Three postcards illustrating the character of the Wren's Nest during the early part of the twentieth century. The views show (**a**) the abandoned trenches pre-1909; (**b**) the Seven Sisters caverns pre-1909; and (**c**) a slightly later picture showing a smartly dressed lady enjoying the spectacular views of the wooded trenches.

(c)

LIMESTONE CLIFF, WREN'S NEST, DUDLEY.

Fig. 1. (*Continued*).

Although it was of scientific and educational value, it was relatively dangerous and unstable due to its industrial past and was very close to an urban population which used the site for recreation. Thus, during the 1940s, as government was developing a national statutory approach to nature conservation, the Wren's Nest had already acquired the combination of characteristics that would later provide the challenges to place it at the cutting-edge of geological conservation. In short, the Wren's Nest offered internationally important geology but in unstable rocks, educational opportunities but on dangerous and unstable land and opportunities to create an interesting visitor experience but on the doorstep of a relatively deprived urban population who regarded the site, quite rightly, as their recreational area.

The origins of the National Nature Reserve

On 27/28 September 1956, the Wren's Nest was declared as a National Nature Reserve (NNR) with the Mons Hill extension being added on 2 March 1957. The foundation for the declaration was arguably the work of Butler (1939), which provided an up-to-date description of the stratigraphy of the site (the remains of his 'type section' still exist today) to compliment the internationally famous and widely distributed collections of fossils from the Wren's Nest. In effect, Butler reminded the geological community of the site's geological as well as its palaeontological importance.

Just three years after Butler's paper, the first steps were taken towards establishing a national scheme for nature conservation with Scott (1942) in his report to government, recommending that 'the Central Planning Authority, in conjunction with the appropriate Scientific Societies, should prepare details of areas desired as nature reserves (including geological parks) and take the necessary steps for their reservation and control ...' By 1945, thinking had progressed further, and the Geological Reserves Sub-Committee of the Nature Reserves Investigation Committee, had been set up. In its report (Chubb 1945), it listed sites meriting permanent protection and included Wren's Nest, under its list of 'conservation areas'. The entry reads 'Wren's Nest, Dudley, (Preferably the whole hill, but at least the square of land bounded on the south by a line running west for 250 yards from Wren's Nest farm, which includes the main type sections of the Staffordshire development of the Wenlock Limestone)'. The report describes 'conservation areas' as large scale physiographic features and areas containing many items of geological interest and notes that they are mostly characterized by striking scenic beauty. This small Sub-Committee included Oakley who had led the Geologists' Association trip with Butler (Butler & Oakley 1936) and Whitehead of the Geological Survey, who was working on the Dudley and Bridgnorth memoir (Whitehead & Pocock 1947) at the time. Both men would have been very aware of the merits of the Wren's Nest.

By 1947, the national conservation effort was taking shape and a Wild Life Conservation

Special Committee had been established to provide advice in the lead up to passing nature conservation legislation. In its report (Huxley 1947), the Wren's Nest was recommended to be a National Nature Reserve as 'a sequence of Upper Silurian facies and fossil faunas; of structural and scenic interest; includes the main type sections of the Staffordshire development of the Wenlock Limestone'.

In 1949, the *National Parks and Access to the Countryside Act* was passed creating a statutory nature conservation framework, including the power to declare NNRs. At the same time, a conservation body, the Nature Conservancy (NC), was created to implement the legislation. Natural England's national site files for the Wren's Nest include correspondence relating to the Wren's Nest from 1950 onwards and are the source of all quotations used below in exploring the origins of the Wren's Nest as an NNR.

The first record of the NC taking an interest in the Wren's Nest is by W. A. Macfadyen (the NC's Chief Geologist). His note of 9 August 1950 records his visit to the site on 1 and 2 August 1950, and comments that 'There seems no reason why we should not claim the whole hill as a geological reserve as Butler wanted . . .' In a later letter he notes that no tipping is currently taking place and that official tipping is now completed. However, by autumn 1950, the concept of uncontrolled access to a potential NNR was causing him some difficulties and on 23 November 1950, he wrote to the regional government planning office stating that 'Since completely unfettered access to the area is to be allowed to all and it is to be a playground for children in a thickly populated district, it is felt that its designation as a NR would be inappropriate'.

By 1951, the local planning authority (henceforth referred to as 'Dudley') was also thinking about NNR status and on 12 November 1951, it approached the NC exploring the use of 'Nature Reserve' status as a means of securing funds to manage the site.

Over the next few years, reconciling designation as a NNR, with uncontrolled public access in an urban setting, continued to be a major issue. The following correspondence illustrates this.

16 November 1951: Macfadyen's internal file note: 'Owing to the proposed usage of the site, it does not appear that designation of the Wren's Nest as a NR would be appropriate, a NR implies some limitation of access by the public in the interests of the wild life or geology, whose safeguarding is the aim of the Reserve. This condition is incompatible with the proposed usage.'

25 August 1953: Macfadyen writes to Professor F.W. Shotton, University of Birmingham: 'This site is presenting us with a headache with regard to the most suitable means of safeguarding the geology and we would be very grateful for your views? . . . To establish an NNR under the terms of the 1949 Act would probably be impossible, since it seems very unlikely that Dudley would let

us have it as such. They have it and want it as an open space and I think that there is no question of them surrendering it; to declare a place a NNR over which the public have unrestricted access would not appear appropriate.'

8 September 1953: Shotton replies to Macfadyen: 'Sections at Castle Hill and the Wren's Nest are world famous, and now that Castle Hill is inaccessible through its conversion to a Zoo, it is more than ever important that the Wren's Nest should be preserved. It has yielded fossils unrivalled for their variety and preservation so that geologists throughout the World know the name of Dudley . . . . Sections are regularly visited by students and geological experts, and always will be.'

4 November 1953: Macfadyen writes to the NC's Committee for England: 'Recent enquiries were made of Dr Pugh (Geol. Survey), Mr Butler (Geol. Survey), Prof. Shotton (Birmingham University) and Dr Oakley (BM(NH)) as those most qualified to give an opinion. They are unanimous in its importance. They consider we should take some positive action for the preservation of the geological interest, and should not leave it merely as a SSSI'. He goes on to add that: 'It has been urged in the past that to declare this site as an NR would be inappropriate in consequence of its use as a public playground, geological opinion is that to leave the site as a SSSI would give insufficient permanent protection. It is suggested that the right course is that as it is evident that the area is geologically of national importance it should be established as a National Reserve'.

Management challenges

The Wren's Nest, with its industrial background, its caverns and trenches and its urban setting (Fig. 2) was always likely to prove difficult to manage as a nature reserve. Site files demonstrate clearly that complex management issues relating to instability, public safety, tipping, vandalism, vegetation management, irresponsible fossil collecting, people management and community engagement have continually provided management challenges on a scale much greater than on other NNRs (Table 1). The files record numerous crisis meetings regarding safety concerns, discussion of engineering options to address instability, and discussions about the need to increase the number of wardens to manage the site and deal with

Fig. 2. The Wren's Nest illustrating its truly urban setting. (Photograph: Michael Murphy.)

Table 1. *A chronology of the Wren's Nest National Nature Reserve*

Year	Significant events and milestones
1924	Mineral extraction on the Wren's Nest ceases.
1935	The Geologists' Association led by Butler (Geological Survey) and Oakley (British Museum (Natural History)) visit. Council housing estates abutting the Wren's Nest are built around this time.
1939	Butler publishes a detailed description of the stratigraphy of the Wren's Nest.
1945	The Geological Reserves Sub-Committee of the Nature Reserves Investigation Committee publishes a list of recommended Geological Reserves. This includes the Wren's Nest.
1947	Whitehead and Pocock of the Geological Survey publish *Geology of the Country Between Dudley and Bridgnorth*. The Wild Life Conservation Special Committee (England and Wales) publish *Conservation of Nature in England and Wales*, recommending the Wren's Nest as a National Nature Reserve (NNR).
1949	The *National Parks and Access to the Countryside Act* passed, initiating a statutory system for nature conservation and giving the Nature Conservancy (NC) the powers to declare National Nature Reserves.
1950	1–2 August: Macfadyen, Chief Geologist for the Nature Conservancy, visits the Wren's Nest and is optimistic it can be designated as a National Nature Reserve.
	23 November: Macfadyen writes to the regional planning officer expressing concern that the existing open access to the site means Nature Reserve status is inappropriate.
1951	12 November: The local planning authority, Dudley, approach the NC regarding using Nature Reserve status as a means of securing funds for site management.
	16 November: Macfadyen remains concerned that open access and nature reserve status is incompatible.
1953	25 August: Macfadyen seeks the views of Professor Shotton, University of Birmingham, re: suitability of the site for Nature Reserve status.
	8 September: Shotton replies to Macfadyen expressing the view that the site is very important and should be made into a NNR.
	4 November: Macfadyen writes to the NC's Committee for England recommending the site is declared as an NNR.
1956	27/28 September: Wren's Nest declared as a NNR. Press release produced for publication on 28 September.
	22 October: Macfadyen is starting to deal with fossil collecting issues and states: 'I am beginning to have applications for permits ... to visit the NR and collect fossils ... There is really no point in issuing official permits for either casual visitors, including school children and parties of students who wish to study the sections and collect such fossils as they may find. The site is an open playground for the surrounding housing estates and all have hitherto had free access, and still have ... I think that it is really only for more serious scientific study coupled perhaps with excavations, that we should issue official permits'.
1957	2 March: Mons Hill extension added to the NNR.
	30 May: The first specimen collecting policy produced.
	20 June: The first Management Plan for the reserve is produced.
1960	2–9 April: The Conservation Corps visit the site for the first time, clearing trees and scrub from around the Seven Sisters caverns. This work is described as part of a long-term task to upgrade the Wren's Nest into a geological park.
	27 July: The tragic death of Royston Bate (14) who accidentally falls into the underground workings.
	10 August: A meeting takes place to consider implications of the Royston Bate accident.
	6 September: It is reported that the first draft of a planned Wren's Nest booklet is complete.
	30 September: Two full-time wardens have been recently appointed. NC will pay £125 towards their salaries. Desirability of erecting signs discussed but is difficult due to hooligans.

(Continued)

Table 1. *Continued*

Year	Significant events and milestones
1961	9 June: Black (the new NC Chief Geologist) visits to consider the implications of proposals to secure the shaft which was the scene of the fatal accident in 1960.
1962	March: Engineering Company Johnson, Poole and Bloomer report on what needs to be done to make the caverns safe.
1964	24 March: Stubbs (the NC's Deputy Geologist) records that work on construction of the proposed geological trail is underway.
1965	PhD research 'Petrology of the Wenlock Limestone of Wren's Nest' by Peter Oliver commences in summer 1965 and involves revisiting Butler's trench for new sampling.
1966	23 November: The Wren's Nest booklet is now planned for 1967/8 at cost of £500. Contributions were collected 10 years earlier and authors are now 'alternately irate and despondent at the continued multifarious delays'. Bollards marking the trail have been in place for 18 months but numbering soon vandalized.
1967	31 March to 6 April: The Conservation Corps visit again.
	April: Plans emerge to use explosives to collapse the caverns for safety reasons.
	5 May: A draft report by Professor Potts on the condition of the limestone caverns leads to further efforts to keep visitors out of the caverns. These include erection of fences and signs and the writing of letters to schools.
	5 May: Dr Poore, Director of the NC, visits to see management, safety, vandalism and litter issues. His itinerary refers to 'Problems of managing Britain's most disreputable Nature Reserve.'
	22 September: The first Wren's Nest booklet (Warwick et al.) is published at 3s 6d. It is widely reviewed as a pioneering publication but some letters complaining about the font, long words, illustrations etc are received.
	6 November: The trail is launched by the Mayor of Dudley and Professor Shotton.
	24 November: Black notes that 1176 copies of the booklet have been sold.
	Paper by H.D. Brook 'Wren's Nest Research' published in the Proceedings of the Birmingham Natural History Society, **21**, 56–68. Describes the Wren's Nest Research Group, formed by the Birmingham University Extra Mural Department in 1963 and then taken on, in 1966, by the Birmingham Natural History Society. Its aims being to work on mapping, cataloguing, and investigating species and palaeoecology.
1968	19 September: An important meeting takes place between the Mayor of Dudley, Dudley's engineer, the engineers Johnson Poole and Bloomer, Shotton, and the NC to discuss concerns about potential collapse of the caverns.
	1 October: It is recorded that since early 1966 there has been increasing concern about the safety of the underground caverns. Up to 200 houses are potentially threatened and instrumentation to give early warning is needed. Infilling may be needed but would be very expensive.
1969	19 March: Stubbs reports that some local geologists don't think the pillars will last more than 1–2 years.
	October: Stability and safety issues are still being discussed, in particular the need to protect roads and houses. Three options, fencing, blasting or infilling are considered. Filling would be needed in order to retain the Seven Sisters caverns which were recognized as being unstable in 1968, with further roof-falls in autumn 1969 (one of 100 tons). The routes of the trails are now unstable. It is planned to replace the two trails with one new one with different text for different interest levels. The original 3100 print run of the booklet is nearly sold out.
	3 November: A revised trail route is suggested.
1970	A report on the work of the Conservation Corps 1966–1969 describes their work building steps, clearing faces and digging drainage. Stubbs describes the Wren's Nest as 'The most urban of all NNRs' and states that without the Conservation Corps, 'it is doubtful if Britain's first permanent geological nature trail would have come into being'. Visitor numbers estimated at 20000 per year.
	16 March: 'Crowning-in' occurs the rear of 46 Hillside Road, near the junior trail.
	27 April: The Conservation Corps are congratulated by Dudley's engineer on their work done in bad weather over Easter to re-route the trail.
	12 May: Rumours emerge that a motorway may be routed through the Wren's Nest. NC staff wonder if a motorway roundabout may solve access problems to the reserve by enclosing it!
	16 September: Phase One of an infilling programme due.
	20 October: Descriptions of the Trail need re-writing following recent re-routing. Black will produce a new booklet to do this.

1971

Peter Oliver's PhD, started in 1965, completed. Representative hand specimens and thin sections for the whole succession produced and are now stored at University of Birmingham.

September: A progress report notes that infilling of workings with sand slurry commenced late 1970 and continued right through into 1971. A tall steel fence was erected around the Seven Sisters but was damaged by a rock fall in January 1971. The re-routed trail is not yet opened, and is waiting for the new booklet which is now nearly ready for printing. There are still two wardens with the NC contributing £125 towards their cost.

4 November: A party from the council, examining the infilling operation, leap for their lives and the Town Clerk twists his knee when a roof-fall results in rock rolling towards them.

1972

29 March: Continued roof-falls reported.

31 March–7 April: The Conservation Corps on site.

9 August: Phase One of infilling with sand and gravel slurry nearly complete; Phase Two being planned. Dudley's engineer doubts if the Seven Sisters can be saved except at considerable cost.

11 September: The new booklet is still awaited. Infilling continues with 220000 tons of sand and gravel slurry already pumped underground.

December: More reports and proposals emerge with regard to filling the caverns and trenches.

1973

17 January: Continuing safety concerns mean that the new booklet can not be launched.

12 February: Dudley wishes to blast and collapse areas including the Seven Sisters. The NC demands the preservation of the pillars but would compromise at Cherry Hole and Devil's Mouth.

26 March: A crisis meeting involving Dudley's engineers, the Chair of Dudley's Underground Workings Sub-committee, the NC, Shotton and Warwick takes place to discuss what action to take.

29 March: Dudley Museum estimates that there have been 20000 geological visitors per year post publication of the booklet.

4 June: Safety concerns lead to a letter being written to schools and universities halting organized visits for the next year.

28 June: Dudley produces a draft press release announcing that collapse by blasting will take place on 3 July, but it doesn't happen.

4 July: Dudley confirms that it plans to go ahead with blasting.

July 1973: The Geological Society of London, Shotton and others in the geological community write to the Secretary of State for the Environment (SoSE), Denis Howell, expressing concern over Dudley's plans to blast the Seven Sisters.

9 October: The successors to the NC, Nature Conservancy Council (NCC) write to Dudley refusing permission for blasting to go ahead.

16 October: The Department of Trade and Industry expresses a view that blasting could be hazardous.

12 December: Dudley's Underground Workings Sub-committee puts-off blasting but wants the NCC to pay for infilling.

14 December: It is reported that blasting has been actively opposed by the NCC since April 1972 and it is hoped that this option will now be dropped and that infilling will continue. Some access is now being allowed to geological visitors.

1974

17 March: The Conservation Corps clears rubbish from the reserve. There are gypsy caravans at the southern end of the site, and trenches are being used by the local community as refuse tips.

28 March: The restrictions on group visits are lifted, providing they stay away from certain areas.

10 April: A site visit takes place to walk the trail in advance of launching the new booklet. Rubbish and gypsy camps are noted and the need for a full time warden, jointly funded by Dudley and the NCC is identified.

8 June: The Conservation Corps makes new steps on the geological trail.

November: The new booklet (Anon. 1974) is finally published with a 15000 print run.

16 December: A working party, including Council Members of the NCC visit to see problems for themselves. They record a squalid appearance due to tipping and vandalism and lack of sign boards due to 'savage vandalism'.

19 December: A deputation from Dudley meets the SoSE, Denis Howell. Dudley agrees to abandon the blasting option but can not continue to infill without money from the Department of the Environment and the NCC.

(Continued)

Table 1. Continued

Year	Significant events and milestones
1975	The formation of the Black Country Geological Society (BCGS), whose members spent considerable time at Wren's Nest in the early years of the Society. Their influence resulted in the appointment of a geological curator at Dudley Museum in 1987 and a raised profile of the NNR with Dudley Council.
	19 March: Duff (a new geologist at the NCC) visits and reports gypsies, debris and rubbish, rags and old clothes, domestic refuse, unfriendly dogs and an unpleasant atmosphere. The bollards are still in one piece but are subject to attacks by heavy objects.
	19 May: The Department of the Environment record that in February 1975, Dudley formally agreed to abandon the idea of blasting but require funds for infilling.
	6 June: Black states that spending £50 000 to £70 000 to save the Seven Sisters is not worthwhile and could be better used in other ways.
	July: The NCC decides it can not afford to help financially with infilling.
	18 November: An initiative to clear litter is called for.
1976	26 April: Options to replicate the scientific interest of the Seven Sisters elsewhere on the reserve are considered. Dudley has given up on blasting the Seven Sisters, can not afford to fill them and, therefore, has decided to fence-off the caverns and allow them to collapse naturally.
	28 April: The SoSE, Denis Howell, and a number of his officials meet with the Director of the NCC and Black to discuss the difficulties at the Wren's Nest. It is agreed that the NCC will excavate a new geological section elsewhere on the Reserve to replicate the scientific interest in the pillars at the Seven Sisters. This will 'achieve spatial separation between NCC and Dudley interests', retaining the scientific interest but allowing safety work to take place.
	3 June: A list of potential alternative sites is identified.
	23 June: The costs of cutting of new section would be £2500, and having a safe section would allow greater educational use of the site.
	30 June: The Black Country Geological Society contacts the NCC offering to help record sections on Mons Hill.
	29 July: The NCC, in a letter from the Director, agrees that 'no attempt should be made to preserve the scientific interest contained in the west of the nature reserve which is in danger of subsidence but that such interest should be replaced by new exposures to be excavated elsewhere in the reserve'.
	10 August: Barker, from the NCC's regional office, is keen to see the unstable bits of the reserve remain as a wildlife sanctuary—the first recognition of a wildlife interest at the site.
	3 September: The specification for the new trench excavation is prepared. No explosives to be used as the school and houses are too near.
	9 September: A meeting between Dudley and the NCC takes place to discuss the new sections and to explore how the reserve can be improved as an amenity and educational resource.
	29 September: A planning application to cut the new sections is submitted to Dudley.
	14 October: The BCGS asks to be involved in the future of the Wren's Nest.
	25 October: The Institute of Geological Sciences (now the British Geological Survey) asks to have White, one of their officers, present to collect material when the sections are cut.
1977	10 December: Hayden and sons tender to undertake the excavation is accepted.
	10 January: Work on the section is due to start and a press release on 12 January leads to newspaper and radio coverage.
	7 February: Work was reported as completed at a cost of £2800.
	2 March: The first mention is made of a possible reserve base. The Wren's Nest school is suggested as a location.
	2 June: Use of wardens is discussed. Since 1959, the NCC has paid £125 per annum to support patrols by 'parks' staff from Dudley who report major damage, subsidence etc. In recent years, patrols have been cut-back and salaries have risen etc. With the site now enhanced by the new sections, more effective use of wardens is needed. The NCC would pay £2000–3000 to cover half of a warden's salary.

1978   17 April. The new guidebook (Hamblin et al. 1978) to accompany the revised trail guide is now available and costs 20p. Dudley writes to complain that they were not aware it was being produced.

2 May: Black notes that 'we are all familiar with the Wren's Nest, I think it must rank as the site most visited by Geology and Physiography personnel [the Geologists working for the NCC] anywhere in Britain'.

29 May: More subsidence is reported (on the optional part of the trail).

14 June: The NCC writes to Dudley, offering money for a warden, and supporting Dudley in its view that a permanent reserve base is needed. A warden's job description is drafted. Candidates must be 26 or older, and the salary will start at £3500. The idea of an estate management team as part of the Manpower Services Commission is floated.

1979   19 March: Barker writes to Black informing him that he has just heard that Dudley has the authority to appoint a warden, set up a reserve base and to tidy the place up. The reserve base is to be a 'mobile classroom unit' attached to the school.

3 April: A meeting takes place to discuss problems with funding the reserve base. It is agreed that a base is needed if the warden is to be successful and that the warden's duties will be primarily educational.

18 April: The new 'alternative section' is accidentally hydroseeded!

June: A site visit takes place to consider the BCGS's concerns about the state of the reserve and to explore the role they could play in helping to manage the reserve.

15 June: The lack of funding for, a warden is still a problem, oil company sponsorship is being considered.

14 September: The use of Mons Hill School is suggested as a reserve base.

1980   28 November: A planning application is submitted. This is to landfill part of the site, the Upper Quarried Limestone trench on west of the site, to prevent illegal tipping. The NCC objects.

1981   The first issue of The Black Country Geologist includes paper by Oliver on the 'Lithological groups within the Wenlock Limestone (Silurian) at Wren's Nest, Dudley'.

1983   Wren's Nest identified as a 'contender' for the disposal of medium level nuclear waste (article in New Scientist). BCGS point out that there is water in the canal beneath the hill and that the Wren's Nest marks the water shed between the Trent and the Severn.

Funding for a full-time reserve warden further discussed. There is a 'ranger service' workshop and there is also a 'Nature Centre' within the school grounds.

1984   18 June: Dudley Orbital Road proposals may affect the reserve.

30 August: Purchase of the Caves Pub to establish a reserve/interpretation centre is briefly discussed but thinking soon returns to the original idea of building a visitor centre within the Mons Hill School.

1985   3 April: The Castle Hill project study into tourism regeneration takes place. Proposal includes realizing the potential of the Wren's Nest (landscape and caves to rival Wookey Hole) as well as initiatives such as the development of the Black Country Living Museum and making more innovative use of government funds to overcome the problems of old limestone workings.

As part of a newly signed Nature Reserve Agreement a full-time on-site warden (Lee Southall) is appointed with 50% NCC and 50% Dudley funding.

November: The Wren's Nest Visitor Centre buildings offered at the school are not suitable.

1986   A viewing platform for the Seven Sisters caverns is constructed.

1987   Colin Reid appointed as the first Keeper of Geology at Dudley Museum and Art Gallery.

14 September: The Limestone Strategy report makes recommendations to stabilize areas underlying parts of the Reserve by pumping slurry into underground cavities. This will be funded through a Derelict Land Grant.

November: The warden role is redefined as 'Senior Warden' and Nick Williams is appointed.

1988   6 November: A revised fossil collecting policy statement is issued.

(Continued)

Table 1. *Continued*

Year	Significant events and milestones
1990	May–September: Mons Hill School building may become available—proposal to convert part of it into a Wren's Nest field centre but Dudley College take over the buildings instead. A feasibility study for a new visitor centre identifies six possible locations.
	1 June: A new geological trail and guide launched (Cutler *et al.* 1990).
1991	January: The adjacent Bluebell Park is identified as the preferred location for a visitor centre. Also explored are links with the museum and re-opening of the Wren's Nest Canal Tunnel.
1992	30 April: English Nature (the successor to the NCC) offer a grant of £100 000 for interpretation in the proposed visitor centre which is planned to open in August 1993.
	28–29 November: The first Dudley Rock and Fossil Fair takes place.
1993	January: Dudley suffers financial problems and cut backs mean that plans for the visitor centre are shelved.
	27 July: The International Conference on Geological and Landscape Conservation held in Malvern visits the Wren's Nest NNR and Black Country Living Museum.
1994	The Black Country Nature Conservation Strategy, undoubtedly influenced by experience at the Wren's Nest, is published.
	July: The adjacent Parkes Hall Pool, Donkey Pool and Bluebell Park Local Nature Reserve is declared, recognizing the wildlife and community value of the area.
	September: Alec Connah takes over as Senior Warden.
1995	10 April: A new crown-hole (130 ft deep) opens up near Cherry Hole.
	July: Development of interpretative displays and equipment for the new warden's base is funded by English Nature and the Geologists' Association.
1996	25 October: The Wren's Nest classroom within the warden's base is opened.
1997	February: Bat survey complete confirming importance of Seven Sisters as a bat summer roost and for winter hibernation.
	April: Phase 1 of a new set of large-scale mine remedial works is initiated.
	April: An updated fossil collecting code is agreed.
	12 December: There is a major rock fall south of Seven Sisters damaging 10 m of fence.
1998	April–July: The steeply inclined 'ripple beds' are starting to slip and the path below them is closed and fencing is erected.
	April: Phase II of the remedial works is now underway.
	November: A bid to secure World Heritage Site Status for the Wren's Nest and Castle Hill fails, and Geopark status is suggested as an alternative way forwards.
1999	An article exploring the challenges of managing the Wren's Nest *Between a rock and a hard place* is published by Connah. (Enact 7(3)15–18, Autumn Issue).
	November: Dudley consultation on draft Unitary Authority Development Plan which includes geological policies.
	23 December: There is a 10–15 ton rock fall in the Seven Sisters.
2000	February: Graham Worton is appointed Keeper of Geology at Dudley Museum and Art Gallery.
	April: Jo Naden is appointed as artist in residence.
2001	7 March: A major collapse of the 'ripple-beds' takes place.
	22 October: A further major collapse at the Seven Sisters takes place.
	November: Stabilization of Seven Sisters caverns is discussed in detail. The principle of infill with retention of surface features is established.
2002	17 March: Jo Naden's exhibition opens at Dudley Museum and Art Gallery.
	25 May: The 'Waves project', a school arts project drawing inspiration from the Wren's Nest, is launched.
	April: Dudley presented award for their management of the Wren's Nest and the adjacent Local Nature Reserve.

2003	August: The infilling of the Seven Sister's caverns, with a temporary infill, and a bat hibernaculum commences.
	November: Anna Coward (née Gorski) appointed as Senior Warden.
2004	April: The Bramford Wildlife Watch group is established.
	Autumn: The temporary infilling of Seven Sisters is completed.
	21 October: The industrial archaeological importance of the Wren's Nest is recognized through the confirmation of Scheduled Ancient Monument status.
2005	July: Trevor Conroy receives an award for 15 years of voluntary service on the reserve.
2006	The 50th anniversary celebrations take place throughout the year and include a conference and community events.
	January: The Friends of Wren's Nest Group holds its inaugural meeting.
	July: The Black Country Local Geodiversity Action Plan, a reflection of the increasing interest in the area's geological heritage is launched.
	27 September: The Wren's Nest NNR is 50 years old. The Bramford School mosaic is unveiled.
	October: A new visitor centre, to be located within the Seven Sisters caverns and with underground access from the Wren's Nest Canal Tunnel, is proposed as part of major Heritage Lottery Bid.
	The existing reserve base is burnt to the ground in an arson attack. (Following the fire, the wardens were relocated to the Dudley College campus.)

This chronology is based largely upon the Natural England national site files for the Wren's Nest, held in Peterborough, England, but also draws on published articles and papers.

tipping and litter etc. To illustrate the latter point, the note of a Nature Conservancy Council (NCC) site visit in March 1975, describes '...gypsy encampments ... large quantities of domestic refuse, rags and old clothes, various mechanical and electrical appliances, ...large and unfriendly dogs and an unpleasant atmosphere'.

The history of the NNR is one of a continual battle between progress and setback, advances taking a long time to be implemented and then often being knocked back before eventually prevailing. For example, a progress report written in 1971 records that safety fencing eventually erected at the Seven Sisters caverns was almost immediately damaged by a rock-fall, whilst the long awaited geological trail established and launched in 1967, had become dangerous and unstable by 1969 and had to be re-routed. The large and complex geoconservation workload at the Wren's Nest is well illustrated by a statement made by Black (who replaced Macfadyen in the NC) who wrote on 2 May 1978, that 'we are all familiar with the Wren's Nest, I think it must rank as the site most visited by G&P [Geology and Physiography staff at the NCC] personnel anywhere in Britain'. The challenges posed, however, have meant that many innovative and technically robust management solutions have been developed on the Wren's Nest.

Management solutions

Instability

Instability has been a major issue throughout the history of the reserve, but particularly so following the tragic death of 14-year-old Royston Bate, on 27 July 1960, who fell into the underground workings. This led to many local authority proposals and schemes to either blast and collapse, or infill the caverns. Although these would have addressed safety concerns on the reserve they would have damaged the geological interest the reserve had been set up to conserve. A controlled blast was undertaken in 1960 to the NW of the Seven Sisters to close an opening know as the 'Devil's Mouth' but this did not damage the geological interest of the reserve. Since 1960 the need to address instability without damaging the geological interest has been very great. In 1969 for example, it was argued that that the geologically important pillars of the Seven Sisters caverns would probably only last another 1–2 years, there were regular reports of roof-falls and crowning-in, and the relatively new geology trail needed to be re-routed due to instability. These continual management challenges led to a great deal of technical, political and practical thought going into developing management solutions. Management options such as blasting, infilling or fencing the underground caverns were considered in great depth and led to practical and politically acceptable solutions balancing conservation requirements with those of public safety. These have included managed infilling and fencing of the caverns rather than blasting, the creation of alternative sections in safe areas, the use of trails to keep visitors away from dangerous areas, occasional temporary exclusion of geological parties on safety grounds (see Table 1, June 1973) and petitioning of senior politicians (see Table 1, July 1973) when it was felt that geoconservation was not being given enough weight. The management

solutions that have been developed on the reserve have ensured that the key geological features still exist today and strong partnerships are in place with plans to enhance and promote the reserve to an even greater extent in the years ahead.

Alternative sections

The fact that some of the Wren's Nest's most scientifically important geological sections occur in the unstable pillars of the Seven Sisters caverns has provided a continual management challenge, balancing geoconservation against calls, on safety grounds, to blast or infill the caverns (Fig. 3). The

Fig. 3. Managing geological features in unsafe and unstable areas has always posed a serious management challenge. (Photograph: Colin Prosser.)

solution to this seemingly irreconcilable problem of retaining something on geoconservation grounds that needs to be destroyed or buried on safety grounds, was the creation of an alternative section. The aim, as described in April 1976, being 'to achieve spatial separation between Nature Conservancy Council (NCC, replaced the NC in 1973) and Dudley interests'. This pioneering approach was a consequence of stability problems that had been of growing concern since 1967 (see Table 1) and which came to a head with a crisis meeting in March 1973. Dudley's plans to blast and collapse the caverns met strong opposition from the NCC and the geological community and by 28 April 1976 (Table 1) the issue had developed to a point where a proposal to create an alternative section, exposing the same strata as the Seven Sisters, but elsewhere on the reserve, was discussed and approved at a meeting involving the Secretary of State for the Environment, Denis Howell. In order to implement this decision, alternative sections were identified and then cut in January–February 1977 at a cost of £2800 (Fig. 4). This solution provided an innovative way of balancing geoconservation and safety works, providing a new section in a safe part of the reserve whilst leaving the pillars to collapse naturally behind safety fencing. Although not as extensive or petrologically variable as the Seven Sisters, these alternative sections are still managed and used today, providing an educational and scientific resource which forms part of the current geological trail.

Ironically, the Seven Sisters pillars did not collapse naturally in the 1960s or 1970s as was predicted, and in 2004, the caverns were filled with a temporary aggregate, thus supporting and retaining the pillars until funds are in place to remove the aggregate, strengthen the pillars, and re-establish them as a key part of the reserve.

Trails, education and a visitor centre

Education has always been seen as an important part of the Wren's Nest experience and despite the problems of providing trails and interpretation in an urban area periodically subject to high levels of vandalism, the Wren's Nest has played a pioneering role providing education through geological trails. The large nearby population of the Birmingham area, and the spectacular geological and mining heritage, undoubtedly served as a driver for this and work started on drafting a trail guide soon after the NNR was declared in 1956 (Table 1, November 1966).

The first trail was officially opened on 6 November 1967 (Fig. 5), by the Mayor of Dudley and Professor Shotton, with the associated booklet (Warwick *et al.* 1967) published a couple of months earlier. The trail was marketed as the first permanent geological trail and was generally very well received although one letter on file expresses the concern that the advice to visitors is too 'terror-inspiring' to let mums allow their 'little angels' to visit. However, the trail was a big success; the 3100 print run sold out within two years and

Fig. 4. The 'alternative sections' being cut in January 1977. (Photograph: Nature Conservancy Council.)

Fig. 5. The Mayor of Dudley opens the 'first permanent geological trail in Great Britain',
November 1967. (Photograph: Nature Conservancy.)

visitor levels immediately after the trail launch
recorded as 20 000 per year. Instability problems
were caused difficulties within two years, and the
trail had to be re-routed. By November 1974, a
new trail guide (Anon 1974) describing a replacement trail was published and in April 1978 a new
handbook (Hamblin *et al.* 1978) to support the
trail was also available. Twelve years later, a completely revised trail and associated publication
(Cutler *et al.* 1990) was published which remains
in use to the present day (though it too is currently
being revised).

The establishment of a geology trail provided a
more structured basis for educational visits to the
reserve directing visitors towards safer, more interesting and less sensitive areas. Conducting guided
tours and visits and, in particular, encouraging
school use has always been an important part of
the warden's service although it was 1985 before
the use of full-time on site wardens was established.
In 1987 Dudley's first keeper of geology was
appointed and quickly became involved in the promotion, education and conservation activities at the
NNR closely working with the Black Country Geological Society to provide significant support to the
wardens. It wasn't until 1996 with the establishment
of a more substantial warden's base that a permanent Wren's Nest classroom was established that
included a 'time clock', a mural illustrating the
changing geological environments of the reserve

and various 3D models and hands-on activities. In
2000, a Wren's Nest educational pack was created
by Dudley's education department, the keeper of
geology at the museum, local primary school teachers and the wardens providing activities directly
linked to the National Curriculum. In 2006 a
number of interpretation panels linked to the geological trail were erected.

As well as a warden's base it has always been a
desire to have a visitor centre at the Wren's Nest.
Various proposals have been made with locations
including a centre within the Dudley College
campus (in the middle of reserve), the purchase
and conversion of the Cave's Pub (again centrally
placed within the reserve) and the construction of
a purpose-built facility. The realization of a
purpose-built visitor centre came very close in
1992 when a development was agreed on the adjacent Bluebell Park. Despite securing a £100 000
grant from English Nature (English Nature replaced
the NCC in 1991) the proposal collapsed in 1993
due to severe financial cut-backs within Dudley.

Sadly, the warden's base and classroom were
destroyed in an arson attack in 2006 and the
wardens are now temporarily housed in Dudley
College. Plans for a visitor centre still remain a
core objective for the reserve and innovative plans
to develop a visitor centre within the Seven
Sisters caverns are currently part of a substantial
bid to the National Lottery's Heritage Lottery Fund.

Volunteers and wardens

Volunteers and wardens have been central to the success of the Wren's Nest as an NNR, however, it was not until 1985 that the first full time on site warden was employed with funding shared 50% by Dudley and 50% by the NCC. Until then, management work on the reserve was undertaken by volunteers and as part of the work of rangers operating across Dudley (the NC and then NCC contributed £125 per annum towards patrols by Dudley Parks staff from 1959 up until the mid 1970s).

In the early years of the reserve, volunteer effort was crucial. In 1959, under the leadership of Brigadier Armstrong, a national Conservation Corps was established with the aim of involving volunteers (usually senior school children and university undergraduates) in practical conservation work. In April 1960 the Conservation Corps visited the Wren's Nest for the first time (Fig. 6) in what was to become an annual event up until the mid 1970s by which time the Conservation Corps had become the British Trust for Conservation Volunteers (BTCV). Their long-term task was to 'upgrade the Wren's Nest into a Geological Park'. Their work typically included scrub and vegetation clearance around important sections, improving and maintaining paths and constructing steps. A considerable amount of work was undertaken to establish and maintain the geological trail—a legacy which is still very visible today.

However, the presence of a permanent warden was essential, not only to deal with the day-to-day management tasks of the reserve but also to provide a strong link with the local communities of the Wren's Nest. Since 1985 there have been four wardens/ senior wardens at the NNR (Lee Southall from 1985, Nick Williams from 1987, Alec Connah from 1994 and Anna Coward ((née Gorski) from 2003 to present) and today there are also two assistant wardens and a number of dedicated long-term volunteers.

The establishment of an on-site warden provided a focus for community engagement and although the relationship between reserve managers and the local Wren's Nest and Priory estates has ebbed and flowed, it has progressively strengthened, particularly recently, with the establishment of a Friends of the Wren's Nest group in 2006 and the Bramford School Wildlife Watch group in 2004.

Day-to-day management

In a very simple way, Macfadyen established the basis for reserve management in 1953 when he stated that the geological requirements were 'To leave the place as it is, with the provision for cleaning up some of the best sections from time to time'. Subsequent management has been guided by this simple advice and a succession of reserve

Fig. 6. Conservation Corps, on their first visit to the Wren's Nest in April 1960, take a break from scrub clearance to have lunch in the Seven Sisters Caverns. (Photograph: Birmingham Post and Mail.)

management agreements and plans. Not only has this involved the maintenance of the geological exposures, and all the challenges that this brings, but also the management of woodland and grassland habitats, and a range of species including badgers and a nationally significant bat population in the caverns. An adjacent Local Nature Reserve, declared in 1994 to recognize the importance of some of the wildlife rich green space to the local community, is also managed by the NNR wardens.

One of the unique attributes of the Wren's Nest is the superbly preserved Silurian reef fauna. Since the earliest quarrying the hill has been famous for its fossils and fossil collecting has been an established activity at least since the mid-nineteenth century. From the earliest days of the NNR it was quickly recognized that fossil collecting was an important part of the experience gained by people visiting the reserve. When, in 1957, concern was expressed about whether permits and restrictions should be in place for collecting, Macfadyen stated that 'Geological specimens may well be collected, to yield the collector at least something of scientific or educational value, and the Conservancy should not discourage this'. This approach has been the basis of collecting policy ever since and today a 'no hammering' policy and the collecting of only small representative samples from loose material is encouraged. Research is encouraged with any site-based work being agreed in advance. Importantly, the many visitors to the reserve can still enjoy the experience and excitement of collecting and new discoveries are made each year.

An extensive photographic record has been amassed over 50 years of management. In the last 20 years, fixed point photography has been used with great success to demonstrate both the changing nature of the geological exposures and a range of vistas on the reserve. This has provided invaluable documentation of the progressive collapse or slipping of the surface layers of features such as the 'ripple beds' as well as how vegetation and trees have encroached upon what were once more open views.

Wider influence

The geology of the Wren's Nest and Castle Hill has influenced both the industrial and cultural heritage of Dudley and lent much to the town's character not the least of which is the appearance of the 'Dudley Bug' in the town's coat of arms.

In 1984, the Castle Hill Project explored tourism regeneration and made a number of proposals (including the establishment of the Black Country Living Museum) that included a wider recognition of the potential of the NNR in the regeneration of the area (suggesting the landscape and caves as a

rival to Wookey Hole). In 1987, after considerable pressure from the Black Country Geological Society, Dudley appointed the first Keeper of Geology at Dudley Museum and Art Gallery and almost immediately a strong link was made between the reserve and the museum which housed many of the specimens collected from the Wren's Nest.

Initiatives driven by the museum make strong links to the reserve. For example, in 2000, the work of artist in residence Jo Nadin, drew inspiration from visits to the Wren's Nest and from Wren's Nest fossils, regular guided trips to the reserve are lead by the museum and educational initiatives such as the museum's 'Waves Project' has used the geology of the reserve and museum collections as a starting point for an outreach arts programme.

Although not directly attributable to the Wren's Nest NNR, the long experience of managing this site and the values that have developed around this have undoubtedly influenced what has happened elsewhere in the local authority and beyond. For example, in 1994 the Black Country Nature Conservation Strategy (one of the first of its kind in an urban area) was published and included explicit policies for local geology. More recently, Dudley has become the first local authority to adopt comprehensive and detailed Supplementary Planning Guidance for its geological heritage and in 2006 the wider 'Black Country Geodiversity Action Plan' (Fig. 7) was launched encompassing the boroughs of Dudley, Sandwell, Walsall and the City of Wolverhampton.

In 1998, the Wren's Nest, together with Castle Hill, were proposed for inclusion on the World Heritage Site nomination list; however, this bid was not successful. The Wren's Nest is now central to both a Black Country European Geopark application and a multi-million pound National Lottery Fund bid to develop innovative access and visitor facilities within the reserve as part of the much wider Black Country 'Urban Park' regeneration programme. (The initial National Lottery Fund bid was unsuccessful but has been resubmitted to the Regional Heritage Lottery Fund and to a new 'Access to Nature' grant scheme.)

50 years on

In 2006, the Wren's Nest celebrated 50 years as an NNR. During its fiftieth year it became clear that the value placed on the NNR after 50 years is even greater than when it was first declared. Dudley and English Nature (Natural England from October 2006) supported a range of celebratory events and activities on and off the reserve. These

The Black Country Geodiversity Action Plan

"making a positive contribution to the enrichment of the Black Country environment and quality of life by conserving, enhancing and managing the region's geological heritage and diversity for the benefit of all"

Fig. 7. The Black Country Geodiversity Action Plan, a consequence of how the Wren's Nest experience has raised awareness amongst local decision makers of the potential of the geological heritage to enrich and enhance the local environment.

included a 50 years conference, the installation of interpretive boards and a celebration mosaic on the reserve (Fig. 8), a 'memories project' that gathered reminiscences and old photographs of the reserve and the production (and consumption) of a commemorative beer—the 'Trilobitter'.

When it was created the Wren's Nest NNR was, geologically, one of the most important places in the UK with a global reputation. This remains; but what has been added is unprecedented experience

in geological conservation that has had to overcome the challenges of urban reserve management, physical instability and acceptance within the local community. Over the 50 years of the NNR there have been setbacks; however, persistence and the willingness to adapt have, on the whole, overcome these problems. Innovation and experience from the Wren's Nest NNR goes beyond the reserve and has been central to the establishment of Dudley as a leading local authority in geological

Fig. 8. Local school children and the mayor with a mural created by the children to celebrate the fiftieth year of the NNR. (Photograph: Jonathan Larwood.)

conservation and the Wren's Nest as a nationally and internationally cited example of what can be achieved in geological conservation.

The authors are grateful to P. Oliver (Herefordshire and Worcestershire Earth Heritage Trust) and G. Worton (Dudley Museum and Art Gallery) for reading and commenting on this manuscript.

References

ANON, 1974. *Wren's Nest National Nature Reserve: Geological Trail.* Nature Conservancy Council, 1–23.

BUTLER, A. J. 1939. The stratigraphy of the Wenlock Limestone of Dudley. *Quarterly Journal of Geological Society of London*, **95**, 37–74.

BUTLER, A. J. & OAKLEY, K. P. 1936. Report of "Coral Reef" Meeting. The Dudley District. *Proceedings of the Geologists' Association*, **47**, 133–136.

CHUBB, L. 1945. *National Geological Reserves in England and Wales.* Report by the Geological Reserves Sub-Committee of the Nature Reserves Investigation Committee. Conference on Nature Preservation in Post-War Reconstruction. Natural History Survey of Great Britain. The Society for the Protection of Nature Reserves. British Museum (Natural History), London. September 1945 (2nd edn, November 1945).

CUTLER, A., OLIVER, P. G. & REID, C. G. R. 1990. *Wren's Nest National Nature Reserve Geological Handbook and Field Guide.* Dudley Metropolitan Borough Council and the Nature Conservancy Council, 1–28.

HAMBLIN, R. J. O., WARWICK, G. T. & WHITE, D. E. 1978. *Geological Handbook for the Wren's Nest National Nature Reserve.* Nature Conservancy Council, 1–16.

HUXLEY, J. S. 1947. *Conservation of Nature in England and Wales.* Report of the Wild Life

Conservation Special Committee. HMSO, Cmd. 7122, 1–139.

JUKES, J. B. 1859. *The Geology of the South Staffordshire Coal-field.* 2nd edn. Memoir of the Geological Survey of Great Britain, 1–241.

MIKULIC, D. G. & KLUESSENDORF, J. 2007. Legacy of the Locust—Dudley and its Famous Trilobite *Calymene blumenbachii. In:* MIKULIC, D. G., LANDING, E. & KLUESSENDORF, J. (eds) *Fabulous Fossils—300 Years of Worldwide Research on Trilobites.* New York State Bulletin, **507**, 141–169.

SCOTT, L. 1942. *Report of the Committee on Land Utilisation in Rural Areas.* Ministry of Works and Planning. HMSO, Cmd. 6378, 1–138.

WARWICK, G. T., STRACHAN, I. & STUBBS, A. E. 1967. *The Wren's Nest National Nature Reserve: A Guide to the Geology and Economic History of the Wren's Nest National Nature Reserve including the Nature Trails.* Nature Conservancy, 1–31.

WHITEHEAD, T. H. & POCOCK, R. W. 1947. Geology of the country between Dudley and Bridgnorth. *Memoir of the Geological Survey of Great Britain,* 1–226, HMSO, London.

A history of geoconservation in the Republic of Ireland

M. A. PARKES

Natural History Division, National Museum of Ireland, Merrion Street, Dublin 2, Ireland
(e-mail: mparkes@museum.ie)

Abstract: Geoconservation in the Republic of Ireland has had only a short history and very few champions within the geoscience community or from the wider population. An early start came with a listing of Areas of Scientific Interest (ASIs) by An Foras Forbartha in 1981. A successful legal challenge to this scheme required a back-to-the-drawing-board approach but the exclusion of geological heritage from subsequent nature conservation assessment was a major setback. Only persistent effort from within the Geological Survey of Ireland has allowed a belated integration of geoconservation into the work of other state agents over the last eight years, through the Irish Geological Heritage Programme. Today, a twin track approach operates with statutory Natural Heritage Areas and non-statutory County Geological Sites. There are broad parallels to UK geoconservation strategies, but having also amalgamated much best practice from European and other countries, through involvement with ProGEO. The existing geoconservation programme is very modest through lack of human resources; it has evolved rapidly to exploit the available opportunities, especially in conjunction with local authorities. It has worked to keep geoconservation in the public consciousness in parallel with the work through official channels.

The development of geoconservation in the Republic of Ireland has had only a short history, initially as part of general nature conservation initiatives of the 1980s. The first coherent published attempt to identify and protect Earth science sites in Ireland came with the publication by An Foras Forbartha (1981) of a listing of Areas of Scientific Interest (ASIs). A successful legal challenge rendered this unworkable and in the early 1990s the National Parks and Wildlife Service (NPWS) developed Natural Heritage Areas (NHA) as a new alternative. European Union LIFE money which was provided to survey these proposed NHA sites formally excluded geology. NPWS approached the Geological Survey of Ireland in 1992 to seek its technical input to develop a replacement listing of geological heritage sites, even though no extra staff or financial resources could be provided. As a consequence, little could be done, although GSI continued to strive to develop geoheritage conservation.

Irish geological heritage then languished for a few years until the efforts of the Geological Survey of Ireland (GSI) returned it to the nature conservation agenda in Ireland. The Irish Geological Heritage (IGH) programme commenced in 1998 (Parkes & Morris 1999a). A comprehensive outline of the operation of that programme was defined in Parkes & Morris (2001), although a brief summary is included here. This paper principally summarizes the evolution of the IGH programme to date, its current status as well as other recent geoconservation initiatives in Ireland. A very brief, mainly pictorial outline of geoconservation in Ireland was recently published in *Earth Heritage* magazine (Parkes 2005), but this paper provides a more comprehensive overview.

The earliest efforts at geoconservation

It can be argued that geoconservation in Ireland started with concerns over the impact of visitors to the Giants Causeway in County Antrim as far back as 1896 (Wyse Jackson 1987; Doughty 2008). However, setting aside these historical examples, there was virtually no structured interest or concern for geoconservation until the early 1980s, neither officially from state bodies nor from the geological community. These are isolated examples of individual geologists attempting to preserve sections or fossil localities, but there are nearly as many stories of prominent geologists (generally from outside of Ireland) ransacking important mineral localities and so on. In all such cases the factual records are sparse.

In the wake of Irish independence, the new social and economic realities accompanied by religious influence that prevailed over political life unfortunately acted to exclude geology from mainstream awareness. Although archaeology prospered, with significant political backing, the natural world in general was low on political agendas and geology, in particular, became marginalized. The fortunes of the geology collections of the National Museum of Ireland are symptomatic of the political and cultural indifference to geological heritage. In 1924, the Geological Survey of Ireland lost its museum space

From: BUREK, C. V. & PROSSER, C. D. (eds) *The History of Geoconservation.*
Geological Society, London, Special Publications, **300**, 237–248.
DOI: 10.1144/SP300.18   0305-8719/08/$15.00 © The Geological Society of London 2008.

within the curved corridor adjoining the Natural History Museum to Leinster House, to make room for more government offices. Those important collections remained in crates and inaccessible until the 1970s (Parkes & Sleeman 1997). Furthermore, the Fossil Hall and other geology display rooms within the Natural History Museum buildings were demolished in 1962, to allow for government expansion, but with no alternative provision. Many of the collections removed from that site remain in crates today. To further exemplify the lowly status of geological heritage, when a new building space was assigned for Earth science within the National Museum of Ireland in the 1980s, various staffing embargoes meant that the Curator was often unable to open displays to the public.

Things were also at a low point in university geological museums. The Natural History Museum in University College Cork was closed in the 1960s, to make room for administrative offices. Much of the geological material was used as hard core for new paths around the campus. More interesting specimens found their way into the homes of staff (Bettie Higgs pers. comm.). This, and other events, are part of the geoconservation history of Ireland—showing the climate at the time. Later, in the 1990s the government provided funding for conservation work in the small geological museum on campus. The geological museum in TCD, however, fared better, although it now occupies only one attic room of a building that was the University of Dublin's Museum when it was built (Wyse Jackson 1994).

Areas of Scientific Interest: An Foras Forbartha's list

The first strategic geoconservation initiative was An Foras Forbartha's (1981) publication of the ASI in Ireland. Aubrey Flegg of the Geological Survey of Ireland was on the working party, but several other geologists such as George Sevastopulo, John Jackson, Willie Warren, Francis Synge and Loreto Farrell contributed to the process. This list of ASIs was on a county basis, with each site categorized as of international, national, regional or local importance. Fifty-one sites were recognized as being of international importance. Biodiversity and geodiversity were integrated and of equal value—making this a ground-breaking development.

Unfortunately a legal challenge over the designation of a bog site in County Galway rendered the scheme unworkable, and it fell by the wayside, although the site data remained a valuable resource to underpin the successor programme. The NPWS, then a part of the Office of Public Works, developed NHAs as a replacement scheme for

ASIs. EU LIFE programme funding supported a resurvey of NHAs but specifically excluded geological and geomorphological sites, thereby artificially severing the potential integration of biodiversity and geodiversity. Within the national press this anomaly pertaining to geological sites was noted by some correspondents and a plea made to the Government to enact legislation that would protect geological and geomorphological sites (Wyse Jackson 1987). In 1992, the Geological Survey of Ireland was invited to survey geological sites for NHA status, and agreed a partnership in 1994. Unfortunately, lack of resources, including staff, precluded the GSI from implementing this scheme until mid-1998.

The establishment of the Irish Geological Heritage programme

The early promotion of peatland geological heritage by Donal Daly (1989, 1994a, b) was important in establishing the geoconservation agenda in the GSI. John Morris picked up the baton, and consulted extensively to shape the IGH programme on best practice followed in Scotland, England and Wales, as well as in Northern Ireland. Although the IGH programme was agreed within GSI in 1994, government embargoes on employment of staff meant that nothing could start, unless staff were redeployed from other programmes. Eventually, by 1998, with some relaxation in the restrictions hindering commencement of the IGH programme, it was possible to appoint a short-term contract geologist. This author was fortunate in securing that post. In June 1998 work began, initially on two themes. From the beginning, the author took a long-term view for the programme, and initiated the establishment of working practices and procedures for geoconservation in Ireland. Whilst short-term delivery of objectives such as thematic reports may have suffered, the overall strength of the work and security of the programme within GSI, and the inclusion of broader outreach activities into the programme and into GSI has benefited from the long-term approach.

An early project that can be regarded as a geoconservation success was the development of the Valentia Tetrapod trackway in Co. Kerry (Parkes & Morris 1999b; Parkes 2000, 2003a, 2004b). This site exhibits the oldest in-situ footprints of an amphibian venturing onto a land surface known anywhere in the world. A special case was made for the state to buy the site and it is now protected and accessible to the general public, with local interpretation (Fig. 1). It is effectively the Irish state's first geological national monument. The preservation and presentation in situ, rather than collection and removal to the National Museum

Fig. 1. The Valentia Tetrapod Trackway Site. (**a**) A view from the pathway over the viewpoint and interpretive panel mount. (**b**) The interpretive panel with real section of trackway and rippled siltstone for touching. The panel faces in the reverse direction from the main view over the trackway as a safety measure to prevent it being used as a step up and over the fence. It was originally intended to be part of the wall structure. (*Continued overleaf*).

of Ireland was the preferred option here, reflecting the view that if local people value their heritage they will act as effective custodians for it. Experience to date indicates that this has been the case, and that there is considerable local pride and careful attention amongst Valentia Island's people.

Irish Geological Heritage programme procedures

The total geology and geomorphology of Ireland is encompassed in sixteen themes (comparable to the more than 100 blocks of the Geological

(c)

(d)

Fig. 1. (*Continued*) (**c**) A view of the pathway down to the viewing point closest to the actual trackway itself. The chain and post fence is a safety measure to protect the trackway from visitors and the visitors from slippery rocks and the sea itself. (**d**) The trackway can be viewed safely from as close as 1–2 m away.

Conservation Review (GCR) in Britain (Ellis 2008)). The smaller number of themes in Ireland reflects the smaller geographical area and lower level of geodiversity, particularly with regard to Mesozoic geology. The sixteen themes are listed below:

- IGH1 Karst
- IGH2 Precambrian to Devonian palaeontology

- IGH3 Carboniferous to Pliocene palaeontology
- IGH4 Cambrian–Silurian
- IGH5 Precambrian
- IGH6 Mineralogy
- IGH7 Quaternary
- IGH8 Lower Carboniferous
- IGH9 Upper Carboniferous and Permian
- IGH10 Devonian
- IGH11 Igneous intrusions
- IGH12 Mesozoic and Cenozoic
- IGH13 Coastal geomorphology
- IGH14 Fluvial and lacustrine geomorphology
- IGH15 Economic geology
- IGH16 Hydrogeology.

After some trialling of different approaches with the operation of IGH1–3, a pattern of operation has now been established for the selection of geological NHAs. For all themes an expert panel was established. Each panel attempted to include all national experts in the particular theme, but many also included international experts. This proved highly effective in reaching consensus about which sites were genuinely important at a national or international scale, even where long held academic rivalries were known to exist. The formulation of this system drew heavily upon examples of good practice in many other parts of Europe, in particular in the former Yugoslavia.

For each theme, an indicative list of all sites which merited consideration for NHA status was developed. This list allows the final selection for designation after more detailed assessment and comparison of sites. Designation can only be legally authorized by the NPWS—the statutory authority for designations. The NPWS is also responsible for the designation of Special Areas of Conservation (SAC) for the Natura 2000 programme, and other European-wide instruments of nature conservation. These are generally driven by ecological concerns either for endangered species, or for particular habitats, but there is an underlying protection for the geological foundation of many of the habitats. Although it is not clearly expressed in the process, both landforms and rock exposures and their subsoils are the fundamental basis of the habitats. Therefore it is unsurprising that many sites of geological importance coincide with land already identified for nature conservation through SAC, NHA, SPA or other instrument (Fig. 2).

Three criteria were used in the selection:

- representativeness;
- unique or exceptional character; and
- international importance.

Subject to meeting the above criteria, multiple sites with the same geological interest were then included in the indicative site lists or not, by a weighting of other factors. These included a significant research history or recognized potential for future research. Further factors applied were educational value, public amenity value, ecological or other special value etc. It is intended that only the minimum number of sites should be designated to provide a full representation of the scientific interest of the theme, thereby ensuring their defensibility and integrity against many other pressures and competing interests in society. Selected NHA sites must be able to stand up to a landowner's legal objection to their designation.

In the course of building each thematic indicative site list a second category of site, the County Geological Site (CGS) was also included (see below for full explanation). It was initially apparent that many sites under consideration by the expert panels were of this lower importance. These were included in the indicative site lists so that local authorities could be supplied with appropriate information. However, any sites under consideration for NHA status that do not eventually get selected as such will continue to be comprehended as CGS.

Consolidation

Subsequent to completion of the indicative site lists for the sixteen themes, a lengthy process of consolidation occurred. Almost 1500 sites had been identified across the themes, and inevitably there were sites that were listed under two or more themes. The areas of such sites often overlapped but did not coincide exactly, due to the diversity of geological and geomorphological scientific interests. Often the exact limits of a site were poorly understood and necessitated a detailed report from an expert panel member who had researched the site, or else a lengthy literature research or a new site visit. To add to the complexity, the same site may have been known by different names to different expert panels, compounding the difficulty in consolidation of the indicative site list.

It was essential for the future simplicity and integrity of the NHA designation process, as well as for the legal defensibility, that sites be as clearly and as tightly defined as possible. In view of the sensitivity of the Irish population to any designation of their land, and potential restriction on landowner rights, it was especially important to consolidate the list for the minimization of total land areas to be proposed for any designation. It was also important to avoid any serial or sequential designation of adjoining or overlapping areas, as would have occurred had each theme been dealt with separately without looking for multiple appearances.

The consolidation process required extensive consultation with active contributors to the expert panels, literature and field research and more recently, extensive use of digital colour aerial photos available

Fig. 2. This limestone pavement at Sheshymore was considered to be the best Irish example by the karst expert panel, but it is included within a much larger area already identified and designated as Special Area of Conservation under European legislation designed to protect particular habitats. The biodiversity interest, in this case mainly botanical, is founded upon the underlying geodiversity, without adequate recognition of the role of geology and geomorphology in forming ecological habitats.

in a GIS, as they have become available. Informed decisions have been made on naming of sites and on the delimitation of normally one single site containing multiple thematic interests, where such overlaps occur (e.g. Fig. 3). The process was also helped by the individual county projects (see below) that took place during the consolidation period.

Completion of individual themes

Aside from the first three themes which have been completed, the full IGH programme process needs further steps to complete the remaining themes, based on the sites included in the indicative site list. Each theme now requires the commission of desk-based site reports from members of each expert panel, according to their knowledge of sites, and their individual availability. In this way it is intended that the IGH programme gets the best quality reports feasible. It will be then the responsibility of the GSI staff to visit sites for a particular theme in order to check their current status, make a photographic record and define a site boundary which will be surveyed according to NPWS protocols. In addition, if at all possible contact can be made with landowners and the IGH interest in the site can be explained.

Once individual site reports have been completed a theme report which links the sites into a coherent framework and provides the overview of the topic must be written. This will normally be done by one or more members of the Expert Panel for the theme in question, but with the basic template and considerable editorial input supplied by the IGH programme to maintain a consistent standard across thematic reports. At time of writing, desk-based site report work is well underway with IGH6 mineralogy and IGH7 Quaternary themes.

County Geological Sites

The IGH programme made a proposal in 1999 to establish County Geological Sites (CGS), which

Fig. 3. Castlepook Cave is typical of many sites, having importance in more than one theme. Castlepook Cave was regarded as an excellent example of a maze cave, typical of many Irish lowland karst terrains, by the IGH1 karst expert panel. It also has great importance under the IGH7 Quaternary theme, as it contains the richest range of glacial, interglacial and postglacial cave fauna. Many of the animal bones are held in the Natural History Museum and have been radiocarbon dated, and include unique records, e.g. hyaenas. The site boundary is drawn to include land above all known cave passages, as the cave is close to surface, and undisturbed deposits exist which could be excavated in the future. The cave is gated.

was adopted within the National Heritage Plan (2002) A CGS is a non statutory designation for sites which receive a measure of recognition within the planning system by being included within each local authority's County Development Plan. They may thus also be protected and promoted, but most importantly geological heritage sites listed as CGS may not be destroyed through a lack of awareness. However, there is much work to be done in monitoring and raising awareness of this situation within local authorities. As recent destruction (May 2007) of a classic geological section at Inch in County Kerry shows, information does not always flow to those it should reach. At Inch, works to stabilize a cliff below a coastal road have covered a critical section with two Old Red Sandstone conglomerates, providing understanding of the Devonian geology of the Dingle Peninsula. The engineers were unaware of the geological importance of the beach and cliff section,

and could perhaps have modified the design of works to maintain access to exposures if they had.

Most local authorities have now been supplied with some provisional data on CGS in their counties as each County Development Plan has been revised (usually on a five year cycle). However the completion of the indicative site list for all themes permits more comprehensive lists of CGS to be supplied from now on.

In many respects the NHAs are analogous to Sites of Special Scientific Interest (SSSI) in the UK (Prosser 2008). Similarly the CGS could be viewed as comparable to the UK Regionally Important Geological/geomorphological Sites (RIGS) (Burek 2008a, b). There is one important difference. Although population density in the UK is usually sufficient to sustain RIGS groups (Burek 2008b), who take an active interest and participate in selecting and managing sites, with local authority or statutory agency support, it is highly unlikely that the CGS scheme could operate in this way. Population density is inadequate to sustain sufficient active groups.

Even amongst geoscientists, who would normally be the nucleus of RIGS groups, there is very sparse interest in geoconservation apart from those working directly in the field (about 3–4 people, only one full-time) and the active participants of expert panels. With a small number of notable exceptions, their direct involvement (other than through desktop site reports) has generally not extended any further to date.

The most comparable equivalent of a RIGS group is the organization ES2k (Earth Science 2000). This organization was originally set up as a response to closure of the Geology Department at Queen's University Belfast and loss of geology teaching in schools. It has matured into an energetic group campaigning on several fronts to raise awareness of geology in formal education and in the wider community. From a Northern Ireland base it has developed to have an all-Ireland (Northern Ireland and the Republic of Ireland) vision and activity.

The ES2k organization has examined the RIGS concept and is wrestling with the issues, at time of writing. It may be feasible to develop a RIGS group in a Northern Ireland context, since there is an active membership base in that region, and where the existence of the network of sites identified and documented by the Environment and Heritage Service is a major advantage. Areas of Special Scientific Interest (ASSI) in Northern Ireland have been documented though not all have yet been designated. This compares closely with the Geological Conservation Review process in England, Wales and Scotland (Ellis 2008). In the current debates over setting up of RIGS groups in Northern Ireland or in the Republic, it has apparently been forgotten that there is no tradition of such activity, nor likely to be a groundswell of new people interested in such an initiative. It will be the same people involved, should it happen, who are already fully engaged with promoting Earth science and geoconservation through their jobs or in ES2k or other groups.

The identification and documentation of geological heritage sites in the Republic of Ireland (as NHAs or CGSs) lies several years behind that of Northern Ireland. With a major struggle to get even CGS into County Development Plans and to be accepted by planners and local authority officials, the formal setting up of RIGS groups is unlikely to succeed and could potentially confuse the geoconservation debate with new and unfamiliar terms, which have no legal standing in the Republic of Ireland. Outside Dublin, the population base is very dispersed. However, the Cork Geological Association, in the south of the country, is a large,

active group and could feasibly undertake activities, at least in part, within a CGS scheme, such as siteworks, leaflets and guides or public promotion. Similarly the Irish Geological Association, with a national membership could undertake some specific activities in a CGS scheme. However, both associations have as their aims the provision of lectures and fieldtrips for their members. A significant shift of mindset would be required to expand into physical management of sites and creating promotional materials. Some members are priming themselves to take up such a challenge.

In general, the author would argue, that encouragement from England and Wales to set up RIGS groups in Northern Ireland and especially in the Republic of Ireland is probably better diverted into other projects to sustain existing groups such as ES2k. However, the scheme has been successfully applied to Scotland (Burek & Potter 2004). The broadly accepted RIGS structure and support does not translate well into the different social, legal and geographic situation in both jurisdictions of Ireland.

County Geological Site audits

The development of the Heritage Officer scheme, promoted and part funded by the Heritage Council has been strongly beneficial to geoconservation. The success of the scheme has been such that almost all counties now employ a Heritage Officer (only counties Wexford, Carlow and Leitrim out of 31 authorities have failed to start or replace contract staff). Many councils have made the position a permanent one, after a part-funded trial for three years.

Each Heritage Officer is charged with developing a 5-year heritage plan for their county, based upon the deliberations of a Heritage Forum, supported by wide consultation. The author was fortunate in serving on three Heritage Fora, and in being able to make submissions to many others. Since geology is formally listed as one of thirteen elements of heritage in the Heritage Act (1995), the Heritage Officers have generally been grateful to receive guidance as to how to address it. A general purpose document (Parkes 2003b, revised by S. Gatley July 2006 and available on GSI website: www.gsi.ie) was made available to all Heritage Officers, and for other officials, advising simple measures to embrace geology within local authority and Heritage Officer remits.

These collaborations have been valuable in the implementation of IGH programme objectives through the means of County Geological Site audits. Whilst each has varied in the specific detail and full scope, dependent upon local needs, four such audits have been completed (McAteer &

Parkes 2004; Parkes & McAteer 2004; Parkes *et al.* 2005; Parkes & Sheehan Clarke 2005). Fundamental to all of them is an accessible description of all geological heritage sites (possible NHA and CGS sites) in the county. At time of writing, audits of three further counties are nearing completion (Clarke & Parkes 2007*a*–*c*). These seven county audits have been extremely helpful in clarifying the indicative site lists with respect to the counties concerned, and in consolidating the lists. The simple level of assessment provides a line on a map of the geological interest, but has not allowed detailed boundary surveys (to NPWS protocols) of possible NHA sites. Such sites will have to be visited again during future completion of individual theme reports. However, for the majority of sites visited, which are only CGS, and not considered as NHAs the audits have provided definition of the limits and boundary of each site.

Current status of the IGH programme

The author took up a permanent position in the Natural History Museum, Dublin in November 2005, as Assistant Keeper, responsible for the Earth science collections. In this museum, the portable elements of geoconservation from sites of importance are cared for—the fossil, mineral and rock collections from well over 200 years of collecting in Ireland, and all around the world. The promotion of awareness and understanding of Earth science is an essential ingredient of the work, exactly as it was in the IGH programme, for if people do not value and realise the importance of sites and collections, they will not sustain them in society against competing concerns.

The author, as a scientific visitor in the GSI, continues to take an active interest in the progress of the IGH programme as well as assisting with the ongoing care of the fossil and other collections in GSI. In this way the natural and existing good relations between the Natural History Museum and the GSI have been strengthened and enhanced.

In GSI, the programme is in the capable hands of Sarah Gatley, who is Head of the Irish Geological Heritage Programme, after many years working in the Bedrock Programme. This ensures that it is an element of the core work in GSI and cannot be regarded as a marginal activity compared to other programme areas, as it had been perceived.

Current work targets aside from managing the county audits and addressing the extensive enquiries, are completion of the desk study site reports by the expert panels in IGH6 mineralogy theme, and making inroads into the IGH7 Quaternary theme. The latter theme is a priority against the extreme pressure on sand and gravel resources from the extractive industry, for concrete, construction and civil engineering projects such as motorways. Eskers are an iconic Irish landform, and they are numerous through the Irish Midlands. The international name for these linear ridges of sand and gravel deposited by sub-glacial meltwater streams has been derived from the Irish language *eiscir* (along with other landforms such as turloughs and drumlins). However, they and their associated outwash features are under particular threat. Establishing the most important examples for protection has a great urgency.

The IGH programme participants, including the author, are currently in co-operative discussions with the Irish Concrete Federation, which represents the main companies involved in legal quarrying. The aim is to develop and promote guidelines for the industry operators as to best practice in geoconservation. This area of activity provides a significant hope for establishing geodiversity and geoconservation as readily understood terms in a country with no tradition of either, and with a lobby group comprising only a handful of practitioners.

INTERREG projects (European Union inter-regional funding scheme)

In parallel to the thematic network of sites approach for geoconservation described above, John Morris (a Principal Geologist in GSI), the original architect and overall manager of the IGH programme, has invested great energy in developing and running a series of complementary projects. These have involved partners including GSI in a host of different countries and organizations and have sometimes themselves required their own separate staffing structures. Almost all have received funding through INTERREG 3 applications. Various other European funding programmes have also been sourced, such as LEADER, CULTURE 2000 and cross border, peace and reconciliation funding.

The individual projects are only listed here, as more detail is outside the scope of this paper:

- MINET—A European network of mining heritage centres.
- EUROPAMINES—a successor project to MINET to build and develop the network and share knowledge.
- Copper Coast IIIC, IIIB—intra-Geopark projects to build practical skills and resources as well as on site works.
- Green Mines—A European mine rehabilitation project focused on Silvermines in North Tipperary, for the Irish partners.

- Celtic Copper—one of several mine rehabilitation, mine heritage and acid mine drainage/fisheries projects at Avoca Mines in County Wicklow.
- Breifne Mountains Project—a cross border, multi county project inspired by Spanish Cultural Parks, aimed at developing sustainable tourism based on the geological and other heritage of an upland landscape in NW Ireland.

The Mining Heritage Trust of Ireland

Most of the projects noted above have involved a significant component of mining heritage. Mining heritage plays an important role in connecting people to geodiversity within their social and historical fabric. As the basis for NHAs, under the Wildlife (Amendment) Act 2000 has to be scientific, many aspects of mining heritage fall outside this scheme. So although mineralogy is covered along with a limited range of deposits of metallic and industrial minerals as economic geology sites, the whole industrial archaeology and social history of mining are not.

Through the endeavours of John Morris, with support of a group of members of the Shropshire Caving and Mining Club, the Mining History Society of Ireland (MHSI) was established in 1996. As a founding member, the author has seen this group establish itself as an important player in a neglected area of heritage. By becoming a charitable company limited by guarantee, and forming the Mining Heritage Trust of Ireland (MHTI) in 2000, major projects became feasible for financial support. The conservation of the Man Engine House at Allihies in West Cork was an important first success. Supported by the GSI, the Heritage Council and many others, MHTI now manages several mining heritage projects and is widely recognized as the expert body in Ireland for such heritage. It therefore complements the purely scientific and very restricted geoconservation work of the IGH programme relating to mining by dealing with all other aspects.

Integration of Irish geological conservation with European and international initiatives

ProGEO

Ireland is a member of ProGEO, the European Association for the Conservation of the Geological Heritage. This organization, founded in 1993, has a very broad-based membership drawn from Iceland to Ukraine and from Norway to Greece. Ireland's participation in meetings and working groups has benefited the IGH programme significantly through experiencing best practice in other countries and exchange of knowledge. ProGEO members had experience of Irish geoconservation within a conference held in 2002 on 'Natural and Cultural Landscapes—the geological foundation', organized through the Royal Irish Academy's National Committee for Geology. The Proceedings (Parkes 2004a) provided a record of many ProGEO member actions in different countries.

Geosites

The Geosites project (Wimbledon 1996; Wimbledon et al. 2000) is run by the Global Geosites Working Group of the International Union of Geological Sciences, with the support of UNESCO (United Nations Educational, Scientific and Cultural Organization). In Europe the ProGEO regional working groups are the mechanism for achieving the aims of the project. These include the compilation of a Global Geosites inventory and database of key sites and terrains, to facilitate provision of advice to IUGS (International Union of Geological Sciences) and UNESCO on Earth Science World Heritage Sites (Boylan 2008).

A provisional listing of frameworks (geological categories or thematic types) of sites (Wimbledon et al. 1998) included a very preliminary Irish contribution. Now that a full indicative site list is available for Ireland, the comparisons between adjoining territories can be made to refine the frameworks and hence determine where sites must be included (Wimbledon et al. 1995). Collaboration with Northern Ireland and the UK is the first step in this process. This falls within the work of the Northern Europe Regional Working Group within ProGEO. By identifying common frameworks and comparing similar sites between countries and regions, the best sites on a regional basis can then be identified for potential geological World Heritage Sites.

UNESCO and European Geoparks

The geopark concept is now well established within geoconservation practice, and supported by UNESCO (Jones 2008). A growing number of geoparks, particularly in China and in Europe are recognized by UNESCO as Global Geoparks. In Europe, the European Geoparks Network (EGN) is recognized by UNESCO as the assessor and arbitrator of applications to join the network. In Ireland there are presently two European and UNESCO Global Geoparks—the Copper Coast Geopark in Co. Waterford and the Marble Arch Caves

Geopark in Co. Fermanagh, which collaborate very well on a range of projects.

There are several community and local authority groups with genuine ambition to enable local areas to become geoparks, such as the Burren in Co. Clare, the Carlingford/Mourne Mountains igneous province, the Sneem-Valentia area of the Iveragh Peninsula in Kerry and possibly the esker landscapes of Westmeath and Offaly in the Irish Midlands. The stated community-based dimension of the EGN is apparently less favoured than structures and finance underpinned by local, regional or national government. Initiatives and desire for geopark status from a local community driven base may not be able to meet the strict standards of the EGN. Given some inherent problems in the present system, including a general philosophical shift away from bottom-up community driven geoparks, to larger state-based National Park scales, it is possible that the German geopark model would suit Ireland much better in the future. A recognizable network of Irish Geoparks working to an Irish national structure, without full endorsement of UNESCO or the European Geoparks Network is likely to have much more impact with the general public, and with the communities seeking to sustain their landscape and tourism infrastructure.

Summary

A personal appraisal of the history of geoconservation in Ireland has been presented by the author who feels privileged to have been working at a central position within the developments of the last decade. Although various criticisms could be made of an apparent lack of progress in designation of sites as NHAs, many positive steps on a long road have been taken. It has largely been achieved with a staff of, at times, only one person, sometimes with temporary assistants for specific projects, alongside broad outreach responsibilities.

The completion of the indicative site list for all themes, the essential completion of three full theme reports and some progress on several others, represents a considerable advancement of geoconservation in Ireland. In addition there have been seven county heritage audits which have established a pattern for local authority fulfilment of their geological heritage responsibilities under different legislation. The Valentia Tetrapod Trackway site has also been developed for public access and interpretation as a national treasure. In a country where there has been no modern tradition of geoconservation, this all represents a firm foundation for rapid future advancement with completion of further themes and expected designation of some geological NHAs in the near future.

The content and opinions expressed are those of the author, who thanks J. Morris, S. Gatley and N. Monaghan for their constructive comments and discussion. The referees B. Higgs and P. Wyse Jackson are both thanked for considerable improvements to the paper. The following temporary Geological Assistants made important contributions to different projects and are thanked for their input: E. Roche, S. Preteseille, P. Coffey, C. McAteer, A. Clarke, F. McGrath and S. Engering.

R. Meehan, M. Simms, I. Enlander, M. McCabe, M. Feely, G. Sevastopulo, R. Unitt, E. Bird and many others unnamed, are all thanked for their advice and input in many different ways, often above and beyond the call of duty. All the expert panel members have played an important role, but are too numerous to list here. K. Higgs and B. Higgs are thanked for their most active interest in geological heritage in general and considerable assistance in sites where issues have arisen over recent years.

References

AN FORAS FORBARTHA. 1981. *Areas of scientific interest in Ireland national heritage inventory*. An Foras Forbartha, Dublin.

BOYLAN, P. 2008. Geological site designation under the 1972 UNESCO World Heritage Convention. *In*: BUREK, C. V. & PROSSER, C. D. (eds) *The History of Geoconservation*. Geological Society, London, Special Publications, **300**, 279–304.

BUREK, C. V. 2008*a*. The role of the voluntary sector in the evolving geoconservation movement. *In*: BUREK, C. V. & PROSSER, C. D. (eds) *The History of Geoconservation*. Geological Society, London, Special Publications, **300**, 61–89.

BUREK, C. V. 2008*b*. A history of RIGS in Wales: an example of successful cooperation for geoconservation. *In*: BUREK, C. V. & PROSSER, C. D. (eds) *The History of Geoconservation*. Geological Society, London, Special Publications, **300**, 147–171.

BUREK, C. V. & POTTER, J. 2004. *The application of Local Geodiversity Action Planning to Scotland—Learning from the Cheshire region experience. In*: BROWNE, M. & McADAM, D. (eds) *Good Practice. RIGS groups working in partnership*. Proceedings of the 6th Annual UKRIGS conference, Ecclesmachen, Broxburn, West Lothian, Scotland 24–26 October 2003. UKRIGS, Chester, 33–37.

CLARKE, A. & PARKES, M. A. 2007*a*. *The Geological Heritage of Kilkenny*. Geological Survey of Ireland.

CLARKE, A. & PARKES, M. A. 2007*b*. *The Geological Heritage of Meath*. Geological Survey of Ireland.

CLARKE, A. & PARKES, M. A. 2007*c*. *The Geological Heritage of Fingal*. Geological Survey of Ireland.

DALY, D. (ed.) 1989. *Extended summaries of lectures at Symposium on Conservation of Earth Science areas of scientific interest*. Geological Survey of Ireland, 1 December 1989.

DALY, D. 1994*a*. Earth science conservation in the Republic of Ireland. *Mémoires de la Société Géologique de France*, **165**, 129–133.

DALY, D. 1994*b*. Conservation of peatlands: the role of hydrogeology and the sustainable development principle. *In*: O'HALLORAN, D., GREEN, C., HARLEY, M., STANLEY, M. & KNILL, J. (eds) *Geological and Landscape Conservation*. Proceedings of the Malvern International Conference 1993. Geological Society, London, 17–21.

DOUGHTY, P. 2008. How things began: the origins of geological conservation. *In*: BUREK, C. V. & PROSSER, C. D. (eds) *The History of Geoconservation*. The Geological Society, London, Special Publications, **300**, 7–16.

ELLIS, N. 2008. A history of the Geological Conservation Review. *In*: BUREK, C. V. & PROSSER, C. D. (eds) *The History of Geoconservation*. Geological Society, London, Special Publications, **300**, 123–135.

JONES, C. 2008. History of Geoparks. *In*: BUREK, C. V. & PROSSER, C. D. (eds) *The History of Geoconservation*. Geological Society, London, Special Publications, **300**, 273–277.

MCATEER, C. & PARKES, M. A. 2004. *The Geological Heritage of Sligo*. Geological Survey of Ireland.

NATIONAL HERITAGE PLAN. 2002. Government of Ireland.

PARKES, M. A. 2000. The Valentia Island tetrapod trackway—a case study. *In*: BASSETT, M. G., KING, A. H., LARWOOD, J., PARKINSON, N. A. & DEISLER, V. K. (eds) *A Future for Fossils*. National Museums and Galleries of Wales, Cardiff, Wales, Geological Series, **19**, 71–73.

PARKES, M. A. 2003*a*. *The Valentia Tetrapod Trackway*. Dúchas, Dublin.

PARKES, M. A. 2003*b*. *Geology in Local Authority Planning. An outline guide to inclusion of geological heritage in County Development Plans and Heritage Plans*. Irish Geological Heritage Programme, Geological Survey of Ireland, Dublin.

PARKES, M. A. (ed.) 2004*a*. *Natural and Cultural Landscapes—the Geological Foundation*. Proceedings of a Conference 9–11 September, 2002, Dublin Castle. Royal Irish Academy, Dublin.

PARKES, M. A. 2004*b*. *The Valentia Tetrapod Trackway*. Geological Survey of Ireland.

PARKES, M. A. 2005. Geological heritage in the Emerald Isle. *Earth Heritage*, **25**, 20–22.

PARKES, M. A. & MCATEER, C. 2004. *The Geological Heritage of Carlow*. Geological Survey of Ireland.

PARKES, M. A. & MORRIS, J. H. 1999*a*. The Irish Geological Heritage Programme. *In*: BARETTINO, D., VALLEJO, M. & GALLEGO, E. (eds) *Towards the Balanced Management and Conservation of the Geological Heritage in the New Millennium*. Sociedad Geológica de España, Madrid, 60–64.

PARKES, M. A. & MORRIS, J. H. 1999*b*. The Valentia Island tetrapod trackway. *In*: BARETTINO, D., VALLEJO, M. & GALLEGO, E. (eds) *Towards the Balanced Management and Conservation of the Geological Heritage in the New Millennium*. Sociedad Geológica de España, Madrid, 65–68.

PARKES, M. A. & MORRIS, J. H. M. 2001. Earth science conservation in Ireland: The Irish Geological Heritage Programme. *Irish Journal of Earth Sciences*, **19**, 79–90.

PARKES, M. A. & SHEEHAN-CLARKE, A. 2005. *The Geological Heritage of Kildare*. Geological Survey of Ireland.

PARKES, M. A. & SLEEMAN, A. G. 1997. *Catalogue of Type, Figured and Cited Fossils in the Geological Survey of Ireland*. Geological Survey of Ireland, Dublin.

PARKES, M. A., MCATEER, C. & ENGERING, S. 2005. *The Geological Heritage of Clare*. Geological Survey of Ireland.

PROSSER, C. 2008. The history of geoconservation in England: legislative and policy milestones. *In*: BUREK, C. V. & PROSSER, C. D. (eds) *The History of Geoconservation*. Geological Society, Special Publications, London, **300**, 113–121.

WIMBLEDON, W.A. 1996. Geosites—a new conservation initiative. *Episodes* **19**, 87–88.

WIMBLEDON, W. A., BENTON, M. J., BEVINS, R. E., ET AL. 1995. The development of a methodology for the selection of British geological sites for conservation: part 1. *Modern Geology* **20**, 159–202.

WIMBLEDON, W. A., ISHCHENKO, A., GERASIMENKO, N., ET AL. 1998. A first attempt at a Geosites Framework for Europe—an IUGS initiative to support recognition of world heritage and European geodiversity. *Geologica Balcanica*, **28**, 5–32.

WIMBLEDON, W. A., ANDERSEN, S., CLEAL, C. J. ET AL. 2000. Geological world heritage. Geosites—a global comparative site inventory to enable prioritisation for conservation. *In*: Proceedings of the Second International Symposium on the Conservation of the Geological Heritage, Roma June 1996. Memorie Descrittive Della Carta Geologica d'Italia LVI. Servizio Geologica Nazionale.

WYSE JACKSON, P. N. 1987. Saving our rocks. *The Irish Times*, 30 July 1987, p. 11.

WYSE JACKSON, P. N. 1994 (ed.) *In Marble Halls*. Department of Geology, Trinity College Dublin.

History of geoconservation in Europe

LARS ERIKSTAD

Norwegian Institute for Nature Research, Gaustadalléen 21, NO- 0349 Oslo, Norway
(e-mail: lars.erikstad@nina.no)

Abstract: Europe consists of many countries and legislative regimes. Some countries have a long history of geoconservation with well developed strategies and practices and others lack the most basic legislation for this kind of work. Some of the earliest geoconservation is found within Germany such as the conservation and visitor control from 1668 in the showcave Baumannshöle. At the start of the twentieth century basic legislation came into force in several countries with more or less effective potential to protect geology. Geology was also important within the National Park movement at the same time.

Modern legislation, inventories and conservation strategies were developed mostly during the second half of the twentieth century. International cooperation has become more important and the association ProGEO and the European Geoparks Network are important in a European context. World Heritage Sites designated on geological criteria are also increasingly common. Geoconservation has been included as a recommendation by the European Council and has begin to be visible in EU policies (such as the soil strategy). As for the future, the challenge lies in getting organizations and disciplines to work together in ways that develop synergy between existing initiatives, share management experience and collaborate over research.

Geoconservation is a part of nature conservation, and its history runs parallel with nature conservation history. Several authors have described old examples of nature conservation initiatives and concern (Gray 2004), but in a more basic and practical setting it is common to start with the idea of national parks originating in USA about 150 years ago. These were interesting times in the natural sciences in general and especially for geology. Geological understanding was developing fast, and geology as a science had high prestige. This geological prestige was partly reflected in the nature conservation movement. Several geographers and geologists were involved, and landscapes and geological features (especially in a landscape setting) became objects of protection.

Early history

Today the discussion of geotourism is central within geoconservation. We tend to think that this is a modern concept that brings geoconservation into a new era. This may be so, but geotourism is far from new. In the search for early references to European geoconservation initiatives, the oldest I have found is the protection of the cave Baumannshöle in Germany in 1668 (Grube 1994). This showcave was discovered in the fifteenth century; it was first mentioned in literature in 1565 and was the object of guided tours as early as 1646 (Duckeck 2007). It was scientifically investigated in the 1650s and in 1668 was subject to a nature conservation decree by Duke Rudolf August controlling access to the cave.

A cave appears as exiting, mystic, dangerous and special to humans. Entering this dark world and discovering beautiful halls, and speleotems adds to this feeling of excitement, so caves have become one of the favourite geotourism destinations (Fig. 1). The vulnerability of underground formations creates major challenges for their management, at least if sustainability is to be included as an aim. In this way, and based on their long geotourism and geoconservation history, caves represent important early markers in geoconservation history.

A vital part of conservation is evaluation. We select sites and areas that we value, which are threatened and judged vulnerable. The understanding of value has changed over time and this is reflected in changes in strategies and management attitudes toward geoconservation. Early conservation efforts often concentrated on peculiar objects standing out as features of the landscape and often with mythical associations. In 1818, Alexander von Humbolt introduced the term 'Natural monument' and very early German conservation included the small mountains Drachenfels (protected in 1836), Totenstein (1844) and Teufelsmauer (1852) which all represent this sort of protection. Drachenfels is considered to be the oldest nature reserve in Germany (Wiedenbein 1994*a*, Grube 1999).

A Belgian Royal Commission on Monuments and Sites was established in 1827 (Jacobs 1994) and the first Czech geological protection was

From: BUREK, C. V. & PROSSER, C. D. (eds) *The History of Geoconservation.*
Geological Society, London, Special Publications, **300**, 249–256.
DOI: 10.1144/SP300.19   0305-8719/08/$15.00 © The Geological Society of London 2008.

Fig. 1. The fascination of the underworld. Caves have a long history of geoconservation and geotourism. Photo from Fatima, Portugal.

established in Bohemia in 1884 in memory of the French palaeontologist Joachim Barrande (Kriz 1994). The protection of this site might be regarded as a more modern understanding of the term 'Geological monument' and as a 'landmark in the Earth sciences' (Robinson 1989). The late nineteenth century conservation of Upper Carboniferous trees in the UK (Victoria Park in Glasgow and Wadsley Fossil Forest in Sheffield) is in the same category (Thomas & Warren 2008).

National Parks and geological site protection

In contrast to the focus on peculiarities, the National Park movement, originating in the USA in the middle of the nineteenth century, focused on an entirely different and larger scale with a much more holistic approach linked to wilderness concept (Thomas & Warren 2008). This did not exclude the importance of National Parks for geoconservation (Fig. 2) which also is clearly illustrated by Yellowstone, the USA's first National Park (1872) and its geothermal features.

The relationship between specified site-oriented geoconservation and the National Parks approach illustrates a major issue for geoconservation, namely the links between geoconservation and the rest of nature conservation, the cultural heritage and in a wider sense, development planning (e.g. Jacobs 1999; Johansson & Zarlenga 1999; Stevens 1994). Although it is important to recognize geoconservation in its own right as a part of nature conservation with its own value criteria and priorities, it is also important to include geology and landscape within a wider context linked to National Parks and other similar large conservation efforts as well as within cultural heritage and the general planning system.

This dualistic approach is visible within geoconservation in Europe throughout its history. The National Park idea was picked up quickly in the Nordic countries. The well known Swedish Polar researcher Nordenskiöld argued for the idea of National Parks as well as for the need to protect special nature types, including geological features. This was the basis of the first Swedish nature conservation act in 1909 and the establishment of the first National Parks in Sweden the same year.

Fig. 2. The National Parks of Europe is important for geoconservation, and geological heritage contributes significantly to the qualities of the parks. Photo from the lava landscapes of the Teide National Park, Tenerife, Spain.

Norway passed its first act in 1910, which clearly expressing geology as one central justification for nature protection (Erikstad 1984).

In Prussia a first inventory of natural monuments was started in 1906 (Wiedenbein 1994a), with the focus mainly on picturesque remains; modern inventories were started as late as in 1969. A list of geological sites with potential need for protection were published in 1945 in the UK (Prosser 2008). The protection of important geological phenomena was one of the aims of the first nature conservation societies founded in the Netherlands early in the twentieth century (Gonggrijp 1994). Finland passed its first nature conservation act in 1923 (Kananoja 1999) and Poland in 1949 (Alexandrovics 1994). Italy declared its first National Park (Abruzzo) in 1922 (Johansson & Zarlenga 1999) and Austria established special legislation for caves in 1928 (Trimmel 1994).

Europe consists of many countries and many legislative regimes. The development of geoconservation during the twentieth century reflects this diversity. There are large differences between countries and regions both in how well the legislation serves geoconservation, how and when

relevant legislation was established and the form and effectiveness of conservation strategies (Daly et al. 1994).

Collection and conservation

Throughout the history of geology, collection of samples has been an important scientific element for close study and documentation of rocks and minerals. Specimens in museums and other scientific collections represent an important reference for the development of the science. There soon developed a great interest in geological collecting outside the scientific institutions and amateur collection of fossils and minerals is a widespread hobby. As a result, trade with specimens has also developed (Basset et al. 2001). Many European countries have some form of restrictions on collection, trade or export of geological specimens. Italy, for example, established legislation for protection of fossils in 1939. According to this legislation, all fossils are the property of the state which forbids individual collection and trade. The legislation has caused problems for research and has effects opposite to the suppression of illegal trade (Pinna 1994).

The debate concerning collection and conservation is difficult and still relevant; European Geoparks have a strict ban on collection and trade of specimens (European Geoparks Network 2007), whereas some geological World Heritage Sites are more liberal (Basset *et al.* 2001). It is a crucial point that collections are important for science and that amateur collecting is one of the important means of disseminating geological knowledge and inspiring future geologists. The most developed European management strategies on this field are probably those found in the UK (Prosser *et al.* 2006).

International cooperation and networks

One of the first major attempts at identifying internationally important geological sites, raising their profile and the need to protect them was the effort to define stratotypes (International Commission on Stratigraphy 2007). This work began in 1970 and has been of great importance for the awareness of geoconservation in Europe (e.g. Holland 1994; Norman 1994).

In 1969 a Dutch working group on geoconservation was established. It started the work with an inventory for Earth science sites important for science and education. During this work the need for international contacts and discussion became obvious. A workshop was held in Leersum in the Netherlands in 1988 (Robinson 1989) with Gerard Gonggrijp as organizer. A new group (European Working Group on Earth-Science Conservation) emerged from this with Gonggrijp as secretary. This group was later transformed into an association (ProGEO). ProGEO: The European Association for the Conservation of the Geological Heritage (ProGEO 2007).

ProGEO has been holding organized meetings and conferences on a regular basis since it was established, often in co-operation with other international organizations and national groups. The first international symposium was arranged in Digne in 1991. Guy Martini was the main organizer and the venue, Réserve géologique de Haute-Provence (Fig. 3), consisting of a network of protected geological sites (geotopes). This venue

Fig. 3. The ammonite wall in Digne, France. This site was an important element in the Réserve géologique de Haute-Provence; now also a member of the European Geoparks Network.

and the focus of integrated networks of protected geotopes with a high degree of dissemination to the public and use of this dissemination in geotourism is interesting as it forms the basis of both ProGEO and the highly active Geoparks European movement. One of the main achievements was the declaration (The Digne Declaration; Martini & Pages 1994) stating that geoconservation is an important task and that geological heritage is important for many reasons. This is illustrated by the following quotation:

The surface of the Earth is our environment. This environment is different, not only to that of the past, but also from that of the future ... Just as an ancient tree retains the record of its life and growth, the Earth retains its 'memories' of the past, inscribed both in its depths and on its surface, in the rocks and in the landscape, a record which can be read and translated ... Everyone should understand that the slightest damage could lead to irreversible losses for the future. In undertaking any form of development, we should respect the singularity of this heritage. (Martini & Pages 1994)

The second international conference was held in Malvern in 1993, and represented a big step forward. The proceedings (O'Halloran et al. 1994) is a book that has had a international impact. Proceedings from European meetings and conferences over the last 25 years form a large resource of literature summing up European geoconservation experiences (e.g. Marinos et al. 1997; Miidel 1998; Zagorchev & Nakov 1998; Barettino et al. 1999; Gisotti & Zarlenga 1999; Parkes 2004). Added to this is the publication of periodicals of which the UK's 'Earth Heritage' and the ProGEO newsletter (ProGEO 2007) are important examples.

New developments: geodiversity and geoparks

Over recent years new ideas and strategies for geoconservation have evolved in large part due to, and supported by, international cooperation. One significant trend is the attempt to introduce a nomenclature parallel to biological conservation. One example is the use of the term 'geotope' (Wiedenbein 1994b; Stürm 1994). Diversity has long been used as an assessment criterion for value and site selection in geoconservation, but the rise and political importance of the term biodiversity increased the relevance of an abiotic parallel to this term. The term 'geodiversity' has been used in Australia since the early 1990s. It was introduced in Europe in a Nordic Project intended for nature management (Johansson 2000), and for a wider international public by Gray (2004). The term has a great impact as the abiotic input to a more holistic view on natural and landscape values and supports directly the Rio convention's understanding of ecosystem diversity. It has been accepted as a common term in the UK and is integrated in practical nature management through initiatives such as production of Geodiversity Action Plans (Burek & Potter 2004; Burek 2008; McMillan 2008) and it now occurs as a suggested major term in new legislation in Norway. It has also found its way into EU documents concerning soil strategies and a possible new soil directive (European Commission 2007). It pushes the geoconservation work forward and facilitates a constructive integration of geoconservation into nature management and planning on all levels.

The other major new development is the rise of the Geopark movement (Jones 2008), in Europe represented by the European Geopark Network linked to the UNESCO Global Geoparks system. This initiative is a powerful combination of geoconservation, dissemination of conservation philosophy and geological knowledge as well as a contribution to local economic development. It shows great potential in raising awareness for geological heritage and is therefore one of the main new developments in geoconservation today.

World Heritage and the Geosite project

Compared with other parts of nature conservation (as for example migrating birds, wetlands etc) there has been a lack of international site status to facilitate a strong international support for geoconservation. The European Geoparks Network can be understood as such a system, but the selection is as much about tourism, education and management as it is about the more traditional criteria of geoconservation value. The World Heritage List (Boylan 2008) includes natural areas, of which some of the first were landscapes with a clear geological significance, e.g. the Galápagos Islands designated in 1977 and the Grand Canyon in 1978 (see http://whc.unesco.org). The criteria for designation include specific geological criteria and there are some very important geological World Heritage Sites in Europe such as the Giant's Causeway (Northern Ireland), the Messel Pit fossil site (Germany), the High Coast Kvarken (Sweden and Finland), the Dorset and East Devon ('Jurassic') Coast and the West Norwegian Fjords (Fig. 4). When the number of geological World Heritage Sites increases, groups of sites will start to represent a wider history than each site by its own. A good example of this are the Nordic sites that are all linked to the Quaternary geological heritage of the Nordic countries.

However, the World Heritage List and the Geoparks are exclusive labels and will never fulfil the need for a more systematic international network of geosites. There is still a need for a system

Fig. 4. The west Norwegian fjords (two defined parts of Sognefjorden and Nordfjord) are now inscribed in The World Heritage list as a geological site, together with the World Heritage Site, 'the high coast' of Sweden and Kvarken in Finland underpins the Nordic contribution to our Quaternary geological heritage.

below the World Heritage Site listing. This was recognized early and a project linked to the World Heritage system, GILGES (The Global Indicative List of Geological Sites), was launched (Cowie & Wimbledon 1994; Wimbledon 1996). Its aim was to make lists of World Heritage candidates of geological origin. GILGES was later transformed into the GEOSITE project which was started as an IUGS initiative. The aim of GEOSITE is to make international listings of sites with high international values based and selected from national inventories and lists through scientific comparisons between the candidates. This is a very ambitious aim and it will take a long time to achieve, but in Europe several countries have identified their inventory framework (national scientific priorities) (e.g. Wimbledon *et al.* 1998; García-Cortés *et al.* 2001) and made up lists according to this (Satkunas *et al.* 1999). At present the project lacks funding, as IUGS has withdrawn its support, but the work continues in Europe as a ProGEO initiative.

The idea of international designation of sites has been taken forward in different ways (e.g. Alexandrowics & Wimbledon 1994) and the question of a possible international convention on geoconservation was one of the main issues discussed at the Malvern conference. The idea was

included in the declaration from the conference (O'Halloran *et al.* 1994). A special task force to work with this idea was also organized (Moat *et al.* 1994). Even if this work has not produced a concrete result, international co-operation has produced important results for European geoconservation both within existing conventions such as World Heritage, but also through the European Landscape Convention, and political statements such as the recommendation from the Committee of Ministers in the Council of Europe (see https://wcd.coe.int/ViewDoc.jsp?id=740629&Lang=en).

Conclusion

The geoconservation history of Europe reflects a continent with great natural diversity and with many countries and regions with different legislation and management regimes. The diversity in legislation, policies and management has resulted in a broad pool of experience and many initiatives of international interest. As for the future, the challenge lies in the task of making synergies between the initiatives that are running. Europe has a long history of conflicts and it is indeed possible to foresee competition and conflict between geoconservation initiatives as well. Co-operation in management and research, between organizations and between disciplines is the tool that may produce synergy. One of the main issues for geoconservation is to be strong enough to become established in EU policies and management systems. Then geoconservation will mature as a part of European nature management on an equal basis with the biological sciences and cultural heritage to the benefit of all.

References

ALEXANDROWICS, Z. & WIMBLEDON, W. A. P. 1994. The concept of world lithosphere reserves. *Memorie Descrittive della Carta Geologica D'Italia*, **54**, 347–352.

ALEXANDROWICS, Z. 1994. L'état et la conception de la protection du patrimoine géologique en Pologne. *Mémoires de la Société géologique de France*, **165**, 149–155.

BARETTINO, D., VALLEJO, M. & GALLEGO, E. (eds) 1999. *Towards the Balanced Managements and Conservation of the Geological Heritage in the New Millenium*. Sociedad Geológica de España, Madrid.

BASSET, M. G., KING, A. H., LARWOOD, J. G., PARKINSON, N. A. & DEISLER, V. K. (eds) 2001. *A Future for Fossils*. National Museum of Wales, Geological Series, 19, Cardiff.

BOYLAN, P. J. 2008. Geological site designation under the 1972 UNESCO World Heritage Convention. *In*: BUREK, C. V. & PROSSER, C. D. (eds) *The History of Geoconservation*. The Geological Society, London, Special Publications, **300**, 279–304.

BUREK, C. V. 2008. The role of the voluntary sector in the evolving geoconservation movement. *In*: BUREK, C. V. & PROSSER, C. D. (eds) *The History of Geoconservation*. The Geological Society, London, Special Publications, **300**, 61–89.

BUREK, C. V. & POTTER, J. A. 2004. *Local Geodiversity Action Plans—sharing good practise workshop*. Peterborough, 3 December 2003, English Nature Research report 601. English Nature, Peterborough.

COWIE, J. W. & WIMBLEDON, W. A. P. 1994. The World Heritage List and its relevance to geology. *In*: O'HALLORAN, D., GREEN, C., HARLEY, M., STANLEY, M. & KNILL, J. (eds) *Geological and Landscape Conservation*. The Geological Society, London, 71–73.

DALY, D., ERIKSTAD, L. & STEVENS, C. 1994. Fundamentals in earth science conservation. *Mémoires de la Société géologique de France*, **165**, 209–212.

DUCKECK, J. 2007. *Showcaves of the world*. World Wide Web Address: http://www.showcaves.com/

ERIKSTAD, L. 1984. Registration and conservation of sites and areas with geological significance in Norway. *Norwegian Journal of Geography*, **38**, 199–204.

EUROPEAN COMMISSION. 2007. *Soil*. World Wide Web Address: http://ec.europa.eu/environment/soil/index.htm/

EUROPEAN GEOPARKS NETWORK. 2007. World Wide Web Address: http://www.europeangeoparks.org/

GARCÍA-CORTÉS, A., RÁBANO, I., LOCUTURA, J. *ET AL.* 2001. First Spanish contribution to the Geosites Project: list of geological framework established by consensus. *Episodes*, **24**, 79–92.

GISOTTI, G. & ZARLENGA, F. (eds) 1999. The second international symposium on the conservation of our geological heritage. *Memorie Descrittive della Carta Geologica D'Italia*, **54**.

GRAY, M. 2004. *Geodiversity—Valuing and Conserving Abiotic Nature*. John Wiley & Sons, Chichester.

GRUBE, A. 1994. The national park system in Germany. *In*: O'HALLORAN, D., GREEN, C., HARLEY, M., STANLEY, M. & KNILL, J. (eds) *Geological and Landscape Conservation*. The Geological Society, London, 175–180.

GRUBE, A. 1999. Geoconservation in Germany–1996. *Memorie Descrittive della Carta Geologica D'Italia*, **54**, 265–272.

GONGGRIJP, G. P. 1994. Earth science conservation in The Netherlands. *Mémoires de la Société géologique de France*, **165**, 139–147.

HOLLAND, C. H. 1994. Geological conservation: notion, necessity and nicety. *In*: O'HALLORAN, D., GREEN, C., HARLEY, M., STANLEY, M. & KNILL, J. (eds) *Geological and Landscape Conservation*. The Geological Society, London, 503–506.

INTERNATIONAL COMMISSION ON STRATIGRAPHY. 2007. World Wide Web Address: wwww.stratigraphy.org/

JACOBS, P. 1994. The very early years of earth science conservation in Belgium. *Mémoires de la Société géologique de France*, **165**, 83–87.

JACOBS, P. 1999. Provocative thoughts on earth heritage conservation or 'does protection of life form an obstacle to overcome for EHC?'. *Memorie Descrittive della Carta Geologica D'Italia*, **54**, 353–360.

JOHANSSON, C. E. (ed.) 2000. *Geodiversitet i nordisk naturvård*. Nordisk Ministerråd, Nord 2000:8, Copenhagen.

JOHANSSON, C. E. & ZARLENGA, F. 1999. Protection of geosites in Europe. State and trends. *Memorie Descrittive della Carta Geologica D'Italia*, **54**, 13–21. (English short version: http://www.norden.org/miljoe/nfk/sk/GM_ENG.pdf)

JONES, C. 2008. History of Geoparks. *In*: BUREK, C. V. & PROSSER, C. D. (eds) *The History of Geoconservation*. The Geological Society, London, Special Publications, **300**, 273–277.

KANANOJA, T. 1999. Conservation of geosites in Finland—a case study from Helsinki. *Memorie Descrittive della Carta Geologica D'Italia*, **54**, 285–288.

KRIZ, J. 1994. Conservation of geological sites, fossils and rock environmenta in Czechoslovakia. *Mémoires de la Société géologique de France*, **165**, 101–102.

MARINOS, P. G., KOUKIS, G. C., TSIAMBAOS, G. C. & STOURNARAS, G. C. (eds) 1997. *Engineering Geology and the Environment*. Balkema, Rotterdam, **3**, 2921–3316.

MARTINI, G. & PAGES, J.-S. (eds) 1994. Actes du premier symposium international sur la protection du patrimoine geologique. *Mémoires de la Société géologique de France*, **165**.

MCMILLAN, A. A. 2008. The role of the British Geological Survey in the history of geoconservation. *In*: BUREK, C. V. & PROSSER, C. D. (eds) *The History of Geoconservation.* The Geological Society, London, Special Publications, **300**, 103–112.

MIIDEL, A. (ed.) 1998. *ProGEO 97 in Estonia, Proceedings. The Second General Assembly of the European Association for the Conservation of the Geological Heritage. Scientific Conference. Tallinn–Lahemaa National Park, Estonia, June 2–4, 1997.* Geological Survey of Estonia, Tallin.

MOAT, T., LARWOOD, J. G. & KING, A. H. 1994. A holistic approach to conserving the Earth's natural heritage. *Memorie Descrittive della Carta Geologica D'Italia*, **54**, 303–308.

NORMAN, D. B. 1994. Fossil collecting: international issues, perspectives and a prospectus. *In*: O'HALLORAN, D., GREEN, C., HARLEY, M., STANLEY, M. & KNILL, J. (eds) *Geological and Landscape Conservation*. The Geological Society, London, 63–68.

O'HALLORAN, D., GREEN, C., HARLEY, M., STANLEY, M. & KNILL, J. (eds) 1994. *Geological and Landscape Conservation*. The Geological Society, London.

PARKES, M. (ed.) 2004. *Natural and Cultural Landscapes, The Geological Foundation. Proceedings of a Conference and the ProGEO Third General Assembly, 9–11 September 2002, Dublin Castle, Ireland.* Royal Irish Academy, Dublin.

PINNA, G. 1994. La protection du patrimoine paléontologique en Italie. *Mémoires de la Société géologique de France*, **165**, 135–137.

PROGEO. 2007. World Wide Web Address: www.progeo.se/

PROSSER, C., MURPHY, M. & LARWOOD, J. 2006. *Geological Conservation—A Guide to Good Practice*. English Nature, Peterborough.

PROSSER, C. D. 2008. The history of geoconservation in England: legislative and policy milestones. *In*: BUREK, C. V. & PROSSER, C. D. (eds) *The History of Geoconservation*. The Geological Society, London, Special Publications, **300**, 113–121.

ROBINSON, E. 1989. European geological conservation. *Terra Nova*, **1**, 113–118.

SATKUNAS, J., MIKULENAS, V., LINCIUS, A. & BALTRUNAS, V. 1999. List of the Most Representative Geosites of Lithuania. *Polish Geological Institute Special Papers*, **2**, 97–102.

STEVENS, C. 1994. Defining geological conservation. *In*: O'HALLORAN, D., GREEN, C., HARLEY, M., STANLEY, M. & KNILL, J. (eds) *Geological and Landscape Conservation*. The Geological Society, London, 499–501.

STÜRM, B. 1994. The geotope concept: geological nature conservation by town and country planning. *In*: O'HALLORAN, D., GREEN, C., HARLEY, M., STANLEY, M. & KNILL, J. (eds) *Geological and Landscape Conservation*. The Geological Society, London, 27–31.

THOMAS, B. A. & WARREN, L. M. 2008. Geological consevation in the nineteenth and early twentieth centuries. *In*: BUREK, C. V. & PROSSER, C. D. (eds) *The History of Geoconservation*. The Geological Society, London, Special Publications, **300**, 17–30.

TRIMMEL, H. 1994. Sixty–five years of legislative conservation in Austria: experiences and results. *In*: O'HALLORAN, D., GREEN, C., HARLEY, M., STANLEY, M. & KNILL, J. (eds) *Geological and Landscape Conservation*. The Geological Society, London, 213–214.

WIEDENBEIN, F. W. 1994a. German developments in earth science conservation. *Mémoires de la Société géologique de France*, **165**, 119–127.

WIEDENBEIN, F. W. 1994b. Origin and use of the term "geotope" in German—speaking countries. *In*: O'HALLORAN, D., GREEN, C., HARLEY, M., STANLEY, M. & KNILL, J. (eds) *Geological and Landscape Conservation*. The Geological Society, London, 117–120.

WIMBLEDON, W. A. P. 1996. GEOSITES—a new conservation initiative. *Episodes*, **19**, 87–88.

WIMBLEDON, W. A. P. (ed.) 1998. A first attempt at a Geosite framework for Europe—an IUGS iniative to support recognition of world heritage and European geodiversity. *Geologica Balcanica*, **28**. 5–32.

ZAGORCHEV, I. & NAKOV, R. (eds) 1998. Geological Heritage of Europe (Proceedings). *Geologica Balcanica*, **28**, 3–4. Bulgarian Academy of Sciences.

Geodiversity in the wilderness: a brief history of geoconservation in Tasmania

IAN HOUSHOLD[1] & CHRIS SHARPLES[2]

[1]*Resource Management & Conservation Division, Department of Primary Industries and Water, GPO Box 44, Hobart, Tasmania, 7000, Australia*
(e-mail: Ian.Houshold@dpiw.tas.gov.au)

[2]*School of Geography and Environmental Studies, University of Tasmania, Private Bag 78, Hobart, Tasmania, 7001, Australia*

Abstract: Since the early 1980s, conservation-orientated Earth scientists in Australia's island state of Tasmania have developed an approach to geoconservation that places emphasis on geomorphology, soils and landform processes, in contrast to the stronger emphasis in some places on the scientific values of bedrock geological features. Although bedrock geoheritage has not been ignored, this geomorphological emphasis emerged from Tasmania's recent political history, during which the conservation of large areas of wilderness has dominated local political debate from the early 1970s to the 1990s. With the recognition of undisturbed natural landscapes and ecosystems (wilderness) as having conservation value, it was only a short step to valuing natural landforms, soils and ongoing geomorphological processes as the key abiotic elements of that broader focus. With popular and political acceptance during the 1980s and 1990s of the conservation of wilderness values as a legitimate government policy, Earth scientists within Tasmanian state government land management agencies had a mandate to develop and implement geoconservation policies. The optimum strategy for the small community of geoconservation workers in Tasmania has been to focus on developing theoretical, legislative and management tools for geoconservation in public land management agencies. Tasmanian workers found existing theoretical frameworks for geoconservation inappropriate for their needs, and adopted additional concepts to identify, justify and implement geoconservation. The concept of geodiversity has proven to be a powerful framework for developing classification systems which in turn allow thematic, georegional analyses to provide a systematic, objective and scientifically defensible context for identifying well-expressed representative examples of the various elements of geodiversity. This approach has resulted in the adoption of a terminology distinct from that previously used on mainland Australia, which is, however, convergent with terminology now used in Europe.

Tasmania, the island state of Australia, is an archipelago of 334 islands (one considerably larger than the rest), found between 30 km and 1700 km SE of mainland Australia (Fig. 1). With a population of around 500 000, it contains spectacular and varied temperate and sub-Antarctic landscapes. An enormous diversity of landforms, soils and bedrock geology is packed into a land area a little less than the size of Scotland. Rocks representing all major geological periods from Holocene to Precambrian crop out across the main island, and differential erosion of resistant quartzite strike ridges, lutites and carbonates in the west has produced spectacular mountain chains. Multiple Pleistocene glacial advances have carved these ranges and major valleys on at least five occasions, and extensive limestone and dolomite cave systems honeycomb hill-flanks. Hard-rock coasts face the Roaring 40s on the south and west coasts, and rounded headlands divide gracefully curved white-sand beaches on sheltered eastern coasts and Bass Strait islands. Raised beaches and exposed marine surfaces up to 400 m above sea level reflect intraplate neotectonic activity. Mean annual rainfall, varying from 4 m in the west to 400 mm in the centre of the island, drives a wide variety of river systems, and relict inland dunefields reflect arid periods during Pleistocene glacial phases. Many soil systems remain in natural condition, including over 550 000 hectares of the western Tasmanian blanket bog peats (Sharples 2003, p. 110).

Much of western Tasmania has been included on the UNESCO World Heritage list, for both natural and cultural values (Fig. 2). The significance of its rocks, landforms and soils has been recognized under UNESCO criteria for universal value. The Western Tasmanian Wilderness World Heritage Area (WHA) is currently managed as a series of National Parks and reserves totalling 1.38 million ha (20% of the state). Macquarie Island is Tasmania's southernmost island, a tiny, sub-Antarctic speck in the Southern Ocean 1500 km SE of

From: BUREK, C. V. & PROSSER, C. D. (eds) *The History of Geoconservation.*
Geological Society, London, Special Publications, **300**, 257–272.
DOI: 10.1144/SP300.20   0305-8719/08/$15.00 © The Geological Society of London 2008.

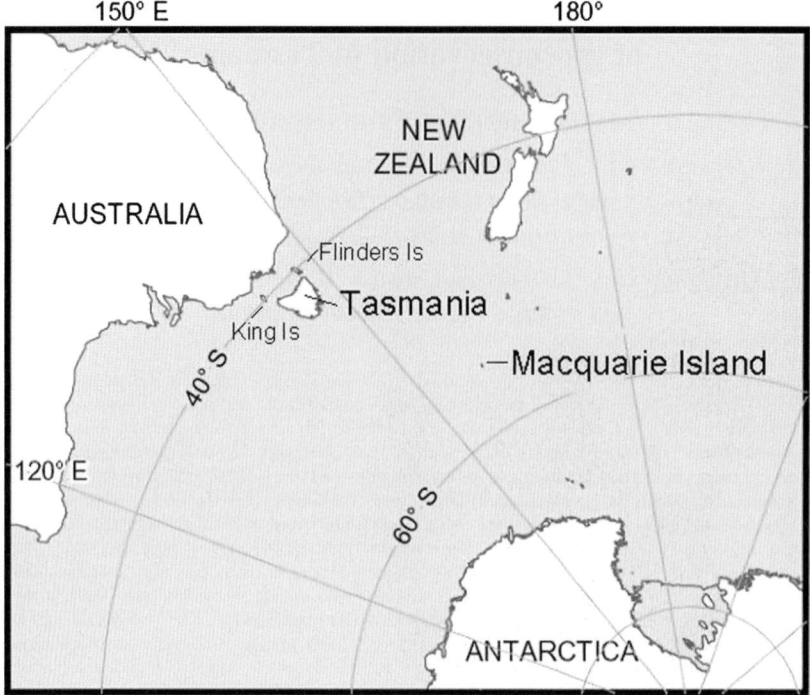

Fig. 1. Map of Tasmania including offshore islands and the Macquarie Island World Heritage Area.

Hobart. Its significance as the world's best exposed terrestrial ophiolite sequence has also been recognized to be of World Heritage significance. The island is one of only a few World Heritage properties listed primarily for its geoheritage values. Little wonder Tasmania has spawned (or attracted) a group of Earth scientists dedicated to conserving its geodiversity!

So how has this interest in geoconservation developed? Tasmania's history as a British colonial outpost, along with its inhospitable but spectacular western and sub-Antarctic landscapes, have combined to inspire todays Earth scientists to both value and protect what remains as a rare gem of natural geodiversity in cool-temperate latitudes.

Colonies and conservation: early perceptions of the landscape

Following at least 40 000 years of exclusive Aboriginal ownership and management, Tasmania was formally settled by British colonists in 1803. Prior to this, much of central/eastern and northern Tasmania was managed, predominantly through the use of fire, by Aborigines. Throughout the Holocene, the intensity of Aboriginal management in

western and SW Tasmania decreased, as rainforest and thick mixed eucalypt forest gradually invaded Pleistocene grasslands. Extensive buttongrass plains and associated blanket bog peatlands of Western Tasmania are at least partly the result of a finely balanced fire-management regime, used to maintain open ground for hunting and communication. Today's SW forest areas retain little surface evidence of Aboriginal use; however, spectacular Pleistocene art in some SW forest cave systems graphically reminds fortunate visitors of the widespread use of these areas during the Pleistocene.

The nuclei of British settlements in Tasmania (then known as Van Diemen's Land) were developed with convicts and free settlers who quickly recognized the value of the land for wool production. Labour shortages meant that increasing numbers of convicts (with increasingly minor offences) became necessary as a free labour source for the wool-growers. The Aboriginal inhabitants were forced ever further into marginal land and within 30 years were re-settled onto remote islands in Bass Strait. The diffusion of European influence was not instantaneous, and the harshness of western Tasmania's terrain and climate retarded the intrusion of European land management systems. Today, some SW Tasmanian

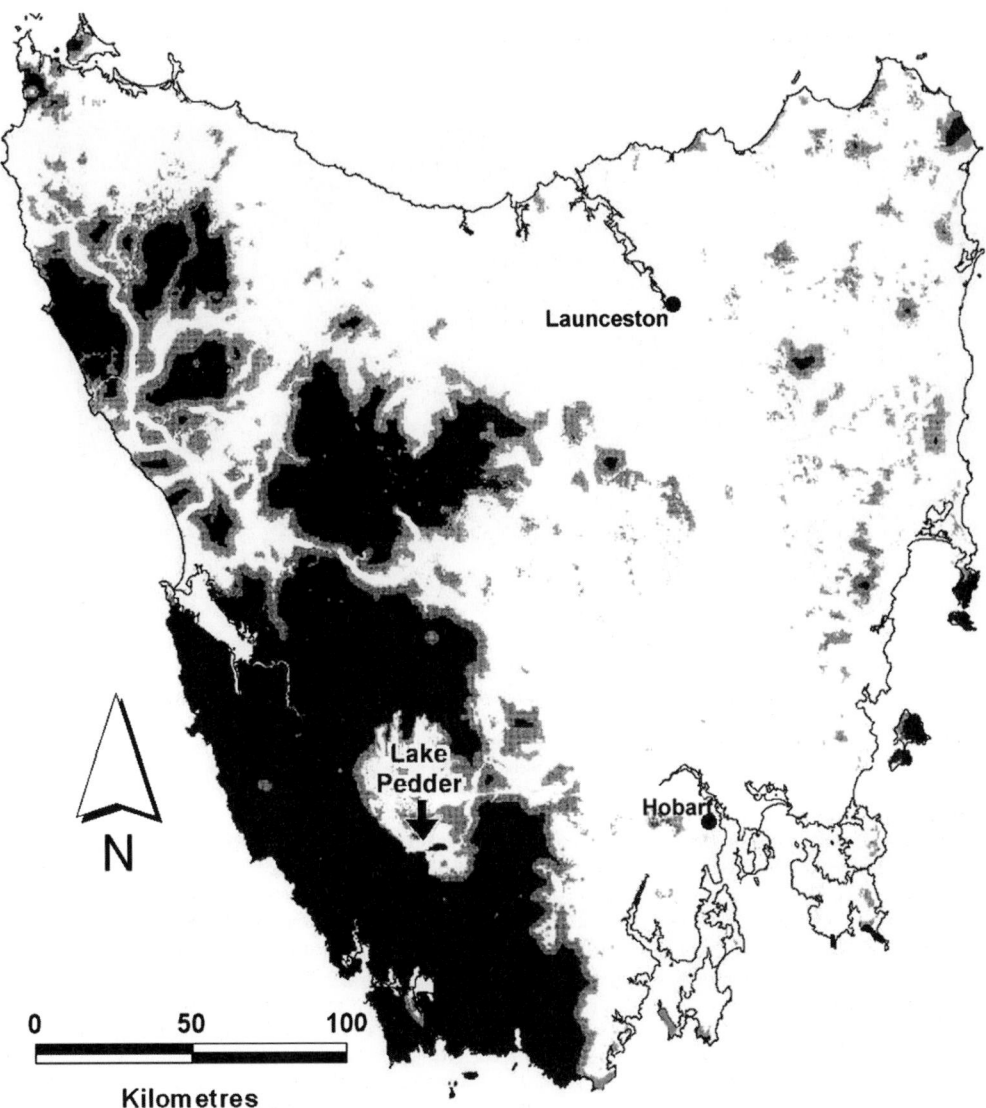

Fig. 2. Map of Tasmania showing extent of high quality wilderness during 1996 (darker tones represent higher quality wilderness). Although many of the lighter areas are agricultural and urban land, the obvious high proportion of wilderness in Tasmania has had a major influence on the state's political history, and was a key driver of geoconservation ideas in Tasmania. This wilderness assessment was prepared using the method of Lesslie *et al.* (1988), and has been adapted from TPLUC (1996). Much of the large area of wilderness in central and SW Tasmania is today encompassed by National Parks and the Tasmanian Wilderness World Heritage Area. The state capital, Hobart, is indicated. Note the large intrusion into the SW wilderness area, which results from roads and the artificial hydro-electric impoundments which inundated the former Lake Pedder in 1972.

valleys and offshore islands remain some of the least intensively managed anywhere in the temperate world. For example, the number of parties that have traversed the length of the New River valley (Fig. 3) in the last 150 years could probably be counted on the fingers of both hands. The entire river basin today is managed as high quality wilderness, with neither formal tracks nor mechanical access permitted. Prior to British settlement, the influence of Aborigines in this heavily forested valley was also likely to have been minimal for many thousands of years.

Fig. 3. The New River basin in SW Tasmania, seen here in a view encompassing most of the catchment basin from its source to the sea, exemplifies the wilderness qualities of large portions of the Tasmanian Wilderness World Heritage Area. This entire river catchment basin of 309 square kilometres contains no roads, human settlements, agriculture or other development, and is entirely mantled by forest and other native vegetation showing no evidence of human disturbance. Apart from potential incipient changes relating to global climate change, fluvial geomorphological processes and landforms throughout this catchment similarly show no evidence of human disturbance, and as such provide a benchmark natural system against which artificially modified fluvial systems can be compared. This catchment also contains extensive limestone and dolomite karst geomorphological systems which have received only just enough exploration to confirm their extensive development (Eberhard *et al.* 1991), and which also remain in as close to pristine a condition as can be envisaged (Photo: Grant Dixon).

In common with other colonized 'new world' nations, and particularly those whose original inhabitants lived lightly on the land, Tasmania retains a significant proportion of its land area in a condition similar to pre-European settlement. In the Americas, parts of mainland Australia, New Zealand and southern Africa the history of management of these 'primitive' areas is comparable. All of these nations experienced a shift in perception of wild lands, from that of a 'frontier to be tamed' to 'a wilderness to be cherished' in the early to mid-twentieth century (Nash 1990; Hay 2002), and a major international movement to conserve them accelerated in the 1960s, alongside the economic prosperity of the years following World War II.

In many respects the history of nature conservation in Tasmania has broadly paralleled its evolution throughout the 'colonized' nations of the western world. Nature conservation in the late nineteenth and early twentieth centuries was focused on the (very useful) concept of 'landscape'. Early Tasmanian Scenic Reserves such as the romantic and

mountainous Mt Field National Park and Cradle Mountain–Lake St Clair National Park (Fig. 4) were declared with an almost subliminal acceptance of the importance of geomorphology. Similarly, some of Tasmania's earliest Crown reserves were cave reserves, specifically created to protect the spectacular underground scenery of karst. Although the terms 'geoconservation' and 'geodiversity' were still decades away, the Tasmanian Scenery Preservation Board recognized the significance of landforms, although mainly at an aesthetic level (PWS 2005). This parallels similar approaches in America, where Yellowstone and Yosemite National Parks, and also New Zealand's Tongariro were all strongly identified with their geomorphological character. In parallel, the Animals and Birds Protection Board was charged with management of Tasmania's wildlife (PWS 2005), and the state's reserve system was managed at the landscape scale by a combination of the two authorities for many decades in the early to mid-twentieth century. This arrangement continued until 1972 when they were disbanded with the establishment

Fig. 4. Cradle Mountain, in the national parks of the Western Tasmania Wilderness World Heritage Area is a glaciated landscape, which is featured as an iconic Tasmanian mountain in tourism literature. This region was one of the first areas of Western Tasmania to be set aside as national park, in the early twentieth century well before the recognition of wilderness values during the 1960s and 1970s. In essence this area was originally reserved for its scenic landscape values; however these values depended primarily on the glacial geomorphology of the region (Photo: Ian Houshold).

of the Tasmanian National Parks and Wildlife Service.

The rise of biodiversity

In the late 1960s and 1970s, a global acceleration in the biological sciences began to deconstruct this 'instinctive' integrated view of conservation at the landscape scale. At the same time as it was becoming clear that many species and communities were facing extinction through environmental pressures, the biological sciences experienced an explosion in quantitative statistical analysis of ecological systems and evolutionary genetics. Researchers and land managers commenced the combined task of documenting the diversity of global biology, based on long-established taxonomic frameworks, and putting in place mechanisms to conserve and manage that diversity.

Unfortunately, the same cannot be said for the Earth sciences. Whilst classification systems had been developed for some bedrock geological features, and various soil classification frameworks were available, no comprehensive typological frameworks were easily available for classifying and characterizing the full diversity of landforms. Traditionally, and particularly so in Australia, both bedrock geology and soil distributions had been mapped and classified predominantly for the purposes of either the mining or agricultural industries. Although suites of geological type sections were being identified in some Australian states well before the 1970s, marking the first moves towards recognition and protection of the representative diversity of bedrock classes, these sites were primarily listed in academic journals and agency reports, and little effort was put into managing or maintaining them. The concept of conserving natural examples of the diversity of soil types was largely unheard of in Australia.

Conservation of some spectacular landforms continued, but at nowhere near the scale of biological conservation. Cave systems provide such obvious examples of complex, beautiful and fragile geomorphological features that they could not possibly be ignored (although numerous examples of inappropriate limestone quarry development can be cited from this period). But other far less obviously attractive landscape elements, such as river reaches, coastal systems and glacial deposits were never given the same priority for conservation in their own right as species of animals and plants. It was generally thought sufficient simply to recognize the importance of geomorphological features, processes and soils for their role in protection of habitat for biological species.

During the 1970s, the Geological Society of Australia (GSA) and the Federal Government's Australian Heritage Commission (AHC) accepted that certain outstanding bedrock features and landforms warranted recognition and protection, although the theoretical support for this was largely limited to recognizing a need to protect scientifically significant features for educational and research purposes. 'Geological Monuments' were defined as reference sites for the geosciences, but were viewed very much as 'outstanding' features or 'oddities' rather than as an attempt to conserve the diversity of features systematically. The first attempts to compile lists of geological monuments (mainly bedrock features) in Tasmania were made by Jennings *et al.* (1974) and Eastoe (1979) for the GSA and AHC, paralleling related geological heritage work on mainland Australia. However, following these initial inventories, geoconservation progress in Tasmania languished until some new initiatives emerged during the 1980s, outside the GSA framework, which were driven by local political factors emerging from the idea of conserving wilderness.

Wilderness politics

One of the great historical trends of the second half of the twentieth century has been the emergence of the concepts of environmental quality and nature conservation as key political issues. The revolution in awareness of environmental issues that swept the western world during the 1960s and 1970s was in many places dominated by concerns over the impacts of industrial and urban expansion on basic environmental values such as water and air quality. However, whilst Tasmania had its share of industrial pollution problems, the global late twentieth century interest in the environment was most prominently expressed in Tasmania through wilderness conservation issues.

'Wilderness' in Tasmania is defined as environments whose landscapes and biophysical processes exhibit no significant alterations owing to the influence of colonial and modern technological society (PWS 1999, p. 92) and which continue to function in ways substantially unaffected by human activities. With large areas of western Tasmania remaining undeveloped to the present, western Tasmania (along with New Zealand's Fiordland and South America's Patagonia) was seen to be one of only three temperate-climate wilderness regions of comparable size remaining in the southern hemisphere (Kiernan 1978). The notion that such wilderness is valuable and worth conserving is perhaps most famously associated with North American advocates such as Henry David Thoreau, John Muir and Aldo Leopold (Nash 1990; Hay 2002). However, it strikes a chord in places such as Australia where, as in North America, large open spaces remain beyond the reach of urban and agricultural development. For those who value it, wilderness is perceived as worth preserving for reasons ranging from protecting ecosystem functions, to a view that we have ethical obligations to permit natural processes to evolve unhindered in at least some places, to a simple romantic desire to preserve places where human egos are not the dominant influence on the landscape.

Historically, these positive views of wilderness have not been dominant, with wilderness having been seen as synonymous with 'wasteland' or as a frontier needing to be tamed. Indeed, the reaction of the Oxford don and expatriate Tasmanian Peter Conrad to the wilderness of his birthplace was to declare: 'The Tasmanian south-west forbids any sense of romantic kinship ... Down here, nature and human affection are incommensurate' (Conrad 1988, p. 58). However, by the time he wrote those words many Tasmanians found Conrad's reaction anachronistic. Instead, the transition from regarding wilderness as something to be feared and conquered to regarding it as worth preserving and cherishing, can be seen as an almost predictable phase in the transition of colonized lands such as North America and Australia from their pioneering stages to more mature established societies. Today, 1.9 million hectares or about 28% of Tasmania is classified as high quality wilderness using a wilderness assessment methodology developed by Lesslie *et al.* (1988; see also RPDC 2003, p. 2.5; TPLUC 1996) (see Fig. 2).

Although the explosion of interest in environmental issues was almost universal in the western world, it was one particular event (the inundation of Lake Pedder) that triggered this interest in Tasmania. Lake Pedder was a large and, to most eyes, beautiful lake of glacial origin set in what had been the heart of the western Tasmania wilderness. However, in the waning phases of Tasmania's post-war economic boom that wilderness was finally starting to be seen as having economic potential for mining, forestry and hydro-electric development. In 1972, Lake Pedder was flooded to form a larger artificially-dammed hydro-electric storage lake with scant consideration for its aesthetic or other natural values. The announcement of plans for this development triggered widespread recognition of the value of Tasmania's wilderness, leading to the most vigorous protests against a government policy that Tasmania had ever seen. Lake Pedder's flooding became a national *cause célèbre* and resulted in the establishment of the world's first 'green' political party (Pybus & Flanagan 1990; Lines 2006).

Although Lake Pedder was ultimately inundated, and a significant portion of the Tasmanian wilderness was scarred by roads, dams and other structures, this loss triggered the rise of an energetic and highly effective environmental activist and political movement in Tasmania (Kiernan 1985). In the course of several decades of loud public debate over the future use of Tasmania's remaining wilderness areas, the environmental ideals fired by the loss of Lake Pedder led to the halting of another planned hydro-electric development on the Franklin River in western Tasmania, and the setting aside of large swathes of western Tasmania as National Parks (1.38 million hectares, equalling 20% of the state). A key milestone was the listing of the Western Tasmania National Parks in 1982 and (expanded) in 1989 as a UNESCO World Heritage Area for reasons which relied to a significant extent on the highly natural (i.e. wilderness) biophysical qualities of the region (DASETT 1989; Sharples 2003). Although not at that time referred to using terms such as 'geoconservation' or 'geodiversity', natural landforms, soils and geomorphological processes were highlighted by the nomination as key values underpinning the integrity of the wilderness, and an appreciation of the value of maintaining 'natural rates and magnitudes of change' in landscape and ecosystem processes developed as an essential element of the protection of wilderness (DASETT 1989).

Geomorphology: the driver of geoconservation

Because geoconservation values were key elements of the arguments put forward to protect these wilderness areas, concerned Tasmanians had a legitimate expectation that these areas would be actively protected as essential natural values of the wilderness. It was obvious that abiotic surface processes (geomorphological and soil processes) underpinned biotic processes, and must form a recognized component of any properly comprehensive nature conservation policy. It was also readily seen that this need *not* form the sole justification of geoconservation. In fact, abiotic features and processes could easily be recognized as having conservation value in their own right, irrespective of their additional significance to broader ecosystem processes (e.g. see Fig. 5). Hence, an approach to geoconservation was developed that explicitly valued landforms, soils and geomorphological processes as key elements of the natural environment.

However, this geomorphological and process-oriented perspective did not sit well within the existing national 'Geological Monuments' approach that had previously been developed by the Geological Heritage Standing Committee of the Geological Society of Australia. Widely seen as the key authority on geoconservation in Australia up to at least the 1980s, the GSA is a professional and scientific organization whose purpose is primarily to promote and support the science and profession of geology in Australia. However, with the mining industry being one of the largest employment and funding sources for the geological profession, conservation has, at times, been viewed with suspicion and mistrust by individuals within the GSA, and geoconservation in particular has sometimes been perceived as a direct threat to the mining industry (see for example Sharples 1999). Indeed, the attendance of the convenor of the GSA's Geological Heritage Standing Committee at a 1991 UNESCO meeting in Paris to discuss possible World Heritage listing of Australian geological sites, and the subsequent publication in the GSA newsletter of the 'International Declaration of the Rights of the Memory of the Earth' adopted at the 1991 Digne International Symposium on Geological Heritage (McBriar 1991), both produced indignant letters to the editor (*The Australian Geologist* No. 81 (December 1991) & No. 82 (March 1992)), which left little doubt that at least some within the GSA found these notions objectionable.

Against this background, the only approach to geoconservation that could be recognized formally by the GSA was one that placed strong emphasis on the research and educational value of geological sites, as a resource whose protection could be justified in terms of protecting sites integral to future advancement of geological knowledge through research and training. However, the broader objective of protecting geodiversity as part of our natural heritage irrespective of its specifically scientific values, or because of the role of geodiversity in the

Fig. 5. Speleothems in Kubla Khan cave, Mole Creek, Tasmania. Well-developed karst systems like this occur widely in Tasmania, and were an early focus of geoconservation activity in Tasmania. Although there are, in fact, links between biological activity and karst geomorphological processes, these are not obvious at sites such as this. Hence, consideration of the obvious natural significance and value of spectacular cave displays such as this supported the development of the idea that non-living geodiversity had conservation values in itself, irrespective of any biodiversity values (Photo: Ian Houshold).

maintenance of natural ecological systems, was not necessarily appropriate within the framework of a professional scientific organization such as the GSA. Thus, the Geological Heritage Standing Committee of the GSA formally stated its policy in the following terms:

Significant geological features (SGF) are those features of special scientific or educational value which form the essential basis of geological education, research and reference. These features are considered by the geological community to be worthy of protection and preservation. (Legge & King 1992)

Whilst protecting the research and educational values of geoheritage form an important part of geoconservation work in Tasmania, it was evident to Tasmanian geoconservation workers during the 1980s and 1990s that the GSA's theoretical framework for recognizing geological heritage values was too narrow to encompass properly the broader focus on geomorphological, pedological and natural process values that had emerged from wilderness and ecosystem conservation ideas in Tasmania.

Tasmanian workers were therefore obliged to develop their own philosophies, conceptual frameworks and assessment methods, and in so doing found inspiration and theoretical guidance in both the broader framework of nature conservation ideas that had developed in western cultures during the twentieth century (see Nash 1990; Hay 2002), as well as in some specifically geomorphological conservation initiatives elsewhere in Australia. Some of the latter which had seminal influences on Tasmanian geoconservation ideas were studies of karst, coastal and volcanic landforms that emphasized fruitful methodological approaches such as the classification of types (diversity), and the selection of best representative examples of each, as keys to identifying significant features (e.g. Davey 1984; Spate & Houshold 1989; Rosengren 1984, 1994).

Agency development of geoconservation practice

By the late 1990s, the establishment of extensive wilderness National Parks in western Tasmania, the halting of planned hydro-electric development, mineral exploration and logging in those parks, and the listing of the region as a World Heritage Area on the basis of its wilderness qualities, all meant that recognition and protecting of wilderness conservation values had officially become government policy (despite considerable opposition from both within and outside parliament). This gave state government land management agencies a mandate to develop policies and management

regimes providing for the protection of natural ecosystem values and processes both within National Parks and on other public lands.

Much of the development of geoconservation in Tasmania therefore took place within government land management agencies, where there was a perceived need to be able to justify geoconservation to a wide range of administrators, land users and managers. Key theoretical issues that were seen to need addressing included the meaning and assessment of geoconservation 'significance', the classification and georegional mapping of geodiversity in order to identify key representative exemplars worthy of conserving, the development of systematic geoheritage inventories, and the differing management implications and requirements of bedrock, soil and landform geoheritage. Consideration of such issues stimulated methodological developments that have been documented by, amongst others, Kiernan (1995), Dixon (1996), Eberhard (1997), Houshold et al. (1997), Jerie et al. (2001, 2003) and Sharples (1995, 2002, 2003). These are outlined below.

Geoconservation and the public land reserve system

In the early 1980s, whilst the 'biodiversity juggernaut' proceeded to dominate nature conservation internationally, Tasmanian government agencies began developing formal processes to manage and protect the biodiversity and geodiversity of the Tasmanian Wilderness World Heritage Area (WHA) following commitments made under the *World Heritage Properties Conservation Act 1983*. Sufficient funds were made available by the Australian Government to employ a geomorphologist in the Tasmanian National Parks and Wildlife Service (PWS) to commence resource inventory and management planning work, and a draft management plan was produced in 1986. Following extensive public consultation and revision the final plan was published in 1992 (PWS 1992), containing detailed provisions for geoconservation. A further complete revision was published in 1999 (PWS 1999).

During research to underpin the WHA plan, significant geomorphological management issues were identified, where detailed investigation and planning would be required in order to protect World Heritage values. The first of these involved the recognition that the wakes of large tourist boats (which carried tens of thousands of visitors per year into Macquarie Harbour and the Gordon River) were causing erosion of fragile bank sediments, with rates of retreat in the order of metres per year occurring at times. A geoscientist was employed to monitor, and remediate these effects by working with operators to design vessels with hulls that would eliminate bank erosion (Bradbury *et al.* 1995). The Earth Science Section of PWS was thus formed, and (although now part of a much larger department, with a current staff of seven conservation geoscientists) remains the only state service section in Australia primarily concerned with management of geodiversity.

Meanwhile, on the SE margin of the WHA a large limestone quarry was removing the upper passages of Australia's most extensive and one of its most significant cave systems. The Exit Cave system, with over 20 km of passages, had previously been assessed to be of World Heritage value by the Helsham Commission (Houshold & Davey 1987; Helsham *et al.* 1988; DASETT 1989). Permission had been granted to continue limited quarrying whilst investigation of the effects of the operation on the cave system were undertaken. Another geomorphologist was employed by the PWS to carry out this research, and following cessation of quarrying, to develop and implement a complex monitoring and rehabilitation programme (Houshold & Spate 1990; Kiernan 1991*b*; Houshold 1992).

Other geoconservation management issues have subsequently been identified, resulting in further employment of relevant specialists, and a permanent core of conservation geoscientists has instigated numerous projects to document the State's geodiversity and produce geoconservation plans for important values on both Crown and private land. Reconnaissance geoheritage inventories have been undertaken for many of the State's reserves, and unreserved public land (e.g. Dixon 1991; Bradbury 1993). Management plans have been produced for geoconservation values in many of the state's National Parks and reserves, the most comprehensive of which was the Mole Creek Karst National Park Management Plan (PWS 2004).

It was thought unnecessary to define a separate reserve class for geoheritage sites, as holistic management of both biological and geoconservation values is supported. All key geoheritage sites protected on Crown land have been gazetted under a suite of classes of reserved land based on IUCN protected area categories, ranging from strict Nature Reserves through National Parks, State Reserves, Conservation Areas and Regional Reserves to Game Reserves and Recreation Areas—each class specifically requiring protective management of geoconservation values under the *Nature Conservation Act 2002*. A Reserve Management Code of Practice (PWS 2003) was developed to provide a clear and consistent set of guidelines for management of all of Tasmania's conservation reserve categories. This code contains many prescriptions relevant to geoconservation, with regard to developments within parks. Karst management, rivers and coastal systems are addressed according to the sensitivity of significant features.

Geoconservation and the forest industry

Not all of Tasmania's old growth forest was protected when the Tasmanian Wilderness World Heritage Area was first listed in 1982. Significant stands of tall eucalypt forest and rainforest in the north, west and SE of the state remained under the jurisdiction of the Forestry Commission (later Forestry Tasmania), and open to both clear-felling and selective logging for sawn timber and woodchips. This had implications for geodiversity, as many significant sites were located in areas potentially subject to logging, and many were susceptible to impacts of timber harvesting activities; karst systems, glacial landforms and deposits, fluvial and relict aeolian landforms are all potentially vulnerable to physical impacts if forest cover is removed.

By the mid-1980s the Tasmanian forest industry was coming under increasing pressure from conservation groups to protect and better manage areas of forested land, outside the World Heritage Area. Significant additions to the reserve system resulted from some intense political campaigns. Large areas of forested land in SE and northern Tasmania were added to the WHA as the result of the Helsham Commission of Inquiry (Helsham *et al.* 1988; DASETT 1989). This inquiry was also an important milestone for geoconservation. Thematic investigations of geomorphological systems were undertaken to provide evidence for the inquiry, the most comprehensive being inventories and significance assessments of Pleistocene glacial landform systems (Kiernan 1987) and karst (Houshold & Davey 1987). The importance of geomorphological systems was recognized at this inquiry, including a unanimous acceptance of the World Heritage value of the Exit Cave system.

Post-Helsham, areas of both public and private land outside the formal reserve system have remained available to the timber industry, but public vigilance has helped to ensure that operations have been undertaken under stringent environmental standards. The Forestry Commission employed a geomorphologist in 1984 to produce a resource inventory and guidelines for management of forested karst systems in the Mole Creek area (Kiernan 1984, 1989). With the instigation of the Forest Practices Unit in 1986, the geomorphologist was then employed to develop provisions for management of landforms in the Forest Practices Code and associated documents (such as the Forest Geomorphology and Forest Sinkhole manuals: Kiernan 1990, 2002). Subsequent revisions of the Code have gradually improved conservation and management of landforms in Tasmania's production forests.

In parallel with the PWS Earth Science Section's inventory programme, projects within the Forest Practices Unit produced reconnaissance inventories of the geoheritage of Tasmania's state forests (e.g. Sharples 1997), following the development of a methods to identify significant geological and geomorphological features which identified the conservation of representative examples of landform diversity as a key goal (Kiernan 1984, 1989, 1990, 1991a; Sharples 1993). One of these studies (Sharples 1993, p. 4) was cited by Gray (2004, p. 6) as the first identified English-language publication of the term 'geodiversity'. However, Sharples does not recall having actually coined the term, whose use probably arose out of office conversations at the time with Kevin Kiernan and other co-workers. Moreover, in Europe the same term was published the same year in German by Wiedenbein (1993) and subsequently in English by Wiedenbein (1994). As often happens in science, similar concepts were simultaneously exercising the minds of separate groups of like-minded workers in Tasmania, Europe and perhaps elsewhere. 'Geodiversity' was a word whose time had come.

In tandem with inventory development, early Tasmanian efforts to develop landform classifications for conservation purposes (e.g. Soutberg 1990) led ultimately to the first comprehensive typology of a class of geomorphological features to be developed in Australia. This was a karst landform classification developed by Kiernan (1995) for an *Atlas of Tasmanian Karst*, itself the outcome of a comprehensive program of karst mapping and inventory work that had commenced with early forestry-driven work at Mole Creek in northern Tasmania. Kiernan (1996, 1997) subsequently produced further typologies for coastal and glacial landform systems. In each case, geodiversity was the rationale for developing a typology or classification. Whilst reconnaissance geoheritage inventories had for some time been compiled on the basis of simply identifying 'outstanding' features, it was recognized that more systematic or thematic inventories would require systematic classifications to enable the identification of representative elements of geodiversity. Thus Kiernan's work (1996, 1997) embodied the idea that by identifying the diversity of karst, glacial and coastal landforms through a suitable classification system, it would be possible to identify well-expressed examples of each element of those typologies as being representative or outstanding elements of geodiversity worthy of conservation management.

The Regional Forest Agreement (RFA), 1997

Ongoing public pressure for protection of natural areas outside the existing World Heritage Area led both State and Federal governments to begin the joint development of a formal agreement which would protect a proportion of remaining wilderness, old growth forests and biodiversity values, whilst setting aside a secure resource for the forest industries. As is fairly normal in Tasmania, neither conservationists nor the timber industry were completely satisfied with the result. Nonetheless, the Tasmanian Regional Forest Agreement (TRFA) led in 1997 to the proclamation of further reserves on public land, along with new initiatives in private forest conservation, which provided funds to purchase and covenant land in order to protect significant forest areas on private tenures.

Although geoconservation values were not a primary consideration in the TRFA, they were regarded as supplementary values, and significant advances were made in the protection of privately owned geoconservation values, particularly in karst areas. The TRFA process also funded a comprehensive audit of all previous geoconservation inventory work, as supporting information to the main biological criteria. This audit produced the first integrated spatial (GIS) database of the state's geodiversity—the Tasmanian Geoconservation Database (TGD; Dixon & Duhig 1996). The TGD remains the state's central repository of information relating to geodiversity, and is managed by an independent reference group composed of specialists in geology, geomorphology and soils.

Additionally, the TRFA provided the impetus to develop defensible means of identifying 'significant' geoheritage features through contextual assessment. Although 'significance' is a subjective value, when making decisions on conservation or development planners and administrators inevitably require some defensible and objective basis on which to weigh up conservation values as against development proposals. As described in Sharples (2002, 2003), the definition of a 'level of significance' for individual features or systems is a commonly used method of doing this. Traditionally, significance level has been measured against a poorly defined spatial reference—a feature may be defined as significant at the local, regional or national level. Such judgements have often been made subjectively, albeit using expert knowledge, however a more objective measure of significance was sought by Tasmanian workers.

To fill this need a system of 'geo-regionalization' was developed, initially with TRFA funding, in order to set a contextual area within which a feature's significance may be judged on further (at least partly objective) criteria

(Houshold *et al.* 1997; see also Gray 2004, pp. 281–283). In essence, contextual regions (or 'georegions') are defined as regions in which similar system controls (climate, lithology, geomorphological history, etc.) have produced similar assemblages of landforms. The significance of any given element of geodiversity (as defined by a classification or typology) can then be more objectively assessed by comparing the quality of available examples (essentially, their condition or degree of artificial degradation, and the excellence with which they exemplify the characteristics of their type) within each objectively-defined contextual region (georegion) containing that class of feature. This method provides a tool to search systematically for the best available representative examples of each landform type, to be recognized as having priority for geoconservation management.

Further development of this idea has been completed subsequently, with the mapping of fluvial georegions for Tasmania (Jerie *et al.* 2001, 2003). Rather than using simple GIS overlays of each system control (as originally proposed by Houshold *et al.* 1997; see also Gray 2004, p. 283), a more powerful spatial synthesis, based on principal components analysis, was used to define areas of similarity in the distribution of the environmental system controls on fluvial geomorphology. A set of 90 fluvial landscape mosaics developed from 14 topographic, climatic, litho-structural and palaeo-environmental inputs now defines the variation in fluvial geomorphological regions across Tasmania. These are being used to identify representative reference stream reaches for listing on the TGD, and to underpin an analysis of the condition of the physical form of Tasmanian streams—the Tasmanian River Condition Index (Houshold *et al.* 2007). It is intended that the same approach should be used for other geodiversity themes such as karst or glacial landforms.

The RFA audit also made it clear that a strategy for geoconservation was necessary across all tenures and agencies. Dixon (1996) produced a comprehensive geoconservation strategy, based on an international literature review, combined with local experience. Many of Dixon's recommendations have now been implemented, and a review of the strategy is currently underway. Key elements of the new strategy include information management (geodiversity classifications, georegionalizations, thematic inventories, GIS databases, web-based public data access), policy and legislation development, implementation of monitoring and management response systems, and programmes for improved resourcing of work. Work on many of these strategy elements is well progressed; however, other strategy elements such as geotourism development and facilitation of improved communication and partnerships with community and industry groups are still in early stages of development.

The TRFA was formalized under an Act of Parliament in 1998 (the *Regional Forest Agreement (Land Classification) Act 1998*). This was the first direct recognition of geoconservation under Tasmanian legislation, although the parliamentary lawyers insisted that the term 'geological diversity' had to be used in the legislation rather than 'geodiversity', since the latter word does not yet appear in widely-recognized English dictionaries. The Act specifically included provisions for the recognition and management of 'geological diversity' across all reserve classes in Tasmania. These provisions were eventually incorporated in the Tasmanian *Nature Conservation Act (2002)* and the *National Parks and Reserves Management Act (2002)*. Other current Tasmanian legislation pertaining to geoconservation in Tasmania includes the *Forest Practices Act (1985)* and the associated Forest Practices Code, the *Mineral Resources Development Act (1995)*, and the associated Mineral Exploration Code of Practice.

Geoconservation on private land

Geoconservation on private land has been relatively slow to develop in Tasmania. Private tenure in Australia is generally less subject to regulation than in more closely settled countries, and private property rights are more jealously held—a hangover from the frontier years. Developments and land uses are generally overseen through local government planning schemes, and many of these have yet to incorporate geoconservation objectives. Tasmanian programmes such as the Private Forests Conservation Program, and the Forest Conservation Fund (both initiated under the TRFA) provide cash and other incentives for private landowners to conserve natural values through conservation covenants and management agreements, but it is unclear how outcomes under these programmes are to be guaranteed and managed into the future.

A current initiative is the Mole Creek Karst Forest Conservation Program, where $3.5 million has been provided by the Australian Government to support landowners to reserve, covenant and better manage land which protects cave and karst systems on their properties. This is Tasmania's first integrated programme to promote geoconservation on private land, and its results will be worth monitoring.

Teaching and research in geoconservation

The School of Geography and Environmental Studies at the University of Tasmania has a long history of analysis and involvement in planning for wilderness and natural area management, at both practical and philosophical levels. Members of the school have been key players in advancing a comprehensive and representative reserve system for the state. For the past four years the school has also employed a geomorphologist to co-ordinate a programme in geoconservation, incorporating second and third year undergraduate courses, an Honours programme, and post-graduate degrees. Links between agency and university staff are maintained through the Tasmanian Geoconservation Database Reference Group, and joint research projects. Graduates are also employed by the agencies to undertake inventory and management planning work.

Both university and agency staff have strong links with national professional organizations. The Australian and New Zealand Geomorphology Group is the peak organization for geomorphologists in these countries, with Tasmanian members strongly represented. The ANZGG conference in western Tasmania (February 2008) included geoconservation as a key theme. The Geological Society of Australia (GSA) continues to represent professionals working in geoscience research and industry, and although still not strongly involved in Tasmanian geoconservation activities, ongoing development of geological heritage strategies is occurring within the national GSA's Geological Heritage Standing Committee. However, the current GSA policy on geological heritage remains the same as in 1992 (see above), being restricted to recognition of scientific and educational values of bedrock geology and landforms, with no reference to soil systems, or to the value of a comprehensive and representative approach to geoconservation.

Terminology

As methodological and legislative developments progressed in Tasmania, an additional priority that became critically evident to local workers was the need for a simple and consistent terminology for geoconservation. By the mid-1990s, a confusing plethora of terms were available to describe aspects of geoconservation, ranging from older terms such as 'Geological Monuments' to newer terms such as 'Significant Geological Features', 'Earth Heritage', 'Geological Heritage', 'Earth Science Conservation', 'Geoconservation', and others. Some of these terms were seen to have restrictive connotations whereas others were simply too clumsy and awkward for regular use. A group of Tasmanian workers including the present writers decided to formalize a standard set

of terms for local use, and embarked on a series of after-work brainstorming sessions at the New Sydney Hotel (roughly half-way between the Forest Practices Unit and Parks & Wildlife Service offices in Hobart). With the assistance of significant quantities of Guinness stout, numerous terms were debated with a view to finding words which best suited our theoretical needs, yet were not unduly long or awkward. The neologisms 'geodiversity', 'geoconservation' and 'geoheritage' were settled on as the best available terms that encompassed all required meanings in a simple concise fashion, and were subsequently formally defined by Eberhard (1997, p. v) and presented in *Earth Heritage* magazine as standard Tasmanian usage by Dixon *et al.* (1997). These terms were not seen as synonyms, rather their intended meanings, in brief, were as follows (see also Gray 2004, p. 6):

- *Geodiversity* refers to the basic *quality* to be conserved;
- *Geoconservation* is the *endeavour* or activity of trying to conserve it; and
- *Geoheritage* is specific *examples* that have been identified as warranting conservation management.

Given the effort that went into isolating these terms as the ones best suiting a broad range of geoconservation purposes, it comes as little surprise to note that the same simple but powerful terms have also emerged as terminological front-runners in Europe (again, partly as a result of brainstorming in a number of forums during the late 1990s–early 2000s; Sue Turner, pers. comm.).

The idea of 'geodiversity' is now well established in Tasmanian geoconservation, as a simple unifying concept which provides a basis for practical methods of identifying key features of geoconservation interest and significance (through classifications, typologies, georegionalizations and thematic inventories). In doing so, it creates a framework within which the significance of geoheritage can be objectively justified to planners, politicians and others.

Some critics have suggested that adoption of the term geodiversity has merely been an attempt to replicate the success of the more widely-known biodiversity concept (e.g. Joyce 1995, Appendix 1, p. A1.6); however, an alternative view is simply that 'diversity' has been recognized as a basic principle underlying all nature conservation, not only bioconservation (Gray 2004, p. 347). The principle was recognized first in bioconservation simply because of the greater attention paid to biological aspects of conservation generally.

Tasmanian experience has been that, as is the case with the biodiversity concept, the geodiversity concept is most fruitful when used as a rationale for classifying the diversity of abiotic things. It is then arguable that a reasonable goal of geoconservation should be to ensure the conservation of adequate representative examples of the comprehensive range of types or classes of abiotic things that exist (as identified by a classification of those things).

Alternative uses of the geodiversity concept have been proposed, for example the idea of measuring the diversity of different things present within specified areas so as to be able to say which areas contain a greater or lesser diversity of things (e.g. Vincent 2004, Vincent *in* Prosser 2002). This usage does not appear to provide a satisfying basis for over-arching geoconservation strategies. For example, if the aim of geoconservation were to maximize the conservation of areas of high diversity, then key phenomena that do not occur in areas of high diversity might be ignored. A geoconservation strategy aimed at protecting areas of high diversity might ignore the repetitive uniformity of the extensive longitudinal dunes of Australia's Simpson Desert, as an area of 'low geodiversity', and as a result no longitudinal dunes might be protected. In contrast, a classification-based approach to using the idea of geodiversity highlights monotonously repetitive longitudinal dunes as one key element in the broader (global) diversity of dunes, and therefore worthy of at least a few good examples of such dunes (and their aeolian process systems) being protected.

Current practice and future directions

Today, geoconservation in Tasmania focuses on developing a 'comprehensive, adequate and representative' approach to the conservation and management of geodiversity. In consequence of our perspective (above) on the appropriate use of the geodiversity concept, and of the Tasmanian focus on geomorphology and landform processes as key elements of geoconservation, much of the theoretical development of geoconservation in Tasmania has been directed to developing geomorphological (landform) classifications to identify the distinctive elements of our geodiversity; georegional mapping to set context areas within which to select significant representative examples of each element of our geodiversity; and to inventories and database development as a means of recording our identified geoheritage for management purposes.

The ideal is that the best (or most outstanding) example of each significant class of natural geological, geomorphological and pedological feature, assemblage of features, processes and systems (i.e. the best exemplar of each element of our

geodiversity) will be identified and managed to conserve its intrinsic, ecological and (non-depleting) instrumental values. Multiple and inter-related representative sites will ideally be recognized, both to maintain the integrity of broader natural systems, and because replication provides a level of insurance against the loss of the best sites.

This process will provide multiple representative examples of each type for the purpose of geoconservation, and also benchmark sites that may be used as references against which the condition of more degraded examples may be compared. These sites also provide models for rehabilitation of degraded landform process systems elsewhere. This process is currently being implemented for Tasmania's rivers, where a project to monitor the physical condition of the state's streams and rivers includes all of these elements. There is no reason why it should not be applied to other aspects of geodiversity.

The same general principles are also recognized to be key drivers of biodiversity conservation, and the present writers consider this to be, not a case of 'copying ideas', but rather a recognition that certain basic principles are common to all nature conservation. All of Tasmania's current programmes to conserve biodiversity are based on comprehensive, adequate and representative reservation and management. Species lists and definition of biological communities parallel typologies of features and assemblages in geoconservation. Selection of key areas for protection of major elements of biodiversity is based on the Interim Biogeographical Regionalisation of Australia (IBRA) which provides a similar level of context setting to georegionalization. When properly integrated, the two should form the basis for a holistic, biophysical approach to nature conservation.

References

BRADBURY, J. 1993. *A Preliminary Geoheritage Inventory of the Eastern Tasmania Terrane*. Report to Parks and Wildlife Service, Tasmania.

BRADBURY, J., CULLEN, P., DIXON, G. & PEMBERTON, M. 1995. Monitoring and Management of Streambank Erosion and Natural Revegetation on the Lower Gordon River, Tasmanian wilderness World Heritage Area, Australia. *Environmental Management*, **19**, 259–272.

CONRAD, P. 1988. *Down Home—Revisiting Tasmania*. Chatto & Windus, London.

DASETT. 1989. *Nomination of the Tasmanian Wilderness by the Government of Australia for Inclusion in the World Heritage List*. Nomination to World Heritage Committee, UNESCO, Paris. Commonwealth Department of the Arts, Sport, the Environment, Tourism & Territories, and the Government of the State of Tasmania.

DAVEY, A. G. 1984. Evaluation Criteria for the Cave and Karst Heritage of Australia. *Helictite. Journal of the Australian Speleological Federation*, **15**, 3–40 (note: issue 15 (2) was dated 1977 but not published until 1984).

DIXON, G. 1991. *Earth Resources of the Tasmanian Wilderness World Heritage Area: A Preliminary Inventory of Geological, Geomorphological and Soil Features.* Occasional Paper No. 25, Department of Parks, Wildlife and Heritage, Tasmania.

DIXON, G. 1996. *Geoconservation: An International Review and Strategy for Tasmania.* Occasional Paper No. 35, Parks & Wildlife Service, Tasmania.

DIXON, G. & DUHIG, N. 1996. *Compilation and Assessment of Some Places of Geoconservation Significance.* Report to the Tasmanian RFA Environment and Heritage Technical Committee, Regional Forest Agreement, Commonwealth of Australia and State of Tasmania.

DIXON, G., HOUSHOLD, I., PEMBERTON, M. & SHARPLES, C. 1997. Geoconservation in Tasmania— Wizards of Oz! *Earth Heritage*, **8**, 14–15.

EASTOE, C. J. 1979. *Geological Monuments in Tasmania.* A Report to the Australian Heritage Commission, by the Geological Society of Australia Inc., Tasmania Division (reprinted 1986).

EBERHARD, R. (ed.) 1997. *Pattern & Process: Towards a Regional Approach to National Estate Assessment of Geodiversity.* 1997 Technical Series No. 2, Environment Australia and Australian Heritage Commission, Canberra.

EBERHARD, R., EBERHARD, S. & WONG, V. 1991. Karst geomorphology and biospeleology at Vanishing Falls, South–West Tasmania. *Helictite*, **30**, 25–32.

GRAY, M. 2004. *Geodiversity: Valuing and Conserving Abiotic Nature.* John Wiley & Sons Ltd, Chichester.

HAY, P. 2002. *Main Currents in Western Environmental Thought.* University of New South Wales Press Ltd, Sydney.

HELSHAM, M. M., HITCHCOCK, P. P. & WALLACE, R. H. 1988. *Report of the Commission of Inquiry into the Lemonthyme and Southern Forests.* Department of the Arts, Sport, the Environment, Tourism and Territories, Australian Government Publishing Service, Canberra, 2 volumes.

HOUSHOLD, I. 1992. *Geomorphology, Water Quality and Cave Sediments in the Eastern Passage of Exit Cave and its Tributaries.* A Report to the Department of Parks, Wildlife & Heritage, Tasmania.

HOUSHOLD, I. & DAVEY, A. 1987. *Karst Landforms of the Lemonthyme and Southern Forests.* Report to the Commission of Inquiry into the Lemonthyme and Southern Forests. Applied Natural Resource Management, Canberra.

HOUSHOLD, I. & SPATE, A. 1990. *The Ida Bay Karst study—Geomorphology and Hydrology of the Ida Bay Karst Area.* A Report to the World Heritage planning team, Department of Parks, Wildlife and Heritage, Tasmania, 2 volumes.

HOUSHOLD, I., SHARPLES, C., DIXON, G. & DUHIG, N. 1997. Georegionalisation—A more systematic approach for the identification of places of geoconservation significance. *In*: EBERHARD, R. (ed.) *Pattern & Process: Towards a Regional Approach to National Estate Assessment of Geodiversity.* 1997 Technical Series No. 2, Environment Australia and Australian Heritage Commission, Canberra, 65–84.

HOUSHOLD, I., GROVE, J., DYER, F. & RUTHERFURD, I. 2007. *Physical form sub-index—River Condition Index for Tasmania.* Report to Tasmania Department of Primary Industries and Water. Earthtech, Melbourne.

JENNINGS, I. B., CORBETT, K. D., LEAMAN, D. E. & BANKS, M. R. 1974. *Tasmanian Geological Features for Preservation or Protection.* Submission to the Committee of Inquiry into the National estate, by the Geological Society of Australia (Tasmanian Division).

JERIE, K., HOUSHOLD, I. & PETERS, D. 2001. Stream diversity and conservation in Tasmania: Yet another new approach. *In*: RUTHERFORD, I. & BARTLEY, R. (eds) *Proceedings of 3rd Australian Stream Management Conference.* CRC for Catchment Hydrology, 329–335.

JERIE, K., HOUSHOLD, I. & PETERS, D. 2003. *Tasmania's River Geomorphology: Stream Character and Georegional Analysis.* Nature Conservation Report 03/5, Nature Conservation Branch, Department of Primary Industries, Water & Environment, Tasmania, 2 volumes + CD-ROM.

JOYCE, E. B. 1995. *Assessing the Significance of Geological Heritage: A Methodology Study for the Australian Heritage Commission.* Report to the Australian Heritage Commission, by the Standing Committee for Geological Heritage of the Geological Society of Australia Inc., Geological Society of Australia, Melbourne.

KIERNAN, K. 1978. World Heritage—One of a Trio. *In*: GEE, H. & FENTON, J. (eds) *The Southwest Book—A Tasmanian Wilderness*, Australian Conservation Foundation, Melbourne, 271–273.

KIERNAN, K. 1984. *Landuse in Karst Areas: Forestry Operations and the Mole Creek Caves.* Report to the Forestry Commission and National Parks and Wildlife Service, Tasmania.

KIERNAN, K. 1985. I saw my Temple Ransacked. *In*: BROWN, B. (ed.) *Lake Pedder.* The Wilderness Society, Hobart, 18–23.

KIERNAN, K. 1987. Geomorphological Reconnaissance of the Southern Forests area, Tasmania. *In*: CANE, S. (ed.) *An Assessment of the Archaeological and Geomorphological Resources of the Southern Forests Area, Tasmania.* Report by Natural Systems Research, Anutech Pty Ltd, Canberra.

KIERNAN, K. 1989. *Karst, Caves and Management at Mole Creek, Tasmania.* Occasional Paper No. 22, Department of Parks, Wildlife and Heritage, Tasmania.

KIERNAN, K. 1990. *Geomorphology Manual.* Forestry Commission and Forest Practices Unit, Tasmania.

KIERNAN, K. 1991a. Landform Conservation and Protection. *CONCOM Fifth Regional Seminar on National Parks and Wildlife Management, Tasmania*, October 1991. Tasmanian Department of Parks, Wildlife and Heritage, Hobart, 112–129.

KIERNAN, K. 1991b. *The Exit Cave Quarry, Ida Bay Karst System, Tasmanian World Heritage Area—A Hydrogeological Perspective.* A Report to the Department of Parks, Wildlife and Heritage, Tasmania.

KIERNAN, K. 1995. *An Atlas of Tasmanian Karst.* Research Report No. 10, Tasmanian Forest Research Council, Inc., 2 volumes.

KIERNAN, K. 1996. *Conserving Geodiversity and Geoheritage: The Conservation of Glacial Landforms.* Forest Practices Unit, Tasmania.

KIERNAN, K. 1997. *Conserving Geodiversity and Geoheritage: The Conservation of Landforms of Coastal Origin.* Forest Practices Board, Tasmania.

KIERNAN, K. 2002. *Forest Sinkhole Manual.* Forest Practices Board, Tasmania.

LEGGE, P. & KING, R. 1992. Geological Society of Australia Inc. Policy on Geological Heritage in Australia. *The Australian Geologist. Newsletter of the Geological Society of Australia Inc.*, **85**, 18–19.

LESSLIE, R. G., MACKEY, B. G. & SCHULMEISTER, J. 1988. *Wilderness Quality in Tasmania.* Report to the Australian Heritage Commission, Canberra.

LINES, W. J. 2006. *Patriots—Defending Australia's Natural Heritage.* University of Queensland Press.

MCBRIAR, M. 1991. Conference Report: The First International Symposium on the Conservation of our Geological Heritage, Digne, Provence, France, June 11–16, 1991. *The Australian Geologist. Newsletter of the Geological Society of Australia Inc.*, **80**, 14–15.

NASH, R. F. 1990. *The Rights of Nature: A History of Environmental Ethics.* Primavera Press Edition, Leichhardt, Australia.

PROSSER, C. 2002. Terminology: Speaking the same language. *Earth Heritage*, **18**, 24–25.

PWS. 1992. *Tasmanian Wilderness World Heritage Area Management Plan 1992.* Parks and Wildlife Service, Hobart.

PWS. 1999. *Tasmanian Wilderness World Heritage Area Management Plan 1999.* Parks and Wildlife Service, Hobart.

PWS. 2003. *Tasmanian Reserve Management Code of Practice 2003.* Parks and Wildlife Service, Department of Tourism, Parks, Heritage and the Arts, Tasmania.

PWS. 2004. *Mole Creek Karst National Park and Conservation Area Management Plan 2004.* Dept. of Tourism, Parks, Heritage & the Arts, Hobart.

PWS. 2005. *Parks and Wildlife Service Tasmania: The Journey.* Department of Tourism, Parks, Heritage and the Arts, Hobart.

PYBUS, C. & FLANAGAN, R. (eds) 1990. *The Rest of the World is Watching: Tasmania and the Greens.* Pan Books, Australia.

ROSENGREN, N. 1984. *Sites of Geological and Geomorphological Significance in the Westernport Bay Catchment.* Department of Conservation, Forests and Lands, Victoria.

ROSENGREN, N. 1994. The Newer Volcanic Province of Victoria, Australia: the use of an inventory of scientific significance in the management of scoria and tuff quarrying. *In*: O'HALLORAN, D., GREEN, C., HARLEY, M., STANLEY, M. & KNILL, J. (eds) *Geological and Landscape Conservation.* Geological Society, London, 105–110.

RPDC. 2003. *State of the Environment Tasmania 2003.* State of the Environment Unit, Resource Planning and Development Commission, Tasmania.

SHARPLES, C. 1993. *A Methodology for the Identification of Significant Landforms and Geological Sites for Geoconservation Purposes.* Report to the Forestry Commission, Tasmania.

SHARPLES, C. 1995. Geoconservation in Forest Management—Principles and Procedures. *Tasforests, Forestry Tasmania*, **7**, 37–50.

SHARPLES, C. 1997. *A Reconnaissance of Landforms and Geological Sites of Geoconservation Significance in the Western Derwent Forest District.* A Report to Forestry Tasmania, Hobart.

SHARPLES, C. 1999. The Dismal Swamp Polje of northwest Tasmania: A Case Study in Geo-conservation. *In*: *Cave Management in Australasia 13.* Proceedings of the Thirteenth Australasian Conference on Cave and Karst Management, Mt Gambier, South Australia. Australian Cave and Karst Management Association Inc., 52–74.

SHARPLES, C. 2002. *Concepts and Principles of Geoconservation.* Parks & Wildlife Service, Tasmania World Wide Web Address: http://www.geoconservation.info/

SHARPLES, C. 2003. *A Review of the Geoconservation Values of the Tasmanian Wilderness World Heritage Area.* Nature Conservation Report 03/06, Nature Conservation Branch, Department of Primary Industries, Water & Environment, Tasmania.

SPATE, A. P. & HOUSHOLD, I. 1989. The significance of the alpine karst areas of mainland Australia. *In*: GOOD, R. (ed.) *The scientific significance of the Australian Alps.* Proceedings of the First Fenner Conference. Canberra, September 1988. Australian Alps National Parks Liaison Committee.

SOUTBERG, T. L. 1990. *Towards a Descriptive Geomorphic Classification System for Nature Conservation Purposes.* Occasional Paper No. 23, Department of Parks, Wildlife and Heritage, Tasmania.

TPLUC. 1996. *Tasmanian—Commonwealth Regional Forest Agreement, Environment & Heritage Report Volume 1, Background Report Part C.* Tasmanian Public Land Use Commission. Hobart, Tasmania Public Land Use Commission.

VINCENT, P. 2004. What's in a Name? Geotopes or Geodiversity? *Earth Heritage*, **22**, 7.

WIEDENBEIN, F. W. 1993. Ein Geotopschutzkonzept für Deutschland. *In*: QUASTEN, H. (ed.) *Geotopschutz, Probleme der Methodik und der praktischen Umsetzung.* 1. Jahrestagung der AG Geotopschutz, Otzenhausen Saarland, 17. University de Saarlandes, Saarbrucken.

WIEDENBEIN, F. W. 1994. Origin and use of the term 'Geotope' in German-speaking countries. *In*: O'HALLORAN, D., GREEN, C., HARLEY, M., STANLEY, M. & KNILL, J. (eds) *Geological and Landscape Conservation.* Geological Society, London, 117–120.

History of Geoparks

CHERYL JONES

*Abberley and Malvern Hills European Geopark, University of Worcester, Henwick Grove,
Worcester WR26AJ, UK (e-mail: c.jones@worc.ac.uk)*

Abstract: The philosophy behind the Geoparks concept was first introduced at the Digne
Convention in 1991 as a means to protect and promote geological heritage and sustainable
local development through a global network of territories containing geology of outstanding
value. In 1997, in direct response to the 'Declaration of the Rights of the Memory of the Earth',
the Division of Earth Sciences of UNESCO introduced the concept of a UNESCO Geoparks
Programme to support national and international endeavours in Earth heritage conservation. In
2000, representatives from four European territories met together to address regional economic
development through the protection of geological heritage and the promotion of geotourism.
The result of this meeting was the signing of a convention declaring the creation of the European
Geoparks Network (EGN). The next significant step for the EGN was the signing of an official
agreement of collaboration with UNESCO in 2001, placing the Network under the auspices of
the organzsation. In 2004 the 17 existing European Geoparks joined with eight new Chinese
national Geoparks to form a Global Network of National Geoparks under the auspices of
UNESCO. This Global Network of National Geoparks has encouraged other countries such as
Iran and Brazil to develop Geoparks programmes. By 2007, European Geoparks were distributed
across 15 European countries. There are 31 members of the European Geoparks Network, bringing
the total number of Global Geoparks to 52. Progress has not always been easy, however, and
finding funding to develop the initiative and secure the future of individual Geoparks remains a
significant challenge.

International programmes for recognizing sites of
geological or geomorphological interest are less
well developed than for biological conservation.
Biodiversity can be protected via a range of designations such as World Heritage, Ramsar and
Man and Biosphere Reserves (MAB). The World
Heritage Convention is capable of recognizing
geological and geomorphological values either
directly or indirectly. Direct recognition of such
values comes through inscription on the basis of
natural features and geological and physiographical formations of outstanding value. However,
the World Heritage process has been criticized by
some Earth scientists because certain important
geological and geomorphological sites are not suitable for inscription as they do not meet the criteria
of outstanding universal value or 'very best of
global sites' (Gray 2004). For example, there is
no explicit provision within the Convention for
the importance of a particular site in relation to
the history or development of geology as a
science. It is likely that the World Heritage List
will not contain more than 150 sites of primary
geological or geomorphological interest. Therefore, the selective nature of World Heritage
listing cannot be regarded as adequate for recognizing global geodiversity (Eder & Patzak 2004;
Gray 2004; Dingwall *et al.* 2005).

The Geoparks concept was developed to
meet the increasing demand from Earth scientists
and non-government organizations for a global
framework to promote and protect geodiversity of
outstanding value (Eder & Patzak 2004). The intention was for Geoparks to represent a global network
of territories that would be complementary to the
World Heritage List, by providing a means of
recognizing internationally important sites which
are of outstanding value, but do not meet the strict
criteria for the World Heritage List. Unlike other
geological designations, the Geoparks initiative
incorporates a highly innovative approach to the
preservation of Earth heritage, by integrating it
into a strategy for sustainable, regional economic
development, primarily through geotourism.

Since the development of the UNESCO Geoparks initiative in 1999, the Geoparks concept has
developed rapidly, with the establishment of both
the European Geoparks Network and the Chinese
National Geoparks Network in 2000. The development of the Global Network of National Geoparks
in 2004 has encouraged other countries such as
Australia, Brazil, Iran, Malaysia and Vietnam to
develop Geoparks Programmes and some of these
areas have successfully achieved Global Geopark
status. By 2007, European Geoparks were distributed across 15 European countries. At the time of

From: Burek, C. V. & Prosser, C. D. (eds) *The History of Geoconservation.*
Geological Society, London, Special Publications, **300**, 273–277.
DOI: 10.1144/SP300.21   0305-8719/08/$15.00 © The Geological Society of London 2008.

waiting there are 31 members of the European Geoparks Network, making the total number of Global Geoparks 52 (http://www.unesco.org/science/earth/geoparks.shtml). This paper reviews the development of the Geoparks concept and the formation of the European and Global Geoparks Network.

The philosophy behind the Geoparks initiative emerged from the Digne Convention in 1991 (Martini 1994). Since 1991, significant progress has been made by individual countries to protect and conserve geodiversity and this work via national programmes aimed towards the protection and promotion of geological sites or areas. One of the recommendations in response to the signing of the 'Declaration of the Rights of the Memory of the Earth' was the creation of a global network of geological territories seeking the support of UNESCO, by integrating geodiversity into its main stream activities.

UNESCO Geoparks initiative

In direct response to this, the Division of Earth Sciences of UNESCO started to develop the concept of a UNESCO Geoparks Programme in 1997 to support national and international endeavours in Earth heritage conservation.

Recognizing that the preservation and international recognition of geological heritage was not covered by any of the existing UNESCO programmes, the proposal for a new initiative to promote a global network of Geoparks was submitted to the Executive Board in 1999. A feasibility study on the development of a Geoparks programme was commissioned to evaluate the need for a new initiative by UNESCO to promote a global network of Geoparks, as well as to examine how this initiative might relate to other relevant UNESCO programmes such as the International Geological Correlation Programme (IGCP), and the Man and the Biosphere (MAB) programme. The call for the Geoparks programme to be integrated into the IGCP and the MAB was rejected by the Executive Board in 2000 but members of the IGCP Board were invited to act as an Advisory Body. Although budgetary limitations were a significant factor there were also serious concerns relating to the overlapping of 'labels' and the downgrading of the biospshere 'label', which had gained wide recognition. The final decision of the UNESCO Executive Board was not to pursue the development of a programme (Patzak 2001); however, it was recognized that UNESCO's role was crucial in enhancing public awareness of Earth heritage issues. Therefore, the Division of Earth Sciences continued to support the general objective 'Education in Earth Sciences' through the promotion of Earth heritage on an *ad hoc* basis when requested by member states (Eder & Patzak 2004). Despite the lack of integration of geodiversity into its mainstream activities, UNESCO has established close collaborative links with the European Geoparks Network and the National Geoparks Network of the People's Republic of China. This collaboration lead to the successful creation of the Global Network of National Geoparks in 2004.

The European Geoparks Network

The success of the Geoparks concept has been focused largely upon the work of European organizations that have worked in close partnership with UNESCO to create the European Geoparks Network (Eder & Patzak 2004). At the 30th International Congress in Beijing in 1996, discussions between Guy Martini (Réserve Géologique de Haute-Provence, France) and Nikolas Zouros (Lesvos Petrified Forest, Greece) considered ways to protect and promote European geological heritage and sustainable regional economic development. This conversation marked the beginning of the early days of the development of the Geoparks initiative which would address the needs of communities located within areas of important geological heritage (Martini & Zouros 2001).

Representatives of four European territories, who had been working on individual programmes, promoting geological heritage and sustainable development, met together in 2000. All four territories were experiencing very similar economic problems such as slow economic development, high unemployment and out-migration of the younger population. These problems were being addressed through projects focusing upon geotourism. The result of this meeting was the signing of a convention declaring the creation of the European Geoparks Network (Zouros & Martini 2003). Using the guidelines proposed by UNESCO in 1997, the following European Geoparks Charter was developed in 2000:

A European Geopark is a territory which includes a particular geological heritage and sustainable territorial development strategy supported by a European programme to promote development. The sites within a European Geopark must be linked in a network and benefit from protection and management measures.

A European Geopark must defend the values of geological heritage conservation and therefore there must be no destruction or sale of geological objects from a European Geopark.

The four founding members of the European Geoparks Network (Réserve Géologique de Haute-Provence, France; Lesvos Petrified Forest,

Greece; Maestrazgo Cultural Park, Spain; and Vulkaneifel, Germany) invited interested regions and organizations from across Europe to join them in learning more about Geoparks and to apply for membership of the new network.

The first European Geoparks Network meeting was held in Molinos Maestrazgo, Spain, in 2000, with the participation of representatives from more than 20 potential Geoparks (Zouros & Martini 2003). The aim of the first meeting was to

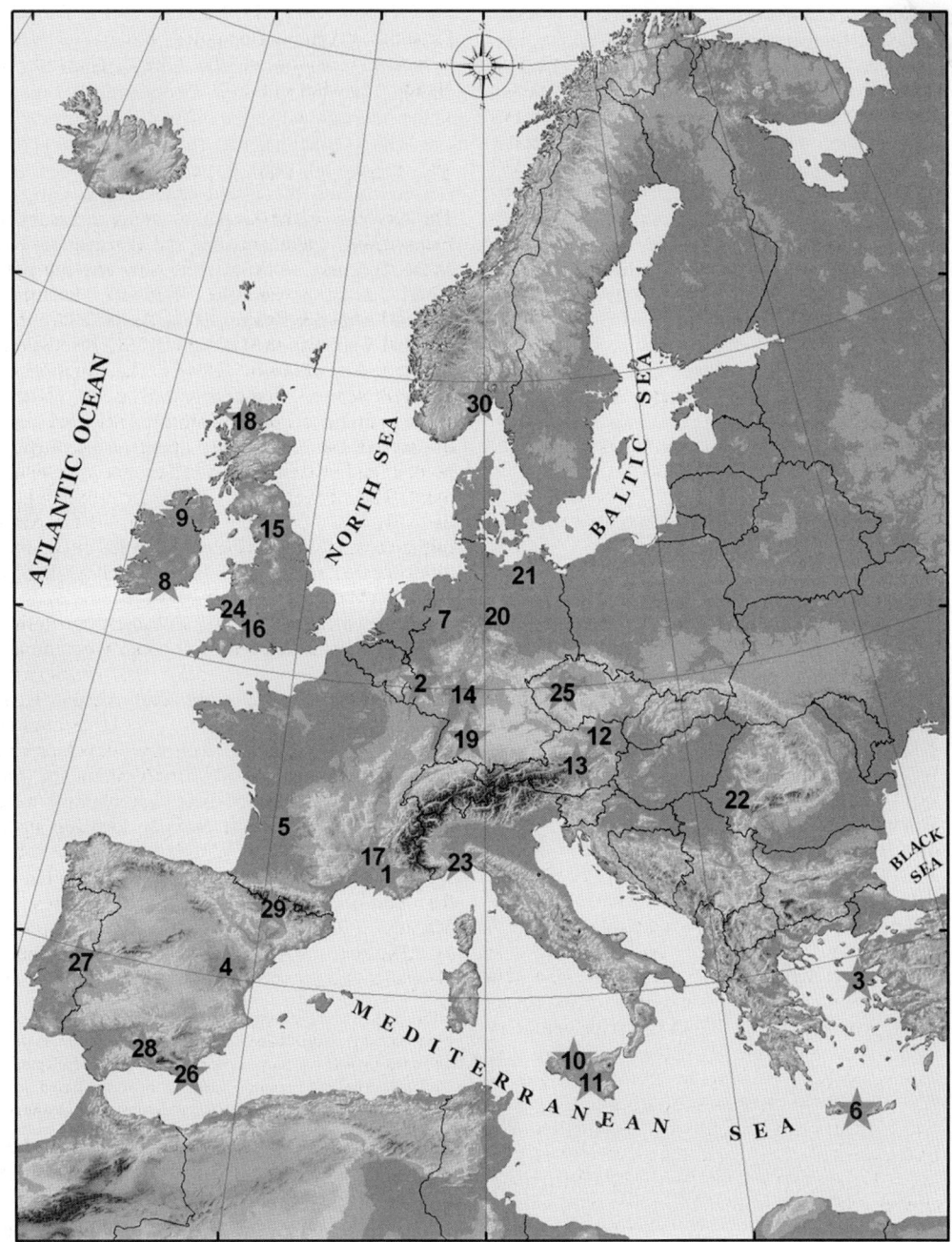

Fig. 1. The distribution of European Geoparks in 2007 (www.europeangeoparks.org).

promote the the European Geoparks Network and to encourage future collaboration with other European partners. By 2001 the European Geoparks Network had increased its membership to 12.

The Geoparks initiative took a major step forward in 2001 with the signing of the Convention of Cooperation between UNESCO and the European Geoparks Network. The convention defined the basis of the partnership between the two organizations and placed the Network under the auspices of the Earth Sciences Division of UNESCO. Since then, UNESCO has played an important role in the development of the European Geoparks Network.

The Network has developed rapidly since its formation in 2000 and membership had reached 30 by April 2007 (Fig. 1). The success of the Network has been due to the dedication of the employees of the individual Geoparks, who actively contribute towards the long term sustainability and the expansion of the Network through joint initiatives, collaborative projects and promotional events, such as European Geoparks Week (Fig. 2). Funding to support these activities has come from European Community Initiatives such as the

LEADER IIC, INTEREG IIIB and INTEREG IIIC programmes (http://www.go-em.gov.uk/european-old/community_initiatives/equal.php).

The Global Network of National Geoparks

In February 2004, representatives from the scientific board of the International Geoscience Programme, the International Geographical Union and the International Union of Geological Sciences along with international experts on geological heritage, conservation and promotion assembled in Paris to discuss the establishment of a Global UNESCO Network of Geoparks and the acceptance of operational guidelines for the designation of Global Geoparks. This meeting saw the merger of the European Geoparks Network and the National Geoparks Network of the People's Republic of China to form the Global UNESCO Network of Geoparks (Zouros 2004). The European Geoparks Network was recognized as the model for the creation of other continental networks of Geoparks (Eder 2004). As a direct result of this

Fig. 2. Trilobite masks. Educational activities during European Geopark Week in the Abberley and Malvern Hills European Geopark. (Photograph by Cheryl Jones.)

meeting the 'First International Conference on Geoparks' was held in Beijing in June 2004. This conference resulted in the acceptance of the Beijing Declaration, which aimed to promote and stimulate the further expansion of the Geoparks initiative across the globe. A World Geopark Office was also opened in Beijing at the same time as the conference (Turner 2006).

In October 2004, a new agreement between the European Geoparks Network and the Division of UNESCO was officially signed. According to the Madonie Declaration, 'A European territory wishing to become a member of the UNESCO Global Network of Geoparks must submit a full application dossier to the European Geoparks Network, which acts as the integration organization into the UNESCO Network for the European continent.' (www.europeangeoparks.org). Further operational guidelines and criteria for the application to the Global Geoparks Network were developed by UNESCO in March 2006 (http://www.unesco.org/science/earth/geoparks.shtml).

Since its formation, membership has now reached a total of 52. Geoparks from Iran and Brazil were accepted as members of the Global Network of National Geoparks in 2006 and the Langkawi Geopark in Malaysia was accepted into the network in 2007. However, it is still early days for the Global Geoparks initiative, and it is envisaged that other countries will develop their own National Geopark Networks and submit applications to UNESCO. For example, Turner (2006) reports that South Australia and Victoria have agreed to support the application of the Kanawinka Geopark for Global Geopark status. If this application is accepted then it will be Australia's first Global Geopark. Furthermore, presentations on the development of new and aspiring national Geoparks were received from India, South Africa, Nigeria, Cameroon and Japan at the Second UNESCO International Conference on Geoparks in Belfast in 2006 (http://82.195.130.20/geo/asp/).

The future of Geoparks

There cannot be any doubt that both the European Geoparks Network and the Global Geoparks Network provide an excellent opportunity for cooperation between experts and practitioners in geological conservation at an international level. Through collaborative projects important geological sites and territories can achieve international recognition and help to redress the imbalance between geological and biological conservation. As the Geoparks initiatives continue to expand there will inevitably be a continual challenge to find the funding needed to secure the long term sustainability of the activities of the networks and its individual members.

References

DINGWALL, P., WEIGHELL, T. & BADMAN, T. 2005. *A Contribution to the Global theme Study of World Heritage Natural Sites*. Protected Area Programme, IUCN.

EDER, F. W. 2004. The Global UNESCO Network of Geoparks. *In*: ZHAO, X., JIANG, J., DONG, S., LI. & ZHAO, T. (eds) *Proceedings of the First International Conference on Geoparks*. 27–29 June 2004, Geological Publishing House, Beijing, 1–3.

EDER, W. & PATZAK, M. 2004. Geoparks—geological attractions: a tool for public education, recreation and sustainable economic development. *Episodes*, **27**, 162–164.

GRAY, M. 2004. *Geodiversity—Valuing and Conserving Abiotic Nature*. Wiley & Sons, Chichester.

MARTINI, G. (ed.) 1994. Actes du Premier Symposium International sur la Protection du Patrimoine Géologique. *Mémoires de la Société Géologique de France*, **165**, 276.

MARTINI, G. & ZOUROS, N. 2001. European Geoparks: Geological Heritage & European Identity— Cooperation for a Common Future. *In*: FREY, M. L. (ed.) *European Geoparks Magazine*. Issue 1, 4.

PATZAK, M. 2001. UNESCO and Geological Heritage. *In*: *Proceedings of the 2nd European Geoparks Network Meeting*. Lesvos, Greece, 3–7 October 2001. Natural History Museum of the Lesvos Petrified Forest, 21–24.

TURNER, S. 2006. Promoting UNESCO Global Geoparks for Sustainable Development in the Australian-Pacific Region. *Alcheringa, Special Issue*, **1**, 351–365.

ZOUROS, N. 2004. The European Geoparks Network. Geological heritage protection and local development. *Episodes*, **27**, 165–171.

ZOUROSA, N. & MARTINI, G. 2003. Introduction to the European Geoparks Network. *In*: ZOUROS, N., MARTINI, G. & FREY, M.-L. (eds) *Proceedings of the 2nd European Geoparks Network Meeting*. Lesvos, Greece, 3–7 October 2001. Natural History Museum of the Lesvos Petrified Forest, Greece, 17–21.

Geological site designation under the 1972 UNESCO World Heritage Convention

PATRICK J. BOYLAN

City University, London, UK (e-mail: p.boylan@city.ac.uk)

Abstract: The World Heritage Convention, 1972, which aims to promote and support the conservation of the world's cultural and natural heritage, can be considered one of the world's most successful international treaties as it has been adopted by 184 states. However, for very many states it is primarily a way in which places that they consider to be their most important heritage sites and monuments can gain additional international recognition through inscription on the World Heritage List. Though geological interest is one of the major criteria for inscription on the World Heritage List, in practice relatively few sites have been inscribed wholly or partly because of their geological or geomorphological importance, (just 72 out of the current 851 World Heritage sites). Further, the present list of geological inscriptions is very uneven and unrepresentative of geological periods and Earth processes, and localities of key significance in the history of geology are almost absent. One weakness is that, although there is some professional geological advice through the World Conservation Union (one of the two official advisory bodies to the World Heritage Committee) there is no mechanism through which international geological science or history of geology organizations can contribute to the development of world heritage policies and the evaluation of nominations to the World Heritage List. However, the greatest problem is that most countries are not evaluating and nominating their national geological heritage. The geological community needs to become much more active in promoting geological conservation and nominations to the World Heritage Committee at the national level.

The World Heritage Convention

On 16 November 1972 the General Conference of UNESCO (the United Nations Educational, Scientific and Cultural Organization) adopted a new international treaty open to adoption by all states, the Convention Concerning the Protection of the World Cultural and Natural Heritage, nowadays universally known as the World Heritage Convention, 1972 (UNESCO 1972).

In terms of the number of States Parties (i.e. countries that have ratified or otherwise formally bound themselves to accept and comply with the provisions of the Convention), this has become much the most widely accepted international cultural or natural heritage treaty in history, with 185 States Parties at 1 December 2007 (compared with 192 member states in membership of the United Nations and of UNESCO itself).

The Convention seeks to ensure the identification and conservation of the heritage of each country at the national level, and also provides a procedure for inscribing the world's most important cultural and natural heritage sites and monuments considered to be of outstanding universal value on a World Heritage List, which now comprises 851 sites and monuments following the most recent decisions (those of the 2007 annual meeting of the World Heritage Committee).

There is very full documentation of all aspects of the progress of the World Heritage Convention from the debates on its adoption during the 1972 General Conference of UNESCO onwards (now available online through UNESCO's World Heritage website http://whc.unesco.org). The details of the earlier history of the emerging plans for such an international treaty are very confused indeed and there are virtually no relevant records in the usually exemplary documentation in the official UNESCO archives in Paris.

Only four years after the adoption of the Convention, Robert L. Meyer's 1976 extensive legal study of the preparatory stages and negotiations had to make heavy use of interviews with, and the private papers of some of the diplomats and experts involved in the negotiations, particularly of several members of the United States delegates to various meetings over a period of about six months in the summer and autumn of 1972, (Meyer 1976). The most authoritative source remains Titchen's 1995 Australian National University Ph.D. thesis on what she termed the *'Construction of Universal Value'* in relation to the Convention, but she too had to rely heavily on secondary sources (Titchen 1995; she also published a journal article on the same theme the following year: Titchen 1996).

Most recently Batisse and Bolla, two of the senior UNESCO Secretariat officers who were

From: BUREK, C. V. & PROSSER, C. D. (eds) *The History of Geoconservation*.
Geological Society, London, Special Publications, **300**, 279–304.
DOI: 10.1144/SP300.22   0305-8719/08/$15.00 © The Geological Society of London 2008.

directly involved in guiding all the negotiations leading up to the final adoption of the present text of the Convention at the 1972 UNESCO General Conference have produced a substantial publication on the emergence of the Convention (Batisse & Bolla 2003), but they confirm what Meyer, Titchen and I (among others) have discovered: that few if any of the preparatory documents and records of negotiations were retained. This very interesting account is therefore based largely on their own recollections 30 years later and on informal interviews with others who had been involved.

Further research on the origins and both intergovernmental and institutional politics of the emergence of the 1972 World Heritage Convention is currently in progress but in summary, despite a long period of preparatory discussions and negotiations, the proposed UNESCO Convention was never intended to cover natural heritage. From 1965 the Culture division of UNESCO had actively supported the establishment of the International Council on Sites and Monuments (ICOMOS) as a global non-governmental organization for specialists in the conservation of cultural sites and monuments, following the recommendations of the 'Venice Charter' on the Conservation and Restoration of Monuments and Sites of 1964.

One of the first major projects of ICOMOS was to develop an international treaty to promote and support the conservation and management of the cultural heritage of architectural and archaeological monuments and sites at both the national and global levels, and it was supported in this by the relevant specialist staff of UNESCO. By July 1971 the ICOMOS-UNESCO draft, covering cultural monuments, buildings and sites, was more or less complete, and was sent to all member states of UNESCO, and the various national responses to this were circulated to states in February 1972. By the spring of 1972, therefore, what was intended to be the final draft text of the proposed World Heritage Convention was submitted by ICOMOS and senior members of the UNESCO Secretariat, particularly Batisse and Bolla, to member states in advance of final consideration by the forthcoming autumn (1972) biennial General Conference of UNESCO member states.

Quite separately, with the support of President Nixon, the United States had been actively promoting the idea of establishing a 'World Heritage Trust' as a free-standing and independent international foundation to provide both private and public support for conservation areas and sites such as national parks. This idea had first gained prominence in 1965 when Russell E. Train, a long-serving President of both the World Wildlife Fund and the USA's premier conservation organization, the Sierra Club, gained the support of a White House Conference on International Cooperation for his plan to establish a non-governmental World Heritage Trust, modelled on the long-established models of the National Trusts of Britain and the USA.

This proposal was taken up by the International Union for the Conservation of Nature (IUCN—now known as the World Conservation Union) in 1968, and it was then incorporated into President Nixon's *Program for a Better Government* submitted to Congress in February 1971, which said 'there are certain areas of such unique world-wide value that they should be treated as part of the heritage of all mankind and accorded special recognition as part of a World Heritage Trust' (Titchen 1995, p. 18). Train's campaign continued with the Sierra Club's 1971 'Action for Wilderness' proposal (Gillette 1972), and the World Heritage Trust proposal became a focal point of the United States' agenda for the United Nations Conference on the Human Environment, held in Stockholm from 5 to 16 June 1972, where it gained very wide support.

The Director-General of UNESCO, Maheu, saw the whole of the Stockholm Conference as a serious threat to UNESCO's position as the designated world intergovernmental body for all the sciences that it had held since the establishment of the United Nations system in 1945–1946. The conference was being organized by the UN directly and not through UNESCO, the UN's official body for the environmental and social sciences, and Maheu's worst fears must have been confirmed when the Stockholm Conference established a new permanent UN agency for the environment totally independent of UNESCO, to continue its work, the United Nations Environment Programme (UNEP).

Though he must have realized that he was losing the wider argument against the Stockholm Conference and the creation of the UNEP, Maheu was determined to block the American proposal for a non-governmental World Heritage Trust, that would be outside UNESCO control. He therefore sent Batisse and Bolla to Stockholm as his personal representatives with a counter-proposal that the draft convention on the world's cultural heritage, the text of which had already been circulated to all UNESCO member states, should be amended to incorporate the natural heritage alongside the cultural heritage (Batisse & Bolla 2003), and hence that both aspects of the world heritage would remain within UNESCO and not be passed to the proposed independent World Heritage Trust. Though there was some support for Maheu's position at Stockholm the battle was far from over, and all then depended on the decisions of the UN member states attending the 1972 UNESCO General Conference less than three months later.

Consequently, as Titchen (1995, p. 40) makes clear: 'it was only in the very final stages of its development that the World Heritage Convention was transformed into an international instrument for the conservation of [both] the cultural and natural heritage'. There were a number of different proposals for amending the already circulated Convention text to bring the natural heritage within its remit, and Meyer (1976) was told by senior members of the United States delegations to Stockholm and to the UNESCO General Conference that as late as mid-September 1972, less than two months before the final adoption of the Convention by the UNESCO General Conference, there were at least five, often very different, drafts circulating within UNESCO itself, while in the same month at the second World Congress on National Parks organized by the IUCN at Yellowstone in September 1972, Train and others continued to press strongly for an independent World Heritage Trust outside UNESCO's control.

In the end the changes to the original text dealing only with the cultural heritage that were finally proposed and adopted during the UNESCO General Conference were minimal and mainly textual, such as adding 'natural heritage' and a definition of natural heritage in appropriate places. There were also amendments adding the IUCN to all references to ICOMOS giving the IUCN a very privileged place within the new world heritage system paralleling directly that already negotiated for itself by ICOMOS in relation to the cultural heritage. However, the United States, among others, continued to press for an independent non-governmental World Heritage Trust rather than a UNESCO treaty though the USA finally acquiesced and indeed went on to be the first country to ratify the World Heritage Convention.

The unusually wide powers that were proposed for the non-governmental ICOMOS in relation to the cultural heritage were confirmed and also granted to the IUCN in relation to the natural heritage. Article 14 of the Convention provides that the Director-General of UNESCO and the World Heritage Committee shall utilise:

... to the fullest extent possible the services of the International Centre for the Study of the Preservation and the Restoration of Cultural Property (the Rome Centre), the International Council of Monuments and Sites (ICOMOS) and the International Union for Conservation of Nature and Natural Resources (IUCN) in their respective areas of competence and capability.

In accordance with this principle, within their respective specializations ICOMOS and IUCN not only advise the World Heritage Committee on all aspects of policy including the *Operational Guidelines* on the preparation and submission by states of their World Heritage List nominations, they also undertake (under contract to the World Heritage Committee) the technical site survey and evaluation of all nominations. Unusually, in relation to international diplomacy and the normal rules of procedure of intergovernmental organizations, under the *Operational Guidelines* countries submitting nominations to the World Heritage List are not permitted to present their own case: instead it is either ICOMOS or the IUCN (as appropriate) that presents and comments on each proposal to the full meeting of the World Heritage Committee.

The Convention defines two categories of heritage in Articles 1 and 2: cultural heritage and natural heritage, and details natural heritage in Article 2 as:

natural features consisting of physical and biological formations or groups of such formations, which are of outstanding universal value from the aesthetic or scientific point of view;

geological and physiographical formations and precisely delineated areas which constitute the habitat of threatened species of animals and plants of outstanding universal value from the point of view of science or conservation;

natural sites or precisely delineated natural areas of outstanding universal value from the point of view of science, conservation or natural beauty.

A third major category of sites, 'mixed sites', recognizes sites which qualify on both cultural heritage and natural heritage grounds.

However, as in most if not all treaty-making negotiations of modern times there was an immediate potential conflict between the international aims and obligations in relation (in this case) to the conservation of the world's heritage, and the over-riding concepts of national sovereignty and non-interference in the internal affairs of a state which are strongly defended by many countries. Becoming a party to any treaty or convention conflicts with the doctrine of near-absolute national sovereignty, so those drafting the World Heritage Convention had to recognize these principles. Consequently, Article 3 provides that:

It is for each State Party to this Convention to identify and delineate the different properties situated on its territory mentioned in Articles 1 and 2 above.

In other words the relevant national authorities of each country have complete control over what should be defined and identified as part of the country's cultural or natural heritage, for the purposes of national conservation law and administrative measures. Similarly, under Article 11 a site or monument can only be considered for inscription on the World Heritage List if it is nominated by the state party. In practice there are very wide variations from country to country over both conservation policy and practice at the national level, and with each country's World Heritage List nominations. Cultural, religious, economic and other national factors seem to be to the fore in some cases: indeed it could be argued that national administrative arrangements, such as which Ministry or Government Agency has the 'lead' responsibility for the World Heritage Convention, can have a very significant effect.

In adopting the convention, countries pledge to take effective measures to conserve and promote the heritage at the national level, as set out in Article 5:

To ensure that effective and active measures are taken for the protection, conservation and presentation of the cultural and natural heritage situated on its territory, each state party to this convention shall endeavour, in so far as possible, and as appropriate for each country:

(a) to adopt a general policy which aims to give the cultural and natural heritage a function in the life of the community and to integrate the protection of that heritage into comprehensive planning programmes;

(b) to set up within its territories, where such services do not exist, one or more services for the protection, conservation and presentation of the cultural and natural heritage with an appropriate staff and possessing the means to discharge their functions;

(c) to develop scientific and technical studies and research and to work out such operating methods as will make the state capable of counteracting the dangers that threaten its cultural or natural heritage;

(d) to take the appropriate legal, scientific, technical, administrative and financial measures necessary for the identification, protection, conservation, presentation and rehabilitation of this heritage; and

(e) to foster the establishment or development of national or regional centres for training in the protection, conservation and presentation of the cultural and natural heritage and to encourage scientific research in this field.

There is currently concern amongst some, myself included, that in practice many countries focus their attention on their national nominations to the World Heritage List, rather than on the country's general obligations towards the whole of the national (and indeed regional) cultural and natural heritage under Article 5. Though Article 29 requires each country to submit two-yearly progress reports on the measures it has taken to comply with the various provisions of the Convention, and in particular Article 5, but compliance with this at best patchy, and many reports are not particularly informative.

The general operation of the World Heritage Convention is monitored and guided by a biennial meeting of all States Parties (held during each General Conference of UNESCO), which elects a World Heritage Committee of 21 states: each for a term of up to six years, one-third retiring at each biennial meeting of States Parties. This is supported by a specialist Secretariat within

UNESCO, the World Heritage Centre, and by three designated advisory bodies: the intergovernmental ICCROM (International Centre for Conservation, based in Rome) and two international conservation organizations, the International Council on Monuments and Sites (ICOMOS) for the cultural heritage, and the International Union for the Conservation of Nature (IUCN—now known as the World Conservation Union) for the natural heritage.

In practice, the main work of the World Heritage Committee is to consider at its annual meeting proposals by states for additional inscriptions on the World Heritage List. The committee also maintains a List of World Heritage 'In Danger', and decides on disbursements from a World Heritage Fund for international assistance of various kinds. This relies mainly on a compulsory contribution by all States Parties levied at the rate of one percent of each country's assessed biennial contribution to UNESCO itself.

Geology and the World Heritage Convention

The definitions of natural heritage within Article 2 of the Convention itself include a specific provision in relation to 'geological and physiographical (i.e. geomorphological) formations', while geological interest is also likely to be a factor within the first defined category of natural heritage, i.e. 'natural features'. However, over the decades these have been expanded and defined more specifically by the World Heritage Committee and the biennial Meeting of States Parties for the purposes of initial nominations by states and the final inscriptions on the World Heritage List within the *Operational Guidelines for the Implementation of the World Heritage Convention*.

Until the end of 2004 the *Operational Guidelines* had two separate lists of World Heritage selection criteria: six for the cultural heritage, and four for the natural heritage. However, the 2005 *Operational Guidelines* (UNESCO 2005) revised these and merged them into a single list, though of course in earlier literature the old criteria designations will still be found. The new natural heritage criteria (with the old equivalents) listed in the current paragraph 77 are as follows:

(vii) contain superlative natural phenomena or areas of exceptional natural beauty and aesthetic importance; *[previously Natural Criteria iii]*

(viii) be outstanding examples representing major stages of Earth's history, including the record of life, significant on-going geological processes in the development of landforms, or significant geomorphic or physiographic features; *[previously Natural Criteria i]*

(ix) be outstanding examples representing significant ongoing ecological and biological processes in the evolution and development of terrestrial, fresh water, coastal and marine ecosystems and communities of plants and animals; [previously Natural Criteria ii] and

(x) contain the most important and significant natural habitats for in-situ conservation of biological diversity, including those containing threatened species of outstanding universal value from the point of view of science or conservation. [previously Natural Criteria iv]

Geological and geomorphological sites are therefore primarily covered by new criterion (viii).

Geological inscriptions on the World Heritage List

The first country to complete its domestic legal procedures and become a party to the Convention was the United States of America in December 1973, followed by Egypt in February 1974 and Iraq in March 1974. The Convention came into force among the original 20 parties on 17 December 1975, three months after the deposit of the notice of ratification by Switzerland had brought the total up to twenty states, the required minimum number specified in the Convention. As each subsequent state completed the necessary national procedures, the Convention came into force in that country three months after the deposit of the required official documentation with the Director-General of UNESCO.

The first twelve sites were inscribed on the World Heritage List during the second statutory meeting of the World Heritage Committee held in Washington D.C. from 5 to 8 September 1978. Following the 32nd session of the World Heritage Committee held in Christchurch, New Zealand, in June–July 2007 (which for the first time removed a site from the World List) there are now a total of 851 inscriptions of all kinds on the World Heritage List across 141 countries, though only 166 (19.5% of the total) are natural heritage inscriptions, compared with 660 cultural sites and 25 mixed sites (UNESCO 2007).

Further, 32 years on from the first World Heritage List inscriptions, there are still only 12 (1.4%) that have been designated solely because of their geological or physiographical importance and the first of these (the Hawaii Volcanoes National Park, USA) was not added to the list until 1987. In practice, geological or wider landscape interest and value is more often cited as an 'added value' in World Heritage List inscriptions and new nominations under other criteria.

Out of the 166 natural heritage inscriptions following the 2007 World Heritage Committee meeting, a further 60 World Heritage inscriptions

cite geological etc. importance alongside other natural heritage criteria. For example, the first World Heritage List adopted in 1978 included four natural sites for which geological significance was cited as one of the factors justifying World Heritage status: the Galapagos Islands (Ecuador), Nahanni National Park (Canada), Simien National Park (Ethiopia) and Yellowstone National Park (USA).

Cowie & Wimbledon (1994) briefly discussed the significance of the World Heritage Convention within the wider context of geological site conservation, and there are a few other studies, but the literature is very limited. In relation to the World Heritage Sites, most of the data is in the official documentation of the World Heritage Committee, which is available on-line at http://whc.unesco.org/.

A full list of the sites inscribed in either the natural sites list or mixed sites list wholly or partly because of their geological or geomorphological interest has been separated out, and is given in the Appendix. In each case a summary of the geological description is given, drawn mainly from the original description in the World Heritage nomination of each approved site prepared by the country presenting the case for inclusion on the World List, and now recorded in the natural site datasheets for each site of the World Conservation Monitoring Centre (WCMC) of the United Nations Environment Programme (UNEP) (WCMC 2006). However, in a number of cases these official geological descriptions are far from adequate in terms of identifying the key geological features, so some additions have had to be made as further clarifications.

The World Heritage Committee, in its Budapest Declaration on World Heritage (UNESCO 2002), argued that one of the four strategic objectives should be to develop a 'Representative, Balanced and Credible World Heritage List' and this objective has been incorporated in the latest Operational Guidelines (UNESCO 2005, paras 54–61). So far the main emphasis in seeking to achieve this ambition seems to be political: to aim at a fairer geographical balance between different countries and regions of the world. (In regional terms the current priorities are seen to be Africa and South America, whereas those countries which already have a substantial number of World Heritage Sites are now restricted to only one new application per year.)

At the Committee level there seems to be little or no serious discussion of other relevant imbalances: not even the widening gap between cultural and natural properties on the World List, let alone the lack of balance between inscriptions reflecting each of the ten different World Heritage criteria, though there have been a few expert studies on representativeness, for example the coverage relating to fossil sites (e.g. Wells 1996), the geological stratigraphical column or to human evolution.

Nevertheless, the situation seems bound to get worse at least in absolute terms. The *Operational Guidelines* call for states to submit 'Tentative List' of potential future national submissions for inclusion on the World List. As of July 2007, 159 states had registered Tentative Lists, which together cover a further 1382 properties over and above the 851 already on the World Heritage List. However, only 300 (21%) of these are natural sites, compared with 905 cultural and 177 mixed sites (12.8%), so the ratio of only one-fifth natural sites is likely to persist far into the future, (UNESCO 2007). Within the natural sites on the Tentative Lists only 97 (7.0%) are included wholly or partly on geological or geomorphological grounds (and five of these are localities which lie close together within the Troodos mountains of central Cyprus and which, if approved, would almost certainly be inscribed as a single World Heritage Site).

This lack of representativeness and balance, and hence ultimately of credibility, is especially marked in relation to the coverage of the world's geological heritage on the World Heritage List. Quite apart from the extremely limited representation of geology overall, even those sites on the World List give a very poor coverage of the range of geological and geomorphological periods, features and interest.

I have carried out a detailed textual analysis by keyword of the terms used in the official descriptions of all geological inscriptions, based on the text of the nomination in the case of those approved and/or the text of the WCMC Datasheet for each World Heritage site (which is usually based on the World Heritage nomination file) and the results are shown in the following four tables.

The coverage of the different standard geological periods is very uneven, as seen in for example the fact that the Quaternary has more than five times the number of sites as the Silurian and more than double those of the Jurassic (Table 1). Further, not a single type locality or region for any of the fifteen traditional geological periods of the Phanerozoic is represented, and out of the 85 named stages recognized by the International Stratigraphical Commission (2004) only one traditional type locality and one designated boundary site are on the World List. These are, respectively, the Kimmeridgian Stage (Upper Jurassic), within the Dorset and Devon Coast World Heritage Site UK (inscribed in 2001), and the designated type locality for the base of the Tremadocian Stage and hence the Cambrian–Ordovician boundary, which is within the Gros Morne National Park, Canada (inscribed in 1987).

Sites of palaeontological importance are poorly represented (Table 2), and one of the few of great

Table 1. *Representation of different geological periods among the existing World Heritage Sites inscribed for their geological value*

Quaternary/Pleistocene: 11 sites
Pliocene: 2 sites
Miocene: 9 sites
Oligocene: 7 sites
Eocene: 7 sites
Paleocene: 2 sites
(Tertiary—undifferentiated in addition): 8 sites
Cretaceous: 12 sites
Jurassic: 4 sites
Permian, Triassic or Permo-Triassic: 11 sites
Carboniferous: 7 sites
Devonian: 4 sites
Silurian: 2 sites
Ordovician: 6 sites
Cambrian: 4 sites
Precambrian: 12 sites

historical as well as continuing contemporary importance, the Cambrian Burgess Shale site in Canada with its unique range of soft-part preservation (inscribed in 1981), largely lost its identity three years later by being subsumed into the huge area of the 'serial' inscription covering most of the Canadian Rocky Mountains National Parks. This has been recognized as the IUCN for some considerable time, and commissioned Wells (1996) to review the approaches to the assessment of World Heritage fossil site nominations.

As can be seen from Table 3, volcanoes and vulcanism (including instrusions and hydrothermal activity) together with sites relating to evidence of plate tectonics etc. represent almost all of the inscriptions. These include the type localities for Vulcanian and Strombolian volcanic eruptions within Aeolian Islands, Italy, World Heritage inscription. The sole meteorite impact site is that of the very large 2 billion year-old ring structure of the Vredefort Dome, South Africa.

Table 4 shows a gross imbalance among the current World Heritage List inscriptions with a very clear bias towards spectacular tourist attractions, particularly glaciers, dramatic glaciated landscapes, river canyons, waterfalls and show caves.

Table 2. *Representation of palaeontological interest among the existing World Heritage Sites inscribed for their geological value*

Fossil mammals: 5 sites
Fossil plants: 4 sites
Reptiles including dinosaurs: 4 sites
Ammonites: 2 sites
Fossil fish: 2 sites
Soft-bodied invertebrates: 1 site

Table 3. *Representation of geological processes etc. among the existing World Heritage Sites inscribed for their geological value*

Plate tectonics, orogeny and subduction: 16 sites
Volcano, volcanic deposits: 12 sites
Intrusion: 11 sites
Hydrothermal and geysers: 5 sites
Rift valley: 2 sites
Meteorite impact: 1 site
Ophiolites: 1 site

Table 4. *Representation of geomorphological processes etc. among the existing World Heritage Sites inscribed for their geological value*

Glaciation: 27 sites
Karst, caves: 15 sites
River erosion, waterfalls, canyons etc.: 13 sites
Glaciers: 9 sites
Coastal processes: 5 sites
Coral reef: 2 sites
Desert processes: 2 sites

More generally, taking into account all four categories, there is no sign, yet, that any re-balancing is on the horizon. For example, two of the three latest (2007) geological additions to the World Heritage List are volcanic landscapes and the third is another major area of karst. All three sites are certainly very important and of world heritage standard, but both these categories are already very well represented in comparison with the many seriously under-represented categories.

Lack of recognition of importance to the history of geology

The position in relation to the history of geology is equally depressing, if not more so. To take a few examples, there are no World Heritage Sites at all associated with key European figures in the early development of the Earth sciences, such as Agricola, Werner, Von Buch, von Humboldt, de Saussure, Hutton, Cuvier, William Smith, Sedgwick, De La Beche, Murchison, Lyell or Quenstedt. Buckland and Conybeare are recognized only in the Dorset and Devon Coast, UK, site added in 2001. Though the Cornwall and Devon Mining District of SW England has been of international importance for mineralogy and mining geology for several centuries the region was inscribed on the World Heritage List in 2006 as a cultural, not geological, site, based on the industrial archaeology of the former mining industry.

Darwin is represented only by the Galapagos Islands, inscribed for their significance in the development of his theory of evolution, but not for his work e.g. the evidence of the uplift of the central Andes such as the 'fossil forest' he discovered high in the Aconcagua range, his confirmation at Queen's Beach, Cape Town, of Hutton's Glen Tilt recognition of the intrusive nature of the granite, nor the key locations for his work on the development of coral reefs and atolls. The IUCN Nomination Report on the Tiede National Park, Spain, inscribed on the World List in 2007, refers specifically to the 'Narices del Teide' (Teide's Nostrils), parasitic cone a little distance below the present summit formed in the 1798 eruption, but makes no reference at all to the great historical importance of this eruption and its effects in the history of science, most notably through the contemporary studies of it by Alexander von Humboldt who hurried to the Canary Islands on hearing the news of the eruption. Similar (indeed longer) lists of highly important sites in the early history of geology could be produced for North America, and for more recent (post *c.* 1850) historic periods.

When cultural heritage nominations are being considered, the historical associations of sites and locations are given very considerable weight within Article 1 of the World Heritage Convention itself, and are highlighted in criterion (vi) under the para. 77 of the 2005 *Operational Guidelines*:

(vi) be directly or tangibly associated with events or living traditions, with ideas, or with beliefs, with artistic and literary works of outstanding universal significance. (The Committee considers that this criterion should preferably be used in conjunction with other criteria.)

For example, it is, because of their extremely important 'intangible' historic associations that the Auschwitz Concentration Camp in Poland and the Robben Island apartheid period prison camp in South Africa have been designated as World Heritage Sites, not because of their intrinsic architectural or similar value.

In marked contrast with this there is no comparable provision for associated historical or other intangible importance within the natural heritage criteria, although there would seem to be no reason why a suitably re-worded version of cultural criteria (vi) could not be added to the natural heritage criteria when the *Operational Guidelines* are next reviewed.

Other weaknesses of the Convention in relation to geological conservation

Although geological heritage interest has been an important part of the World Heritage Convention from the beginning, none of the consultations and discussions up to and including 1972 seem to have involved any geological organizations. The International Union of Geological Sciences (IUGS) was formed at a meeting in the UNESCO headquarters in 1961 (four years before ICOMOS), and has had a formal advisory role with UNESCO since then. Its origins can be traced back to the first International Geological Congress held in Paris in 1878, and by 1972 it had a very important and continuing partnership with UNESCO, for example through the International Geological Correlation Programme (IGCP), now the joint UNESCO–IUGS International Geoscience Programme.

However, there is no evidence of any significant formal consultation with or involvement of either UNESCO's staff Earth scientists or its official partner, the IUGS in any matter relating to the World Heritage Convention either during its drafting and adoption in 1972 or during the subsequent 35 years. The Convention remains located both administratively and, arguably, philosophically within the Culture division of UNESCO, whereas the UNESCO Science division's important environmental and geological conservation programmes, such as Man and the Biosphere and more directly still the Geoparks initiative, appear to be totally divorced from the World Heritage Convention and Centre.

The International Commission of the IUGS on the History of Geological Sciences (INHIGEO) was established in 1967 but the second meeting in Prague in 1968 had to be abandoned following the Russian invasion. However, by 1972 it already had a distinguished world-wide membership of experts who could cover most of the geological heritage, and with relevant interests and expertise would I am certain have been only too willing to advise the emerging world heritage system. For example, at one of its early meetings INHIGEO specifically addressed the influence of place and locality on the early development of geology, (INHIGEO 1978). As with the exclusion of the IUGS overall, it seems particularly strange that INHIGEO, the specialist international body for the study of the geological heritage, was neither involved in any way in the initial development of the World Heritage Convention, nor in any aspect of its subsequent operation since 1972.

The IUCN does have some geologist members of its World Commission on Protected Areas (WCPA) and individual geological site conservation and management experts, including some from this commission, do undertake surveys and reviews of natural heritage nominations with a geological content. No criticism is intended of the individual scientists concerned and their valuable

work, but the IUCN has no specific geological programme or policy within the very much wider role of the WCPA and of the IUCN overall. This is in marked contrast with the provisions in relation to the cultural heritage, where ICOMOS and its more than 6000 members world-wide see the development of world heritage policies (what they term 'doctrines') and their implementation through the World Heritage Committee and the Convention's *Operational Guidelines* as a very central part of the overall purpose and function of ICOMOS.

In fact, under Article 13 (7) of the Convention the World Heritage Committee could invite other expert non-governmental organizations to advise on and participate it its work, so specialist geological knowledge from e.g. INHIGEO or the IUGS could easily be added to the range of expert advice available to the World Heritage Committee, should a majority of States Parties agree to this.

In summary, despite the worthy aims of the original 1972 Convention, geology and geomorphology are very poorly represented on the World Heritage List, and all the signs are that this continues to decline in relative terms in comparison with the cultural heritage, and also to a significant extent compared with other categories of the natural heritage.

The problem ultimately lies with the States Parties and their priorities in terms of their World Heritage nominations. A few countries, such as the United Kingdom, do aim to carry out a wide-ranging periodic review in preparing each revision of the national Tentative List submission. The most recent UK review involved consultations with at least three English ministries and agencies, two each from Scotland, Wales, and Northern Ireland, and the self-governing administrations of the Channel Islands, Isle of Man, and British Overseas Territories and Dependencies, such as Gibraltar, Bermuda, the remaining Caribbean territories, through to the St Helena and Tristan da Cunha, the Falkland Islands and South Georgia and Pitcairn and Henderson Islands. Over recent years following nomination by the UK the completely uninhabited Gough and Inaccessible Islands of the Tristan da Cunha territories off Antarctica have been added to the World Heritage List because of their marine and island ecology including one of the world's largest seabird colonies. Similarly, Australia takes its responsibilities in relation to the Convention very seriously on behalf of the sixteen mainly very small Pacific nations and territories for which it is responsible in terms of international relations.

In far too many cases the emphasis seems to be on the promotion of the country's existing high profile tourist attractions rather than reflecting a considered assessment and prioritizing of the whole of the country's cultural and natural heritage. Ensuring a much higher level of geological involvement within the implementation of the Convention in the World Heritage Committee, and adding recognition of the historical geological heritage to the provisions of the *Operational Guidelines*, could be very significant in the long term.

The current Tentative Lists suggest that there are unlikely to be many major advances in the next 5 to 10 years. Most of the 95 Tentative List proposals based partly (or, much more rarely, wholly) on geological grounds seem to be 'more of the same': yet more spectacular karst landscapes and caves, waterfalls, glaciers, volcanoes of interest to tourists.

However, there are some proposals which, if eventually inscribed, would improve the current state of affairs in relation to the historical and 'classic' geological sites. These include the Carboniferous of Joggins that was so influential on Charles Lyell, the Ediacaran (Pre-Cambrian) fossil site of Mistaken Point (both in Canada); the ophiolites of the Troodos, Cyprus; the Tertiary fossil site of Ipolytarnóc, Hungary; the Campi Flegrei region of Italy, (both known since at least the eighteenth century); and Switzerland's Glarner Hauptüberschiebung nappes region of the Alps. (Most encouragingly, the Tentative List submission for this last region specifically refers to its great significance in the history of science because of the pioneering nineteenth century geological work of Escher, Heim and Bertrand.)

Little real progress will be possible until governments around the world start taking their national geological heritage seriously, and begin to submit more geological nominations to the World Heritage Committee. The geological community in each country needs to become actively involved in promoting geological conservation generally, and in particular World Heritage nominations, at the national level.

Appendix: World Heritage Sites: Natural and mixed lists: geological inscriptions

The geological information presented here is mainly from United Nations Environment Programme's World Conservation Monitoring Centre datasheets; otherwise it is taken from the World Heritage Committee's documentation where WCMC datasheets are not available or appear to have incomplete information. The entries appear in date order of the original inscription. In the years which are omitted from the list, there were no inscriptions.

1978

Galápagos Islands (Ecuador)

The volcanic archipelago of the 128 islands of the Galápagos rises from a submarine platform on the junction of the Nazca and Cocos tectonic plates. The islands are young, formed by moving slowly eastward over a hot spot in the Earth's crust between 3 to 2.4 million years to 700 000 years ago. Most of the larger islands are the summit of a gently sloping shield volcano, some rising over 3000 m from the ocean floor. The western part of the archipelago experiences intense volcanic and seismic activity, culminating in collapsed craters or calderas: in June 1968, the southeastern floor of the Fernandina caldera dropped some 300 m, the second largest caldera collapse since Krakatoa's in 1883. The summits are studded with parasitic vents a few tens of metres high, and frequently flanked by lava flows. Other landscape features include crater lakes, fumaroles, lava tubes, sulphur fields and a great variety of lava and other ejecta such as pumice, ash and tuff.

Nahanni National Park (NW Territories, Canada)

The park is a diverse area of mountain ranges, rolling hills, high plateaus, broad depressions and incised valleys, stretching a succession of intricate terraces around a hot spring, to the deeply dissected sandstone, mudstone, shale and limestone of the Funeral and Headless Ranges and Tlogotsho and Liard Plateaus, and the Mackenzie plain. Large areas of the centre of the park have remained unglaciated for up to 300 000 years. The South Nahanni River runs through the Park covering a seventh of the river's 35 000 sq km watershed which drains via the Liard River into the Mackenzie basin. The South Nahanni and its tributary, the Flat River, are older than most of the mountain ranges through which they cut. Within the reserve the river drops 475 m overall, over Virginia Falls in a spectacular drop 92 m high, then runs for 70 km. through a series of four canyons from 460 to 1200 m deep, and through karst terrain with grottoes, sink holes, labyrinths, closed canyons and an underground river system.

Yellowstone National Park (mainly Wyoming, plus Montana & Idaho, USA)

The park lies in a caldera basin over a volcanic hot spot in the most seismically active region of the Rocky Mountains, that rise up to 4000 m. Crustal uplifts 65 million years ago raised vast blocks of sedimentary rock to form the southern Rocky Mountains. For 25 million years andesitic volcanic ashflows and mudflows were common, covering and petrifying forests: nearly 200 species of petrified plants have been found. A more recent period of rhyolitic volcanism began in the region about two million years ago. During this time thousands of cubic kilometres of rhyolitic magma filled immense chambers under the plateau, then erupted to the surface in three cycles of eruption (dated at 2.2 million, 1.2 million and 630 000 years ago) and produced huge explosive outbursts of ash. The latest eruptive cycle formed a caldera 45 km wide and

75 km long when the active magma chambers erupted and collapsed: the crystallizing magma and injections of new magma are the source of the hydrothermal geysers, hot springs, mud pots and fumaroles. Yellowstone contains more geysers than the rest of the world put together, with more than 300 in all, 200–250 being active, and more than 10 000 hydrothermal features. Most of the area was glaciated during the Pleistocene, and many glacial features remain. Yellowstone Lake, 35 400 ha in area, 2357 m high with a maximum known depth of 122 m, is the largest high elevation lake in North America.

1979

Dinosaur Provincial Park (Alberta, Canada)

The site is an outstanding example of fluvial erosion patterns in semi-arid steppes; slow-moving rivers that emptied into the shallow Bearpaw Sea of the Cretaceous period, 75 million years ago left deposits which have developed into the clay shale and sandstone and have yielded the dinosaur remains for which the park is renowned. Some 38 species of over 34 genera of 12 families of dinosaurs have been found in the park, including specimens from every known group of dinosaurs from the Cretaceous period. The families Hadrasauridae, Ornithomimidae, Tyrannosauridae, Nodosauridae, Pachycephalosauridae and Ceratopsidae are particularly well represented. About 15 000 years ago the area was flat and covered by an ice sheet some 600 m thick. During this ice age, glacial meltwater carved steep-sided channels; ice crystals, wind and flowing water continued to shape these extensive 'badlands' which today display a variety of representative features.

Everglades National Park (Florida, USA)

Everglades National Park is a shallow drainage basin comprised of two broad zones: wet freshwater prairies with forested islets, and coastal saltmarshes, mangrove swamps, estuaries, beaches and dunes. The basin is tilted to the SW draining south Florida in a slow-moving sheet of water 40–80 km wide, depending on rainfall. The area is underlain by extensive Pleistocene limestones with oolitic and bryozoan facies, overlain by variable thicknesses of marl and peat, a thin porous crust which filters the surface water percolating to the aquifer. Florida Bay is has an average depth of 1 m and a maximum depth of 3 m and encloses hundreds of islands. Its substrate is composed of anastomosing mudbanks and unconsolidated calcareous sediments over limestones and is one of the most active areas of modern carbonate sedimentation. The park lies at the interface between temperate and subtropical America, between fresh and brackish water, shallow bays and deeper coastal waters, and protects a complex of habitats which support a high diversity of wildlife. The area of transition from freshwater glades to saltwater mangrove swamps is a highly productive zone that nurses great numbers of commercially valuable crustacea. The Dry Tortugas is an isolated cluster of coral reefs and shoals.

Grand Canyon National Park (Arizona, USA)

The park is dominated by the spectacular Grand Canyon; a twisting, 1.5 km deep and 447 km long gorge, formed during some six million years of geological activity and erosion by the Colorado River on the upraised Earth's crust. On-going erosion by the seasonal and permanent rivers produces impressive waterfalls and rapids of washed-down boulders along the length of the canyon and its tributaries. Exposed horizontal geological strata in the canyon span some 2000 million years of geological history, from Late Precambrian, through the Palaeozoic, Mesozoic and the Cenozoic. The oldest Precambrian strata, the Vishnu Metamorphic Complex, are unfossiliferous. The first fossil evidence appears in the late Precambrian Bass Limestone with remains of early plant forms. The sequence of Palaeozoic strata have both marine and terrestrial fossils demonstrating alternate periods of submergence and uplift.

Kluane/Wrangell-St Elias/Glacier Bay/ Tatshenshini-Alsek (Alaska, USA & Yukon Territory and British Columbia, Canada)

Glacier Bay is a superlative example of the ice-affected landscapes with high mountain ranges, coastal beaches with protected coves, deep fiords, tidewater glaciers, coastal and estuarine waters, and freshwater lakes are characteristic of the region. A large fiord of 105 km in length, the bay has experienced four major advances and retreats of glaciers in recent geological time. Two centuries ago, the bay was completely filled with Grand Pacific Glacier and has witnessed an unprecedented rate of glacial retreat of about 95 km in the past 200 years. As the main glacier has retreated, 20 separate glaciers, many of them tidewater glaciers, have been created. The Tatshenshini-Alsek region contains the largest non-polar ice-cap in the world, with over 350 valley glaciers and an estimated 31 surge-type glaciers. The area is part of the most seismically active region in North America. The Tatshenshini-Alsek rivers and their wide U-shaped valleys are prominent natural features of the park.

Ngorongoro Conservation Area (Tanzania)

The Conservation Area rises 1000 m from the plains of the eastern Serengeti, over the Ngorongoro Crater Highlands to the western edge of the Great Rift Valley. The highlands have four extinct volcanic peaks over 3000 m dating from the late Mesozoic/early Cenozoic periods. The crater is the largest unbroken caldera in the world which is neither active nor flooded, though it contains a saline lake. The formation of the crater and highlands are associated with massive rifting which occurred to the west of the Great Rift Valley. The area also includes Empakaai Crater and Olduvai Gorge, famous for their geology and associated palaeotological studies.

Plitvice Lakes National Park (Croatia)

The Plitvice plateau of Jurassic dolomite to the west of the lakes area lies at 650–700 m between the Licka Pljesevica (1640 m) and Mala Kapela (1280 m) mountains. The upper end of the Korana Valley overlying the dolomite is a wide basin holding the upper lakes while the lower lakes occupy a narrow limestone canyon. The Plitvice Lakes basin is a karst river basin of limestone and dolomite, with approximately 16 lakes, behind dams created during the last 4000 years by the deposition of calcium carbonate in solution by encrustation on mosses, algae and aquatic bacteria. This results in the building, at about 1–3 cm/a, of phytogenetic travertine (calcareous tufa) barriers which have created lakes of various sizes linked by cascades and waterfalls, some up to 25 m in height. These have characteristic strange shapes and contain travertine-roofed and vaulted caves. The carbonates date from the Upper Triassic, Jurassic and Cretaceous ages and are up to 4000 m thick.

Virunga National Park (Democratic Republic of the Congo)

The park [primarily inscribed as one of the last remaining important habitats of the Mountain Gorilla] lies in the western (Albertine) rift valley and adjacent mountains, and it includes the forested granitic Rwenzori and volcanic Virunga massifs and swamp-edged lake. In the south is the Nyamuragira–Nyiragongo lava plateau and the northwestern fifth of the volcanic Virunga massif, shared with Rwanda. The area in the Virungas comprises the flanks of six volcanic mountains, of which two are still very active: an eruption of Nyiragongo destroyed 14 villages and an estimated 40% of the town of Goma on Lake Kivu in January 2002, and Nyamuragira erupted twice later that year. The steep western face of the Rwenzoris is glaciated and shares the third (Mt Ngaliema), fourth and fifth highest mountains in Africa with Rwenzori National Park in Uganda.

1980

Durmitor National Park (Montenegro)

Durmitor National Park comprises Mount Durmitor plateau and the valley formed by the canyon of the River Tara with canyons, mountains and plateaus, ranging in elevation from about 450–2522 m.

Geologically, the park is made up of rocky massifs dating from the Lower Triassic to the Upper Cretaceous, the Cenozoic and the Quaternary. The dominant features are the limestone formations of the Middle and Upper Triassic, the Upper Jurassic and the Upper Cretaceous, especially the so-called Durmitorean flysch. The 16 glacial lakes of the Durmitor and the river canyons rivers were formed during the Quaternary period following the sudden thaw of the glaciers on the Durmitor and neighbouring mountains. There are numerous examples of weathering processes, rock shapes and features characteristic of karst, fluvial, and glacial erosion.

1981

Great Barrier Reef (Queensland, Australia)

This is the world's most extensive stretch of coral reef. Extending to Papua New Guinea in the north, it comprises some 3400 individual reefs, including 760 fringing reefs, which range in size from under 1 ha to over 10 000 ha and vary in shape to provide the most spectacular marine scenery on Earth. There are approximately 300 coral cays. The form and structure of the individual reefs show great variety. Two main classes may be defined: platform or patch reefs, resulting from radial growth; and wall reefs, resulting from elongated growth, often in areas of strong water currents. There are also many fringing reefs where the reef growth is established on subtidal rock of the mainland coast or continental islands.

Los Glaciares (Argentina)

This large mountainous lacustrine area which includes a snow-capped sector of the southern Andean Cordillera with many glaciers derived from the Patagonian Ice Field, which is the largest ice mantle outside Antarctica, occupying about half of the park and feeding a total of 47 glaciers. In addition, there are approximately 200 other glaciers, each of which are less than 3 sq. km, that are independent of the main ice field. Glacial activity is concentrated around two main lakes, namely Argentino and Viedma, which are themselves the product of ancient glacial activity.

Mammoth Cave National Park (Kentucky, USA)

The park contains an area of karst of international importance. The limestone rocks of Upper Mississippian age are highly soluble and contain fossils throughout, including brachiopods, crinoids and corals. The main series, in which the cave systems and karst landscape have developed, are the St Louis, St Genevieve and Paoli limestones of the Meramecian. The Chester Upland is capped by sandstones of the Upper Mississippian–Lower

Pennsylvania periods. The core area is a dissected plateau comprising sandstone-capped ridges which protect the underlying caverns, separated by limestone valleys pitted with sinkholes. It contains the longest cave system in the world, in a many-levelled labyrinth, with known passages extending for over 550 km. Most types of limestone cave formation are found here, including long passages with huge chambers, vertical shafts, stalagmites, stalactites and gypsum flowers and needles. On the surface there is a superb karst topography with largely subsurface drainage, sinkholes, cracks, fissures and springs.

Willandra Lakes Region (New South Wales, Australia) (Mixed Site)

This comprises a system of dry lakes formed during the early Cenozoic, when marine transgressions in the Murray Basin deposited calcareous sand, marl and limestone. These were overlain by sands and dunefields in the Quaternary. The region is characterized by linear dunes, whose west to east orientation reflects the controlling wind system. Although these relict features were stabilized by vegetation, they were reactivated around 18 000 BP to 16 000 BP and were subsequently re-established. The interconnected lake basins were fed by a former tributary of the Lachlan River known as Willandra Billabong Creek. The six main lakes and numerous smaller depressions covered an area of 1088 sq km, and ranged in size from ephemeral ponds to Lake Garnpung, which was over 10 m deep and over 500 sq km. The formation of crescent lunette dunes on the eastern side of the lakes has been dated to at least 40 000 years to about 15 000 years ago.

1982

Río Plátano Biosphere Reserve (Honduras)

The rugged mountains forming 75% of the area which rise to Punta Piedra at 1326 m have many steep ridges, remarkable rock formations such as Pico Dama o Viejo, a 150 m granite pinnacle, and many waterfalls, one 150 m high. Two thirds of the Platano river runs through the mountains, with stretches of white water, and in one cataract disappears under massive boulders in a forested gorge. The remainder of the reserve, a coastal plain up to 40 km wide, rises gradually from the shoreline lagoons and grasslands to 100 m where the foothills begin abruptly. It is partly underlain by a belt of infertile deeply weathered Pleistocene quartz sandy gravels. The river meanders for 45 km through the lowlands forming ox-bow lakes, backwater swamps and natural levees.

Tasmanian Wilderness (Australia) (Mixed Site)

Rocks vary in age from Precambrian to Devonian and have been subjected to two main structural events, the Frenchman and Tabberaberan orogenies. The Precambrian units are widespread and consist of quartzite, schist, phyllite, conglomerate, dolomite, siltstone and sandstone. The more resistant sequences, such as quartzite, form most of the prominent ranges in the area, whereas less resistant schist, dolomite and phyllite underlie many of the valleys and plains. The Permian unit consists of glaciomarine sequences including tillite, sandstone, siltstone, mudstone and limestone horizon, whereas the Triassic unit above this contains banks of sandstone, mudstone, siltstone and coal, probably laid down during a humid, cool climate in swamps, lakes and river channels. The rocks contain rare plant and amphibian fossils. Pleistocene ice caps, cirque glaciers and valley glaciers were generally confined to the high mountains and plateaus, but glacial erosion has contributed to spectacular landform features including horns, arêtes, cirques, U-shaped valleys and rock basins (tarns). Below about 600 m, depositional features are typical including moraines and various other outwash deposits. Other features include cave systems, natural arches, clints and grikes, dolines, karren, pinnacles and blind valleys and a large meteorite impact crater of Pleistocene age in the Andrew River valley is of worldwide significance. (see also Houshold & Sharples 2008)

Tassili n'Ajjer (Algeria) (Mixed Site)

The park comprises two distinct geomorphological units: a sandstone plateau and a mountainous volcanic ridge. The plateau (Tassili) is part of an ancient sandstone layer surrounding the Precambrian granite massif of the Ahaggar, which extends down to a lower plateau edged by a 600 m escarpment with north-facing cliffs cut by several deep gorges and steep-sided watered valleys running northward into sands, and which runs for 700 km in a gentle arc WNW–ESE. The red to black-weathered sandstone has been deeply eroded into forests of 20–30 m pillars like ancient ruins, and rises to the SW-facing escarpment above the shifting dunes of the Erg d'Admer and Erg Tihodaine. The mountain ridge region is of relatively recent volcanic rock, and is part of a continental divide between northward and southward flowing watersheds. The semi-permanent river, Oued Iherir, has secreted travertine deposits which form natural dams and pools that cascade from one level to another.

1983

Great Smoky Mountains National Park (Tennessee & North Carolina, USA)

The topography of the Great Smoky Mountains range comprises moderately sharp-crested, steep-sided ridges separated by deep V-shaped valleys. Many of the mountain ridges branch and subdivide from the central ridgeline, creating a complex of drainage systems with 3057 km of fast-flowing clear mountain streams. The park contains 45 watersheds and the water table is near the surface in almost all sections. Precambrian metamorphic rocks consisting of gneisses and schists, and sedimentary rocks of the Precambrian Ocoee series are predominant, while younger sedimentary rocks are found in the Appalachian Valley.

Gulf of Porto: Calanche of Piana, Gulf of Girolata, Scandola Reserve (Corsica, France)

The Elpa Nera inlet and Scandola peninsula areas are part of a large geological complex that appears to have undergone two distinct cycles of volcanic activity in the Permian. Since then, the area has been subject to alternating cycles of erosion and rejuvenation, comprising porphyry, rhyolites and basaltic pillars, which have all been considerably eroded by wave action. Some ancient metamorphic rocks also occur. The sheer, jagged cliffs contain many caves and are flanked by numerous stacks and almost inaccessible islets and coves. The coastline is also noted for its red cliffs, some 900 m high, sandy beaches and headlands.

Pirin National Park (Bulgaria)

The Pirin Range runs NW–SE and the park is in the north half of the range and has a very varied topography. Much of its very scenic northern quarter is composed of a karst landscape of limestone developed predominantly in Proterozoic marbles. The lower southern three quarters are mainly of South Bulgarian granites and gneiss. Together these form a vast alpine landscape of crags, caverns and waterfalls, with gorges and deep valleys which divide both sides of the mountains into long steep ridges. The high ridges and sharp peaks, 81 of which rise over 2500 m, are the remains of an old Miocene peneplain with lateral ridges of Pliocene age. The area was widely denuded and differentially glaciated in the Quaternary period, with 176 glacial cirque lakes. There are also more than 70 hot springs in the foothills.

Sangay National Park (Ecuador)

The park comprises three geomorphological zones: the volcanic High Andes, the eastern foothills, and an area of alluvial fans. The highlands, of pre-Cretaceous metamorphic and plutonic rocks, rise from 2000–5000 m and are dominated by three strato-volcanoes, of which Tungurahua and Sangay are both still active: Sangay regularly ejects hot rocks and tephra and has the one of the world's longest records of continuous volcanic activity. Tungurahua last erupted violently between 1916 and 1925 and erupted in 2002. The eastern foothills in the NE and SE are low irregular mountains between 1000 m and 2000 m high formed of outcrops of sedimentary rocks. Large east-sloping alluvial fans dominate the east side of the park between approximately 800 m and 1300 m. Younger segments of these fans are only slightly dissected, but older parts are cut into by canyons up to 200 m deep.

Talamanca Range-La Amistad Reserves/La Amistad National Park (Panama & Costa Rica)

The Cordillera de Talamanca is the highest and wildest non-volcanic mountain range in Central America. It was formed by the orogenic activity which created the land dividing the Pacific Ocean from the Caribbean. A long period of marine deposition in the shallow surrounding seas up until the Middle Miocene was followed by a period of marine volcanism, which included the intrusion of a huge granitic batholith and the uplifting of the whole area to some 4000 m above sea level during the Plio-Quaternary orogenesis. The resulting peneplain has been eroded gradually, creating a rugged topography with many slopes inclined at over 60° This was the only area of Central America known to have been glaciated in the Quaternary, as evidenced by cirque lakes and glaciated valleys.

Vallée de Mai Nature Reserve, Praslin Island (Seychelles)

The granitic islands of the Seychelles together form what is in effect a 'microcontinent' that has had quite a different history from the other volcanic or coralline islands in the Indian Ocean. The reserve comprises a valley in the central hills of the island. Two streams originate in the valley, joining the sea in the west and east. The other principal river in the park flows westward into Baie Sainte Anne.

1984

Canadian Rocky Mountain Parks (Canada)

The central Rocky Mountains are a high massif of sedimentary rock dating from the Precambrian to Cretaceous periods, oriented NW–SE along the Continental Divide. The Main Ranges form the Continental Divide and comprise limestone, dolomite, sandstone and shale and the parks include nearly all the highest mountains in Canada, with five exceeding 3600 m. Within the Cambrian Stephen Formation in the Main Ranges of Yoho Park are the world-famous Burgess Shale fossil beds. First discovered in 1909 by the Smithsonian Secretary and palaeontologist, Charles Walcott, this site (first registered as a World Heritage Site in its own right in 1981) contains unique and exceptionally well preserved fossils of soft-bodied marine organisms that lived in mid-Cambrian seas around 515 million years ago. The diverse and sometimes bizarre animals preserved in the shales, many of them unique to this formation, represent a complete ecosystem that existed for only a very short time, relatively speaking, after the first explosion of multicellular life on Earth. Since their discovery, the fossils of the Burgess Shale have provided scientists with a wealth of information about the variety of Cambrian life forms, some of which are difficult to classify within the established modern or fossil taxonomy. Active glaciers and icefields still exist throughout the region, particularly in the Main Ranges, and the Columbia Icefield is the largest in North America's subarctic interior. There are numerous lakes in Mount Assiniboine Park, most of which are located in broad alpine valleys and plateaus in glacially scoured depressions in the limestone bedrock.

Yosemite National Park (California, USA)

Yosemite is dominated by the Sierra Nevada; granite underlies most of the park and is exposed as domes, partial domes, knobs and cliffs. Glaciation has influenced the topography over most of the area including the Yosemite Valley, a 914 m deep cleft carved by glaciers through a gently rolling upland. The valley is a widened portion of the prevailing narrow Merced River canyon which traverses the southern sector of the park from east to west. The massive sheer granite walls present a freshly glaciated appearance with little post-glacial erosion. This area also contains many waterfalls and some 300 lakes.

1985

Huascarán National Park (Peru)

The park has a great diversity of geomorphological features. Situated in the Cordillera Blanca which is the highest tropical mountain range in the world, the park has 27 snow-capped peaks above 6000 m, 663 glaciers, 296 lakes and 41 rivers discharging into the Santa, Pativilca and Maranon watersheds. The lowest point in the reserve is Grand Cataract, near the northern boundary. The base rock consists principally of marine Upper Jurassic and Cretaceous sediments and Cenozoic volcanic deposits which form the Andean batholiths.

1986

Central Eastern Rainforest Reserves (New South Wales & Queensland, Australia)

This is a very large and complex World Heritage Site, which is difficult to characterize and summarize. Originally inscribed in 1986, it was greatly extended in 1994, and now covers 51 separate protected areas in eight broad regions spread along more than 600 km of the mainly mountainous country of the Great Divide, separating the humid Pacific coast from the arid interior of Australia. Within the main range to the west of the watershed of the Great Divide is a region of 160 sq km of strongly dissected tableland of late Oligocene to early Miocene basalts and associated volcanics. In the Focal Peak region lies the Mount Barney Intrusive Complex, a shield volcano predominantly of granophyre approximately 24 million years old, with erupted basalts and rhyolites. The Border Ranges has one of the world's best preserved and largest Cenozoic shield volcanoes, with a major erosion caldera

at Mount Warning and the isolated plug representing the original neck of the volcano at Mount Warning. In marked contrast the coastal Iluka Peninsula Reserve is characterized by a series of siliceous sand dune ridges overlying Triassic and Jurassic sedimentary rocks. The Washpool and Gibraltar Ranges are a series of high ridges, plateaus and sharply dissected valleys formed in middle Palaeozoic metasediments, with a late Permian volcanic complex and a Permo-Triassic granite, in places weathered into spectacular tors. The New England/Dorrigo region is notable for its Permo-Carboniferous metamorphic and sedimentary rocks, and basaltic plateaus and outcrops of Cenozoic volcanoes. In the Barrington Tops area to the east of, but partly connected to, the Great Divide, early Cenozoic volcanic activity has produced massive basalt lava flows over a basement of steeply dipping Palaeozoic sediments.

Giant's Causeway and Causeway Coast (Northern Ireland, UK)

The Causeway Coast has an unparalleled display of geological formations representing volcanic activity during the early Cenozoic period some 50–60 million years ago. Cenozoic lavas of the Antrim Plateau, represent the largest remaining lava plateau in Europe. The 6 km stretch of coastline comprises a series of headlands and bays, the former consisting of resistant lavas. The geological succession during the Cenozoic, consists of: the Lower Basalts, where about six of the eleven lava flows are 67 m thick; the Interbasaltic Bed exposed along extensive sections of the cliffs east of Giant's Causeway; and the Middle Basalts, which are thick flows ranging from 30 m to over 150 m. Specific sites of interest include the Giant's Causeway itself (a sea-level promontory of approximately 40 000 almost entirely regular polygonal columns of basalt. This exposure of columnar basalts in perfect horizontal sections at such a scale creating a pavement, is arguably unique in the world. The coastline is cut through by olivine and tholeiite dykes.

Škocjan Caves (Slovenia)

Škocjan is a shallow limestone canyon in the Dinaric karst with an associated underground river and cave system featuring four deep and picturesque chasms. It is a classic example of contact karst between the limestone and impermeable rock and is the type location for both the landforms and the terms 'karst' and 'doline' (swallowhole). The subterranean passages carved by the Reka River are dramatic examples of large-scale karst drainage. Its caves are the beginning of a system of subterranean passages from their source to the Adriatic coast near Trieste, Italy. Over 5 km in total length with a maximum depth of 230 m, Skokian has one of the largest underground canyons in the world: 2 km long, up to 100 m wide in places, up to 148 m high, with stalactites and stalagmites, 25 cascades including a 163 m waterfall, and a flow rate that can reach 300 $m^3\ s^{-1}$.

1987

Gros Morne National Park (Newfoundland & Labrador, Canada)

This park is geologically very diverse with areas of Pre-cambrian granite and gneiss, Cambrian and Ordovician sedimentary rocks, with an unusually complete palaeonto-logical sequence including the officially adopted world stratotype for the base of the Tremadocian stage and hence the Cambrian–Ordovician boundary. There are also extensive Palaeozoic serpentinized ultra-basic rocks, gabbros and volcanic rocks in the SW, together with abun-dant evidence of recent glacial activities. Exposed oceanic crust, mantle, a section of ancient Mohorovicic discontinu-ity, and other distinctive geological features demonstrating the effects of plate tectonics are also found, and are important in determining the evolution of the North Atlantic Basin. The park comprises two distinct physiographic components: coastal lowlands with late Quaternary piedmont moraines, and an alpine plateau. The shoreline features beaches, steep cliffs of unconsoli-dated deposits, and dune formations up to 30 m in height which extend inland for some 1.6 km in a number of places, whereas meandering creeks, eutrophic bog lakes, dead ice moraine deposits, erratics and small patches of iso-statically raised beach deposits are found on the plain. The heavily glaciated upland alpine plateau of the eastern part of the park has perched lakes, bare rock and valleys.

Hawaii Volcanoes National Park (Hawaii, USA)

The park extends from the southern coast to the summit calderas of Kilauea and Mauna Loa volcanoes. Mauna Loa is a massive, flat-domed shield volcano built by lava flow layers and is considered to be the best example of its type in the world. It extends from 6096 m below sea-level to a maximum of 4103 m above sea-level. These are among the world's most active volcanoes and con-stantly exhibit changing features, especially from the two principal rift zones which feature extensive recent flows. The Halemaumau fire pit was a continuously active lava lake into the early 1900s and others existed along the East Rift. Eruptive activity has been almost con-tinuous along the area's East Rift Zone, and has produced extensive new lava flows and a 300 m high cinder cone.

1989

Mosi-oa-Tunya/Victoria Falls (Zambia)

Since the uplifting of the Makgadikgadi Pan some two million years ago, the Zambezi River has been cutting through the basalt plateau, exploiting east–west trending fissures in the basalt, and forming a series of retreating falls. Below the present Victoria Falls the river enters a zigzag series of narrow gorges, relics of seven past water-falls, and Devil's Cataract in Zimbabwe is the start of the cutting back to an eighth waterfall that will eventually leave the present crest high above the river in the canyon below. Sixteen kilometres of the Batoka gorges border the parks and these continue for some 100 km to the east, being 140 m deep at one place. The park com-prises the banks of the Zambezi River above Victoria Falls and a series of deep gorges below them. The falls are at the heart of the park, and when the Zambezi is in full flood and is 2 km wide (in February and March) they form the world's largest sheet of falling water. During these months some 540 million cubic metres of water per minute pour over the falls, which are 1690 m wide and drop 108 m at Rainbow Falls. The spray plume which may obscure the view of the falls in the rainy season can rise 500 m and be visible 30 km away.

1990

Te Wahipounamu (South Island, New Zealand)

SW New Zealand (Te Wahipounamu) lies across the boundary between the eastern, Pacific plate and the Indo-Australian plate to the west and is one of the most seismically active regions in the world. The mountainous character of the area results from tectonic movement over the last five million years. A detailed history of uplift over almost a million years is recorded in a flight of 13 or more marine terraces on the south coast of Fiord-land and the contiguous Waitutu area. The terraces were formed by marine erosion at the coast, but are now found at up to 1000 m above sea level. The rocks of Fiordland are generally crystalline, dominated by a wide range of plu-tonic types such as granite and diorite, and metamorphic gneisses. In the extreme SW there are unmetamorphosed sedimentary rocks. In the NE, the Fiordland block abuts a set of north–south trending volcanic and sedimentary rocks of mainly Permian age. The Dun Mountain Ophiolite Belt is the key unit, comprising a slice of oceanic crust and the underlying mantle. Eastwards, a Permian terrane of greywacke sandstone becomes progressively more highly metamorphosed to become schist which forms the Southern Alps contained within Mount Aspiring National Park. This band of schist narrows as it extends further NE, paralleling the Alpine Fault on its southeastern side. On its eastern margin in Mount Cook National Park, the schist gradually changes back into Permian–Triassic grey-wacke of a separate terrane. The uplifted mountains have been very deeply excavated by glaciers, resulting in high local relief. West of the Alpine Fault, the rocks of south Westland consist of a basement of Ordovician greywacke with some high temperature metamorphic rocks and gran-ites, and minor areas of younger Cretaceous and Cenozoic sedimentary rocks along the coast. Severely eroded by Pleistocene glaciers, these now generally form blocks of rugged hill country or isolated hills standing above post-glacial alluvium and lagoon-infilling sediments. Pleisto-cene moraines and outwash form extensive areas of subdued hill country and low plateaus.

Tongariro National Park (North Island, New Zealand) (Mixed Cultural and Natural Site)

The park lies at the southern end of a discontinuous 2500 km chain of volcanoes which extends NE into the Pacific Ocean. This chain corresponds with the subduction of the Pacific Oceanic plate beneath the Indian–Australian continental plate. The volcanoes in the park, which are predominantly andesitic in composition, fall into two groups on the basis of location, activity and size. The northern group of volcanoes and their associated vents, domes, cones and craters have not been active for between 20 000 and 230 000 years. Glacial activity 100 000–14 000 years ago has rounded the profiles of this group. The active volcanoes group comprises the Tongariro complex comprises recent cones, craters, explosion pits, lava flows and lakes superimposed on older volcanic features, Mount Ngauruhoe a composite andesite cone of interleaved pyroclastic material and lava which may be as little as 2500 years old and is still building, with violent ash eruptions that occur at approximately nine-year intervals. The SE of the park is dominated by the active volcano Mount Ruapehu. Volcanic activity commenced approximately 500 000 years ago and tephra deposits indicate a peak of activity 10 000–14 000 years ago, and the current active vent lies beneath the 500 m diameter Crater Lake at an elevation of 2550 m. In addition, the park contains other extinct volcanoes, lava and deposits from a glaciation that peaked around 14 700 years ago, and some small valley glaciers. Marine Miocene–Pliocene mudstone and sandstone form two hilly areas in the west.

1991

Shark Bay (Western Australia, Australia)

Shark Bay comprises a series of north–south facing peninsulas and islands which separate inlets and bays from each other and the Indian Ocean. The coastline is 1500 km long and includes some of the highest cliffs of the Australian coastline. There are three distinct landscape types. Gascoyne-Wooramel province comprises the coastal strip along the eastern coast of the bay and is a low-lying plain backed by a limestone escarpment; Peron province which comprises the Nanga/Peron peninsulas; Faure Island/sill comprising undulating sandy plains with gypsum pans or birridas, and ancient interdune depressions filled with gypsum. The seaward margin of this province terminates in a scarp 3–30 m high and narrow sand beaches; Edel province which comprises Edel Land peninsula and Dirk Hartog, Bernier and Dorre Islands, is a landscape of elongated north-trending dunes cemented to loose limestone. The province terminates to the west as a series of spectacular cliffs. The basement rock in the area is Late Cretaceous Toolonga limestone and chalk. The most extensive younger rocks are Peron sandstones and Tamala limestones. These rocks are often overlain by a series of longitudinal fossil dunes accumulated during the middle to late Pleistocene. The extensive supratidal flats are comparable to the coastal sabkhas of the Arabian Gulf, with gypsum forming through evaporation of saline groundwaters. The inland terrestrial landscape of Shark Bay is predominantly one of low rolling hills interspersed with birridas (inland saltpans). There is also much palaeontological interest in Shark's Bay's famous benthic microbial communities, dating from c. 4000 BP to the present, which have mineralized to form stromatolites. These appear to be very closely comparable with ancient stromatolites up to 3 billion years old, which are some of the world's oldest known fossils.

1994

Australian Fossil Mammal Sites (Riversleigh/Naracoorte) (Queensland & South Australia, Australia)

These two very widely separated (and significantly different) fossil localities have been grouped together under a single World Heritage List inscription. The Cenozoic fossil fields of Riversleigh, Queensland, are on the watershed of the Gregory River within the Karumba Basin in the Gulf of Carpentaria. The Cenozoic deposits of Riversleigh occur as inliers within eroded areas of the extensive, flat-lying, Cambrian Thorntonia limestone. This in turn surrounds less common remnants of Proterozoic sediments. The Cenozoic sediments can be categorized into four groups: Oligo-Miocene alluvial and lacustrine deposits; Oligo-Miocene karst and fissure fills; Pliocene cave sediments; and Quaternary fluvial and cave sediments. Naracoorte, South Australia, is a region of flat covered karst, punctuated by a series of stranded coastal dune ridges that run parallel to the present coastline. The caves of the Naracoorte Caves Conservation Park are formed in a ridge of Oligo-Miocene Gambier limestone capped by the Naracoorte East Dune. In the late Pleistocene the caves were open to the surface allowing sediment and bones to accumulate in their entrances and dolines, the most significant of these accumulations being those of Victoria Fossil Cave. Riversleigh's Oligocene–Miocene faunal assemblages around 15 million years old include ancestral forms of most present-day and recently extinct marsupials, and the 35 bat species identified make this one of the richest occurrences of bat fossils in the world. The Pleistocene fossil vertebrate deposits of Victoria Fossil Cave at Naracoorte are Australia's largest and best preserved, in terms of both volume and diversity, and one of the richest deposits in the world, with tens of thousands of specimens representing at least 93 vertebrate species including many examples of the Australian Pleistocene megafauna, with marsupials up to the size of a buffalo.

Canaima National Park (Venezuela)

Canaima includes the uplands of the Gran Sabana and the eastern table mountains (tepuis) of the Roraima Range, as well as the sandstone plateau of Chimantá and Auyántepui and the NW Canaima lowlands. It comprises Precambrian rocks which have been subjected to 600 million years of erosion to form a spectacular landscape. It is composed mainly of horizontal sandstone and lutite strata with intruded igneous rocks (notably diorite dykes). There are three disjunct physiographic units: undulating lowlands between 350 and 650 m; the flat plateau of the Gran Sabana (800–1500 m); and the tepui summits (2000–2700 m). The summits reach 1000–2000 m above the surrounding plateau and their surfaces are often scarred by gullies, canyons and sinkholes of several hundred metres depth. Water drains from the flat summits forming hundreds of waterfalls. The Río Caroní, with its many tributaries arising within the park, supplies the Guri dam which provides electricity to large areas of the country. There are many waterfalls in the park including Angel Falls, the world's tallest at 1002 m.

Ha Long Bay (Viet Nam)

Ha Long is a large bay on the Gulf of Tonkin with a multitude of limestone rocks of two kinds: the Cát Bà limestone of the early Carboniferous period (450 m thick); and the Quang Hanh limestone ayer of the middle Carboniferous to Permian (750 m thick), both with many caves and other karst features. These two limestones form the bedrock of most of the islands of the Bay, but others islands have exposures of some Lower Palaeozoic schists and other metamorphic rocks. The caves and other karst features developed in the middle to upper Pleistocene, and the whole area has been modified by coastal erosion processes following the flooding of the bay with the rising sea level from the early Holocene. In total, there are 1600 islands and islets, and numerous caves and grottoes are found, many with stalactites and stalagmites.

1995

Carlsbad Caverns National Park (New Mexico, USA)

The park overlies a segment of the 560 km long Permian fossil reef (Capitan Reef) which surrounds the Delaware Basin of western Texas and SE New Mexico. Several deep canyons have been eroded in the SW-trending reef revealing cross-sections of other geological formations. Subterranean formations have also been exposed as an extensive cavern system has developed within the 610 m thick reef complex. The most notable example of this can be found within Lechuguilla Cave where five formations (Yates, Seven Rivers, Queen, Capitan Reef and Goat Seep) have been identified. Fossils preserved within the exposed rock formations include bryozoans, bivalves, gastropods, echinoderms, brachiopods, fusulinds, sponges, trilobites and algae. Unlike many caves which were caused by carbonic acid dissolution, Carlsbad Caverns developed as hydrogen sulphide gas from

underlying oil and gas deposits seeped upwards and combined with freshwater to form sulphuric acid which then eroded the limestone. Carlsbad Cavern is the largest of 81 known caves within the park; Lechuguilla Cave is not only the deepest (477 m) and longest (133 km), but contains the largest collection of hydromagnesite balloon-like formations and subaqueous helictite formations. Gypsum has been deposited in a variety of forms throughout many of the caves and ranges from thin crusts to beds of more than 30 m thick. Calcite speleothems include stalactites, stalagmites and columns and sulphate mineral deposits.

Caves of Aggtelek Karst and Slovak Karst (Hungary & Slovakia)

This is the most extensively explored hydrothermal karst area in Europe. Within the fossiliferous Middle Triassic reef limestones 712 caves have so far been identified. Many of the younger caves which have formed at the plateau edges are on several levels and contain extensive dripstones. The most notable of these is the Baradla-Domica cave system which is 21 km long and connects Hungary with Slovakia. It has a cavern capable of holding 1000 people, a 13 m long stalactite and the underground river Styx. These caves are also noted for having the world's highest stalagmite (32.7 m), aragonite and sinter formations and an ice-filled abyss, which considering the territory's height above sea-level, is a unique phenomenon for central Europe. All these karst landforms are the result of long-term geomorphological processes typical of this temperate climatic zone. Hydrological conditions are characterized by a lack of surface streams, except between mountain basins, and the complex circulation of underground water.

Messel Pit Fossil Site (Hessen, Germany)

Messel Pit, studied continuously since the late nineteenth century, is the richest site in the world for understanding the living environment of the Eocene. It provides a unique record of the early stages of the evolution of mammals and includes exceptionally well-preserved mammal fossils, ranging from fully articulated skeletons to the contents of stomachs of animals of this period, as well as rich fossil reptile and insect faunas and fossil floras. The pit is now approximately 1000 m long by 700 m and the deposits, mainly of oil shales, range through virtually the whole of the Eocene, from around 57 million years to 36 million years. The deposits formed in a mainly lacustrine environment in basins, watercourses and subsiding hollows in the Old Red Sandstone (Devonian) country rock, which in turn is underlain by crystalline magmatic older Palaeozoic rocks. Plant fossils include early species of club mosses, royal ferns, grass ferns, cypress, plum yew, swamp cypress and the walnut tree. The abundant fauna of around 40 species of mammal includes Eocene species of opossums, pangolins, anteater, scaly-tailed hedgehogs, forty specimens of the primitive and miniature Messel horse *Propalaeotherium parvulum* (whose skeleton measures approximately 50 cm in length), a bat and a large rodent, together with crocodiles and bird fossils including ancient ostriches, woodpeckers, falcons and rails. The fish are also indicative of the evolution of modern bony fish types, including garfish, eels and perch; insects are the most numerous invertebrates found at the site, with several specimens having very well preserved structure and metallic colourings. These include click beetles, weevils, jewel beetles, dung beetles, stag beetles, ground beetles, water beetles, longhorn beetles and rove beetles.

1996

Lake Baikal (Russian Federation)

Lake Baikal is the seventh largest (636 km long by 27–80 km wide, a 2100 km coastline and covering 23 000 km^2) and the deepest (1182 m below sea level) lake in the world and contains over a fifth of the world's unfrozen surface freshwater. It is walled in by mountains and fed by 335 rivers flowing from these, with only the Angara River, a tributary of the Yenesei, flowing out of it. The lake is of tectonic origin, situated in an active rift complex system of block-faulted depressions and consists of three deep basins resting on 7 km of sediments. The present lake is around 25 million years old (late Oligocene), and is the oldest large lake in the world, but it occupies a lake depression that formed through the Palaeozoic, Mesozoic and earlier Cenozoic eras. Hydrothermal vents 400 m deep at Frohlika Bay are evidence of the ongoing tectonic activity of the area.

Volcanoes of Kamchatka (Russian Federation)

The 1200 km-long Kamchatka peninsula runs north–south between the north Pacific and the sea of Okhotsk. Its southern half is formed mainly by two parallel mountain ranges, including a 700 km long volcanic belt, which is the surface expression of the northwesterly subduction (by 8–10 cm a^{-1}) of the Pacific Ocean plate under the Eurasian plate and shows a complete range of the vulcanism characteristic of the Pacific 'Ring of Fire'. The peninsula has in total some 300 volcanoes, most of them basaltic composite stratocones and andesitic stratovolcanoes, though with some shield volcanoes in addition, of which 33 are currently active. Most of the active volcanoes are of explosive character, and since 1690 some 200 eruptions have been recorded. There are also calderas, scoriae cones, lava streams, cinder fields, over 160 thermal and mineral springs, geysers, solfataras, mud pots and many other volcanic features. Thirty-four of the volcanoes, including the 13 most active, are within the World Heritage site. Kamchatka is also a major centre of glaciation, with 47 glaciers covering 269 km^2, the largest

being the Erman glacier which continues to advance at 30–50 m per year. Two periods of Pleistocene glaciation have influenced much of its landscape, creating cirques, hanging valleys, U-shaped valleys, moraines and glacial till and almost all the types of ice formation common in volcanic areas.

Laponian Area (Northern Sweden) (Mixed Site)

Inscribed on cultural heritage grounds as the largest remaining area in the world that is the home and preserves the traditional way of life of the Saami or Lapp people, this Arctic region, covering four national parks, also has outstanding geological features illustrating both historical and current geological processes associated with recent and contemporary glaciation. The World Heritage area covers two landscape types: an eastern lowland area of Archaean geological origin, and a western mountainous landscape, covering two-thirds of the area. The Nordic alpine landscape of Sarek and Stora Sjöfallet national parks has high, steep mountains, deep valleys and powerful rivers, with more than 200 peaks over 1800 m and 100 glaciers. The important geomorphological features include monadnocks, kursu valleys, sandurs, boulder hollows, tundra polygons, U-shaped valleys, glacial cirques and moraines, talus accumulations, drumlins, weathering phenomena and palsa bogs.

1997

Heard and McDonald Islands (Australian Territories in the Southern Ocean)

Heard Island and the McDonald Islands lie 4100 km SW of the Australian mainland and 1700 km north of Antarctica. These are the only volcanically active sub-antarctic islands, which are formed on the submarine Korgulen Plateau of middle Eocene to early Oligocene oceanic limestones. The islands are essentially accumulations of Oligocene to present-day oceanic volcanic rocks rising 3700 metres above the adjacent seafloor, with some Eocene–Oligocene to upper Miocene marine limestones. The main body of Heard Island is roughly circular, with a diameter of about 25 km. Topography is dominated by Big Ben massif, with the volcanically active Mawson Peak (the only active volcano in Australian territory) and both karst and volcanic landforms occur. The latter include extensive areas of lava tunnels. To the east, a narrow sand and shingle spit extends approximately 10 km out into the southern ocean. About 80% of Heard Island is glaciated, with ice up to 150 m deep and glaciers extending from 2745 m to sea level. Ice cliffs form a high percentage of the coastline. The glaciers appear to be fast-flowing as a result of the steep slope and high precipitation, and are likely to be particularly sensitive to climatic fluctuations. Measurements between 1947 and 1980 suggest that glacial retreat has been marked on Heard Island, particularly on the eastern flanks. The McDonald Island group, all of which ice-free, is composed of basaltic lava and tuffaceous material, resulting from eruptions of volcanic vents near sea level, and the rocks are compositionally distinct from those of Heard Island.

Lake Turkana National Park (Kenya)

Lake Turkana occupies the beds of two grabens at the northern end of the Kenyan Great Rift valley in barren desert country. It is the largest and most northerly of all the Rift Valley lakes, with a delta extending into Ethiopia and measures 249 km by 48 km at its widest. Rich fossiliferous deposits are found around the lake for 60 km north from Allia Bay and up to 20 km inland. The plains are flanked by volcanic formations including Mount Sibiloi, where there are the remains of a petrified forest of the upper Miocene. To the north of Alia Bay, the extensive Koobi Fora palaeontological finds have been made including hominid remains, beginning with the 1972 the discovery of the first fossils of *Homo habilis*. These are evidence of the existence of a relatively intelligent hominid two million years ago and reflect the change in climate from moist forest grasslands towards the present hot desert. The human and pre-human hominid fossils include the remains of four species, the most important being the 1999 discovery of 3.5 million year old *Kenyanthropus platyops*. Other important palaeontological findings include ancestral forms of several modern animal species.

Macquarie Island (Tasmania, Australia)

This World Heritage site is an island 34 km long and up to 5 km wide on the exposed crest of the Macquarie Ridge Complex. This component of the oceanic crust, formed at a spreading ridge, in the early or middle Miocene in water between 2 km and 4 km deep. It has been raised to its present position as the Indian–Australian tectonic plate interacted with the Pacific plate, squeezing the deposits upwards through the ocean floor by as much as 6 km, and finally began to emerge above sea level around 600 000 years ago. It is considered to be the least disturbed and best preserved section on the globe of oceanic crust formed in deep water and now exposed above sea level. Volcanic rocks, mainly pillow lavas with varying proportions of rare massive lava flows, basaltic dykes and various sediments now comprise about 80% of the island. The northern part of the island comprises mainly intrusive rocks apparently derived from deeper crustal levels than the southern section. Dolerite dyke swarms are extensive in the northern region and also around Lusitania Bay and Sandell Bay in the south. Besides the dyke swarms, the northern section is composed mainly of serpentinized peridotite and gabbro masses, although there are small areas of extrusive volcanic rocks. The main landscape feature is a central rolling plateau 250 m–300 m above sea level, bounded on all

sides by steep cliffs, from the foot of which extends a coastal platform up to 800 m wide. Glacial drift up to 20 m thick covers much of the plateau and there are several lakes with a combined area of more than 200 ha. Numerous smaller lakes, tarns and pools are found both on the plateau and on the raised beach terraces.

Morne Trois Pitons National Park (Dominica)

Dominica is a relatively young island which began to emerge through extensive volcanic activity during the Miocene. Morne Trois Pitons is the dominant natural feature of island, and is the name of the basaltic remains of a former volcano rising to approximately 1300 m, within eight kilometres of the sea. The landscape is characterized by volcanic piles with precipitous slopes, and deeply incised valleys (glacis slopes). There is also a fumarole known as Valley of Desolation (or Grand Soufrière), with fumaroles, hot springs, mud pots, sulphur vents and the Boiling Lake, which is the world's second largest of its kind. This is surrounded by cliffs and is almost always covered by clouds of steam. Other outstanding features in the area include the Emerald Pool, Stinking Hole (a lava tube in the middle of the forest) and the Boeri and Freshwater crater lakes, estimated to have formed around 25 000 to 30 000 years ago.

Pyrénées–Mont Perdu (France & Spain) (Mixed Site)

This Mesozoic limestone massif along the crest of the French–Spanish frontier, is a mountain landscape with lakes, waterfalls, rocky outcrops, glacial cirques and canyons. The four glacial cirques are located to the north; to the south there are three canyons and a gorge. Three distinct geomorphological regions are found. In the north, three convergent valleys are surmounted by crests oriented north–south comprising schists and sandstone. The cirques of Estaubé and Troumouse are separated in the SE by a crest dominated by Munia Peak (3133-m). The second region comprises a line of steep limestone steps stretching for 20 km, most of the summits of which are higher than 3000 m. Third, high sandstone and schist plateaus, at about 2000 m are found to the SW of the 'Tres Serols'.

1999

Lorentz National Park (Province of Papua, Indonesia)

The park can be divided into two very distinct zones: the swampy lowlands and the high mountain area of the central mountain ranges with folded and metamorphosed Cretaceous ocean sediments around 100 million years old, though with covers of Eocene to Miocene deposits in places. The extremely rugged, folded, and in places heavily mineralized mountains of the central cordillera are the result of the collision between two continental plates, which continue to cause the mountain range to rise. Heavily glaciated in the Pleistocene, and with extensive karst development with four major caves with important Pleistocene plant and animal fossils, the Lorenz area and the Jayawijaya Mountain Range still retain small ice caps. This is one of only three equatorial highlands around the world that is of sufficiently high altitude to retain permanent ice; however, the Lorentz glaciers are the smallest ice caps are all receding very rapidly indeed at the present time. By 1992 there was only 3.3 km^2 of permanent ice remaining, compared with 13 km^2 in 1936 and 6.9 km^2 as recently as 1972.

Miguasha National Park (Gaspé Peninsula, Quebec, Canada)

The extremely rich Devonian fossil deposits at Miguasha (dated at 350–375 million years old) has attracted international interest since the mid-nineteenth century and is now recognized as of paramount importance in having the greatest number and best preserved fossil specimens found anywhere in the world of the lobe-finned fishes that gave rise to the first four-legged, air-breathing terrestrial vertebrates—the tetrapods. The park covers 3 km of the 8 km exposure in coastal cliffs of the fossiliferous deposit, near the mouth of the St Lawrence. The Devonian sequence comprises grey sediments, mainly of the Escuminac Formation, which is composed of alternating layers of thick sandstone, silt, and calcareous schists, and is overlain by the distinctive red Carboniferous Bonaventure Formation. The fossils identified and described include vertebrates, invertebrates, plants and spores of the Devonian period, many of them unique. The very early fossil fish include agnathids, acanthodids, and actinopterygians, from which 90% of all present-day fish today have evolved. They include the first jawed fish to evolve (*Cheirolepis canadensis*) and a coelacanth morphologically identical to the surviving 'living fossil' *Latimeria*. Other key faunal remains include *Scaumenacia curta*, a transitional fish form with both lungs and gills, and a Crossopterygian species exhibiting several features comparable to the first tetrapods. The flora is an important indicator of the palaeoenvironment, and includes *Archaeopteris halliana*, a precursor to modern-day gymnosperms.

2000

Gunung Mulu National Park (Northern Sarawak, Malaysia)

The World Heritage site lies in the NW Borneo geosynclinal belts. There are a wide range of land forms, including steep ridges and escarpments, karst phenomenon (towers, caves, terraces and floodplains), together with hot springs and many waterfalls. The early Cenozoic Mulu Formation, a 4000–5000 m thick series of shales and sandstones, covers the whole SW part of the park, including the Gunung Massif, which rises to 2377 m. This is overlain by the pure white or grey upper Eocene to lower Miocene limestone of the Melinau Formation, up to 1500 m thick, and the later Miocene Setap Shale Formation. This is exposed mainly in the NW part of the park as clay-shales in the valleys and siltstones and quartzite sandstones on the scarps. During a period of geological uplift between 5 and 2 million years ago there was very large-scale development of caves in the Melinau limestones, and these and other karst features have now been exposed through river erosion and down-cutting. The caves that have been created are some of the largest found anywhere in the world, and are superb examples of tropical multi-level river caves with flood in-cuts, extensive clastic sediment deposits and elliptical tubes linking the different cave levels. Over 295 km of the caves in the park have so far been explored and mapped, and a number of them are unique. Deer Cave is the world's largest natural cave passage measuring 120–150 m in diameter, Sarawak Chamber is the world's largest natural chamber, measuring 600 m long, 415 m wide and 80 m high, and the 'Clearwater Cave System' measures 108 km in length and is believed to be the eleventh longest cave system in the world, as well as containing the longest cave in Asia.

Ischigualasto/Talampaya Natural Parks (Argentina)

The Ischigualasto–Talampaya region is a desert area forming the western border of the Sierras Pampeanas of central Argentina, and the two parks cover almost the entire sedimentary basin known as the Ischigualasto–Villa Union Triassic basin. This formation consists of continental sediments deposited by rivers, lakes and swamps during the entire Triassic period. The river deposits include large areas of flood plains with over-bank and crevasse splay sediments that indicate rapid flooding, probably after monsoon-type storms. Lake and swamp deposits contain large amounts of fossil plants, some of them forming coal seams and others in very rare three-dimensional preservation of the actual plants. Of the six geological formations that make up the Triassic basin, five have abundant plant fossils and three are also noted for their equally abundant vertebrate fossils. They contain some of the oldest known dinosaur remains and document the transition from Early Triassic mammalian ancestors to the age of dinosaur dominance in the Late Triassic. The frequent layers of volcanic ash yielding radiometric dates enable the close recording and interpretation of these critically important faunal changes.

Aeolian Islands (Italy)

The Aeolian Islands, separated from the island of Sicily by waters 200 m deep, provide an outstanding record of volcanic island-building and destruction, and ongoing volcanic phenomena. Studied intensively since the eighteenth century, the islands have provided the science of vulcanology with the historic and contemporary type localities for two major kinds of eruption: Vulcanian on Vulcano, and Strombolian at Stromboli. The islands are all of volcanic origin and are of great importance for their geodynamic, volcanic and archaeological natural and ethno-anthropological features. They include a recent volcanic system of seven volcanoes, which began to form approximately 1 million years ago. The islands' volcanic landforms represent classic features in the continuing study of vulcanology and the development of landforms worldwide, as well as in geological and geomorphological education.

Kvarken Archipelago/High Coast (Finland)

The geology and geomorphology of the World Heritage area are both of great interest. In the SW, the well exposed underlying bedrock is ancient granite and gneiss, with magmatic rocks in the north and east, including reddish rapakivi granite, gabbro and anorthosite, whereas the Baltic sea floor is mainly an ancient sandstone base, overlain by younger rocks, including Ordovician limestones. Throughout the entire area are intrusive layers of finer-grained diabase, and in some areas large slabs of diabase overlay ancient sandstone. The Scandinavian Peninsula has been affected by three major ice ages: the Elster, Saale and Weichsel. During the most recent of these, the Weichsel, a vast glacier/ice-cap was centred upon, and had its greatest volume directly over, the High Coast area. This ice cap had its greatest extent around 18 000 years BP. Immediately following the retreat of the ice, the land was uplifted, with initial rates of elevation of 100–150 mm a^{-1}. It is estimated that the total uplift since this time is around 800 m, possibly the highest uplift of any area in the world in recent geological history. The final retreat of the ice from the area of the High Coast occurred some 9600 years ago, and at this time the land was still some 285 m lower than its current position. Rates of uplift are currently about 8 mm a^{-1}. The geomorphology of the region is significantly shaped by the combined processes of glaciation, glacial retreat and the emergence of new land from the sea. The process continues today. Glaciation has greatly affected the landscape. Glacial flow was from the NW, and these sides of the mountains are the most heavily worn. Valleys are orientated NW–SE with the SE side of the mountains being generally far steeper. Deep grooves have been worn into the bedrock in many areas. There are a number of steep faults and fissures, many gouged out by glacial erosion and the activities of freezing and both fluvial and glacial erosion. Though much of the area must have been covered by moraines and other glacial deposition as the ice retreated, these only survive at higher levels: below the original Late Glacial sea level, now raised to approximately 285 m, many areas were washed clean by the sea, and now have redeposited post-glacial fine clays, silts and sands in the valley bottoms and gravels and larger rocks deposited in more exposed areas.

2001

Dorset and East Devon Coast (UK)

The Dorset and East Devon Coast displays a remarkable combination of internationally renowned geological features, and is considered to be one of the most significant Earth science sites in the World. It comprises a near-continuous sequence of Triassic, Jurassic and Cretaceous rock exposures that represent almost the entire Mesozoic Era: nearly 190 million years of Earth history. Additionally the coast contains 'text-book' geomorphological features which are amongst the finest of their kind, including landslides, beaches, the Fleet Lagoon, cliffs and raised beaches. Several fossil localities within the site could merit World Heritage Site status in their own right. The rock strata dip gently to the east, the oldest rocks, the Triassic, are found at the west and of the site, with younger strata of the Jurassic and then Cretaceous outcropping to the east. Together the succession reveals a complete section through the Wessex basin, one of the best Mezozoic–Cenozoic intra-plate sedimentary basins in Europe, which has been studied continuously since the early nineteenth century, and which provides the world type localities for the Kimmeridgian stage of the Jurassic. The nominated site includes a range of internationally important fossil localities that provide excellent evidence of life during Mesozoic times. Numerous vertebrate, invertebrate and plant fossils have been discovered, as well as fossil footprints and tracks. The Jurassic fossil fauna within the nominated area is considered to be the some of the most abundant and diverse anywhere in the world and specimen quality is exceptional, with well-articulated skeletons and soft-part preservation of features such as skin and stomach contents. The coast is notable for its invertebrate fossils, particularly ammonites which have been used to zone the Jurassic. In addition exceptionally well preserved remains of a late Jurassic fossil forest, estimated to be over 140 million years old, are exposed on the Isle of Portland and the Purbeck coast. Considered to be one of the most complete fossil forests of any age, many of the trees are preserved *in situ* with soils and pollen.

Jungfrau-Aletsch-Bietschhorn (Switzerland)

The region provides an outstanding example of the formation of the High Alps which resulted from uplift and compression during the Cenozoic, 40–20 million years ago. Within an altitude range from 900 m to 4274 m, the region displays 400 million-year-old crystalline rocks thrust over the younger *in-situ* calcareous sediments due to the northward drift of the African tectonic plate. In addition to the record of the processes of mountain building is the great variety of geomorphological and glaciological features found in the site. The summit ridge is one of the great watersheds of Europe with many peaks above 4000 m. and this is the most heavily glaciated area in the Alps. Classic examples of glacial action are found in abundance, including U-shaped glacial valleys, cirques, horn peaks, valley glaciers and moraines, and the Aletsch glacier is the largest and longest in western Eurasia. The geomorphology of the area reflects its geological constitution, in particular its petrography and tectonic structure. It is dominated by the crystalline Aar Massif and extends as far as the Helvetic nappe system in the Wengernalp region. The massif is made up of two units: old metamorphic rocks formed during the Caledonian orogenesis, 400–450 million years old and granitic intrusions formed during the Hercynian orogenesis, 300–350 million years ago. During the Mesozoic period, the Aar Massif was covered by a tropical sea for approximately 200 million years. Sediments formed a thick horizontal layer of rock above the crystalline complex, measuring several kilometres in depth. During the Cenozoic formation of the Alps, 40–20 million years ago, these were subjected to severe compression, uplift and metamorphism, though not thrusting or other dislocation. The landscape of the upper slopes is dominated by glacial processes. Below these valleys the landscape is shaped by rivers, but with many glacial depositional features such as moraines and glaciated forelands.

2003

Monte San Giorgio (Switzerland)

Monte San Giorgio is the single best known location recording marine life of the Triassic period, though with important terrestrial fossils as well, many of exceptional completeness and preservation. This largely forested pyramidal low mountain rises 826 m directly from Lake Lugano and the adjacent valleys. The gently dipping geological formations are fossiliferous Triassic carbonate formations which crop out between both older volcanic and more recent sedimentary formations of the Southern Alpine Series. Permian andesites and rhyolites of volcanic origin are exposed on the north face, and Jurassic limestone formations occur on the lower southern slopes, dipping down the edge of the mountain and under the sediments of the Po valley. The Triassic sediments, predominantly limestones, are more than 1000 m thick and record 15 million years of submarine tectonic activity and marine sedimentation under varying conditions. The different environments and deposits of successive transgressions and regressions include conglomerates and sandstone, reef limestone, dolomites and the bituminous shales of the Besano Formation are the richest fossil-bearing horizons. Within this there are at least five distinct, regularly superimposed, fossil beds containing exceptionally rich,

rare, well-preserved fossils of the Middle Triassic period (245–230 million years ago), which have yielded more than 10 000 fossil remains: 30 species of marine and terrestrial reptiles, 80 different species of fish, hundreds of invertebrate species, ammonites, echinoderms, crustaceans, bivalves, cephalopods, insects and terrestrial plants. The intercalated layers of volcanic ash provide a built-in time radiometric scale.

Phong Nha-Ke Bang National Park (Viet Nam)

The National Park has a complicated geological structure, beginning in the Ordovician period. The topography and geomorphology include: non-karst landforms of low, round-top, mountains with planation surfaces and abrasion-accumulation terraces along and at the margins of the central limestone massif, ancient tropical karst landforms in Mesozoic limestone, and Cenozoic karst, which covers around two-thirds of the park. The karst formation process has resulted in many features such as underground rivers, dry caves, terraced caves, suspended caves, dendritic caves, and intersecting caves. There are 17 active cave systems, of which the most famous is the Phong Nha Cave, which connects with the Son River, and has a currently surveyed length of 44.5 km. The caves demonstrate discrete episodic sequences of events, leaving behind various levels of fossil passages, formerly buried and now uncovered palaeokarst (karst from previous, perhaps very ancient, periods of solution); evidence of major changes in the routes of underground rivers; changes in the solutional regime; deposition and later re-solution of giant speleothems and unusual features such as sub-aerial stromatolites.

Purnululu National Park (Western Australia, Australia)

The park comprises four major ecosystems: the Bungle Bungle Mountain Range, a deeply dissected plateau that dominates the centre of the park; wide sand plains surrounding the Bungle Bungle Mountains; the Ord River valley to the east and south of the park; and limestone ridges and ranges to the west and north of the park. The Bungle Bungle Mountains are an unusual and very dramatic plateau of Devonian quartz sandstone (approximately 360 million years old), created through a complex process of sedimentation, compaction, uplift (caused by the collision of Gondwanaland and Laurasia approximately 300 million years ago and the convergence of the Indo-Australian plate and the Pacific plate 20 million years ago), as well as long periods of erosion. The landscape comprises a mass of conical towers with regularly alternating dark grey bands of cyanobacterial crust and the plateau is dissected by 100–200 m deep, sheer-sided gorges. The sandstone karst is of great scientific importance in demonstrating so clearly the process of cone karst formation on sandstone—a phenomenon recognized by geomorphologists only over the past 25 years and still incompletely understood, despite recently renewed interest and research.

Three Parallel Rivers of Yunnan Protected Areas (Lijiang, Yunnan and Tibet Autonomous Prefecture, China)

This World Heritage site comprises a group of four parallel north–south trending mountain ranges stretching 310 km from north to south and 180 km from east to west and reaching heights in excess of 4000 m above sea level. Three great rivers: the Yangtse, Mekong and Salween flow through steep parallel gorges, which are in places 2000 m deep, and 310 km long in the case of the Mekong. The site is of outstanding value in understanding important events in the geological history of the last 50 million years in the evolution of the land surface of Asia: the collision of the Eurasian Plate and the underlying Indian Plate which is being subducted along the line of the Lancang River fault creating vast thrust-nappes, the closure of the ancient Tethys Sea, and the uplifting of the Himalaya Range and the Tibetan Plateau. There is a wide range of rock types which provide evidence of the past marine evolution under the Tethys and neo-Tethys seas separating the landmass of Laurasia in the north from Gondwanaland in the south. Some of the results are visible in complex patterns of folded rock and unusual mineral formations, with four dominant types of igneous rock: ultrabasic, basic, intermediate acid and acid rock as well as ophiolite, in association with deep-water silicalite. There are also an excellent representatives of alpine landscapes and their evolution. The eastern mountains, plateaus and valleys are covered with meadows, waterfalls and streams and hundreds of small glacial lakes left by glacial erosion processes. More than 424 glacial lakes, glacial moraines and other glacial land-forms remain, and a variety of alpine karst features exist within the protected area.

2004

Ilulissat Icefjord (Greenland, Denmark)

The Greenland icecap, 1.7 million square kilometres in area, is the only remnant in the Northern Hemisphere of the continental ice sheets of the last Quaternary Ice Age. The Ilulissat Icefjord is the sea mouth of Sermeq Kujalleq, one of the few glaciers through which the ice of the Greenland ice cap reaches the sea. The Icefjord is a tidewater ice-stream located 1000 km up the west coast of Greenland, and drains into the bay of Disko Bugt (bight) which is partially blocked by the large island of Disko. It is the second fastest and most prolific ice-calving tidewater glacier in Greenland, and produces a constant procession of icebergs while still actively eroding the fiord bed. The surrounding country is of heavily glaciated Precambrian gneiss and amphibolite rocks extending some 50 km inland to the ice cap with flanking lateral moraines and ice-dammed lakes; also lakelets, glacial striations, roches moutonées, and perched erratics typical of glaciated landscapes. Though the Greenland icecap formed during the middle and late Pleistocene the oldest surviving ice is around 250 000 years old, though around Ilulissat Icefjord, the evidence of glaciation is mainly from the last 100 000 years. This culminated in the 'Little Ice Age' 500 to 100 years ago when the ice expanded in pulses to a maximum during the nineteenth century. A marked glacial recession has occurred during the twentieth century: between 1851 and 1950 the glacier had retreated by 26 km.

Pitons Management Area (Saint Lucia)

The Lesser Antilles are the island peaks of a 700 km-long volcanic arc of 18 volcanoes, overlying a tectonic plate subduction (under-thrusting) zone. The Pitons Management Area of the island of St Lucia contains the greater part of a collapsed stratovolcano contained within the volcanic system: the Soufriere Volcanic Centre. Prominent within the volcanic landscape are two eroded remnants of lava domes, Gros Piton and Petit Piton. The Pitons are two steep forested cone-shaped mountains rising side by side from the sea on the SW coast of Saint Lucia with spectacular abruptness. Gros Piton is 3 km wide at the base, Petit Piton is 1 km wide and is linked to it by the high Piton Mitan ridge. The peaks are the degraded dacitic cores of two lava-dome volcanoes probably formed on the side of a collapsed andesitic stratovolcano. There are a variety of other volcanic features including cumulo-domes, explosion craters, pyroclastic deposits (pumice and ash), and lava flows. Collectively, these fully illustrate the volcanic history of an andesitic composite volcano associated with crustal plate subduction. Although the volcano has been dormant for at least 20 000 years, current geothermal activity is seen at Sulphur Springs: a solfatara with sulphurous fumaroles and hot springs is surrounded by a variety of other apparently recent volcanic features, including explosion craters, lava flows and deposits of pumice and ash.

2005

Vredefort Dome (Northwest and Free State Provinces, South Africa)

Vredefort Dome is the oldest (2023 million years), largest (190 km radius), and most deeply eroded (about 38 km deep) meteorite impact structure in the world. As such it is the site of the world's greatest single, known energy release event, and it contains high quality and accessible geological (outcrop) sites which demonstrate a range of geological evidences of a complex meteorite impact structure. A comprehensive comparative analysis with the 200 or so other known meteorite impact structures demonstrated that Vredefort is the only example on Earth providing a full geological profile of an astrobleme below the crater floor, thereby enabling research into the genesis and development of an astrobleme immediately post impact. Located 120 km SW of Johannesburg and covering over 30 km^2 this site is a representative part of the whole meteorite impact structure, and covers part of the ring structure and a cross-section of the geological formations and structures that provide evidence for the impact. On the ground, the magnitude of the diameter of the multi-ring structure and of the forces which contributed to forming the overturned, steeply dipping and highly faulted hills of the Vredefort Dome can be best appreciated at a landscape scale from a number vantage points within the World Heritage site.

Wadi Al-Hitan (Whale Valley) (Faiyum, Egypt)

The three Eocene formations of the Tethys Sea that are visible in this World Heritage site in the Western Desert are: the Gehannam Formation (c. 41–40 million years old) consisting of white marly limestone and gypsum shale and yielding many fossils of primitive whales (Archaeoceti), sirenians, sharks, turtles, and crocodilians, the Birket Qarun Formation; the middle unit, consisting of sandstone, clays and hard limestone, which also yields whale skeletons; and the Qasr El-Sagha Formation of Late Eocene age, about 39 million years old, which is rich in an invertebrate fauna indicating a shallow marine environment. Wadi Al-Hitan has for many years been famous for its invaluable fossil remains of the earliest, and now extinct, suborder of whales, the Archaeoceti, which represents one of the major steps of evolution: the emergence of the whale as an ocean-going mammal from a previous life as a land-based animal. The number (over 400 individuals), concentration and quality of such fossils here is unique, as is their accessibility and setting in an attractive and protected landscape. The presence of many baby skeletons suggests that the place was a shallow and nutrient-rich embayment frequented for calving. The fossils of Al-Hitan show the youngest archaeocetes, in the last stages of losing their hind limbs. They already display the typical streamlined body form of modern whales, whilst retaining certain primitive aspects of skull and tooth structure. In total, the fauna includes invertebrates such as nummulites, molluscs, gastropods, bivalves, echinoids and crabs, and plant remains: at least 25 genera of more than 14 families in addition to the four classes of vertebrates.

West Norwegian Fiords: Geirangerfjord and Nærøyfjord (Norway)

These dramatic fiords are the grandest landscapes in a country of spectacular fiords, and are unlike many others have not been modified for hydroelectric power developments. Each is at the upper end of a major fiord system that developed along faults and fracture zones at right angles, giving them a characteristic zigzag form. Both fiords are submarine hanging valleys, which have floors between 300–500 m deep in ice-scoured basins, the floor of Nærøyfjord ending 1000 m above the floor of Sognefjord. Geomorphologically, the areas are extremely well-developed examples of fiord landscape and excellent examples of young active glaciation. Relatively recently, in geological terms, the products of glacial weathering were removed, leaving ice- and wave-polished surfaces on the steep fiord sides which provide superbly exposed and continuous three-dimensional sections through the bedrock. In Geirangerfjord these are Precambrian gneisses of the West Gneiss Region, a world-class example of deeply subducted continental crust and of well preserved high-pressure rocks. In Geirangerfjord there are outcrops of peridotite and serpentinite in the predominantly gneiss bedrock. In Nærøyfjord, the underlying rocks are anorthosite and gabbro, with softer phyllite in Aurlandsfjord. The Geirangerfjord area is 60 km inland at the end of Storfjord. It branches into two: Sunnylvsfjord of which Geirangerfjord is a branch, and Norddalsfjord of which Tafjord is a branch. Its fiords are 1–2 km wide and their sides reach a height of 1300 m in places with old transhumance farms in the hanging valleys. These mountains are more alpine in character than those of the more southerly Nærøyfjord, where block fields are more prevalent and permafrost and glaciers persist on the highest summits.

2007

Jeju Volcanic Island and Lava Tubes (Republic of Korea)

Jeju Volcanic Island and Lava Tubes comprises three sites that together make up 18 846 ha, 10.3% of the surface area of Jeju Island, the southernmost territory of the Republic of Korea. It includes: Geomunoreum, regarded as the finest lava tube system of caves anywhere, with its multi-coloured carbonate roofs and floors, and dark-coloured lava walls; the fortress-like Seongsan Ilchulbong tuff cone, rising out of the ocean, a dramatic landscape; and Mount Hallasan, the highest in Korea, with its waterfalls, strange rock formations, and crater lake.

Jeju Island is a shield volcano about 1.2 million years old, characterized by a thick sequence of basalt lava flows, surmounted by a trachyte dome. The island originated as underwater hydromagmatic eruptions on the continental shelf, which were then overlain by basalt lavas erupting from about 360 subsidiary cones, mostly scoria cones with tuff cones on the coast. The basalt flows were tube fed, forming extensive lava tube caves of which 120 are known today.

The Hallasan Natural Reserve comprises a substantial part of the summit area of the primary volcano. The diverse volcanic landscape includes a 1.6 ha crater lake, 550 m in diameter and 108 m deep, a younger (c. 25 000 year old) intruded trachyte dome, and a series of columnar-jointed basalts forming prominent cliffs.

The Geomunoreum Lava Tube System contains five lava tubes in lavas that erupted from the Geomunoreum scoria cone 300 000 to 100 000 years ago. Formed by differential cooling within the lava field, the lava tubes are elongated tubular cave structures varying in length, configuration and composition. The Seongsan Ilchulbong Tuff Cone is a hydroclastic volcanic feature on the coastal flank of the Jeju volcano. Composed of a mix of breccia, lapilli tuff, stratified tuff and bedded tuff, it was formed by a Surtseyan-type (Icelandic) eruption from a shallow sea bed in the late Pleistocene epoch (120 000–40 000 years ago). It is a 179 m high castle-like feature with a bowl-shaped summit crater 570 m in diameter. Wave erosion has exposed the internal sedimentary structures and stratification.

South China Karst (China)

The South China Karst region extends over an area of half a million square kilometres lying mainly in Yunnan, Guizhou and Guangxi Provinces. South China is unrivalled for the diversity of its karst features and landscapes. The site presents a coherent serial property comprising three clusters: Libo Karst, Shilin Karst and Wulong Karst. South China Karst represents one of the world's most spectacular examples of humid tropical to subtropical karst landscapes. The stone forests of Shilin are considered superlative natural phenomena and a world reference. The cluster includes the Naigu stone forest occurring on dolomitic limestone and the Suyishan stone forest arising from a lake. Shilin contains a wider range of pinnacle shapes than other karst landscapes with pinnacles, and a higher diversity of shapes and changing colours. The cone and tower karsts of Libo, also considered the world reference site for these types of karsts, form a distinctive and beautiful landscape. Wulong Karst has been inscribed for its giant dolines, natural bridges and caves.

The nominated property contains a cross-section of key features of the regional geology of the area including the deposition of carbonates up to the Triassic period (250 million years ago) and the subsequent tectonic evolution of the area including three phases of evolution during the Quaternary period. The geological histories of the mature karst and the palaeokarst landscapes are 'intact' as they were little affected by glaciation. The great variety of karst landscapes in the South China Karst is attributed to: the age of the thick accumulations of limestone which resulted in relatively hard limestone and more stable and massive landforms; and the influence of several phases of tectonic uplift (including a major recent phase associated with the Himalayan orogeny and associated with the uplift of the Tibetan plateau) causing folding and faulting of the rocks and permitting access

of water to corrode and erode the limestone to form the current karst forms.

The nomination notes four landscape types as outstanding. These have considerable internal landscape diversity, but can be summarized as:

- *Fengcong karst (cone karst)*: characterized by linked conical hills and depressions, valleys and gorges;
- *Fenglin karst (tower karst)*: comprising isolated cones or towers on broad plains;
- *Stone forests*: with a wide diversity of closely spaced pinnacles and towers; and
- *Tiankeng karst (giant dolines)*: massive circular collapse structures often in close proximity to spectacular gorges, decorated caves and where cave/doline collapse can create natural rock bridges.

Teide National Park (Tenerife, Canary Islands, Spain)

Situated on the island of Tenerife in the Canary Islands, Teide National Park covers 18 990 ha and features the Teide–Pico Viejo stratovolcano that, at 3718 m, is the highest peak in Spain. Standing 7500 m above the ocean floor, it is regarded as the world's third tallest volcanic structure and is situated in a spectacular environment. The visual impact of the site is all the greater due to atmospheric conditions that create constantly changing textures and tones in the landscape and a 'sea of clouds' that forms a visually impressive backdrop to the mountain. Teide is of global importance in providing evidence of the geological processes that underpin the evolution of oceanic islands, complementing those of volcanic properties already on the World Heritage List, such as the Hawaii Volcanoes National Park (USA).

Tenerife is composed of a complex of overlapping Miocene–Quaternary stratovolcanoes that have remained active into historical times. The dominant feature of National Park is the Teide–Pico Viejo stratovolcano. Examples of relatively recent volcanism include the Fasnia Volcano (1705) and the eruption of the parasitic 'Narices del Teide' (Teide's Nostrils in 1798). The older and more complex crater of Pico Viejo dates from the Pleistocene. The stratovolcano is located in the centre of a large depression known as Las Cañadas Caldera, which is delimited to the east, south and part of the west by abrupt escarpments of up to 650 m that display the geological history of the area along their 25 km length. In the east the Las Cañadas escarpment comprises alternating layers of lava and explosion debris, followed by an arc of pumice deposits and, finally, outflow deposits.

The landscape continues to develop through active erosion and deposition as exemplified by features such as the Corbata del Teide torrent and the talus slopes of the Las Cañadas wall. To the north and NW of the stratovolcano the wall of the caldera is absent apart from a limited escarpment at La Forteleza. This is considered by many to reflect the lateral collapse of a proto-volcano via massive and complex avalanche-like collapses in the direction of Icod and Oratava. Between the base of the stratovolcano and the foot of the wall is an extensive field of lavas (including obsidian) and recent pyroclastic material. This area also contains numerous medium and small forms including ridges, cones, craters, volcano fields, domes, fissures, blocks, needles, tubes, channels, badlands and lahars. The geology of the National Park represents the entire range of the magmatic series, with a large amount and variety of fully differentiated acid (felsic/phonolitic) volcanic as well as basaltic materials.

References

BATISSE, M. & BOLLA, G. 2003. *L'invention du 'Patrimoine Mondial'*, Association des anciens fonctionnaires de l'Unesco, Club Histoire, Paris. Website: http://unesdoc.unesco.org/images/0013/001317/131732f.pdf/

COWIE, J. W. & WIMBLEDON, W. A. P. 1994. The World Heritage List and its relevance to geology. *In*: O'HALLORAN, GREEN, C., HARLEY, M., STANLEY, M. & KNILL, J. (eds) *Geological and Landscape Conservation*. Geological Society, London, 71–73.

GILLETTE, E. R. (ed.) 1972. *Action for Wilderness*. 12th Wilderness Conference, Washington D.C., 1971. (Sierra Club, San Francisco).

HOUSHOLD, I. & SHARPLES, C. 2008. Geodiversity in the wilderness: a brief history of geoconservation in Tasmania. *In*: BUREK, C. V. & PROSSER, C. D. (eds) *The History of Geoconservation*. The Geological Society, London, Special Publications, **300**, 257–272.

INHIGEO. 1978. *VIII Symposium. Münster and Bonn, 12–24 September 1978*. International Commission for the History of Geological Sciences, Münster.

INTERNATIONAL STRATIGRAPHICAL COMMISSION. 2004. *Overview of Global Boundary Stratotype Sections and Points (GSSP's)* http://www.stratigraphy.org/gssp.htm/ (Last accessed 10 March 2007).

MEYER, R. L. 1976. *Travaux Preparatoires* for the UNESCO World Heritage Convention. *Earth Law Journal*, **2**, 45–81.

TITCHEN, S. M. 1995. On the construction of universal value: UNESCO's World Heritage Convention *(Convention concerning the Protection of the World's Cultural and Natural Heritage, 1972)* and the identification and assessment of cultural places for inclusion in the World Heritage List. Ph.D. thesis, Australian National University, Canberra, Australia.

TITCHEN, S. M. 1996. On the construction of 'outstanding universal value': Some comments on the implementation of the 1972 UNESCO World Heritage Convention. *Conservation and Management of Archaeological Sites*, **1**, 235–242.

TURNER, S. 2006. Rocky road to success. A new history of the International Geoscience Programme (IGCP), *In*: PETITJEAN, P., ZHAROV, V., GLASER, G., RICHARDSON, J., DE PARIDAC, B. & ARCHIBALD, G. (eds) *Sixty Years of Science at UNESCO 1945–2005*. UNESCO, Paris, 297–314.

UNESCO. 1972. *Convention concerning the Protection of the World's Cultural and Natural Heritage, 1972*. Website: http://whc.unesco.org/en/conventiontext/

UNESCO. 2002. *Budapest Declaration on World Heritage, adopted by the 26th Session of the World Heritage Committee*. Website: http://whc.unesco.org/en/budapestdeclaration/

UNESCO. 2005. *The Operational Guidelines for the Implementation of the World Heritage Convention*. Website: http://whc.unesco.org/en/guidelines/

UNESCO. 2006. *World Heritage Centre—Tentative Lists*. Website: http://whc.unesco.org/en/tentative-lists/ (Last accessed 02/03/2007).

UNESCO. 2007. *World Heritage List*. Website: http://whc.unesco.org/en/lists/ (Last accessed 03/10/2007).

WCMC. 2006. *Protected Areas and World Heritage Datasheets* (Cambridge: World Conservation Monitoring Centre—UNEP). Website: http://www.unep-wcmc.org/sites/wh/ (Last accessed 20/03/2008).

WELLS, R. T. 1996. Earth's geological history—a contextual framework for assessment of World Heritage fossil site nominations. *In*: IUCN. *Global Theme Study of World Heritage Natural Sites No. 1*. IUCN, Gland, Switzerland.

Index

Figures are shown in *italics*; tables in **bold**.